Standard Application of Electrical Details

Standard Application of Electrical Details

Jerome F. Mueller, P. E.

MCGRAW-HILL BOOK COMPANY
New York St. Louis San Francisco Auckland Bogotá
Hamburg Johannesburg London Madrid Mexico
Montreal New Delhi Panama Paris São Paulo
Singapore Sydney Tokyo Toronto

Library of Congress Cataloging in Publication Data

Mueller, Jerome F.
Standard application of electrical details.

Includes index.
1. Electric drafting—Detailing. I. Title.
TK431.M84 1984 621.3′022′1 83-782
ISBN 0-07-043961-3

1 2 3 4 5 6 7 8 9 0 HAL/HAL 8 9 8 7 6 5 4 3

ISBN 0-07-043961-3

The editors for this book were Harold Crawford and Frances Koblin, the designer was Naomi Auerbach, and the production supervisor was Sally Fliess. It was set in Times Roman by University Graphics, Inc.

Printed and bound by Halliday Lithograph.

Contents

List of Drawings

Figure No.	Title	Page No.
14-3	MATV system schematic diagram	195
14-4	Community center part plan	196
14-5	Conduit routing details—exterior of apartment building	197
14-6	MATV details	199
14-7	Television riser	200
14-8	Typical empty conduit TV raceway	201
15-1	Expanded stage view looking toward audience	205
15-2	Section A-A	206
15-3	Stage lighting	207
15-4	Small control console dimmer	208
15-5	Small control console dimmer—side view and plug counter detail	209
15-6	Small control console dimmer—front view	210
15-7	Typical supporting bracket for studio lighting	211
15-8	TV control room lighting	212
16-1	Heat tracing for fuel transport line	214
16-2	Typical insulated wire layout to protect a roof edge and downspout	216
16-3	Typical cable installation in concrete slab	217
16-4	Typical arrangement—MI heating cable in asphalt	218
16-5	Typical schematic diagram of heater with fan relay	219
16-6	Disconnect switch	220
16-7	Typical wiring for resistance heater	221
16-8	Disconnecting contactors	222
16-9	Disconnect switch	223
16-10	Electric heating panel for wet plastered ceiling	224
16-11	Electric heating cable in concrete slab	225
17-1	Air terminal roof installation detail	228
17-2	Typical class I protection schematic	229
17-3	Typical residential cable and grounding	230
17-4	Typical multiple terminal grounding rod	231
17-5	Typical terminal grounding rod	232
17-6	Lighting protection Class II system	233
17-7	Grounding details	234
17-8	Grounding details	235
17-9	Stack protection	236
17-10	Lightning mast ground connection details	238
18-1	Typical sacrificial anode installation	240
18-2	Anode arrangement for multiple tank installation	241
18-3	Quay wall sheet piling cathodic protection	242
18-4	Impressed current protection with high silicon cast iron anodes	244
18-5	Cathodic protection for hot water storage tank	245
18-6	Cathodic protection system for elevated water tank	246
18-7	Electric service and rectifier arrangement	247
18-8	Anode suspension arrangement	248
18-9	Cathodic protection system for oil storage tank	249
18-10	Cathodic protection system for land side of sheet piling	250
18-11	Cathodic protection for water side of sheet piling	251
18-12	Cathodic protection for heating conduit system	252
18-13	Cathodic protection system for gas main	253
18-14	Typical graphite anode installations for cathodic protection	254
18-15	Various soil treatments for graphite anode ground rod cathodic protection	255
18-16	Cathodic protection—magnesium anode installation and magnesium block anode installation for tanks	256
18-17	Sectionalized anodes in cable installation cathodic protection	257
18-18	Hand hole and anode suspension detail for elevated water tank	257
18-19	Riser anode suspension detail for elevated water tank	258
18-20	Typical rectifier installations	259
18-21	Cathodic protection miter gate	261
18-22	Electrical connection to steel pipe	262
19-1	Diagram for Class II power installation	265
19-2	Diagram for Class II lighting installation	266

Preface

In writing a book of electrical details the task is at once both simple and exasperatingly difficult. It is simple in that, as a practicing consultant, I have literally hundreds of potential electrical details available for this book—a not uncommon consultant situation. The problem is to select the details that might assist someone in the daily task of assuring that the quality of the job at hand is maintained right down to the simplest, most obscure detail on the plan.

There is a school of thought which says that if you show every possible detail that might ever be used under any circumstance you cannot go wrong, since at least you have covered all of the areas. This is never going to be true because there will always be a new situation where no detail truly applies. However, there are few times in one's career where a standard detail of some sort will not fit the problem at hand with some modification. The general premise of this book is to provide a basic collection of details that cover all of the bases, so to speak. While the details may not cover your requirement totally, we hope that between the details and the accompanying text there are sufficient data for you to be able to use the depicted details with some modification.

Another major purpose of the book is to provide the details and explanations that customarily are given to the new employee in the department who does the detailing. And, in the smaller office, to assist the hard-pressed designers who do not have a backup detailer and frequently are forced to make very quick decisions on how to present a certain situation which, although no one has told them, they instinctively know should be detailed.

The problem of detailing invariably comes down to explaining what should be on the detail and repeatedly explaining what you want on the detail is what you do want on the detail. This sometimes tedious series of explanations can take considerable time and effort by the principals of the smaller firm who reluctantly may take less time than they should, but more time than they think they ought to, in this sort of explanation. Toward the solution of this problem, the text that accompanies each detail explains what is needed to the detailer so that many questions can be answered before you are pressed for a further explanation. If this task is accomplished, this fact alone will make this book a considerable bargain.

Undoubtedly many readers will feel that there should have been a greater emphasis on design. The author simply felt that the book was a book on details and that a true book on design would require a considerably larger amount of text which would not be of much value to the junior detailer and the beginner. Secondly, a truly usable book on design covering all the areas in this book would require a series of separate textbooks, with each one covering perhaps one or two chapters of this book. The purpose of this book is to supplement the many fine design texts that are available with a collection of supporting details which are available only in part in these books.

Anytime an author presents a book of drawings, he or she is immediately open to criticism from others who feel that the drawings lack quantity, quality, clarity, completeness, and a host of other seemingly indispensable virtues. The author makes no apologies for any failings in this area. On the contrary, it is felt that this collection of details represents a step forward in the electrical design profession in the supplying of support design material for many personnel and others involved in the field.

JEROME F. MUELLER, P. E.

Standard Application of Electrical Details

How to Use This Book

The material contained in this book is intended to provide details which supplement and clarify the basic electrical system design. The objective of detailing is to clarify the special points of connection of the systems shown on the plans and/or described in the specification. The detail presumes that the plan and specification descriptions, either together or individually, are the basis of the overall design.

Whether you are designing structural, mechanical, or electrical systems, the design customarily begins with the collecting, arranging, and totaling of the loads that will affect the design solution. For the electrical design to which these details apply, this basically means totaling the lighting and power requirements of the individual spaces that make up the total design.

Since the normal electrical design begins with individual area lighting and power requirements and continues back to the private or public electrical power source, the chapters in this book follow the same sequence. Depending on the number of design personnel involved, the various system and control designs can either be performed simultaneously or as follow-on work. This book simply presents them in the normal follow-on sequence usually encountered in the small, busy firm.

The details presented are those that would most likely be used to clarify the design development drawings. The book is not intended to be a design manual but rather a design detail reference manual.

The book does not, for example, cover outlets and many other areas in the depth of detail noted in the National Electrical Code. Nor does it cover electrical power design in the depth contained in available design manuals. The lighting system, interior and exterior, which is under the control of the electrical designer is more thoroughly covered.

Further, the Illuminating Engineering Society (IES), with their two-volume 1981 edition of the IES *Lighting Handbook,* is the primary, if not "the," source of lighting design information. A number of lighting-fixture manufacturers also provide excellent lighting design information.

In our opening chapter we cover, in abbreviated fashion, the essence of the lighting calculations. Commonly, engineers and designers, trained in older systems and through long experience with personal design approaches, tend to use shortcut rule-of-thumb calculations. While frequently no real harm is caused by the use of these methods, it does give the beginning draftsperson, designer, or detailer the wrong impression of design, and in these days of energy conservation, such lack of awareness can defeat or reduce energy conservation schemes.

For particular detail requirements there is no need to peruse the entire book starting with the opening chapter. However, some chapters do require reference to preceding details. Thus it would be well to carefully read the chapter you select in its entirety.

Proper design depends on proper approach. The designer must be familiar with all the background factors of design; he or she must have depth in engineering knowledge, and, over and above all of this, must understand electrical systems both in their broad outlines and in their detailed applications. Electrical design is not a mechanical procedure of filling in formulas and adding pieces together. It involves a clear understanding of past procedures as well as new techniques. Finally, a factor that is important to both the designers and to the detailers in the organization—department heads must have a complete understanding of the total design to provide the necessary instructions for its proper development.

The age-old question of what should be detailed is never-ending. Primarily, the electrical system has five basic areas of concern. These are:

1. Safety
2. Capacity
3. Flexibility excess
4. Accessibility
5. Reliability

Safety demands a thorough knowledge of all electrical codes that apply, including those of the National Electrical Code (NEC), the National Fire Protection Association (NFPA), and the Underwriters' Laboratories (UL). Hospital codes, fire alarm codes, and other specialized city, town, state, and federal codes must be carefully reviewed.

Capacity is primarily an engineering function. The proper design of a system will be one that has not only adequate capacity for the present tasks but proper capacity, or proper means of acquiring additional capacity, for future tasks.

Flexibility is almost obvious. To be flexible in any system or in any area means to be able to move from one situation to another without undue effort.

Accessibility, because it is also obvious, is something that is frequently overlooked. Things that are seemingly accessible on a plan are often not accessible in fact when installed.

Reliability is primarily a quality of design based again largely on the written work of the specification.

At this point it might be added that this discussion is on design in a book devoted to detail. However, before closing this short chapter, the emphasis must be on the direction that the design implies and designer imparts to the detailer. Based on the above statements, the question logically arises, "What should the design detail cover?" A seemingly redundant yet very necessary answer is that the design details should cover those areas required to be clarified for the proper installation of the electrical system as designed. This answer is of course not satisfactory, but it is, in essence, the answer. Another answer perhaps would be to peruse the table of contents of this book to become aware of the major areas of detailing. However, every job does not involve all of these areas. To be more specific, there are four major areas of detail design, investigation, and possible requirements:

1. The lighting area, including fixtures and all of their connections
2. The power area, including motor loads, power layouts, control equipment, protection devices, power switchgear, and electrical service
3. Distribution, including the distribution panelboard, bus duct, subdistribution, load centers, substations, transformers, primary and secondary feeders, types of systems, actual or schematic service layouts, and connections to alternative sources of power such as emergency generators
4. Systems and controls: By systems we mean all forms of communication, emergency lighting, signaling, clocks, programs, sound, stage, television, and other specialized system; controls imply primarily temperature controls and specialized equipment control systems.

No list or collection of lists can provide the head of an electrical group with cookbook-type procedures that will show which systems require what details. The design itself is, in the end, the determining factor. What this book does is to provide the details in such a fashion that custom tailoring will be limited to minor modifications. Also the explanations that accompany these details will greatly reduce the time to explain why, what, and how design elements should be arranged and portrayed.

ONE

Lighting Fixtures

General Although the purpose of this book is not to be an authoritative design text, the subject of installing electrical fixtures is directly related to the one area the electrical designer truly controls—lighting. The following discussion is a condensed version of lighting design calculations, which are based on the latest methodology of the Illuminating Engineering Society as outlined in the two-volume 1981 edition of the *Lighting Handbook*. Only the most common, frequently used calculations are covered.

It is felt that this discussion is particularly important as the performance of lighting calculations by lower-level employees and the frequent use of rules of thumb and company shortcuts tend not only to demean this important basic task but also to create solutions that do not maximize energy conservation and fail to minimize the quantity and output of lighting fixtures.

Lighting Design Lighting output is commonly referred to in footcandles. Perhaps the use of the word "footcandles" could be considered as a way of dating the knowledge of the user. Today the word is "illuminance." *Illuminance* is defined as luminous flux divided by area. If the area is measured in square feet, the illuminance is expressed as footcandles or lumens per square foot. If the area is measured in square meters, the illuminance is defined as *lux,* which is lumens per square meter.

The formula for calculating illuminance (footcandles or lux) is well-documented and described in the reference volume of the IES *Lighting Handbook* and in some manufacturers' literature. It is important that the formula be followed in your calculations. The basic formula is

$$\text{FTC} = \frac{N \times L \times \text{CU} \times \text{LLF}}{A}$$

where FTC = illuminance measured in footcandles or lumens per square foot
N = number of luminaires to be supplied in the area
L = lumens supplied by each luminaire
CU = coefficient of utilization (provided by the manufacturer of the lighting fixture)
LLF = light loss factor
A = area

The light loss factor is a description that used to be called the maintenance factor. It is the product of luminaire ambient temperature, voltage to luminaire, ballast factor, luminaire surface depreciation, room surface dirt depreciation, lamp lumen depreciation, lamp burnout factor, and luminaire dirt depreciation. Various areas would obviously have large differences in some of these items.

The coefficient of utilization first requires the calculation of two, sometimes three, separate room cavity ratios. Note the relationships shown in Figure 1-1. The formula for calculating the room cavity ratio is

$$\text{CR} = \frac{5 \times h \times (l + w)}{A}$$

where CR = cavity ratio (room, ceiling, or floor)
h = height of cavity (see Figure 1-1 and use h_{rc}, h_{cc}, or h_{fc} as required)
l = room length
w = room width
A = room area ($l \times w$)

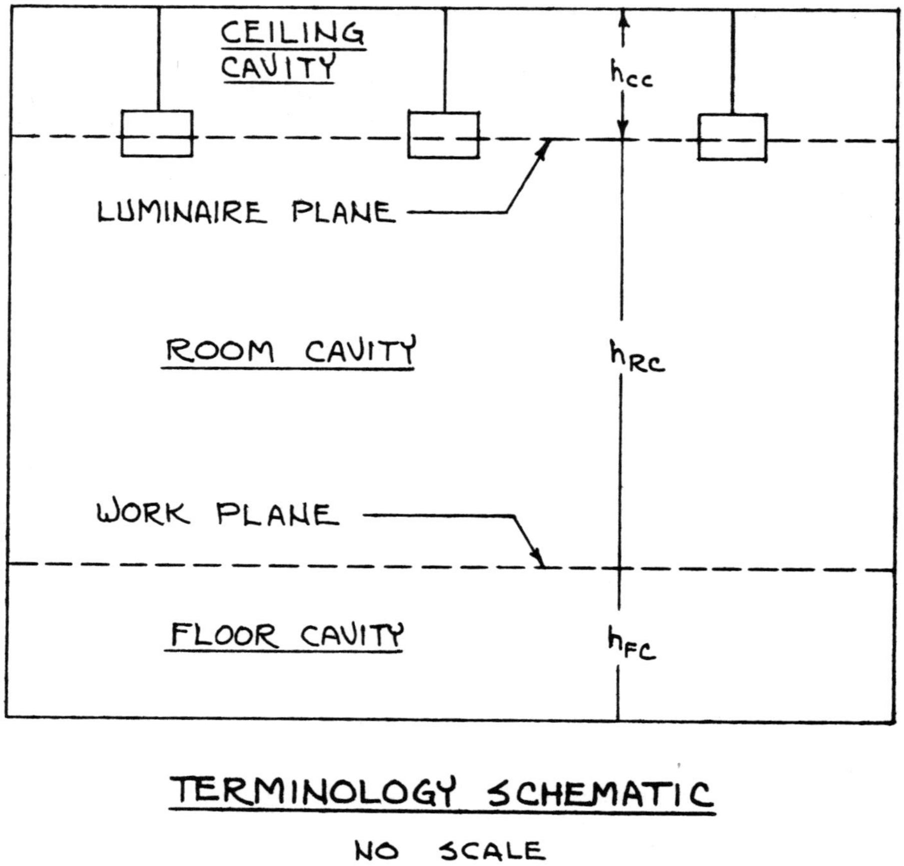

FIGURE 1-1

The manufacturers' catalogs contain tables to convert these items into effective cavity reflectances for both the room and floor cavity. With this information one can then go to the IES *Lighting Handbook,* and, depending on the particular fixture, obtain the proper coefficient of utilization.

As stated in our opening description, How to Use this Book, this discussion is not intended to equal or repeat what is already thoroughly covered in the IES *Lighting Handbook.* Instead, this is an abbreviated version designed to call your attention to the importance of this task.

While the light loss factor is a subjective judgment, it is important to realize that it is the product, not the sum, of a number of judgments that you must make. For example, if fixture depreciation results in 90 percent of initial output available and dirt results in 80 percent availability, the net result is 90 percent × 80 percent, or 72 percent.

Properly selected, the coefficient of utilization requires floor, room, and ceiling cavity ratios to be calculated or picked from a table. With these data you can then, from another table, find ceiling and floor cavity effective reflectances. Armed with these data you can use the manufacturer's (or the IES *Handbook*) data to find the correct coefficient of utilization.

Since the basic illuminance (footcandle) formula is relatively simple and straightforward, the difference between casually obtained and correctly obtained values of coefficient of utilization and light loss factor can make large differences in expected footcandles and energy consumption. It is no secret that many buildings today are turning off large numbers of light fixtures since the tasks performed in these areas *do not,* and never did, require all the fixtures installed. Whatever the reason for existing design, America can no longer afford to overdesign.

There are also other methods of performing lighting output calculations such as point methods, equivalent sphere, and contrast rendition factor method. The average illuminance calculation method is commonly used for interior spaces.

In lighting calculations there are two other specialized applications that can cause confusion because of their infrequency in the normal work load of most firms. These are covered in abbreviated fashion in the following paragraphs.

Floodlighting There are two commonly used methods for floodlighting calculations, the point method and the beam lumen method. The point method is a good, accurate method, but it is also somewhat complex. The beam lumen method is similar to the interior footcandle calculation and easier to recall and to comprehend. Basically the beam lumen method is based on the following formula:

$$\mathrm{FTC} = \frac{N \times L \times \mathrm{CU} \times \mathrm{LLF}}{A}$$

where FTC = illuminance measured in footcandles
N = number of floodlights
L = beam lumens (lamp lumens times beam efficiency of floodlight)
CU = coefficient of beam utilization (the ratio of the lumens on target to the total beam lumens)
LLF = light loss factor
A = area to be illuminated

Calculating the beam lumens that fall on the target requires that one construct a diagram showing the angles from the lamp to the largest area and various points at different angles off the line from the fixture to the target. Comparing these angles to the grid of the isocandela, a curve of the floodlight in a few sample cases can produce a fairly accurate average coefficient of utilization. A further explanation can be found in the reference volume of the IES *Lighting Handbook.*

Roadway Lighting There is an entire chapter in the *Lighting Handbook* on roadway lighting. For designers accustomed to thinking that the 5- or even 10-fc level is approaching darkness, it may come as a surprise to learn that the top requirement (pedestrian tunnels and commercial walkways) is only 4 fc and the top level for roads (major commercial roadways) is only 2 fc. The commonly used formula in these cases is

$$\mathrm{FTC} = \frac{L \times \mathrm{CU} \times \mathrm{LLF}}{A}$$

where FTC = average illuminance measured in footcandles
L = lamp lumens (luminance output of roadway lamps)
CU = coefficient of utilization (based on a curve supplied by the fixture manufacturer and calculated using a ratio of road width to luminaire mounting height)
LLF = light loss factor (given or estimated dirt depreciation factor)
A = area covered, obtained by multiplying the spacing of the luminaires by the roadway width

Lighting Fixture Details—General Most lighting fixtures are not very heavy and are easily supported, directly or by stems, from a securely supported outlet box from which their electrical power is obtained. Where there are dropped ceiling tile grid systems, the fixtures are commonly selected to rest in and be supported by the grid system. Thus, there is no need to detail this obvious, well-understood method of fixture support.

Given the above facts, the details of fixture installation generally required cover specialized interior lighting, fixtures of considerable weight that are usually hung in high-bay industrial areas, or those that may be abused, such as gym fixtures. Special areas such as underwater pool lighting and stage and television studio lighting are also areas to be detailed. Finally, the interior of a prison or mental institution has certain special security requirements that require detailing.

Exterior lighting of all types frequently must be supported on its own system supports or attached to a structure, pole, or base in special fashion. This, too, requires explanatory detailing.

In all cases the details shown deliberately avoid naming manufacturers who make special supporting devices. It is felt that this avoidance would preclude favoring any manufacturer. Specification or manufacturers should not be a problem for any skilled, experienced engineer.

Process Lighting Frequently the lighting system for an industrial plant or process turns out to be more complicated than was originally thought. While the general illumination of the area may pose no particular design problem, the special lighting needs around a particular process does. The concern of the design becomes how to support and locate the fixture in a position that not only suits the present process but also allows for considerable freedom if the process is altered, relocated, or removed.

One of the basic concepts of good industrial design, which is frequently overlooked, is clearly present in Figures 1-2 to 1-4. In each case the fixture is plugged into an outlet. A good design would include a number of outlets of this type, which may not necessarily all be utilized when the lighting system is installed. Having these outlets in position provides for a great deal of lighting flexibility.

In today's energy-conscious economy one sadly sees fixtures simply locked in place and turned off. Nothing is more ridiculous than a large group of fixtures simply hanging, unlit, in a not-usable-for-anything-else arrangement.

Each of the three figures depicts a simple hook-and-eye arrangement supported from the existing building or process structure in differing fashions. As in all drawings in this book, the arrangements depicted were taken from a real-life, actual project and are currently installed.

The fixture shown is a mercury-vapor type. It could, with slight modification, be a high-pressure-sodium or incandescent type. Typical of most installations of this type is the mounting height of 14 to 18 ft. Depending on the application and project requirements, the mounting height could be any height that suits the design.

The cord and junction box arrangement provides ideal fixture mobility. If the process or lighting requirement changes, the fixture can be easily unplugged and relocated. The fabrication of the eye bolt can be easily and quickly accomplished in any industrial machine shop, or it can be bought in the nearest electrical supply outlet.

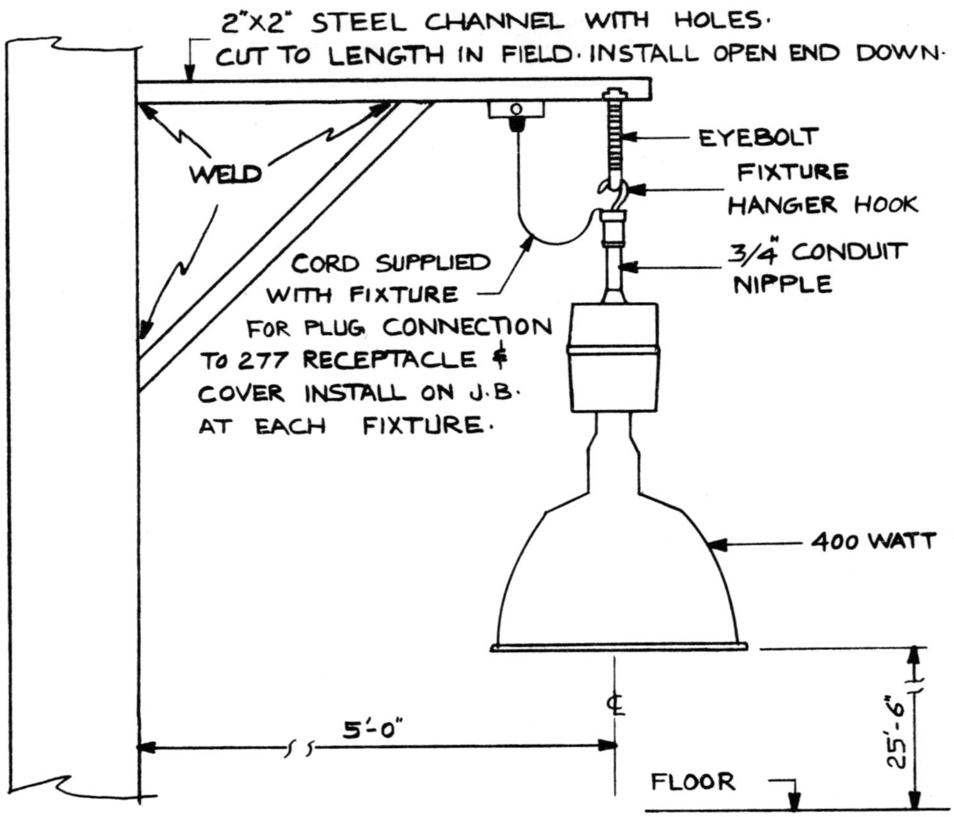

TYPICAL INDUSTRIAL FIXTURE INSTALLATION

NOT TO SCALE

FIGURE 1-2

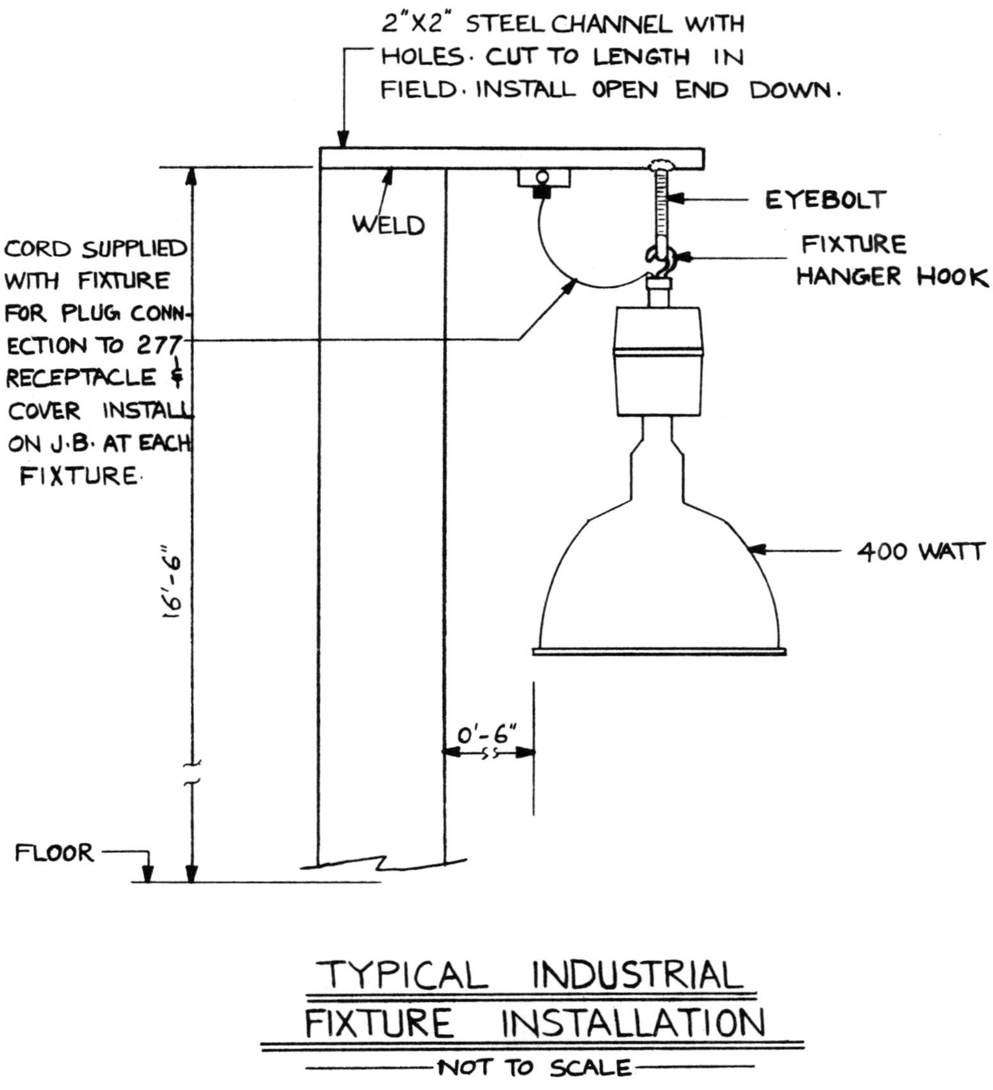

2"X2" STEEL CHANNEL WITH HOLES. CUT TO LENGTH IN FIELD. INSTALL OPEN END DOWN.
WELD
EYEBOLT
FIXTURE HANGER HOOK
CORD SUPPLIED WITH FIXTURE FOR PLUG CONNECTION TO 277 RECEPTACLE & COVER INSTALL ON J.B. AT EACH FIXTURE.
400 WATT
16'-6"
0'-6"
FLOOR
TYPICAL INDUSTRIAL FIXTURE INSTALLATION
NOT TO SCALE

FIGURE 1-3

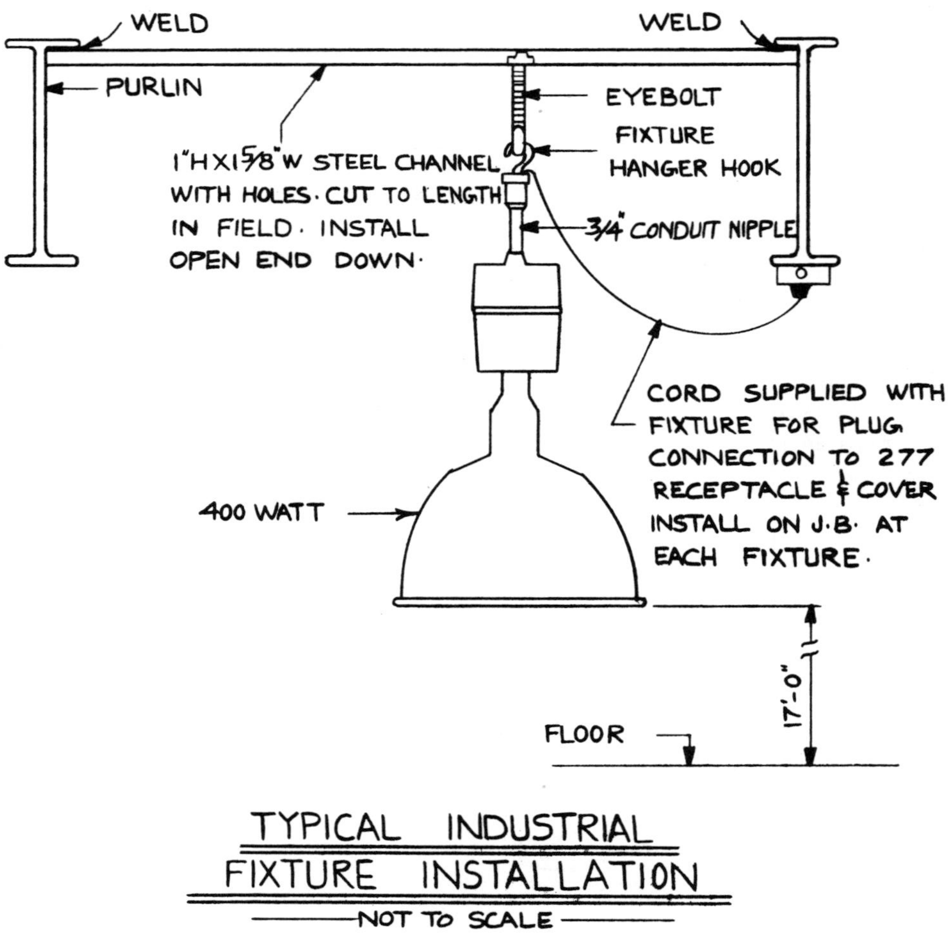

FIGURE 1-4

High-Bay Lighting High-bay lighting is very similar whether it be in an indoor swimming pool, a gymnasium, or an industrial plant. Because several points can be made with one drawing, Figure 1-5 depicts lighting for an indoor swimming pool. It shows a fixture and a power hook assembly that are the standard products of more than one large, well-known lighting fixture manufacturer. If the firm using this detail so chooses, it can easily specify the manufacturer's name and model number on the detail.

Because this simple specifying of standard manufactured items seems so easy and so logical, the question arises as to why show the detail at all. The answer to the question is also very easy. When the installing contractor reviews the costs, this proper attachment is one item that could be omitted, and you, as the designer, will be faced with time-consuming, unending arguments as to why some other, jury-rigged arrangement isn't equally as good, if not better. You can easily multiply by 10 the time spent in showing this detail versus the time spent in proving your written words mean exactly this arrangement and no other.

Frequently, unless the detail appears on your drawings, you will find the simple piece of ¾-in rigid conduit between the ceiling-mounted junction box and the power hook either too short, too long, or not there at all. The detail leaves no doubt as to what you want. That is, a piece of conduit long enough so that the bottom of the fixture is flush with the bottom of the joist.

Designers frequently dimension details unnecessarily. There is no guarantee that the manufacturer's power hooks and fixtures are all identical in length. Unless you want to list a long series of related dimensions for each manufacturer, you are far better off to omit all dimensions. If you wish, you could add to the detail the words "bottom of fixture must finish flush with bottom of joist."

Since this is a pool with a chlorinated water source

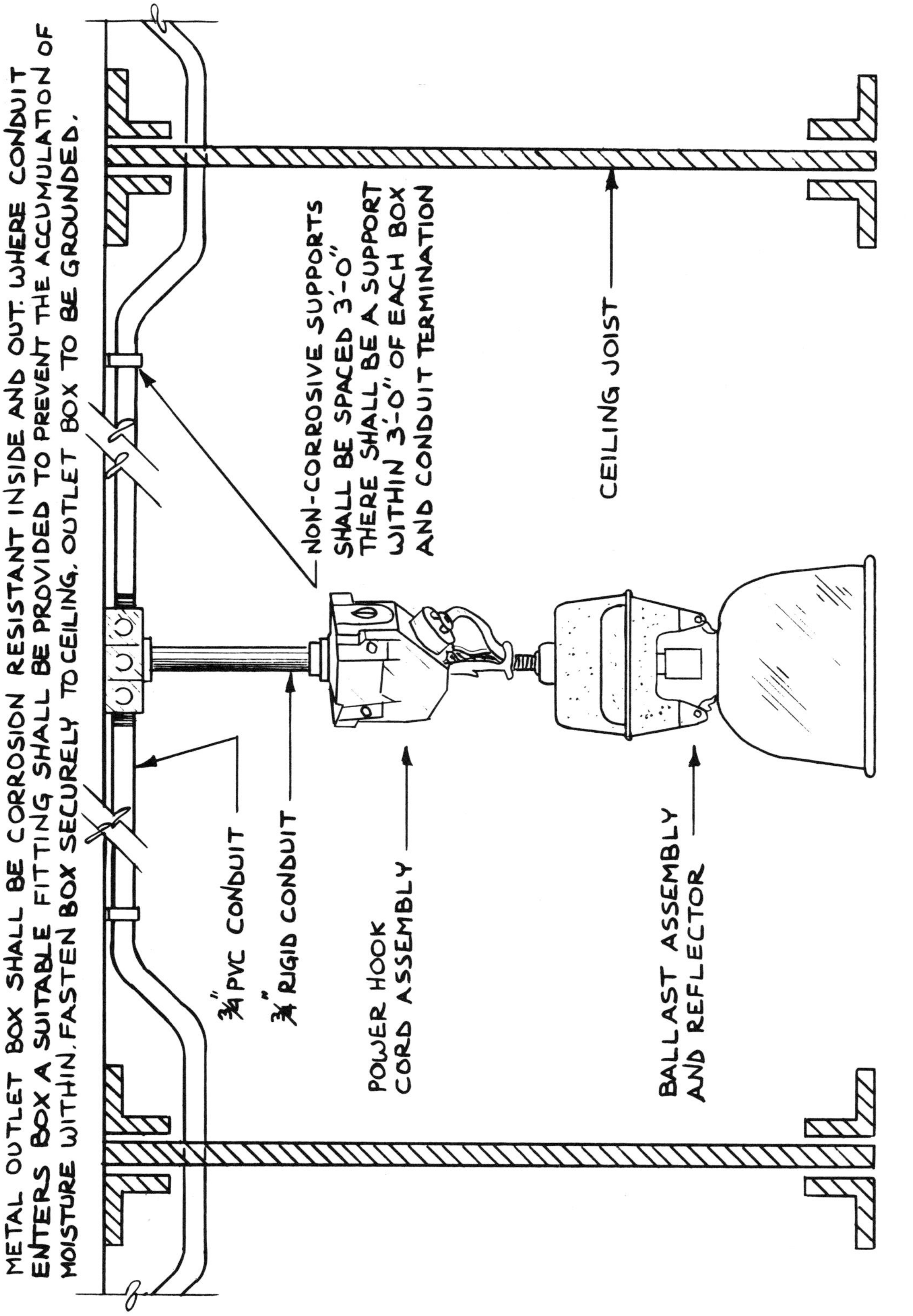

NATATORIUM PENDANT LIGHTING FIXTURE DETAIL

NOT TO SCALE

FIGURE 1-5

which creates a corrosive atmosphere, your attention is called to the type of conduit, outlet box, and supports clearly described on the drawing. Far too frequently these items are overlooked to the sorrow of the design firm involved.

The same design, minus its corrosion notes, could be used for a fixture installed in a high-bay industrial or other area or in a gymnasium. The fixture itself could have a lens or not as required. If it were in a gym, a wire guard should be specified to protect the lamp.

Security Lighting Whether in a prison, as depicted in Figures 1-6 and 1-7, or in a mental institution, the problem of lighting design and detailing is not so much the type of lighting as it is the security of the light source. Some of the points to be made may seem obvious as you study the details, but they can easily be overlooked as you prepare the design drawings. Probably the most important point is "How do you relamp a fixture in a maximum security situation?"

In Figure 1-6, the fixture in the maximum security cell is protected on the cell side by a welded wrought-iron lattice grille and a tempered glass lens. Having gone to that length to seal the fixture, you have to have a way to change lamps without breaking the seal. As can be seen from the detail, the method of getting at the lamp from the back side must be detailed. It is certainly no ordinary recessed box, nor is it inexpensive. Note also the separately circuited night-light.

Also depicted in Figure 1-6 is a detail of a fixture in a medium security cell. It is presumed that the occupant is not the sort who deliberately breaks fixtures that, if broken, will not help in an escape. The lens is high-stress and can survive accidental blows delivered in periods of frustration. With the right tools the fixture can be relamped when the cell is unoccupied. The tamperproof screws are difficult, but not impossible, to remove, if the prisoner tries hard enough. Patients in mental institutions, apparently immune to pain, have removed this type of screw with only their fingernails as a tool.

Figure 1-7 shows two other applications, one for fixtures installed over cell air cores and one for fixtures installed over cell galleries and exercise corridors. Again, it is the type of abuse as well as use that determines the carefully detailed fixture housing.

In a number of prison situations the cell air core fixture could be abused. Its housing is almost like a fort. The enclosure is a 14-gauge casing. Its lens is covered by a steel rod grillwork. It can be disconnected from outside the area. At fixed times it can be opened up and relamped.

The exercise corridor fixture is a fairly common prison-type fixture. It is dual-circuited and very clearly depicts both a regular lamp and a night-light. Generally speaking, light fixtures designed around the ½-in-square tempered lens are best-suited for maximum security applications. The installation arrangements should provide minimum access to persons being detained. The 8½-in size precludes providing an escape route.

Lamp access should generally be as shown on the details. Long-life lamps should be used. Where feasible, long-life mercury-vapor fixtures should be specified. Night-lights should include long-life rectifiers within the fixture to provide sufficient illumination for night security checking and power-failure backup.

If a fixture is fitted with louvers for audio surveillance, then all fixtures in the area, whether used for surveillance or not, should have similar louvers to avoid suspicion of the person surveyed.

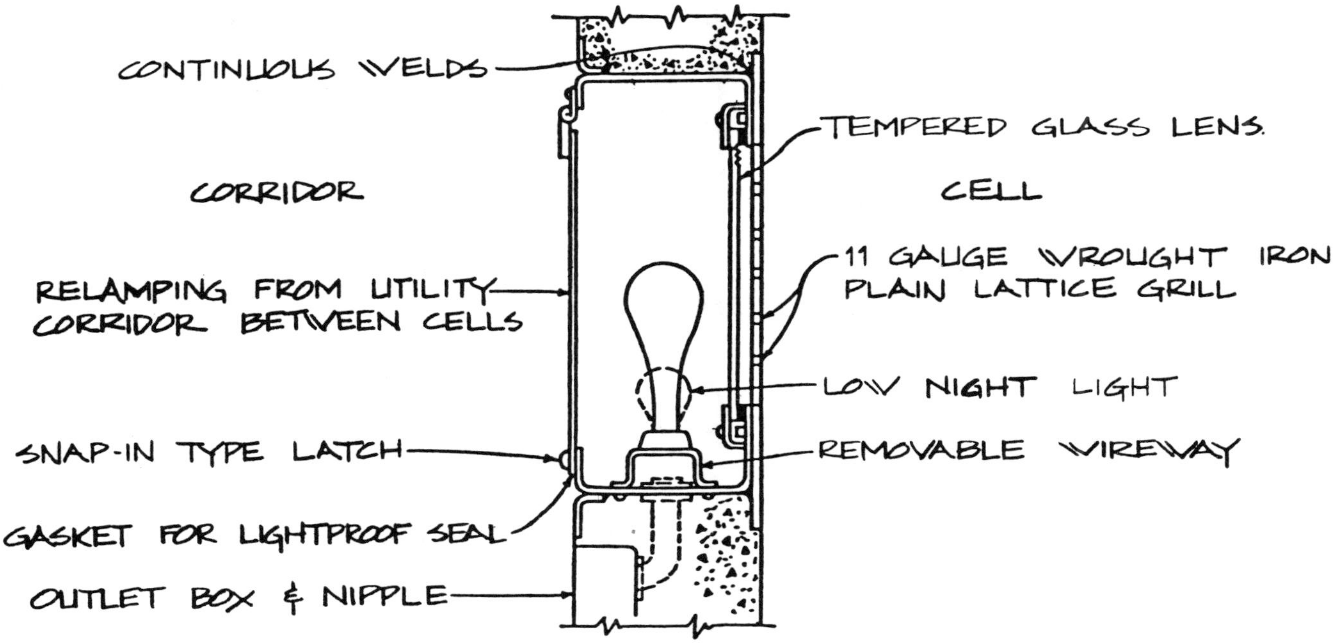

LIGHTING FIXTURE
MAXIMUM SECURITY CELL

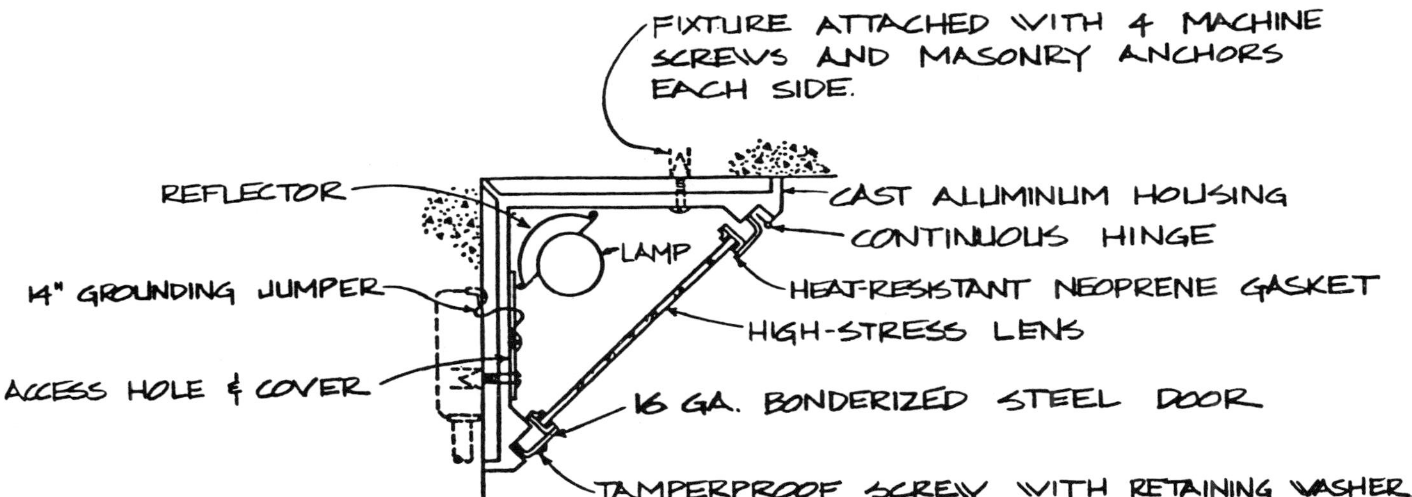

LIGHTING FIXTURE
MEDIUM SECURITY CELL

PRISON LIGHTING DETAILS

NO SCALE

FIGURE 1-6

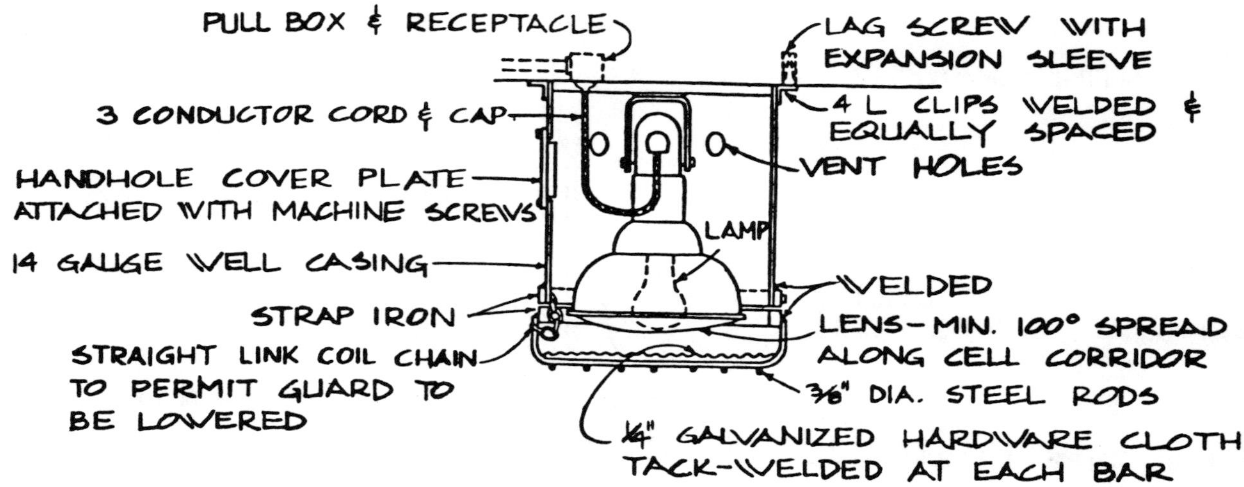

EMERGENCY LIGHTING FIXTURES
OVER CELL AIR CORES

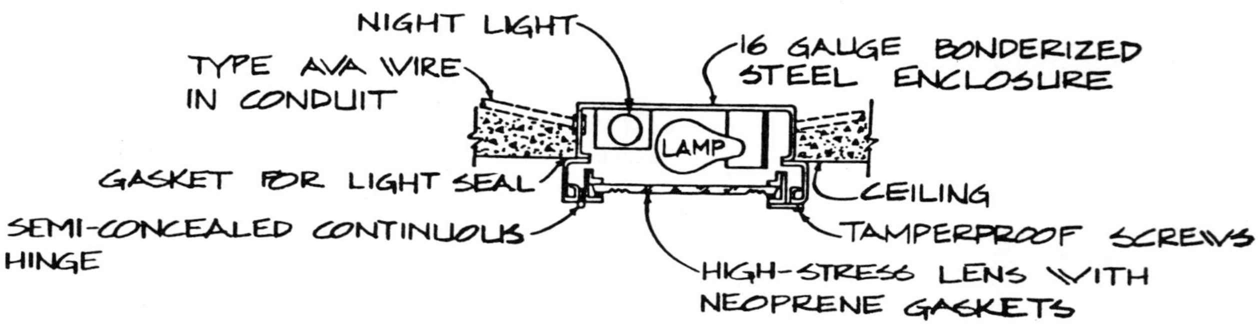

UNIT INSTALLED OVER CELL GALLERIES
AND EXERCISE CORRIDORS

PRISON LIGHTING DETAILS

NO SCALE

FIGURE 1-7

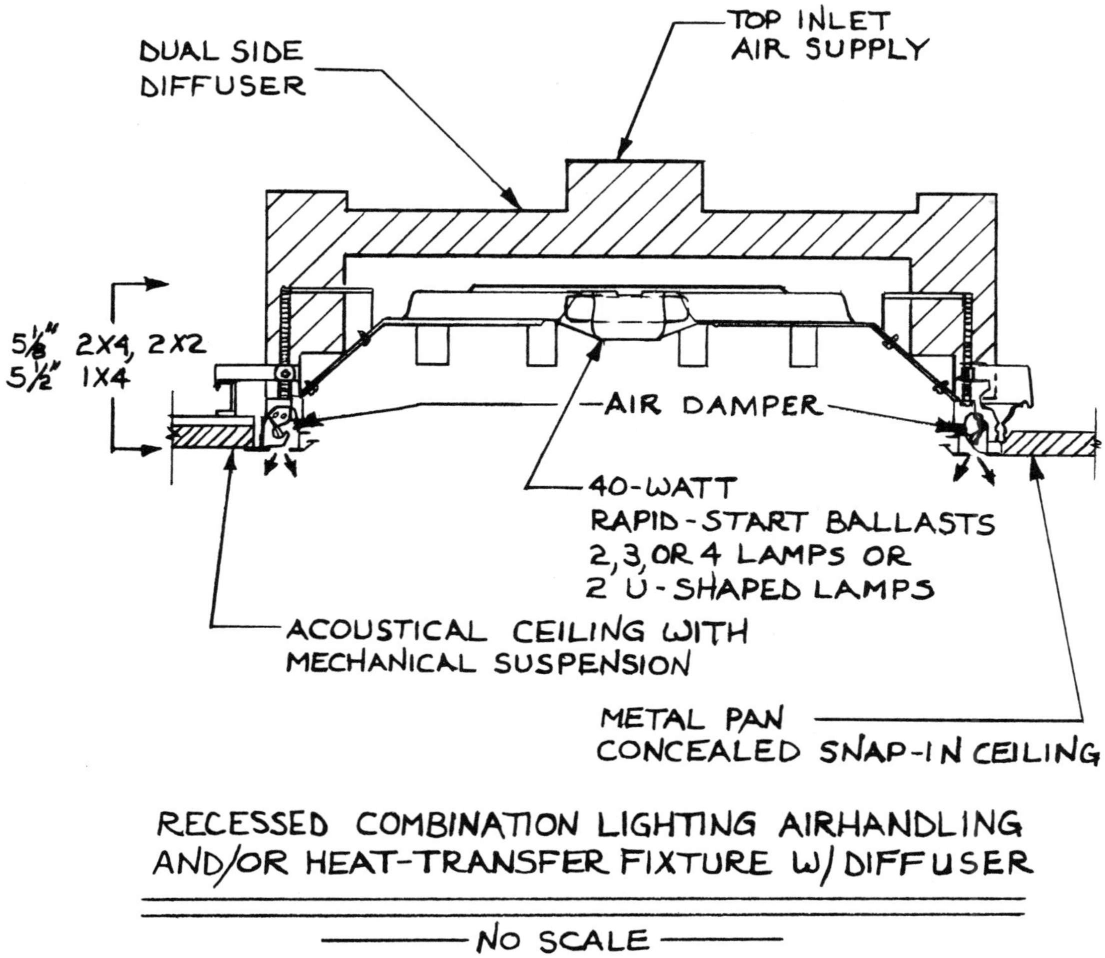

FIGURE 1-8

While steel is usually required, fixture construction finishes may be cast aluminum with baked enamel coating. The steel used in the housing should be at least 16 gauge. Finally, there are few manufacturers who have lines of special, custom, security-type fixtures.

Air-Handling Light Fixture As we pointed out in our opening discussion of lighting fixture details, many fixtures simply are supported by the suspension system of the dropped ceiling. The detail of the fluorescent air-handling fixture shown in Figure 1-8 is in that category. Whether the fixture is square or rectangular, incandescent or fluorescent, the basic type of support is the same.

It is very easy to discover in this application that the correct amount of illumination requires too many or too few fixtures to meet the needs of delivering or exhausting air for the heating, ventilating, and air-conditioning system. Certainly outputs can be adjusted to resolve the differences, but design compromises are frequently less than ideal for the compromisers. If both the illumination and space air-handling requirements can be met without forcing the design of either into peculiar arrangements, the air-handling fixture can be a very pleasant, aesthetically pleasing design solution. The detail illustrates the lighting fixture in a standard format and the air-boot connection as a crosshatched section.

The importance of detailing what is obviously a standard manufactured product lies not in the complication or arrangement of support but rather in the clarification of who does what and what will be the result. Frequently there can be union or local practice disputes over who supplies the air boot. You will find the boot shown in both fixture and in diffuser manufacturers' catalogs. Your detail should very clearly define who supplies this item, what type it is, who installs and connects the item, and who balances the air-supply output.

The ideal solution to this problem, as well as to any light fixture schedule on the plans, is to have any data in either of these cases become a reference to the detail, to have the detail clearly describe each item, including manufacturer and model number, and to have a clear-cut definition of work division and responsibility.

Since this detail affects both the heating, ventilating, and air conditioning, and the electrical divisions of the

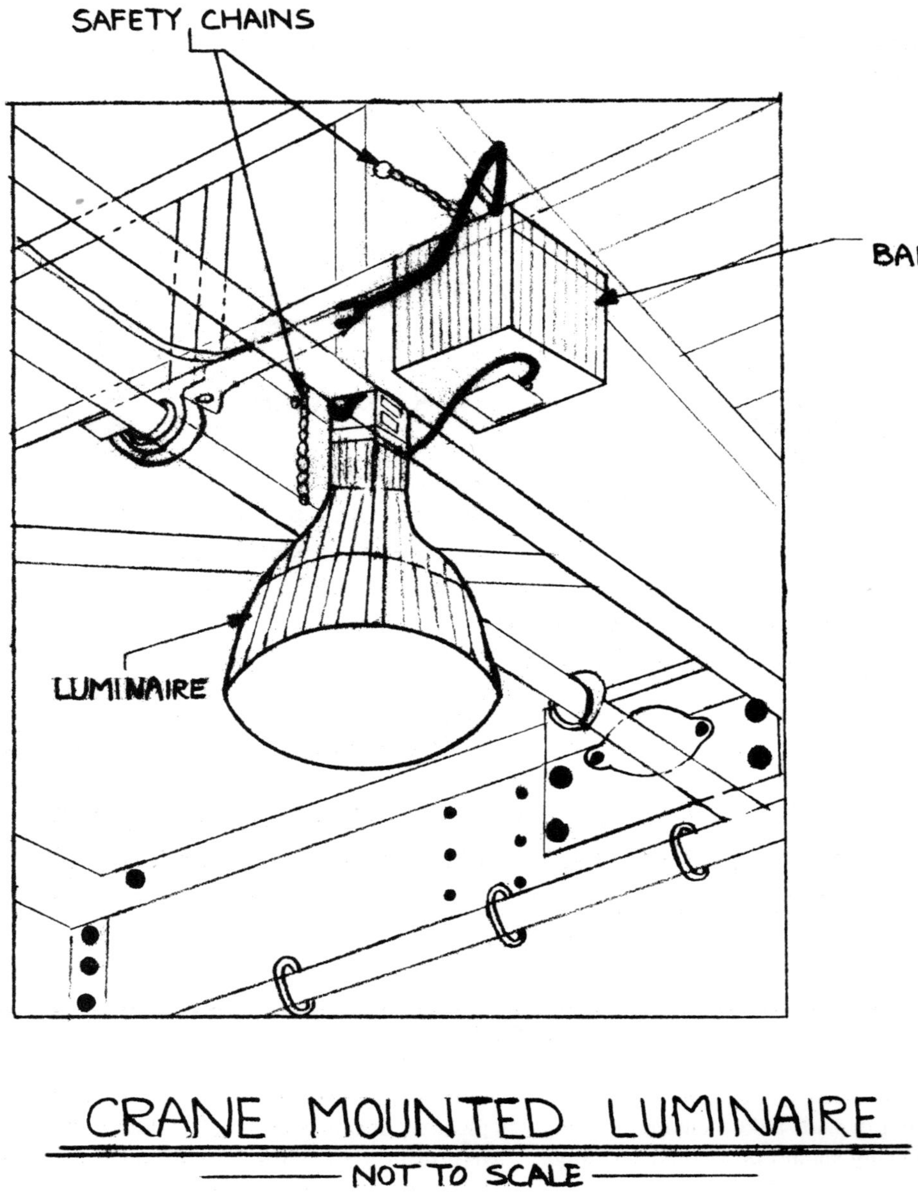

FIGURE 1-9

contract, the *identical* detail should appear in both areas of the drawings or a very clear notation should be made on the heating, ventilating, and air-conditioning drawings referring this contractor to the electrical drawings to find the work to be done.

Crane-Mounted Lighting In lighting an industrial area the solution commonly requires an even light output onto the work surface from a carefully selected and located series of high-bay lighting fixtures. However, there are occasions when the material handled by a moving crane requires lighting directly below the crane. Figure 1-9 is a solution to that problem. *It is not the only solution.*

The fixture and ballast shown must be of rugged, heavy-duty material and solidly built. There is no guarantee that even this type of fixture construction will always absorb the bangs and bumps suffered by the installation. Note that both the fixture and the ballast have safety-chain support. Presumably if either device gets knocked off, shaken off, or just plain falls off, it will not drop on the heads of workers below because of the safety-chain support.

The safety chain should be short so that the fall of the ballast or luminaire does not get much help from the pull of gravity and does not become a swinging, lethal weapon in the work area. A length of 4 ft 0 in is reasonable and 6 ft 0 in is maximum.

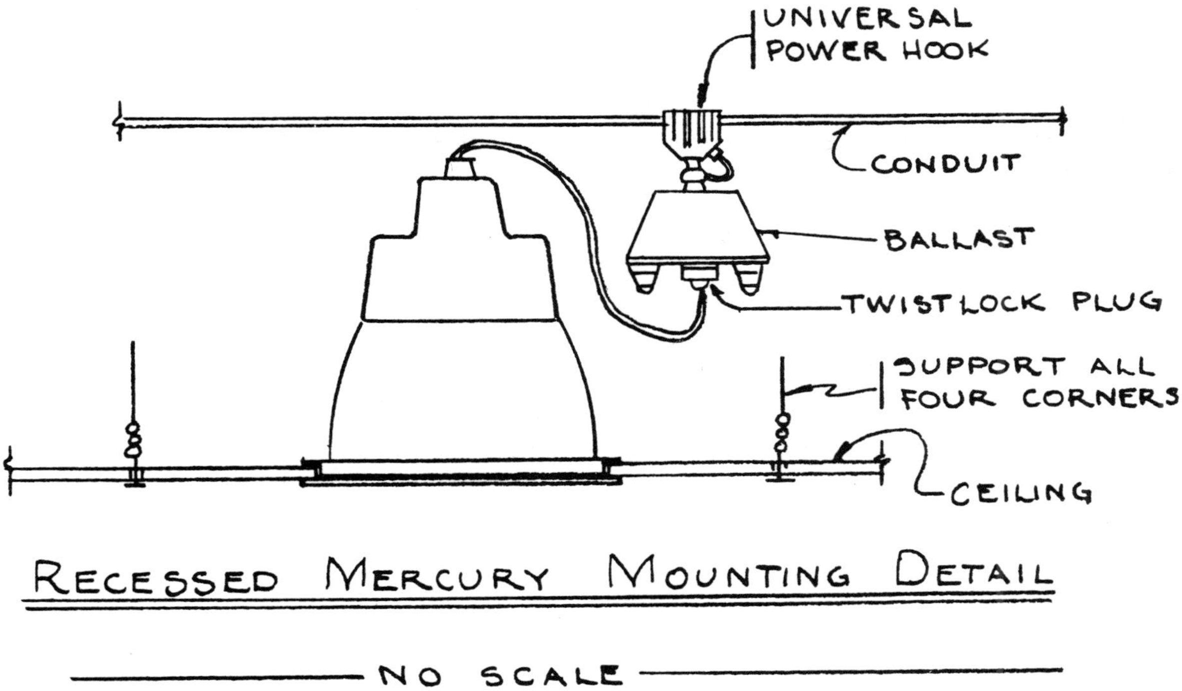

FIGURE 1-10

Both the fixture and the ballast should be carefully specified to include vertical or horizontal surface-mounting brackets. In addition there should be a remote-mounted power cord connector specified for attaching the ballast to the luminaire.

In the detail shown the fixture is bolted to the underside of the crane and the ballast is bolted to the side of the crane. Where required, a properly sized ⅜-in angle bracket can be welded onto the joist and the luminaire and its bracket bolted to the angle, thus raising the luminaire further up and making its bottom more nearly match that of the crane.

This description of fixture and related installation parts and the method of attachment is difficult to describe in words in a specification or light fixture schedule. It would be far better to have the schedule and the specification both refer to the detail. The detail should carry all the installation notes, manufacturers' and model numbers, and other applicable data.

Recessed Large Lighting Fixtures Figure 1-10, depicting a recessed mercury lighting fixture, almost did not make it into this book. It is so simple, so straightforward, and so obvious that showing it hardly seemed necessary But humorously (if your competitor has the problem) or not so humorously (if the problem is yours), the area above the dropped ceiling frequently gets decreased near the end of the design and you are then suddenly faced (if you or your design checker catches it) with a fixture and ballast that just won't fit. To protect against all possibilities, some offices will put this detail on the plans with the notation "Use this ballast and fixture installation arrangement where space conditions preclude fixture-mounted ballast."

For the sake of clarity, the conduit in the drawing is shown above the fixture. It could have been alongside and lower than the fixture. Keep in mind that the minimum clearance is the height of the fixture plus the space necessary for the cable from the ballast to and into the top of the fixture.

The suspension of the fixture itself is not shown, since it varies from manufacturer to manufacturer. The weight of the fixture is such that it cannot simply sit in a ceiling tile opening but must be separately suspended on its own hanger.

Underwater Lighting In far too many underwater pool lighting systems, the electrical drawings show a fixture installation in a simplified, abbreviated detail that is purely schematic. The premise of this approach is that the fixture is very clearly and completely specified and the shop drawing submitted for approval will very clearly cover the details of fixture installation. This is reasonably accurate. Unfortunately, this approach presumes not only the electrical contractor but also the concrete contractor and the general contractor know exactly what their area of work will encompass. Again, this is true provided you are dealing with contractors who have all had previous experience in this type of work.

So far we have outlined all the reasons why you never need to show this detail. Only one point is missing. It is no problem for the electrical contractor to install the fixture in the niche provided, and the junction box and conduit will be installed before any concrete is poured. However, since there was no detail, the concrete and general contractors can, and will, state that no niche, grounding, or fixture attachment system was shown on the bidding documents and that any required will be provided at extra cost. Avoiding this frustrating claim for an extra undoubtedly accounts for the large variety of sketchy details shown on most pool underwater lighting plans. Since it takes very little more effort to produce an argument-free detail whose source is a shop drawing of a standard manufactured product, it hardly seems necessary to point out that this is what should be done.

There is more than one location for the junction box shown in Figure 1-11. It could be mounted as a free-standing item at the same height under the diving board. Finally, the fixture installation has a notation to allow for pool finish—tile, etc.—which can, by itself, save a lot of discussion.

The grounding system for a pool is shown in a separate detail in another section of the book. The fixture is 24 V and is fed from a separately located transformer not shown in the detail.

Pool Overhead Lighting When lighting an indoor swimming pool, the difficulty of the task revolves about the problem of relamping the fixture located over the water. Concurrent with this problem is the avoidance of glare on

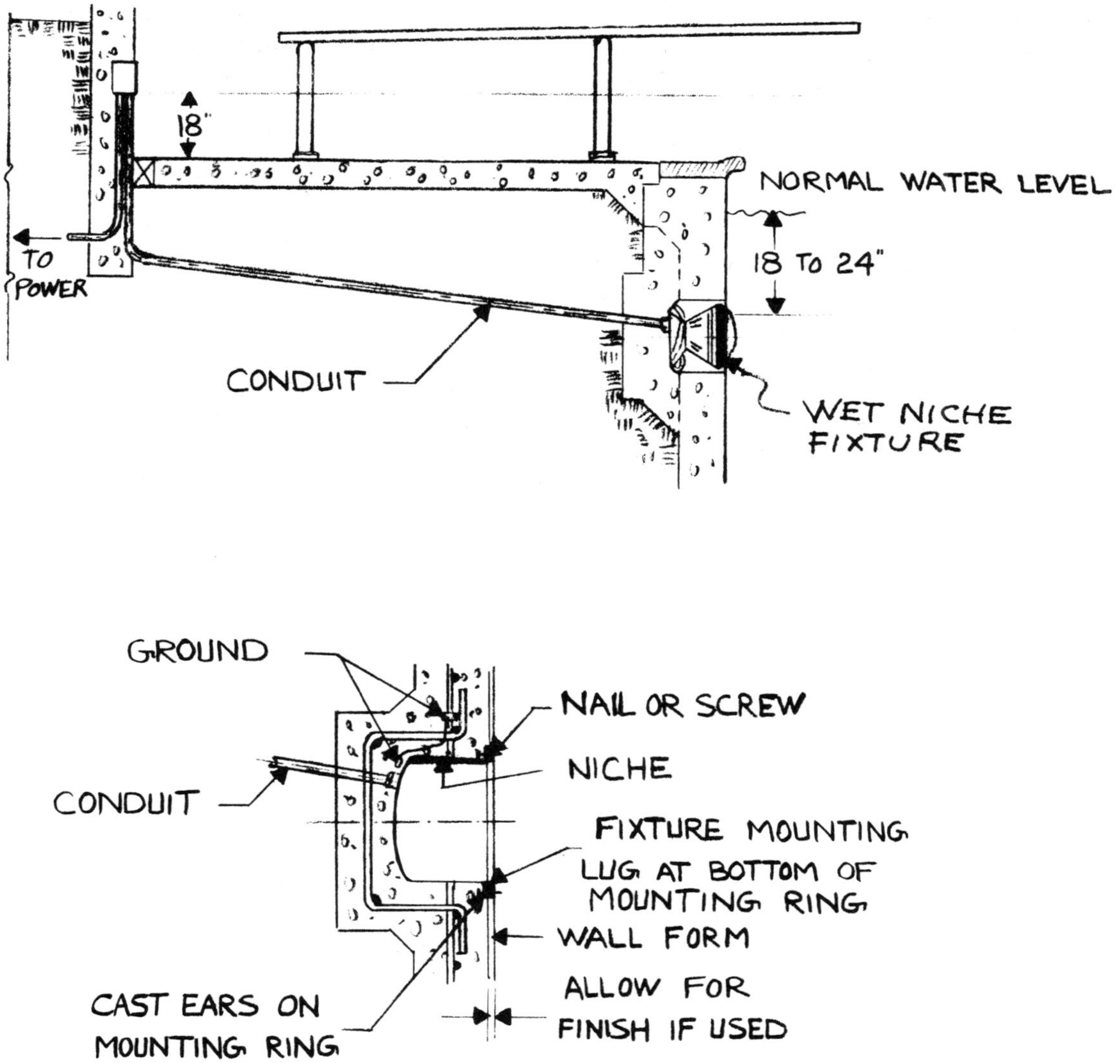

UNDERWATER POOL LIGHTING DETAILS

NOT TO SCALE

FIGURE 1-11

the surface of the water which would prevent seeing swimmers in distress below the surface. Generally the solution most frequently used is the directional lens fixture located above the deck and just sufficiently outside the water area so that a ladder can be placed on the deck and a straight-up climb be made directly to the fixture.

The problem of lighting the pool from a cove by the usual fluorescent lamp is that the amount of footcandles that can be produced is frequently insufficient for pools featuring major athletic events. Figure 1-12 shows a cove utilizing a floodlight-type fixture. Depending on the quantity of illumination required, floodlights can produce a large and varied amount of light output.

There are manufactured systems which can be selected with special tracks to plug in fixtures of this type. The fixture itself can be incandescent, mercury, metal halide, or variations of same. The height of the cove can vary to some degree. The one illustrated, while not dimensioned, has a height of 36 in and the fixtures are mounted on 8 ft 0 in centers.

The fixtures should be the vapor-proof type, and all metal portions of fixture, track, conduit, and junction box should be carefully specified to operate in the chlorine-laden atmosphere of an indoor pool.

The depth of the cove is noted on the detail and is 2 ft 6 in, which is a minimum value for proper fixture concealment.

Pole-Mounted Lighting Many types of outdoor lighting require a complete, independent system to support the luminaire. The most common outdoor lighting installation is the one used for roadways, drives, parking lots, and

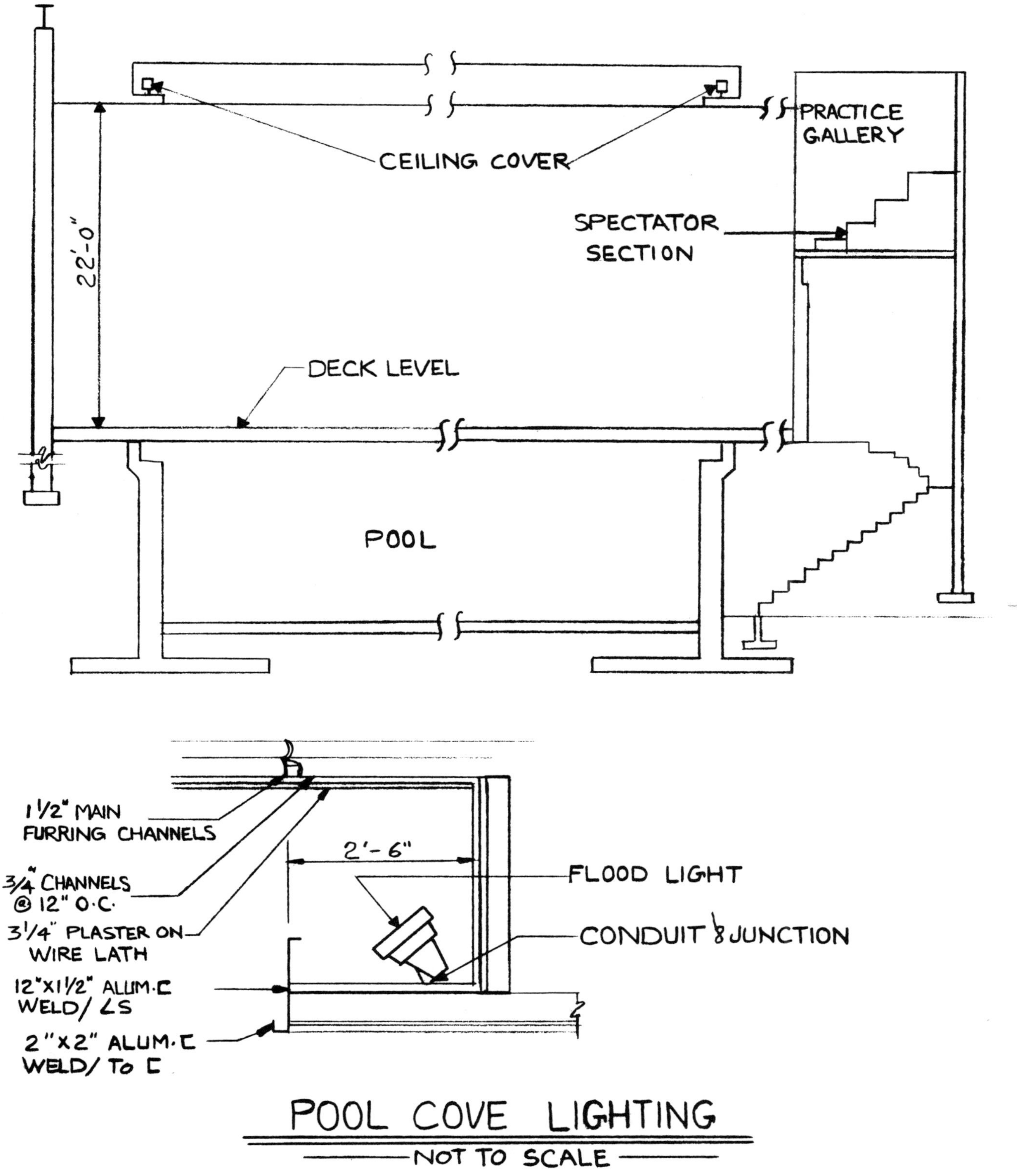
CEILING COVER
PRACTICE GALLERY
SPECTATOR SECTION
22'-0"
DECK LEVEL
POOL
1 1/2" MAIN FURRING CHANNELS
3/4" CHANNELS @ 12" O.C.
3 1/4" PLASTER ON WIRE LATH
12"x1 1/2" ALUM. C WELD/ ∠S
2"x2" ALUM. C WELD/ TO C
2'-6"
FLOOD LIGHT
CONDUIT & JUNCTION
POOL COVE LIGHTING
NOT TO SCALE

FIGURE 1-12

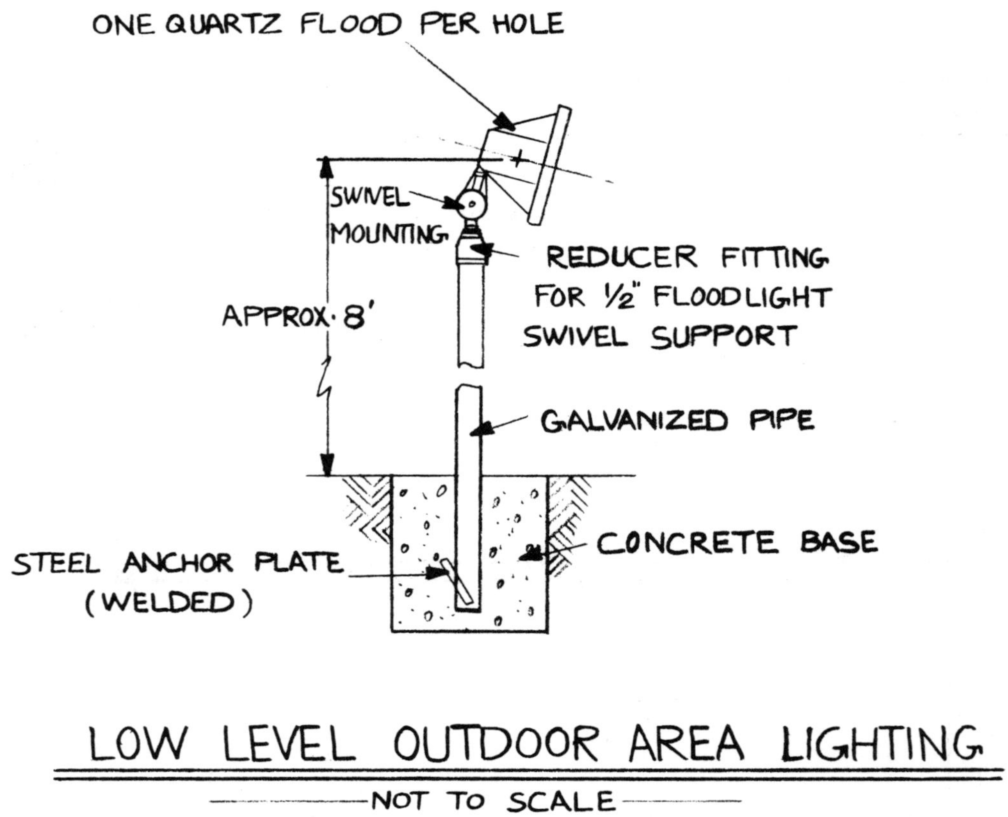

FIGURE 1-13

similar spaces involving vehicular traffic, with and without the accompanying pedestrian traffic.

While the common understanding is that pole-mounted lighting is usually arranged with the luminaire some 20 to 55 ft off the ground, there are visual tasks requiring fixtures at much lower levels. Figures 1-13 to 1-15 illustrate a variety of pole and fixture arrangements.

In Figure 1-13 a quartz floodlight luminaire is connected via a swivel mounting to a galvanized steel pipe. The fixture height is only 8 ft off the ground and the concrete is of simple, standard, 3000-lb construction. Frequently around elevated, narrow loading platforms or in the loading of flatbed trailers or railroad flat cars, the accurate centering of the vehicle or merchandise requires almost daylight conditions at night. This simple arrangement provides an economical solution to the problems which arise.

For the lighting of all types of outdoor parking, driving, freight handling, and pier deck areas, two common fixtures are shown in Figure 1-14. The fixtures could have been high-pressure sodium instead of mercury, and the indication that they are mounted on a deck certainly does not limit their use to a pier deck.

In each case it should be noted that not only is the fixture described but also the lamp, the ballast, the horizontal arm, the pole, the photoelectric control, the transformer base, and the grounding connection. Both the multidirectional and the well-known wide-light fixture enclosures are shown. In using the detail, here again specifying all the parts by manufacturer's name and model number right on the detail with appropriate reference in the specification and light fixture schedule would be welcomed by most suppliers and contractors.

Figure 1-15 shows the typical detail of roadway lighting. Because you are involved with vehicular traffic, the clearance off the road may vary with specific requirements in your particular area. Do not blindly assume that the 8 ft 0 in shoulder dimension is correct. Be certain to have the correct dimension on your project. This also applies and relates to the 6 ft 0 in dimension on the horizontal bracket support of the luminaire.

Because this detail came from a real job, there is a very specific, carefully described reinforced concrete base shown. Note also that the grounding is very specifically shown, with each pole-mounted installation having its own ground rod. The fixture is some 40 ft 0 in off the roadway, as may be noted in the pole description.

The type of pole itself is not left to chance or merely

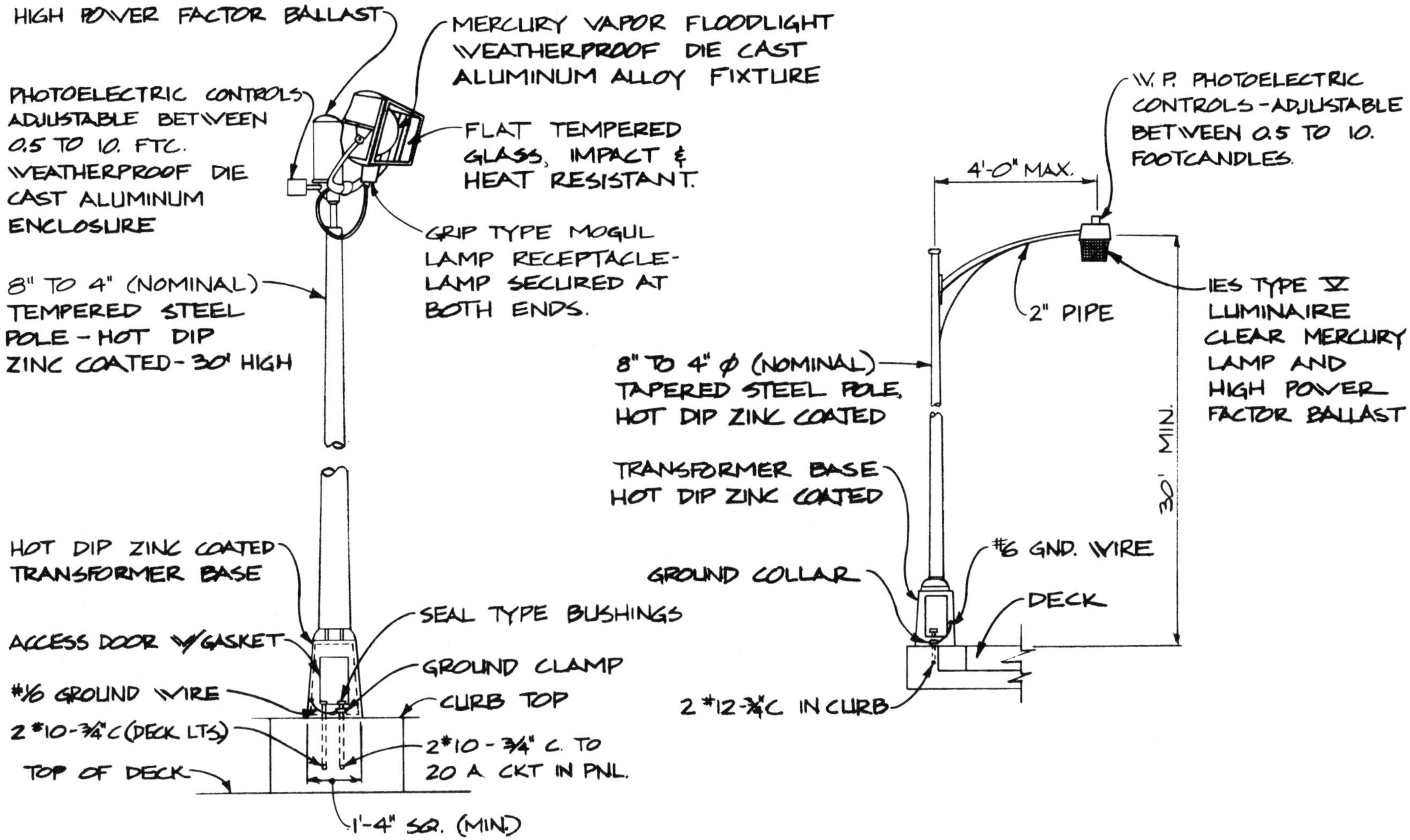

LIGHTING STANDARD DETAILS

NO SCALE

FIGURE 1-14

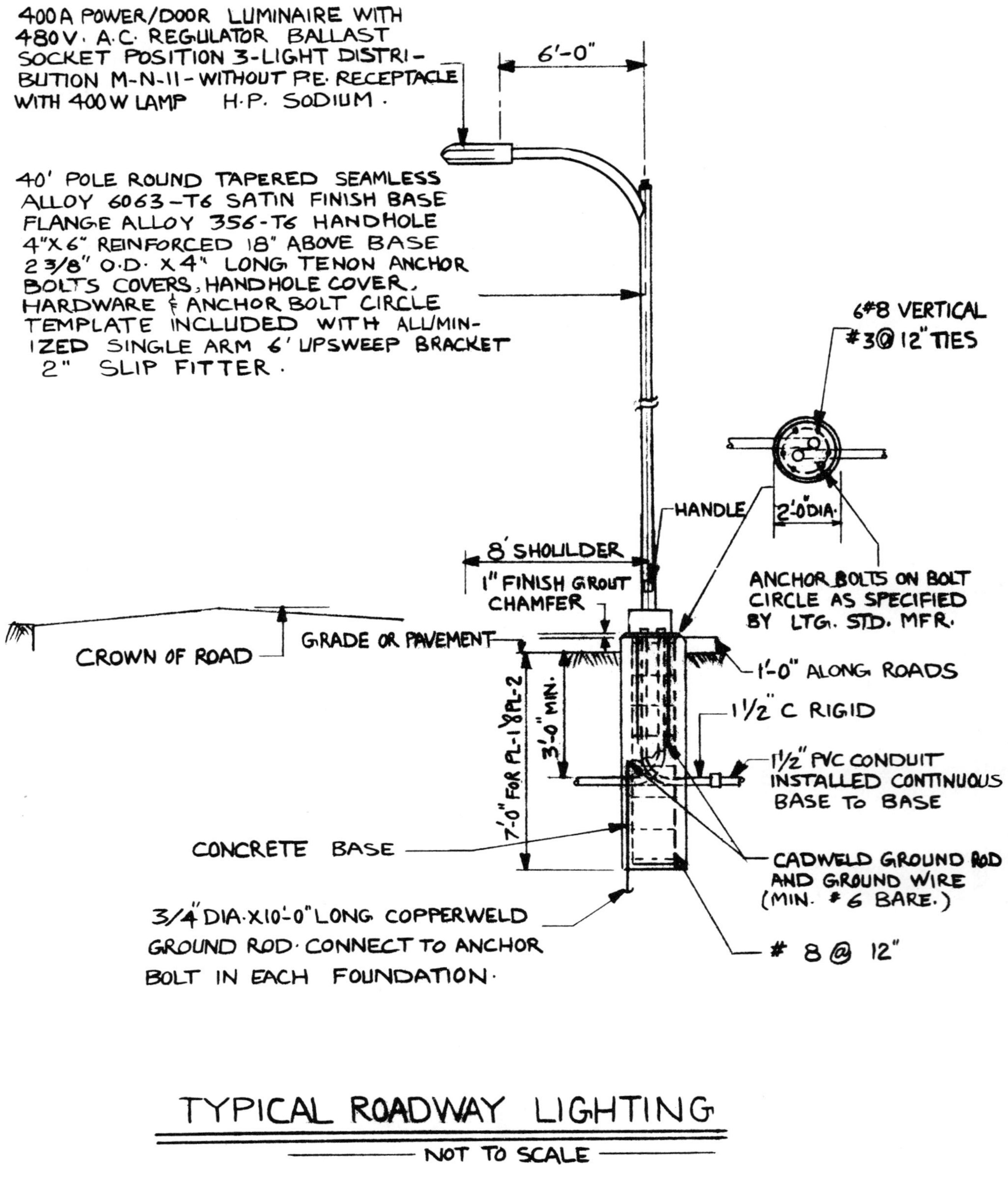
400A POWER/DOOR LUMINAIRE WITH 480V. A.C. REGULATOR BALLAST SOCKET POSITION 3-LIGHT DISTRIBUTION M-N-II-WITHOUT P.E. RECEPTACLE WITH 400W LAMP H.P. SODIUM.
6'-0"
40' POLE ROUND TAPERED SEAMLESS ALLOY 6063-T6 SATIN FINISH BASE FLANGE ALLOY 356-T6 HANDHOLE 4"X6" REINFORCED 18" ABOVE BASE 2 3/8" O.D. X4" LONG TENON ANCHOR BOLTS COVERS, HANDHOLE COVER, HARDWARE & ANCHOR BOLT CIRCLE TEMPLATE INCLUDED WITH ALUMINIZED SINGLE ARM 6' UPSWEEP BRACKET 2" SLIP FITTER.
6#8 VERTICAL
#3@ 12" TIES
HANDLE
2'-0" DIA.
8' SHOULDER
1" FINISH GROUT
CHAMFER
ANCHOR BOLTS ON BOLT CIRCLE AS SPECIFIED BY LTG. STD. MFR.
GRADE OR PAVEMENT
CROWN OF ROAD
1'-0" ALONG ROADS
1½" C RIGID
3'-0" MIN.
7'-0" FOR PL-1 & PL-2
1½" PVC CONDUIT INSTALLED CONTINUOUS BASE TO BASE
CONCRETE BASE
CADWELD GROUND ROD AND GROUND WIRE (MIN. #6 BARE.)
3/4" DIA. X10'-0" LONG COPPERWELD GROUND ROD. CONNECT TO ANCHOR BOLT IN EACH FOUNDATION.
8 @ 12"
TYPICAL ROADWAY LIGHTING
NOT TO SCALE

FIGURE 1-15

described by a manufacturer's name and model number. Note the very detailed description of the pole. The same detail has been applied to the description of the high-pressure-sodium luminaire. Since this fixture is a part of the roadway lighting around an industrial plant, the type of conduit is also described. Note that while the distribution system is PVC, the conduits in and out of the reinforced concrete base and up to the pole are rigid galvanized iron.

There are, of course, other types of poles and pole-mounted lighting, including the low-level, residential, frontyard pole seen in most neighborhoods. These three sets of details do, however, cover the normal range of most pole-mounted lighting systems. Another version, which we have not shown, is the high-mounted cluster arrangement. This is a very specialized, proprietary lighting system offered in many versions by several large lighting manufacturers.

Such a detail should only be put on the drawings after thorough analysis, discussion, and consultation with the fixture manufacturer. The unit consists of a pole 80 to 150 ft in height supported on a specially designed and reinforced concrete foundation. The fixtures, usually 6 to 12 in number, are supported from a specially designed mounting ring that can be raised and lowered by a hoisting motor.

The fixtures are usually metal halide or sodium vapor, shielded, 400 or 1000 W. Poles are normally galvanized or Corten steel. Service hoists, one per pole, consist of head frame, luminaire support ring, and hoisting assembly. This description is only an abbreviated version of a very considerable and technical plan, detail, and specification required for this special lighting system.

Exterior Floodlight While the installation of exterior floodlights mounted on a building, on a wall, or on some other fixed means of support is not very complicated, there are certain basic features of the installation that should be emphasized.

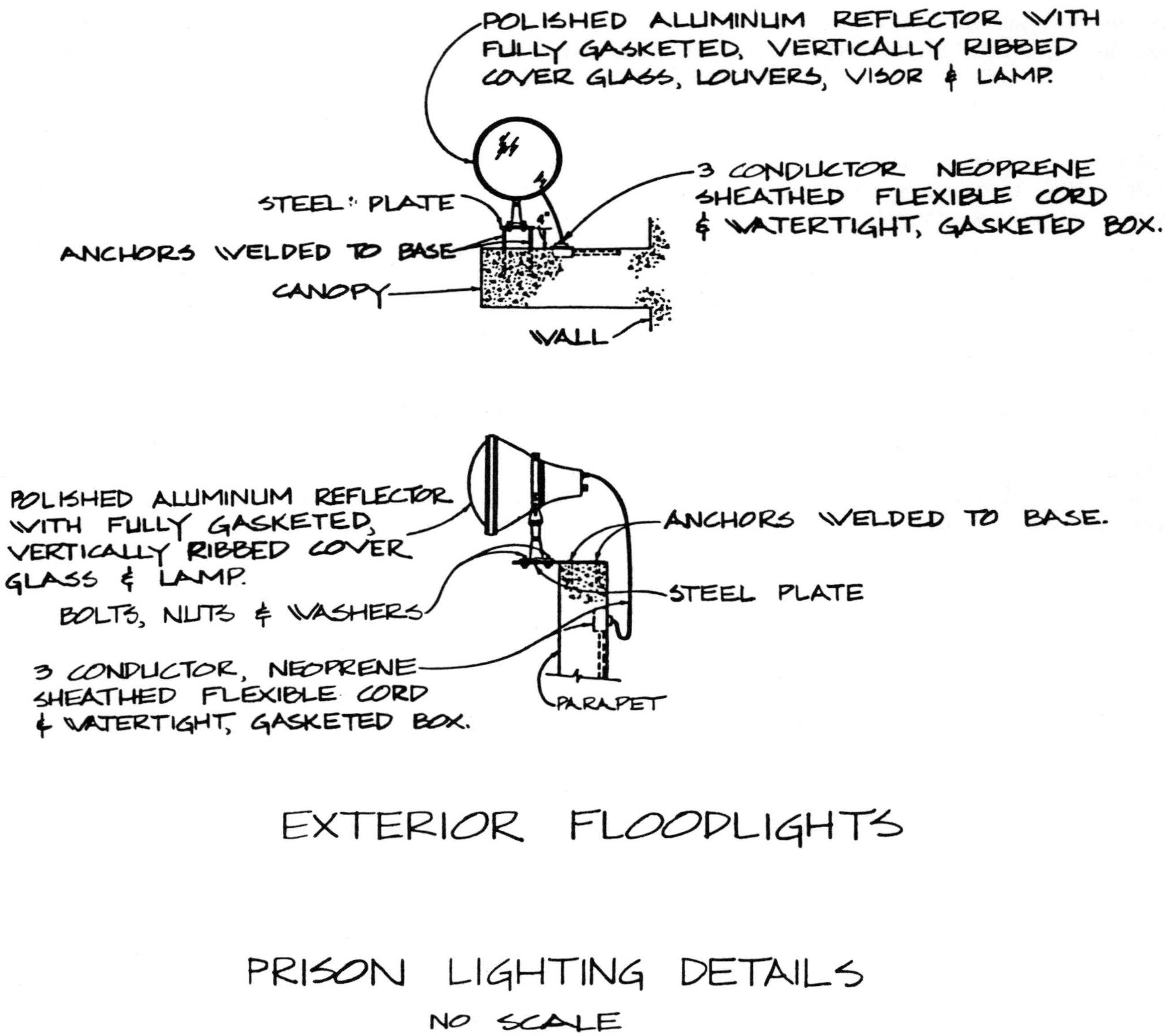

FIGURE 1-16

Figure 1-16 illustrates two details from a prison lighting project, but they could have been located, arranged, and installed on any similar outdoor area. The details were picked because the location, installation, and arrangement are a little out of the ordinary.

In the upper detail the fixture is located on a canopy with an anchored, welded steel base providing the structure to which the fixture is bolted. Keep in mind that this is a prison and fixture attachment security is vital. The fixture support is further anchored into the concrete. While this is not necessary in every project, it is a good idea where there is any concern about the quality of the concrete or where hurricane-force winds may be expected.

In the lower detail the anchors that are holding the plate mounted on top of a parapet wall are not as clearly defined as perhaps they should be. This detail, again designed for a prison, may also be necessary in a normal environment because parapet walls tend to present problems. One point is missing in both cases. We hope you would notice and ask "How thick is the base plate?" The answer is ¼ in.

In both details there is a concealed rigid conduit leading to a gasketed, watertight outlet box. In this outdoor detail, as in many others, thought must be given to fixture replacement. The easiest way to change any outdoor fixture

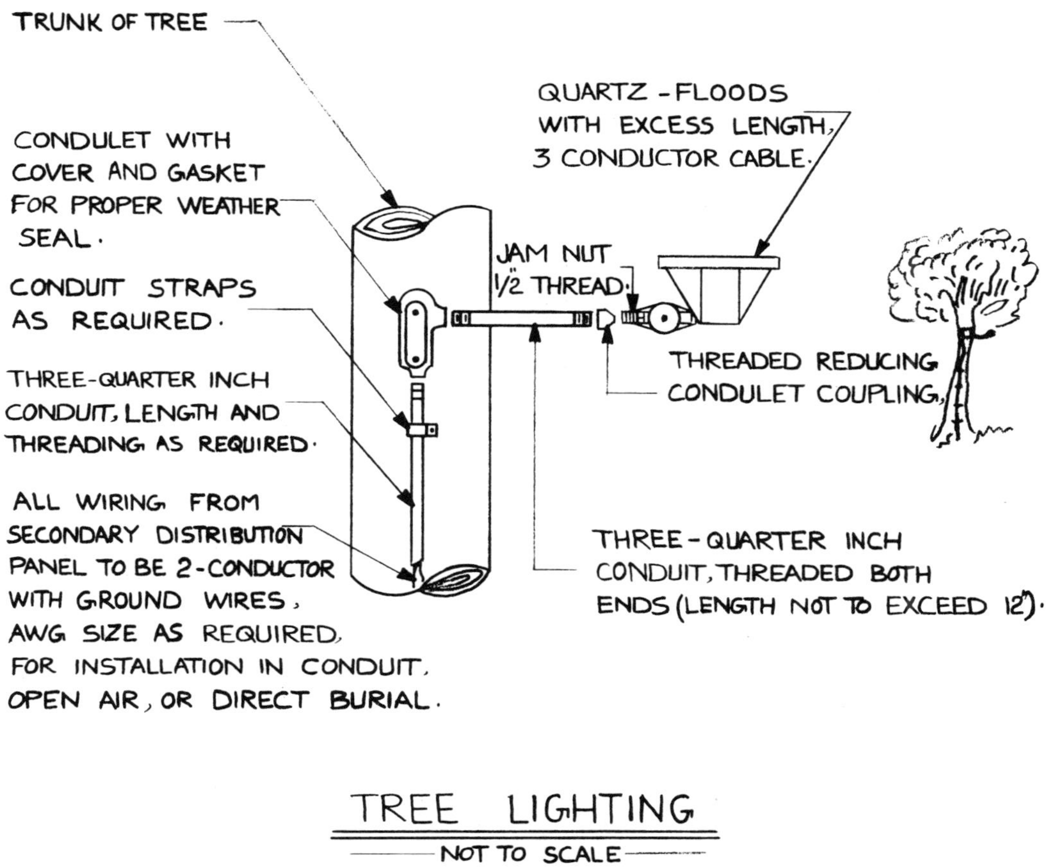

FIGURE 1-17

of this type is to simply unplug it and replace it, which does not disturb the function or operation of the other fixtures.

Decorative Outdoor Lighting While it is not the intent of this book to get into revolving color displays and all the other aspects of outdoor effect lighting, we would be remiss if we did not show at least one detail in this category. Whether the tree, bush, or what have you has one or a dozen fixtures lighting its surface, your basic task is to highlight the object with a safe, logical, electrical installation. A representative example is shown in Figure 1-17. A balanced tree-lighting job should have at least two of the fixtures we show, but that is a design problem. We are concerned with the installation detail.

The quartz lamp will provide more than adequate illumination under and through the tree. Designers in a northern clime, however, would do well to consider how the illuminated tree, neatly shown to the right in our little sketch, will appear when it has no leaves.

Normally the line to the tree will be underground. At the base of the tree the conduit riser should be in an inconspicuous location or painted to blend in with the tree trunk. Generally one or two conduit straps are all that is required to attach the riser to the trunk.

You will undoubtedly have a fixture located at some distance out from the trunk. In our detail we have said you can use a ¾-in rigid conduit for distance up to 12 ft, which should cover most cases. We have shown a swivel-type joint at the fixture so that the luminaire can be aimed as desired. Again, the installation is outside and all fittings, fixtures, and connections must be selected to meet the weather conditions you expect to encounter.

Ground-Mounted Floodlights In Figure 1-17 we showed how to detail the lighting of a tree with the fixture on the tree itself. Probably the thought arose in your mind that this method may be all well and good but the client, the landscape architect, and the project architect do not particularly care to see fixtures bolted on trees, and furthermore there are lots of other shrubbery, statues, and the like which also need illumination, and for which the previous arrangement is unworkable.

Figures 1-18 to 1-20 illustrate some solutions covering the installation of various types of fixtures from a ground-based location. In all of the details there are common, basic items necessary to support and power the fixture. These are discussed in the paragraphs that follow.

The first item to consider in any of these installations (which really are only samples of an almost unlimited variety of possible support systems) is that the fixture must be supported either by a driven stake or by a concrete pad. If the fixture is not going to be subjected to abuse and it is not very large or very heavy, serious consideration should be given to the stake with a separate, hidden, weatherproof outlet. While the pad may not seem like much of an object since it usually is about 1-ft square, it is an unattractive blob of concrete on a green lawn in the daytime.

As the details illustrate, you can put wiring, outlet, and fixture together on one pad and provide a neat, solid, close-coupled installation. However, there will be many times when the separation of the outlet base and fixture really is the best solution.

There also will be cases where a fluorescent fixture is the best solution. These are far fewer than the incandescent or quartz luminaire solutions, but where a fluorescent fixture is required, the ballast may be a problem all by itself. An answer is shown in the separate, partially buried ballast which also acts as a fixture support. Note that this installation still requires the weatherproof junction box.

Since your fixture is very close to the ground and subject to all the vagaries of the weather, you should not only clearly emphasize the weatherproof requirements but also directly name manufacturers, types, and model numbers for all items on the detail. Items that should be clearly specified include fixture, cable, conduit, connector, cap, ballast, tee, coupling, union, and stake. This sort of detail has long been a semistandard in most consulting firms. There are a number of manufacturers who offer similar details and whose products are itemized in applicable fixture catalogs.

Unless there is some compelling reason to do otherwise, a number of outdoor outlets of this type are usually wired to a single circuit controlled by a manual switch or timer. The installation can further be controlled by a photocell arrangement.

Fence-Mounted Lighting In any enclosed, restricted area that houses people or equipment, there is frequently a requirement to mount the fixture on the enclosure. Far too often the engineers and designers view this installation as a simple, obvious problem and merely specify that the fixture be mounted on the enclosure at such and such height above the ground.

It is a simple problem and, with a reasonably cooperative contractor, the installed result is usually reasonably satisfactory. The word "reasonably" is used twice in the above sentence to make the point about this detail being omitted. Like every other part of your design, it is frequently difficult to get exactly what you want where you want it unless you clearly spell it out.

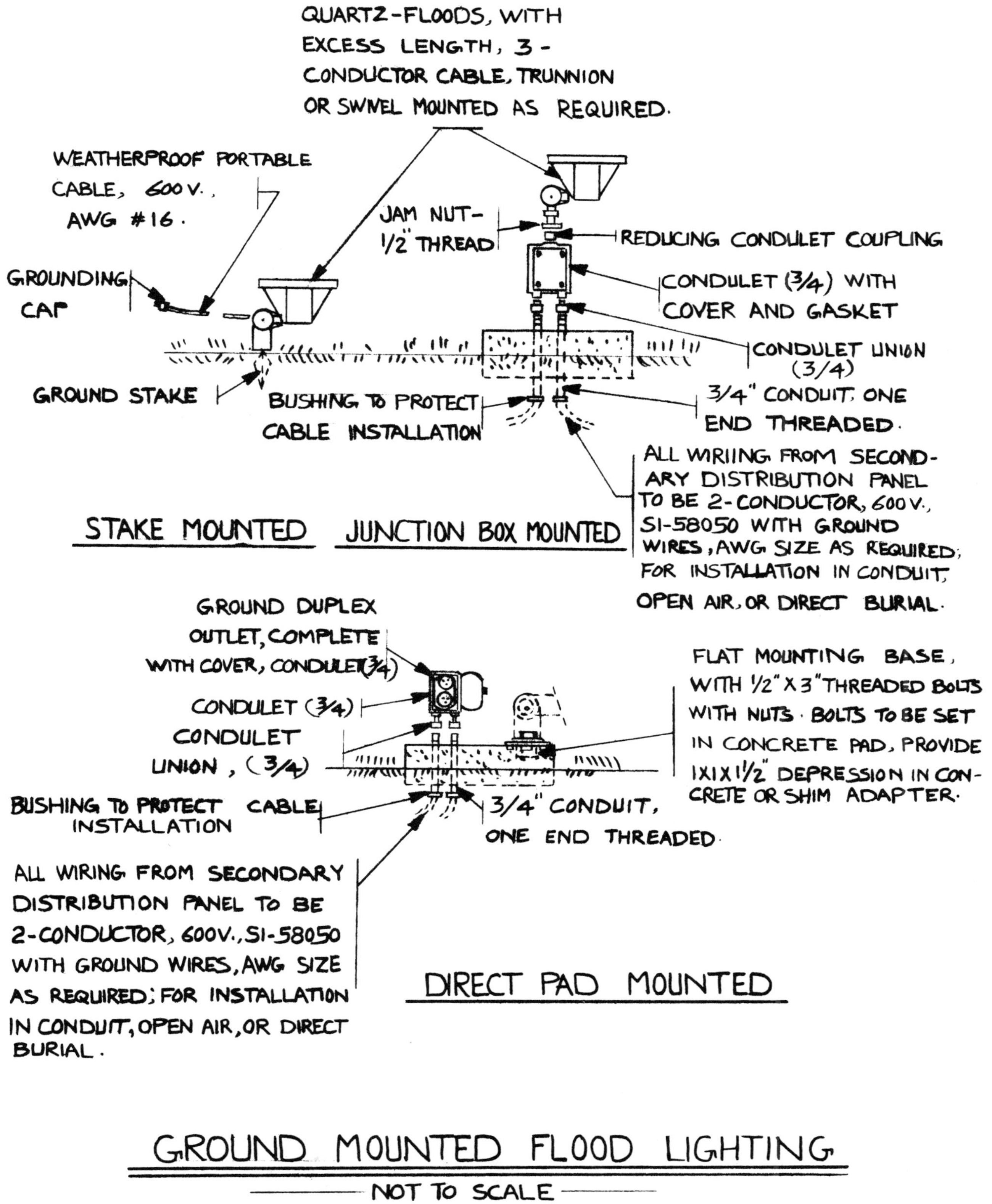
QUARTZ-FLOODS, WITH EXCESS LENGTH, 3-CONDUCTOR CABLE, TRUNNION OR SWIVEL MOUNTED AS REQUIRED.
WEATHERPROOF PORTABLE CABLE, 600 V., AWG #16.
JAM NUT- 1/2" THREAD
REDUCING CONDULET COUPLING
GROUNDING CAP
CONDULET (3/4) WITH COVER AND GASKET
CONDULET UNION (3/4)
GROUND STAKE
BUSHING TO PROTECT CABLE INSTALLATION
3/4" CONDUIT; ONE END THREADED.
ALL WIRIING FROM SECONDARY DISTRIBUTION PANEL TO BE 2-CONDUCTOR, 600 V., SI-58050 WITH GROUND WIRES, AWG SIZE AS REQUIRED; FOR INSTALLATION IN CONDUIT, OPEN AIR, OR DIRECT BURIAL.
STAKE MOUNTED
JUNCTION BOX MOUNTED
GROUND DUPLEX OUTLET, COMPLETE WITH COVER, CONDULET (3/4)
CONDULET (3/4)
CONDULET UNION, (3/4)
BUSHING TO PROTECT CABLE INSTALLATION
FLAT MOUNTING BASE, WITH 1/2" X 3" THREADED BOLTS WITH NUTS. BOLTS TO BE SET IN CONCRETE PAD, PROVIDE 1X1X1 1/2" DEPRESSION IN CONCRETE OR SHIM ADAPTER.
3/4" CONDUIT, ONE END THREADED.
ALL WIRING FROM SECONDARY DISTRIBUTION PANEL TO BE 2-CONDUCTOR, 600V., SI-58050 WITH GROUND WIRES, AWG SIZE AS REQUIRED; FOR INSTALLATION IN CONDUIT, OPEN AIR, OR DIRECT BURIAL.
DIRECT PAD MOUNTED
GROUND MOUNTED FLOOD LIGHTING
NOT TO SCALE

FIGURE 1-18

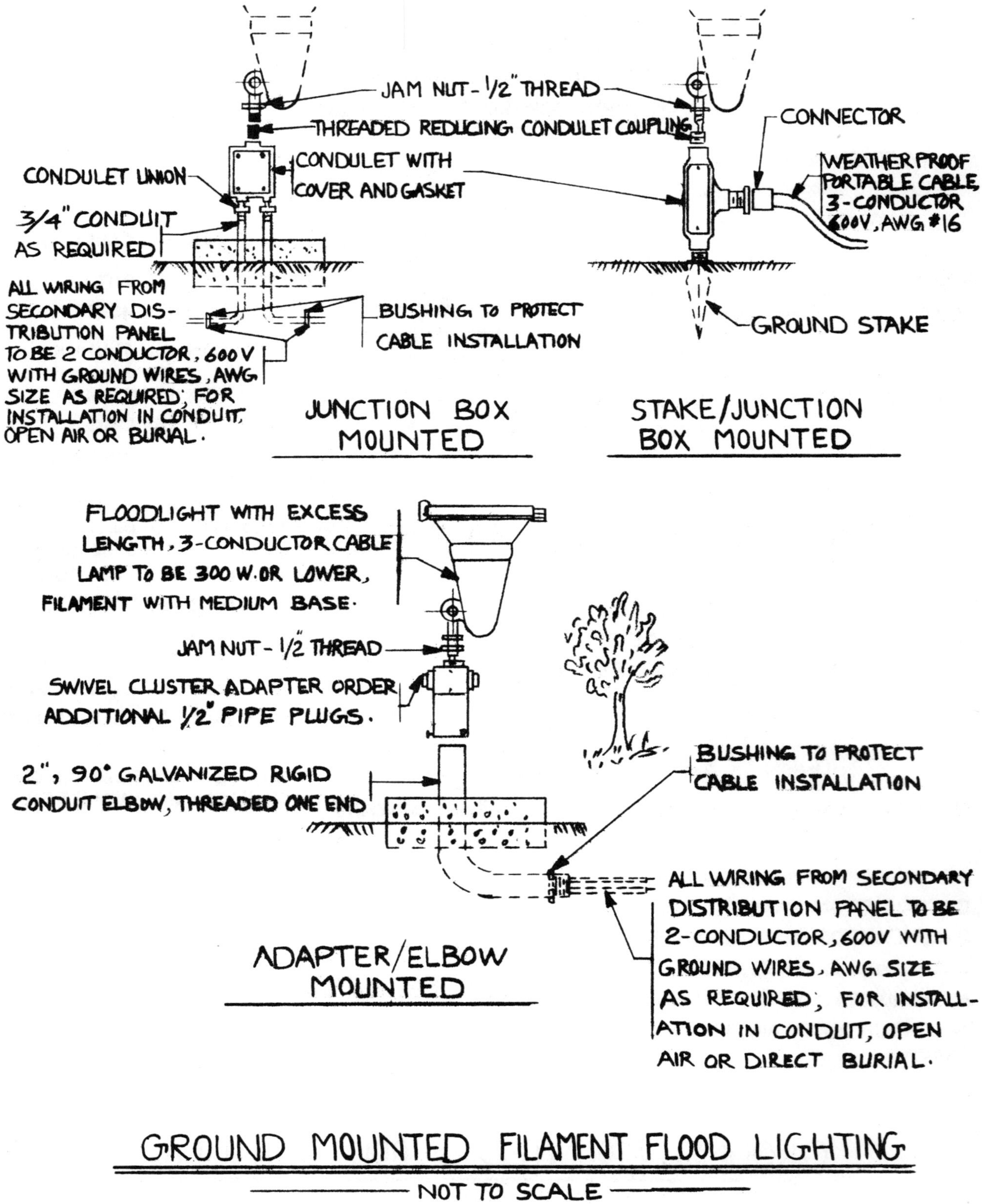
JAM NUT - 1/2" THREAD
THREADED REDUCING CONDULET COUPLING
CONDULET WITH COVER AND GASKET
CONDULET UNION
3/4" CONDUIT AS REQUIRED
ALL WIRING FROM SECONDARY DISTRIBUTION PANEL TO BE 2 CONDUCTOR, 600 V WITH GROUND WIRES, AWG SIZE AS REQUIRED, FOR INSTALLATION IN CONDUIT, OPEN AIR OR BURIAL.
BUSHING TO PROTECT CABLE INSTALLATION
JUNCTION BOX MOUNTED
CONNECTOR
WEATHER PROOF PORTABLE CABLE, 3-CONDUCTOR 600V, AWG #16
GROUND STAKE
STAKE/JUNCTION BOX MOUNTED
FLOODLIGHT WITH EXCESS LENGTH, 3-CONDUCTOR CABLE LAMP TO BE 300 W. OR LOWER, FILAMENT WITH MEDIUM BASE.
JAM NUT - 1/2" THREAD
SWIVEL CLUSTER ADAPTER ORDER ADDITIONAL 1/2" PIPE PLUGS.
2", 90° GALVANIZED RIGID CONDUIT ELBOW, THREADED ONE END
BUSHING TO PROTECT CABLE INSTALLATION
ALL WIRING FROM SECONDARY DISTRIBUTION PANEL TO BE 2-CONDUCTOR, 600V WITH GROUND WIRES, AWG SIZE AS REQUIRED, FOR INSTALLATION IN CONDUIT, OPEN AIR OR DIRECT BURIAL.
ADAPTER/ELBOW MOUNTED
GROUND MOUNTED FILAMENT FLOOD LIGHTING
NOT TO SCALE

FIGURE 1-19

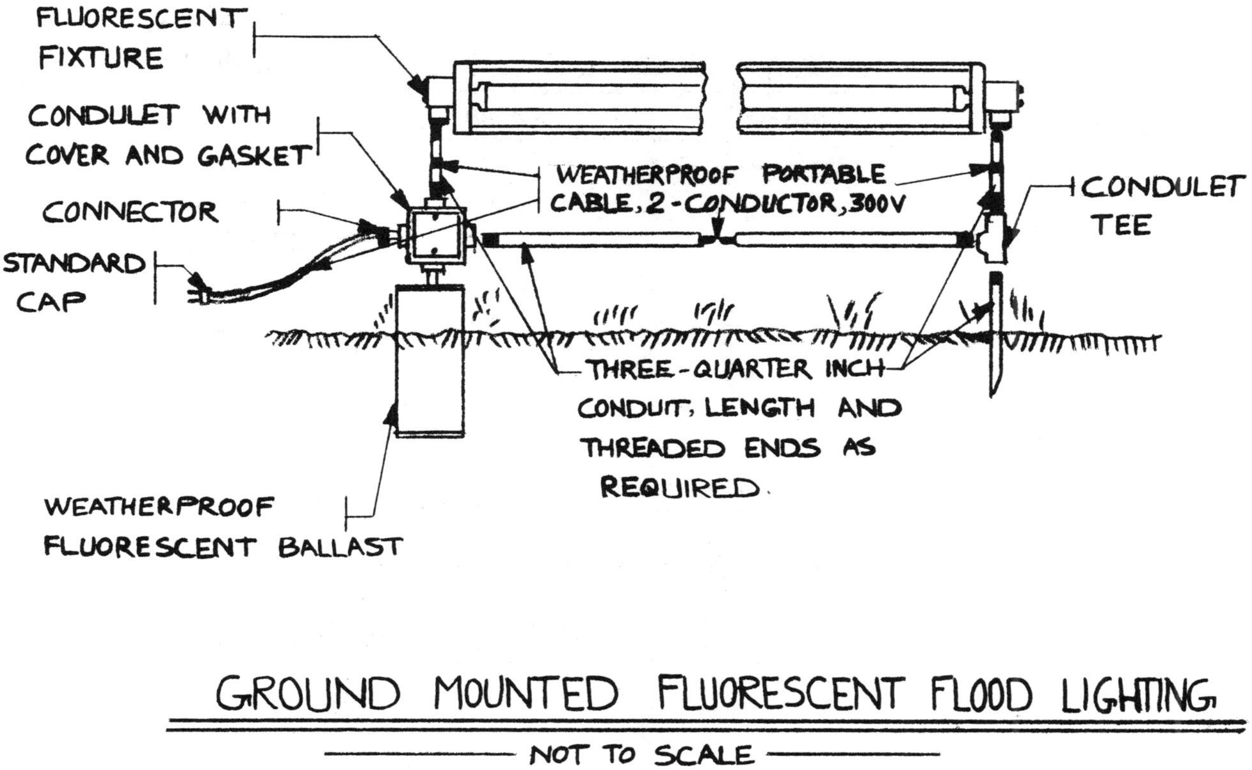

FLUORESCENT FIXTURE
CONDULET WITH COVER AND GASKET
CONNECTOR
STANDARD CAP
WEATHERPROOF PORTABLE CABLE, 2-CONDUCTOR, 300V
CONDULET TEE
THREE-QUARTER INCH CONDUIT, LENGTH AND THREADED ENDS AS REQUIRED.
WEATHERPROOF FLUORESCENT BALLAST
GROUND MOUNTED FLUORESCENT FLOOD LIGHTING
NOT TO SCALE

FIGURE 1-20

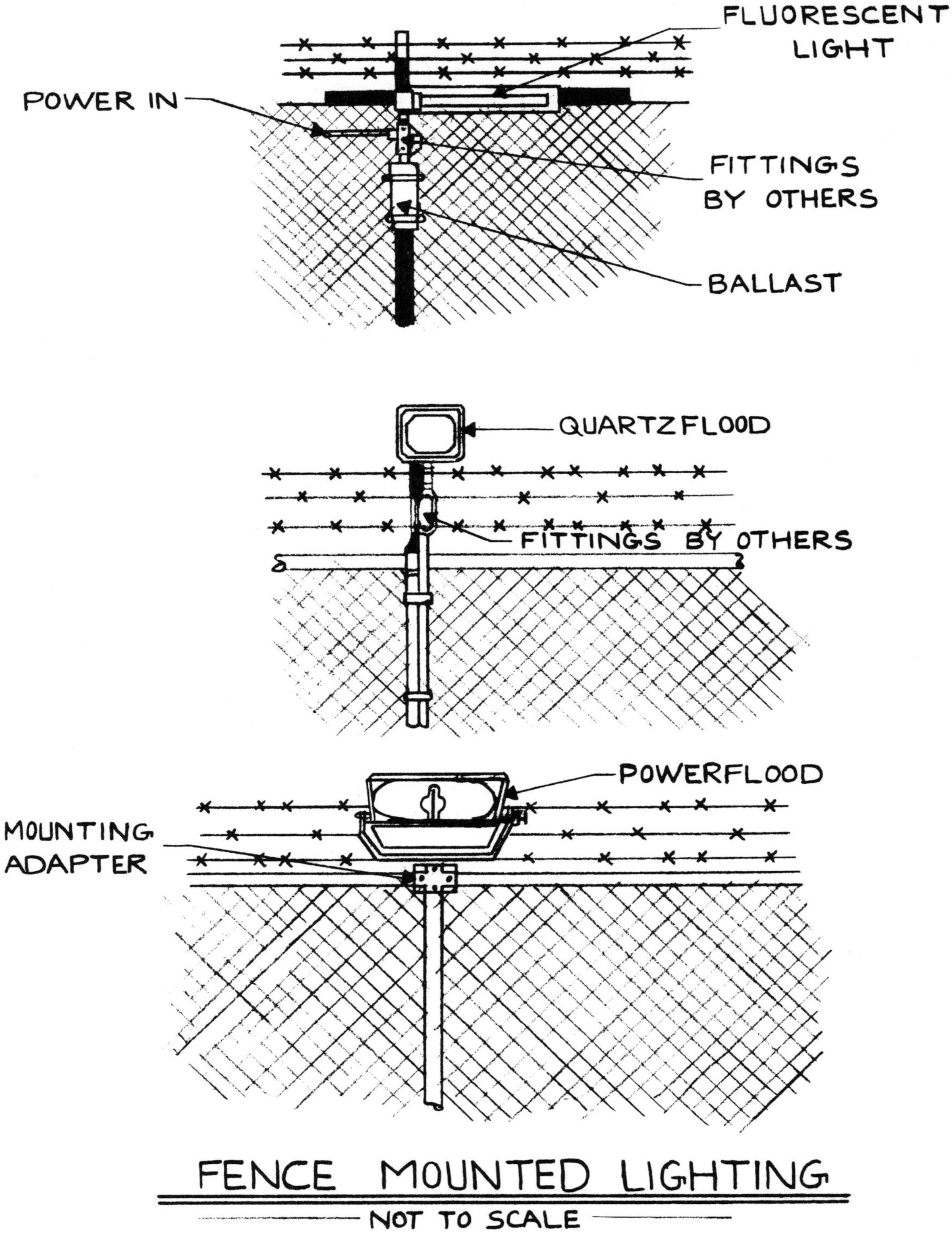

FIGURE 1-21

The details in Figure 1-21 show clearly that the fluorescent fixture goes right at the top of a fence and at the fence joint. On your plan you would show the number and spacing of the fixtures.

The details also show approximately how high the fixture is above the barbed-wire fence enclosure and that the fixture is to be totally within the barbed-wire enclosure. Since there is no way of knowing your exact project requirements, no dimension is given. If it is not important, a dimension is not really needed. You will get the installed result pretty much as shown. But if height above the fence is important, then the fixture detail should have a center-

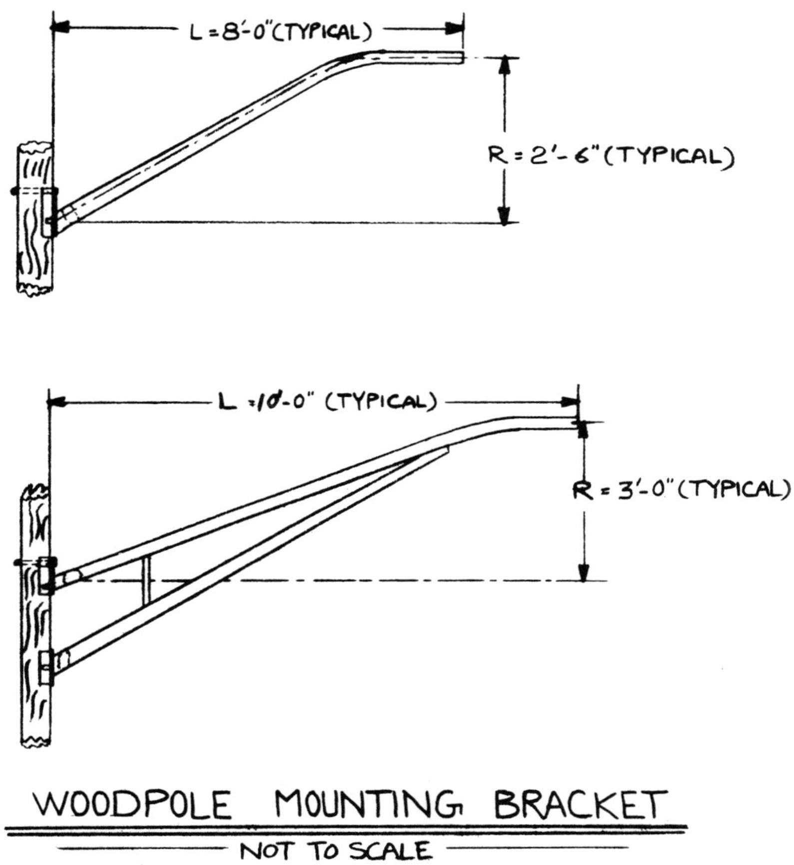

FIGURE 1-22

line dimension from the top of the fence to the centerline of the proposed luminaire.

Special Pole-Mounted Lighting Details Occasionally pole lighting is mounted on an existing wooden pole. Although the length allowed for the horizontal arm may vary slightly from manufacturer to manufacturer, the usual length of the unbraced arm is 8 ft and the braced arm is 10 ft, sometimes 12 ft. Figure 1-22 should be used where you clearly want to designate which type of horizontal arm meets your requirements.

The concrete base for any pole should be either designed

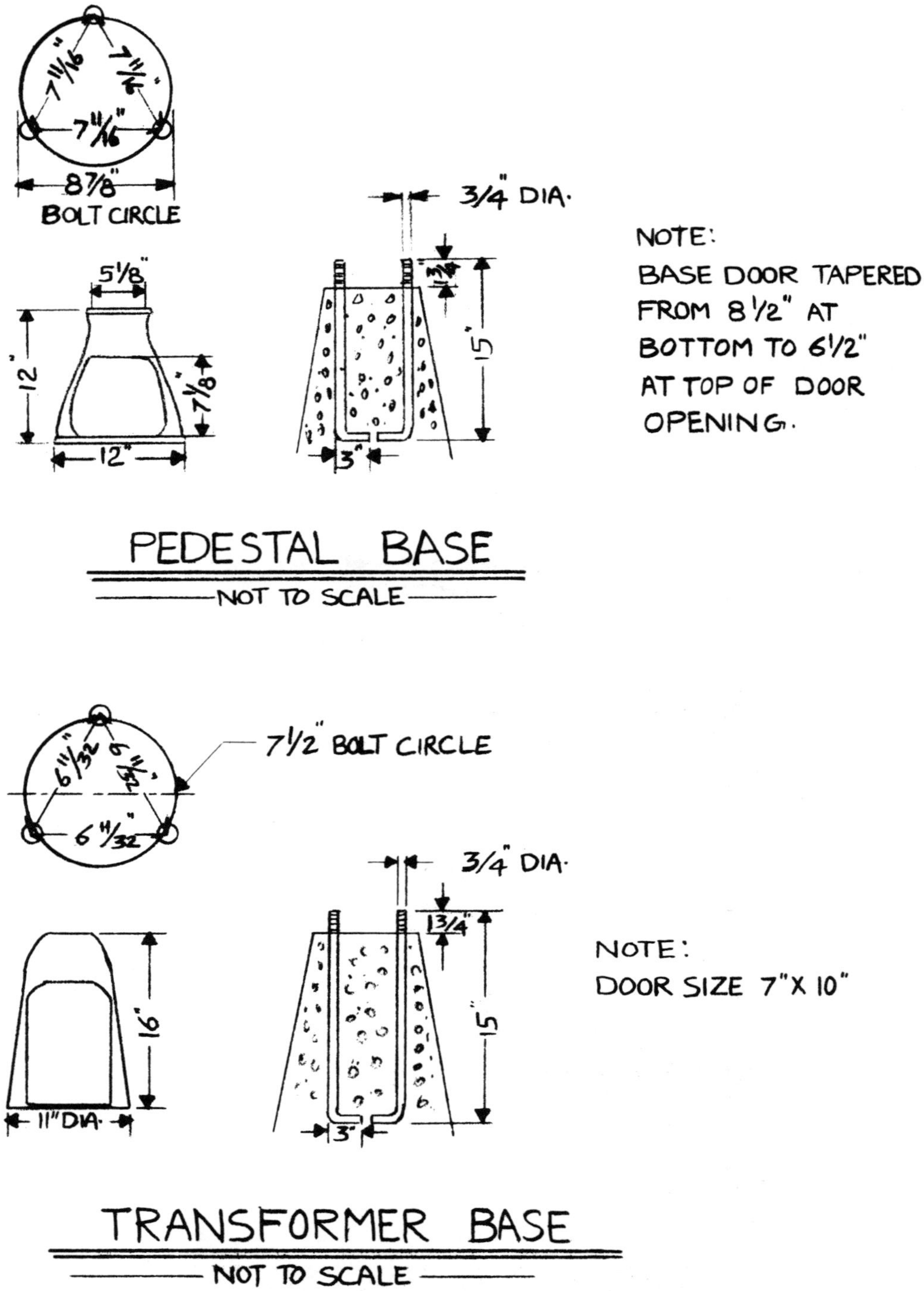

FIGURE 1-23

by a competent structural engineer or taken from a standard base design that is provided by the fixture manufacturer you select. The intent of the base detail shown on Figure 1-23 is not to describe and specify the base itself completely but rather to show the tie-down rods, base plate, and housing of a pedestal-based fixture as compared to a transformer-based fixture.

Not infrequently the pole and luminaire are correctly selected and specified and the base is carefully designed, while the method of attachment to the base and the base plate itself are completely ignored. Sometimes a pedestal base is shown where a transformer base is specified and

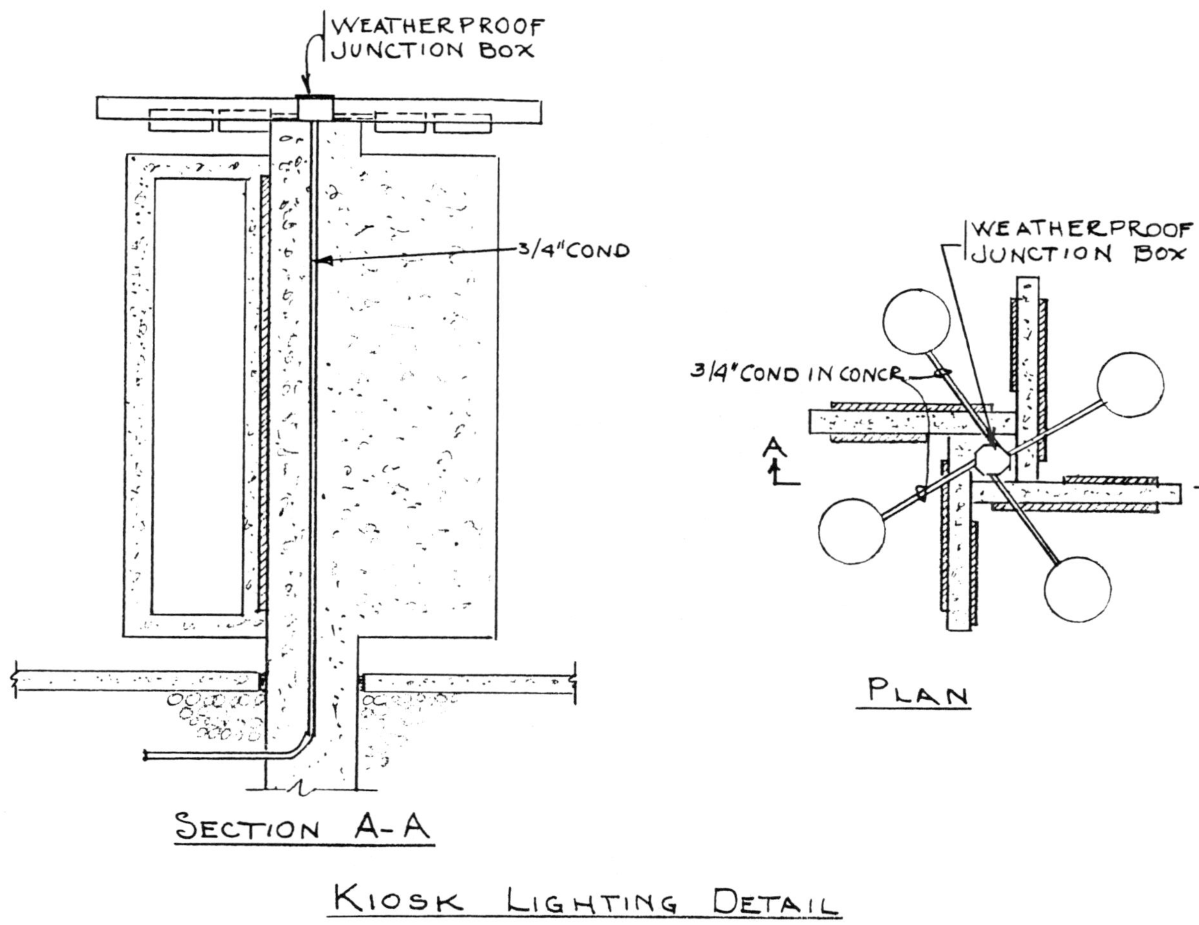

FIGURE 1-24

required. Careful coordination of this simple detail with the specified fixture will not only clarify the requirements but also assure that your fixture gets properly attached to its base.

Kiosk Lighting The kiosk has become an object of the modern landscape architect's site planning. What used to be rather plain metal display board is now an attractive, decorative, concrete structure. It may or may not include the concrete top shown in Figure 1-24. Since the person detailing the lighting of this object may be unfamiliar with the structure, the abbreviated version of a complete lighting detail is also shown.

There are basically two problems if your fixture is semi-flush-mounted in the roof of the kiosk. The first is how to conceal the conduit and the second is where to put the junction box so it does not show. Fortunately all kiosks have to be supported off some kind of vertical pole and usually this is, or can be made into, a passageway for the conduit.

As shown in Figure 1-24, the problem of conduit location in the cleverly designed kiosk was solved by running the conduit inside the hollow center section, effectively creating a precast sort of hollow pole.

This kiosk has a roof and the fixtures are partially recessed there. The weatherproof junction box is accessible from the top, a position seen only by the passing birds. If the kiosk had no roof, some form of junction box could still have been place at the top. Perhaps a ball-type fixture (or any of a number of other types) could have been supported on short runs of conduit or decorative horizontal supports.

Aviation Ground Lighting There are a number of aviation ground-lighting systems and a host of Federal Aviation Administration (FAA) requirements related to airport ground lighting. FAA requirements vary in detail, arrangement, and complexity, depending on the airport and its usage. The differences between a small, private-plane airport and a big international or commercial airport are obviously considerable.

However, in either case there are very few lighting units which protrude above ground. Items such as taxiway cen-

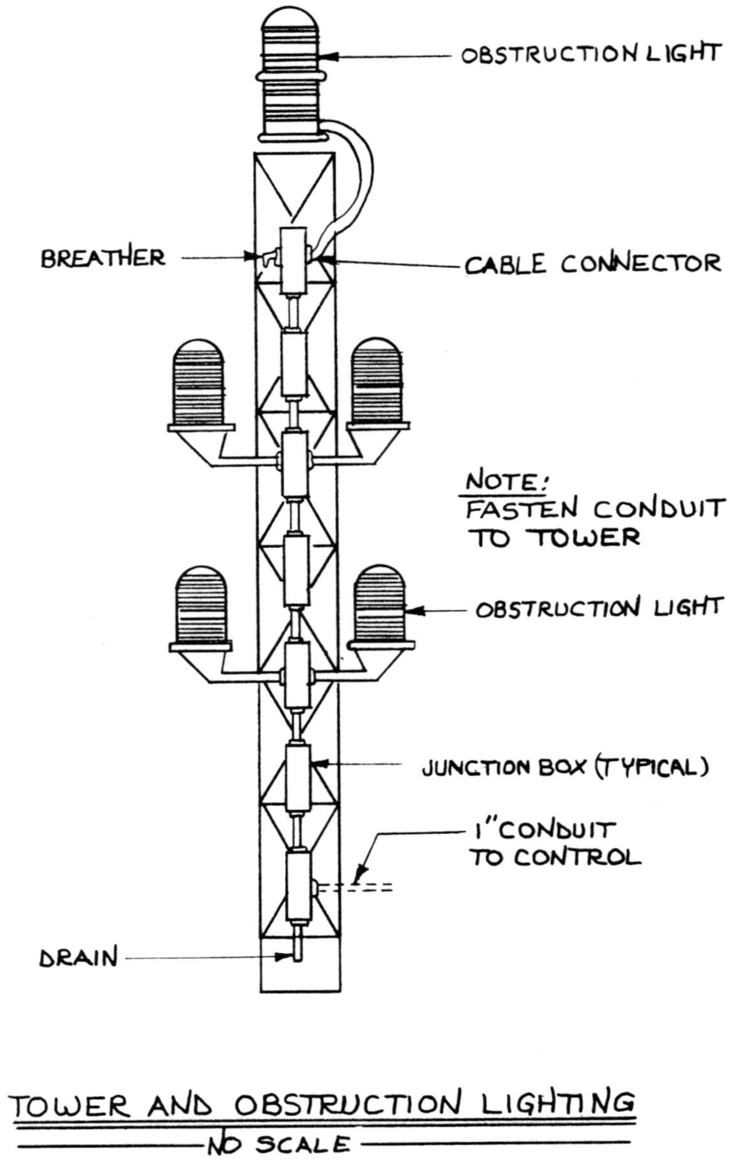

FIGURE 1-25

terline lighting, stop bars, and runway lighting are all designed to fit flush in the pavement. In large airports with instrument landing systems there are approach lighting systems above the ground that are a part of the coordinated instrument landing system.

On most airports that have lighting systems there are above-grade fixtures that mark the edges of taxiways, runways, and the landing threshold. There are also obstacles on and around the field that are lighted, and commonly there is a lighted wind sock for visual detection of prevailing wind direction.

These details are not subject to engineering judgment and are strictly controlled as to size, shape, height, and location. As a result, a small group of manufacturers provide these fixtures per FAA standards. Thus these are the fixtures that must be installed and will be funished, if properly specified, whether detailed or not.

However, there is considerably more work involved in installing an above-ground fixture compared to one that merely sits in a weatherproof enclosure flush with the surface of the runway or taxiway.

The tower or obstruction lighting detail in Figure 1-25

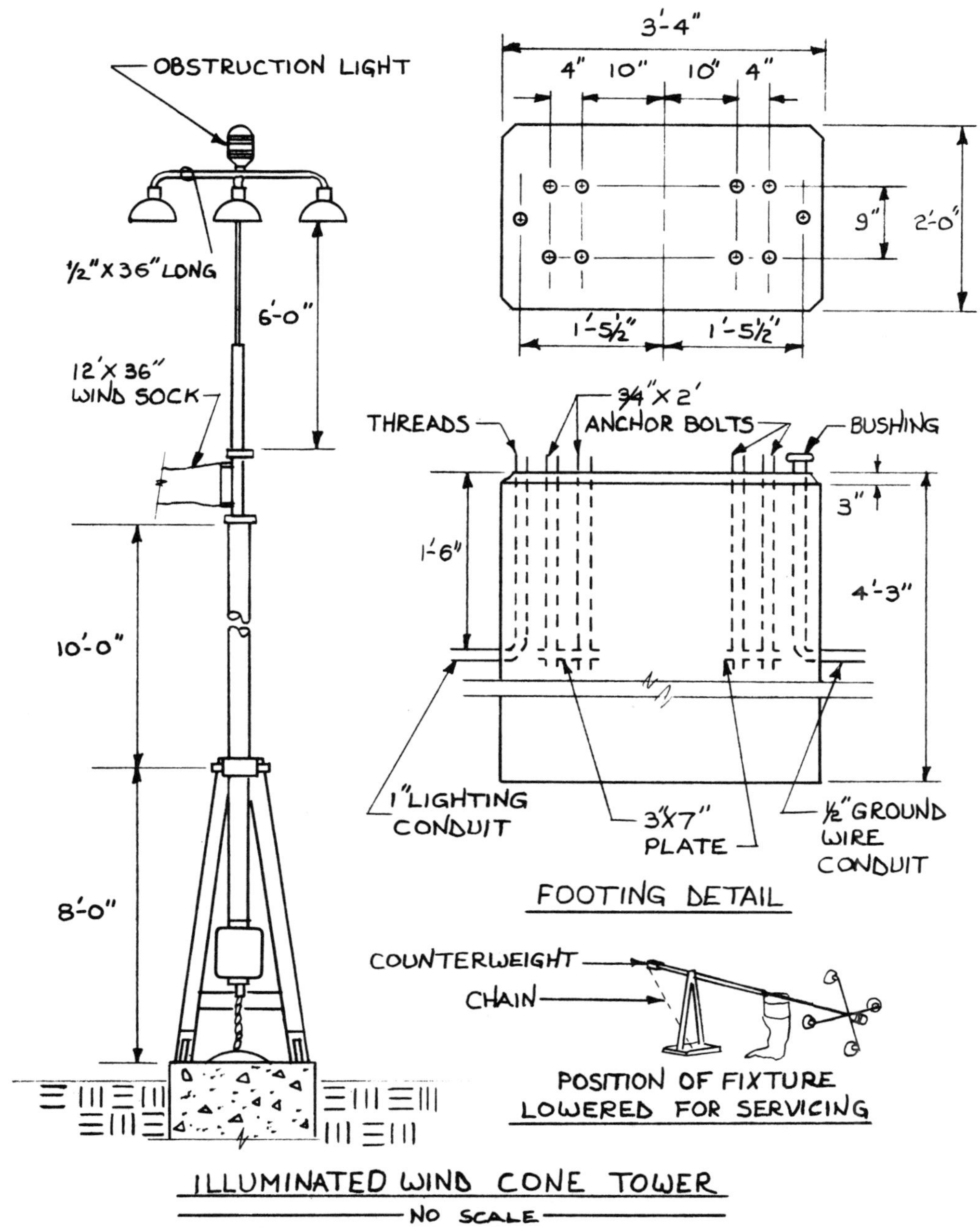

FIGURE 1-26

is typical of any 300 ft obstruction, be it a pole, building, stack, or other structure, within a certain distance of the airport that requires lighting. If the fixture is attached to a tall smokestack, the method of attachment should be detailed.

Figure 1-26 shows a wind sock, which is usually 36 in in diameter by 12 ft long, and has an arrangement of four 200-W fixtures with reflectors and horizontal arms mounted on a pole above the sock. Although not very important, the location of the luminaires at the magnetic north, south, east, and west points would be of some assistance to the pilots. On this type of pole the fixtures are about 25 ft above the ground. For ease of relamping, there should be an arrangement of the type illustrated which is capable of being lowered.

The above-ground taxiway or runway lighting shown in Figure 1-27 may vary if certain special conditions require a change in height. Thus, this detail is fairly obvious.

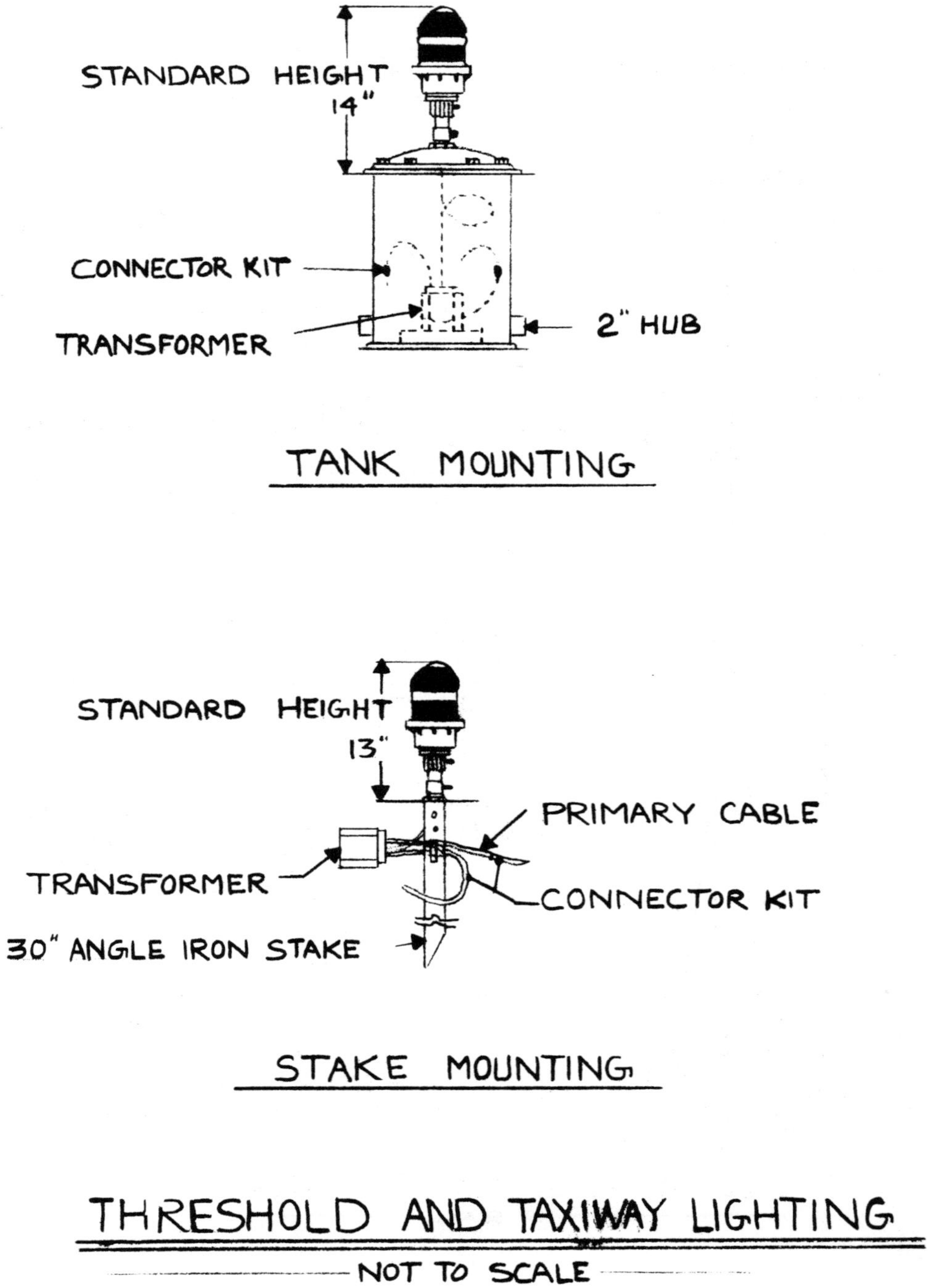

FIGURE 1-27

Understanding a Lighting Fixture Schedule From time to time reference has been made to a lighting fixture schedule. Figure 1-28 shows a blank schedule and Figure 1-29 a completed schedule.

Before you quickly turn to the completed schedule, let us first consider the blank schedule. What is it? What does it accomplish? What doesn't it accomplish? The schedule contains a number of columns with specific headings for very specific reasons. Let us examine each in sequence.

Lighting Fixture Schedule: This is correct use of the language. Actually a "light" fixture is a misnomer. Luminaires provide light and they make up the system of lighting.

Key - Customarily the lighting fixtures are designated on the plans by the letters of the alphabet. If you have more than 26 different types, the additional fixutures

LIGHTING FIXTURE SCHEDULE						
KEY	MFR.	CAT. NO.	DESCRIPTION	LAMP		REMARKS
				QUAN.	TYPE	

FIGURE 1-28

should have double letters; that is 1 to 26, would be A through Z, and 27 to 52, AA through ZZ.

Manufacturer: Normally your preferred manufacturer is named. If the rules under which you operate require three names, then three complete lines of data for each fixture will be required.

Catalog Number: Here you would insert the complete, accurate catalog number of the fixture you have selected.

Lamp: For some offices, the subheadings "Quantity" and "Type," as shown in Figure 1-28, are the preferred method of lamp description, since you then clearly designate number and type of lamps. Other offices feel that using the manufacturer's designation, such as 2F40T12-RS, is much more understandable to contractors and suppliers. Thus, as shown in Figure 1-29, a wider space, simply titled "Lamp," is used. It can be done either way.

Remarks: Frequently, where three fixtures are required, the firm specifies a Smith fixture and then under "Remarks" states that those of similar nature made by Jones and Brown will be considered equal. This is not only inadequate as a description it is also dangerous, since there may be a number of Jones and Brown fixtures that could be considered as equals. The real purpose of this column is to alert the contractor to certain special features of the fixture installation that will be considered.

Description: As you may have noticed, we skipped over this item, which should have come in the middle of our discussion, and deliberately kept it until the end. While the listing of the catalog number and manufacturer really is all that is needed to describe a fixture and the phrase "with lamps" would cover the lamp, the description is intended to alert everyone to the fixture type. Although contractors and suppliers are knowledgeable, they don't necessarily recognize every manufacturer's catalog number instantly. This is their clue column.

The "Description" column can also frequently save you a lot of grief. Presumably as in item I under "Key" of the completed list you do want a recessed, vaporproof, 150-W, incandescent, tamper-proof fixture. Above all you want it recessed, not surface-mounted. You could accidently pick the wrong catalog number and later discover the one you picked was surfaced-mounted. Most likely you would get a call during the bidding period asking for a clarification. But, even if you did not, you would still be on pretty firm ground in demanding what you described.

Some engineers and engineering firms go to great lengths in describing their selected lighting fixtures. They repeat and print a great deal of the material in the fixture manufacturer's catalog. In a few special situations this may be both advisable and necessary, but in most cases this excess verbiage does not help the supplier or the contractor. The contractor knows you are not inventing a fixture. What he or she really wants to know is who makes the fixture you want and what fixtures you will accept as equal. In some cases there are many substitutes; in others there are few. Ideally, specifying three or more fixtures which really are equal and acceptable to you is the preferred solution. If this is the case, you should add a special notation to your fixture list clearly stating that this is a requirement of your contract documents.

The completed list shown in Figure 1-29 was used in a base bid and alternate situation. In that case there were no equals. The contract specifications clearly stated that any substitution to be considered had to be named together with an amount to be deducted from the contract price if its substitution were permitted. Interestingly enough, substitutions for all the fixtures except B were turned down by the owner because of the very minor credit offered. Type B was substituted at a considerable saving to the owner.

LIGHTING FIXTURE SCHEDULE					
KEY	MANUFACT'R	CATALOG NUMBER	DESCRIPTION	LAMP	REMARKS
A	HOLOPHANE	#920-WG-UPH35 PHCA-CR-SC	PENDANT SEALED NON-CORROSIVE	H33GL-T400/DX	SEE DETAIL ON DRAWING E-2
B	MORRIS KURTZON	V-7872-40-SS-GVS	RECESSED VAPOR-PROOF	2T-43-RS	LAY IN GRID
C	MORRIS KURTZON	C-40-L	SURFACE VAPOR-PROOF	2F40 T12-RS	WITH MOUNTING BRACKETS
D	HOLOPHANE	400 (WALLPACKET)	WALL MOUNT WEATHERPROOF	150 W INCAND	WITH GUARD
E	DUAL LITE	NF-500T-12V-12W W/MP 101 MTG PLATE	EMERGENCY LIGHTING	SUPPLIED	2 HEADS REQD. PER LOCATION
F	DUAL LITE	WRWW-PPW	EXIT SIGN	SUPPLIED	
G	LEVITON	9874 (TYPE V-25)	PORCELAIN KEYLESS	100 W INCAND	BARE LAMP PORCELAIN SOCKET
H	DUAL LITE	EXT-124 R	RECESSED	25 W INCAND	12 VOLT
I	MORRIS KURTZON	WL-78G-SS-TP	RECESSED VAPOR PROOF	150 W INCAND	TAMPER PROOF
J	PRESCOLITE	WB-24-WG-6	WALL MOUNT	100 W INCAND	VAPOR-PROOF WIRE GUARD
K	GUTH	APG 6281/120/NAT. MOD	CORNER FIXTURE	2F40	VAPOR PROOF

FIGURE 1-29

TWO

Junction Boxes, Outlets, Connections

In Chapter 1 we began with details applying to lighting fixtures. In this chapter we discuss details that apply to junction boxes, outlets, and connections. We use this approach because the electrical designer generally starts a design with a lighting plan coupled with a plan of electrical convenience outlets, motors, and other items requiring electrical services.

Because many of these connections are presumed to be obvious, Code-required, or self-explanatory, frequently little or no thought is given to appropriate or applicable details. In many cases it is near the very end of the design effort that other designers or design checkers may bring up the need to detail certain connections. There is nothing inherently wrong with this approach. It is simply the hard way to do the required detailing. Or, perhaps better stated, it is the hard way to discover what detailing should be done.

After completing the basic lighting, outlet, and motor-connection phase of the design, a review of these locations and their relationship to structure and site should be made to decide what ought to be detailed. The latest version of the National Electrical Code and one of the several handbooks that interpret the Code are essential for this purpose. They should be part of the basic reference information in every design office. The National Electrical Code is published every three years by the National Fire Protection Association. Two of the better known versions of the handbook are published separately by McGraw-Hill and the National Fire Protection Association. The handbooks provide further explanation of the intent of the Code as well as some details to clarify the text.

Insofar as the subject of outlets is concerned, specific reference is made to Articles 210 through 250, Article 370, and Article 430. Most of this material deals with items obvious and inherent in the design or spelled out in the specifications. However, the Code should be checked to avoid overlooking any item of your layout that might require detailing.

In the detailing that follows, we show some of the unusual features. The simplicity of most pull-box detailing could be used as a serious argument as to why what is intended is not shown—because "a box is simply a box!" What is important to your detailing is to always show a pull box, properly sized, wherein a number of conduits join together and then proceed via a wireway or conduits to a load distribution or motor control center.

In lieu of just showing a box we have chosen a detail of a watertight exterior junction box as an example on the premise that this an exterior location, requiring specialized detailing of the box itself, is a familiar situation.

Watertight Junction Box A standard-sized watertight junction box could easily be specified and not otherwise detailed or noted. A statement such as "All exterior junction boxes shall be watertight" might suffice.

Figure 2-1 covers a relatively nonstandard situation where you want to remove all doubt as to your intentions. In this detail the first question to be resolved is how strong to make the box. Certainly it is one thing to put the box in a lawn where the heaviest weight it will encounter is a human foot or a lawn mower. It is quite another to put the box where it is subject to an unlimited variety of heavy traffic.

The best way to treat the cover plate is to relate its strength to civil engineering highway standards of wheel loading and let the manufacturer calculate the proper strength and resulting thickness of the cover plate. That is exactly what was done in the accompanying detail. The plate and its reinforcement are clearly indicated, but no exact dimensions are given. The specifications for the

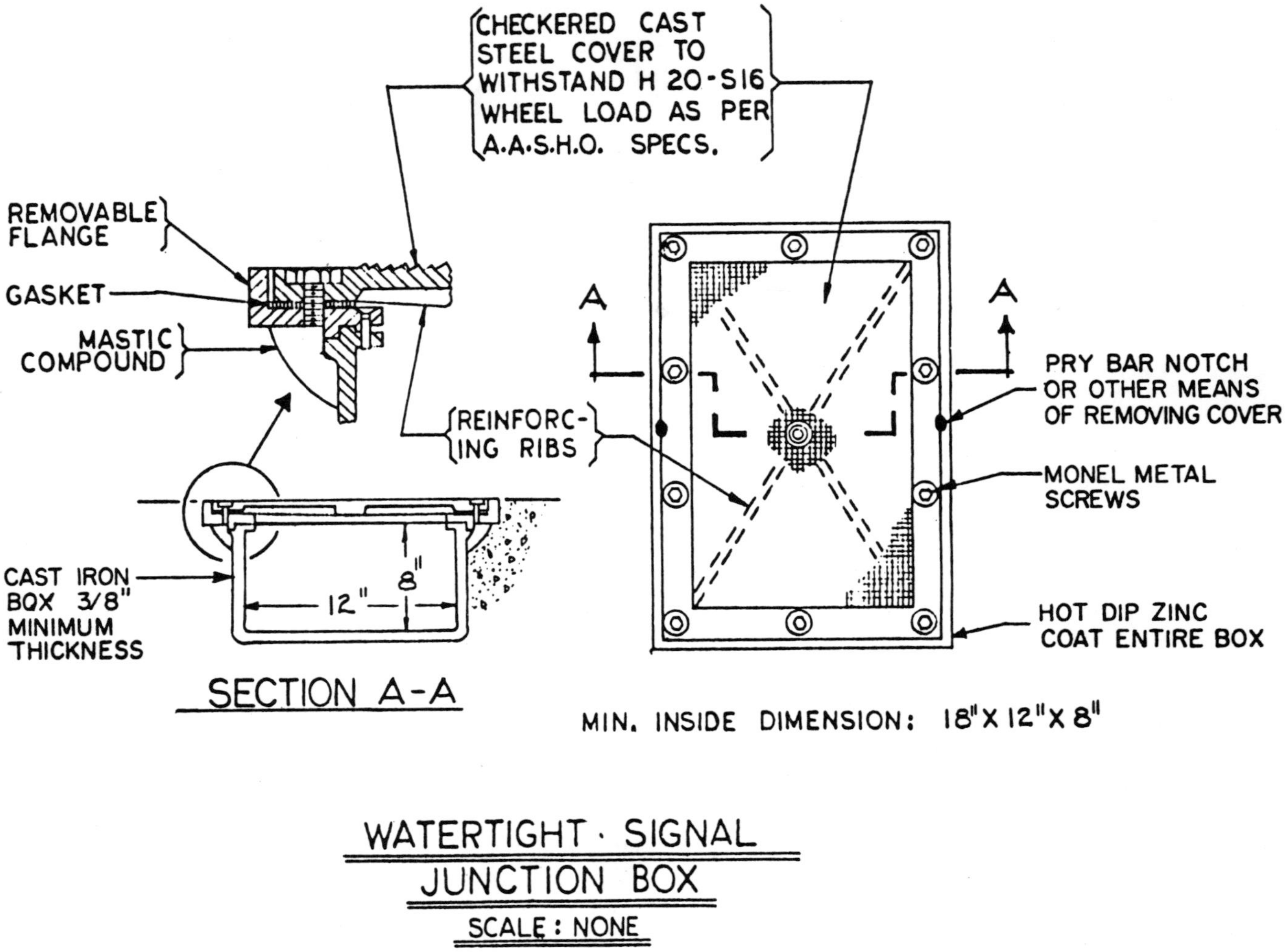

FIGURE 2-1

frame of the box itself should be calculated by a structural engineer. (If you want one that will stand up under almost any conditions, you can copy the one shown).

Since the box is expected to be watertight it must be bolted and gasketed. As you can see in the detail, provision for removal of the cover must be made. In the enlarged detail of the frame and cover several important points are dealt with. First the gasket is at the bottom of the frame. There must always be space between the frame and cover or the cover will not come off. The flush bolts screw into threaded outlets in the frame and pull the cover tight to the gasket, making a watertight seal.

Since the clearance between the edge of the cover and the frame edge is approximately 1⁄16 in, removing the bolts does not help to remove the cover plate. Our detail shows a pry-bar notch. If you want to get fancy, you could depict a handle that rests in its own cast niche as a method of providing a cover lifting device. Monel metal screws are a relatively standard specification.

Do not forget that your installation is outdoors and must be carefully specified to meet the conditions of your area—rain, snow, hail, high winds, and the like.

Specialized Connections In selecting details for this book, or any other book of a similar nature, there is an obvious limitation on the material that can be presented. This is especially true in this chapter. Since this book begins with the sources of electrical loads, and we are currently dealing with one of those sources, we are faced with a decision as to how to cover the specialized areas of outlets and connections. While we could present unusual, seldom used details that would cover many pages and be of little value, we have chosen special situations that illustrate common problems.

Figures 2-2 and 2-3 are both concerned with the installation of electrical power and services on a pier. The installation could just as easily have been a bridge. Our point is that while the consulting firm or individual detailer is not often involved in this kind of work these two details do involve exterior, wet, salty-air conditions, which except for

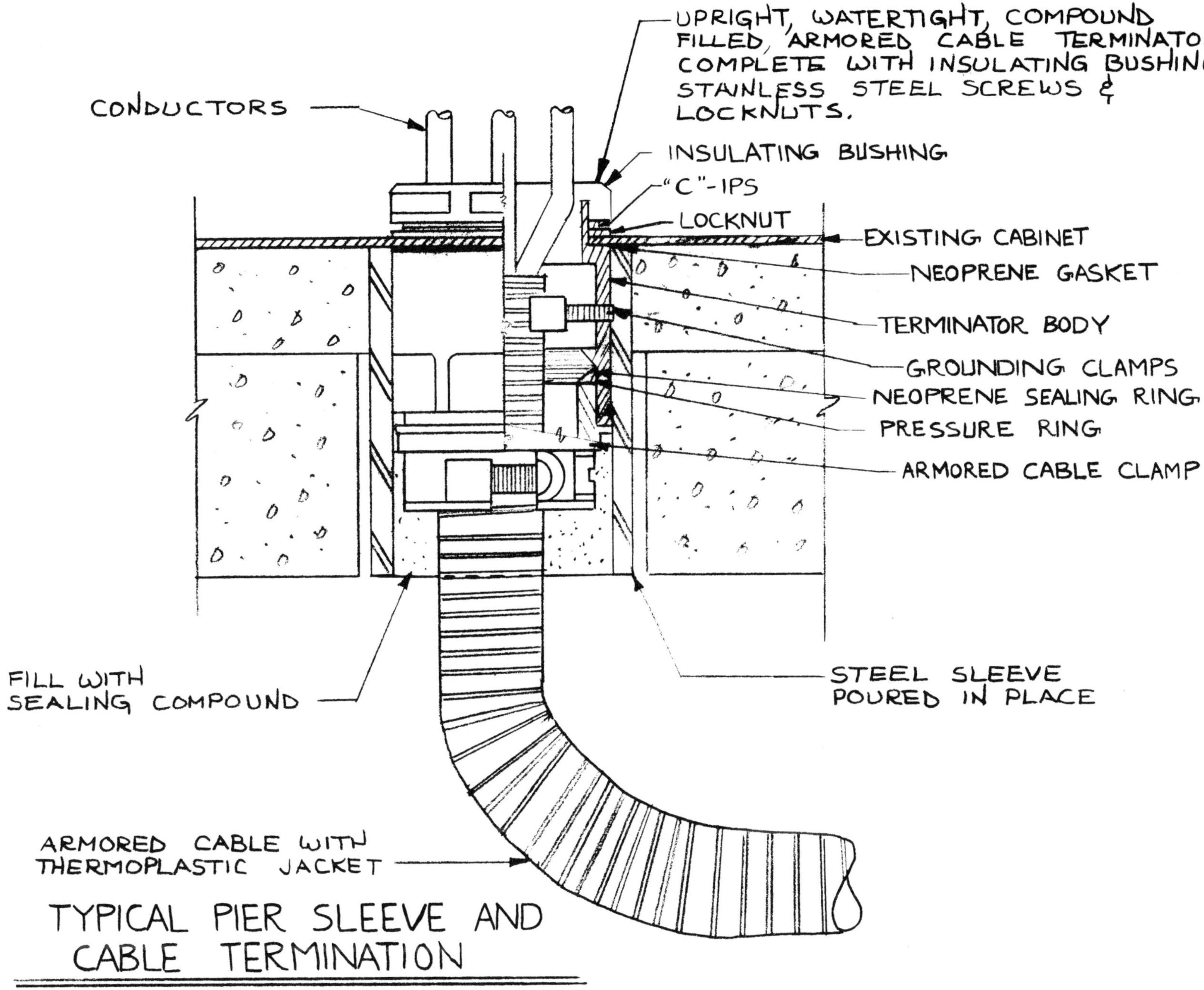
CONDUCTORS
UPRIGHT, WATERTIGHT, COMPOUND FILLED, ARMORED CABLE TERMINATOR. COMPLETE WITH INSULATING BUSHING, STAINLESS STEEL SCREWS & LOCKNUTS.
INSULATING BUSHING
"C"-IPS
LOCKNUT
EXISTING CABINET
NEOPRENE GASKET
TERMINATOR BODY
GROUNDING CLAMPS
NEOPRENE SEALING RING
PRESSURE RING
ARMORED CABLE CLAMP
FILL WITH SEALING COMPOUND
STEEL SLEEVE POURED IN PLACE
ARMORED CABLE WITH THERMOPLASTIC JACKET
TYPICAL PIER SLEEVE AND CABLE TERMINATION

FIGURE 2-2

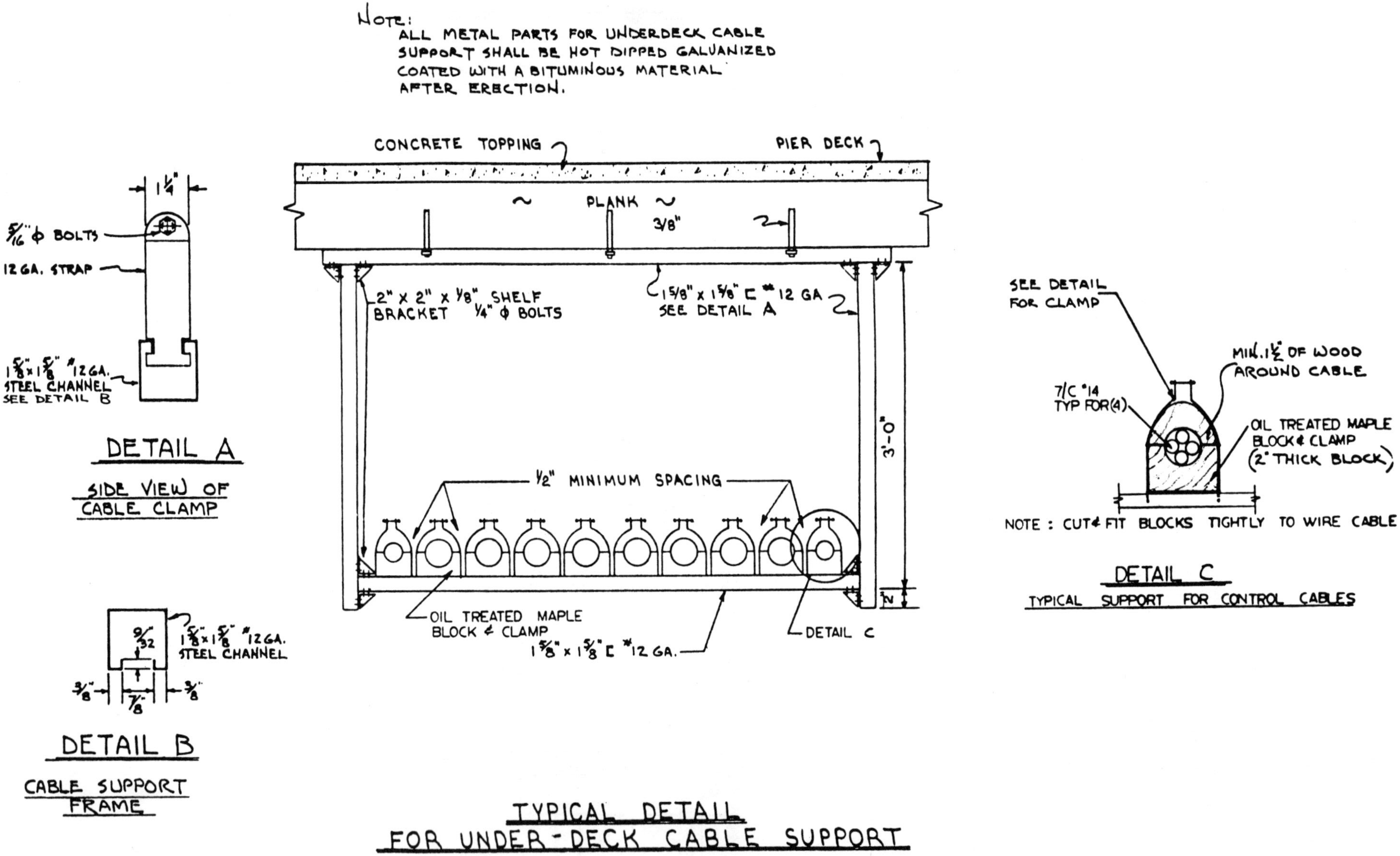
NOTE:
ALL METAL PARTS FOR UNDERDECK CABLE SUPPORT SHALL BE HOT DIPPED GALVANIZED COATED WITH A BITUMINOUS MATERIAL AFTER ERECTION.
CONCRETE TOPPING
PIER DECK
PLANK
3/8"
2" x 2" x 1/8" SHELF BRACKET 1/4" ϕ BOLTS
1 5/8" x 1 5/8" C #12 GA
SEE DETAIL A
3'-0"
1/2" MINIMUM SPACING
OIL TREATED MAPLE BLOCK & CLAMP
1 5/8" x 1 5/8" C #12 GA.
DETAIL C
1 1/4"
5/16" ϕ BOLTS
12 GA. STRAP
1 5/8" x 1 5/8" #12 GA. STEEL CHANNEL
SEE DETAIL B
DETAIL A
SIDE VIEW OF CABLE CLAMP
9/32
1 5/8" x 1 5/8" #12 GA. STEEL CHANNEL
3/8"
7/8"
3/8"
DETAIL B
CABLE SUPPORT FRAME
SEE DETAIL FOR CLAMP
MIN. 1 1/2" OF WOOD AROUND CABLE
7/C #14 TYP FOR (4)
OIL TREATED MAPLE BLOCK & CLAMP (2" THICK BLOCK)
NOTE : CUT & FIT BLOCKS TIGHTLY TO WIRE CABLE
DETAIL C
TYPICAL SUPPORT FOR CONTROL CABLES
TYPICAL DETAIL
FOR UNDER-DECK CABLE SUPPORT

FIGURE 2-3

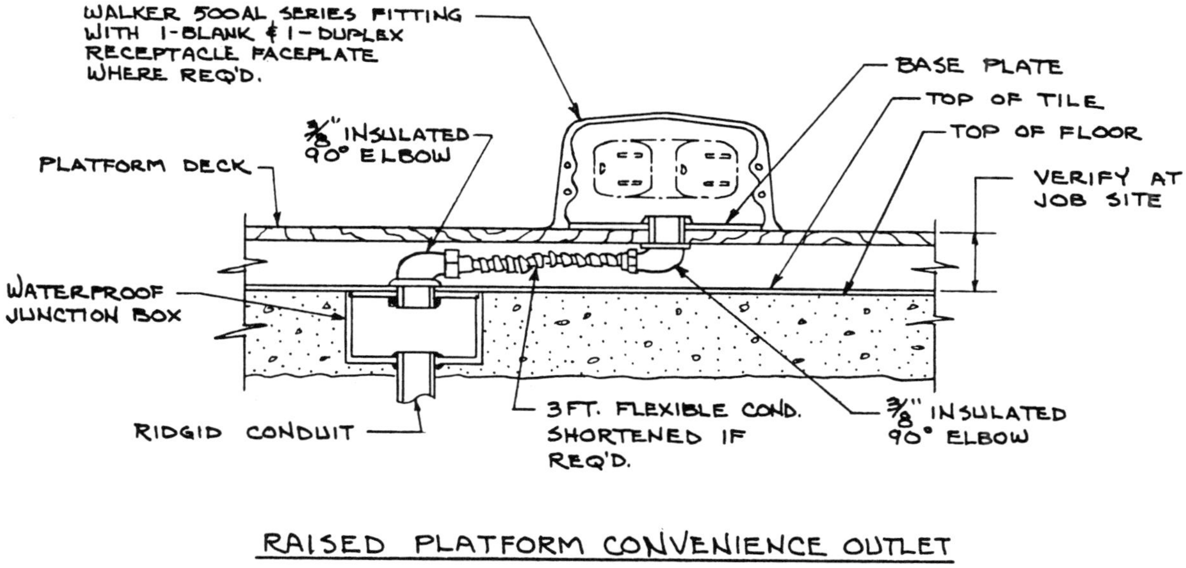

FIGURE 2-4

the salty-air problem are illustrative of a common problem. This is also true of many other special, one-of-a-kind situations.

In the first of the two details we illustrate the proper termination of a cable running under a pier (or bridge) and rising up through a steel sleeve (galvanized schedule 40) to a cabinet sitting on top of the *pier*. Basically it is not an unusual situation. Since we judged that the best overall solution was to run wiring under the pier in an armored cable with a thermoplastic jacket, the *sleeves* become an obvious necessity. The fitting (cable terminator) is a standard item that could have been specified and the detail could have been omitted.

However, this is a serious situation where protection against water is required and should not be left to chance. In your detail the cable should be sized and the sleeve dimensioned to provide minimum clearance between sleeve, cable, and terminator. *The space between these devices must be filled with a waterproof sealant.* The terminator comes with grounding clamps *to be attached to the pier steel* by a suitable connector.

Figure 2-3 is truly unique. It illustrates one solution for supporting 10 separate power and control cables. There were some obvious problems such as sag and movement of the cables caused by wind and water. As you can see this was a pure invention of an oil-treated piece of maple to isolate the cables from rubbing on any sharp-edged steel surface.

This detail is not presented as the only way to accomplish the intended task but rather as a way, invented on the spot, to provide a workable solution. In the actual installation all of the cables were successfully supported for a very considerable distance under the pier by this method with supports on 8 ft 0 in centers.

Our point here is that you will always come across specialized connections which cannot be anticipated. These simply will have to be invented.

Raised Platform Outlet As stated in the opening discussion, there is little that needs to be detailed in outlets. Figure 2-4 presents one of the few exceptions. It is a surface-mounted outlet on a furred or raised wall or floor.

Frequently the wall or floor is furred or raised off of a concrete or masonry surface. If it is a wall, the problem is frequently simple. Conduit, BX, or plastic cable can be fished down or up the furred space and connected to a flush or surface outlet. A raised platform above a floor can be quite a different matter. There may be a problem with water. The concrete floor may be poured and the conduit in place when the outlet is moved for any reason.

Where there is the chance that water may be a factor it is generally safest to have the conduit and wire terminate in a waterproof junction box as shown in the accompanying detail. The reason is obvious: it is better to be safe than sorry.

The real trick in the detail is the use of two elbows—one at the junction box and one at the outlet with a short piece of BX cable in between. Now you have real flexibility. Unless something in the design is really way off, it is reasonable to assume that the outlet location shown on your detail and on your plan is surely within 3 ft of its final location. And, as the detail states, the BX cable can be adjusted to suit the condition. It becomes one of those almost ideal solutions where you can hardly be outfoxed by any location of the outlet. The outlet shown is one of a number of its type made by several manufacturers.

Connecting Aluminum Conductors to Copper Steel In modern electrical wiring the mixing of copper and aluminum is becoming more and more prevalent. Be-

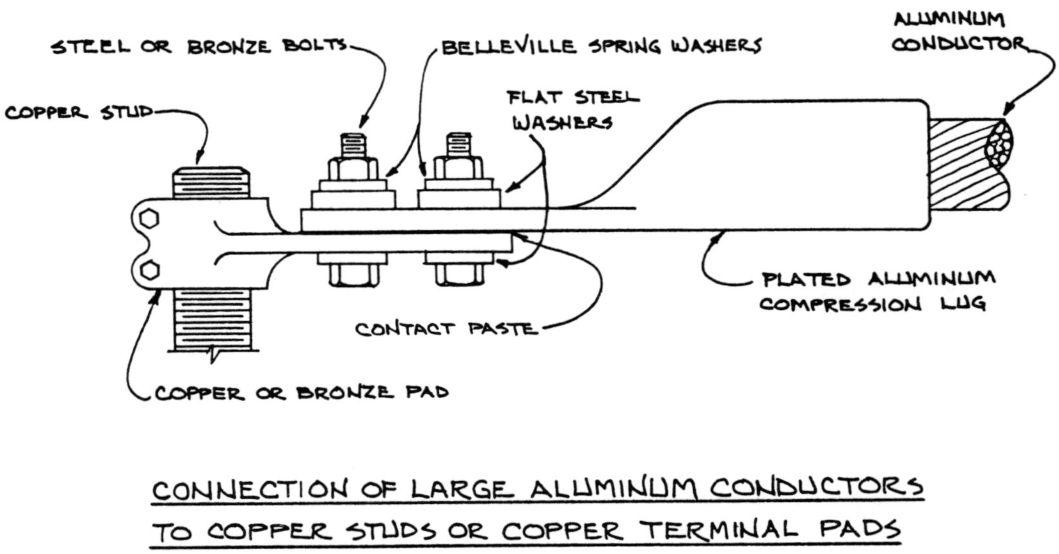

FIGURE 2-5

cause of the rash of fire claims and innumerable articles, discussions, and Code meetings on the subject, there are few, if any, reliable contractors making improper connections. Generally speaking, the chances of a good contractor making an improper connection between these two metals are very slight.

However, nobody is perfect, and people, well-intentioned or not, do make mistakes. Figure 2-5 shows a standard that is common in the industry. Putting it on your drawing may almost seem like a waste of time. But, if this connection is not shown on your drawing and a claim is made that the connection was done improperly, whether this claim is true or not, you will have a very difficult, if not impossible, task of trying to convince a judge and a jury that omitting the detail was a minor matter.

If there is a key item in this drawing, it is the use of contact paste plus steel washers topped off by spring washers. And do not forget to note that the aluminum conductor has an *aluminum* compression lug and the copper stud has a *copper bronze pad.*

Expansion Connections Frequently one sees the interconnection of outlets or the home run of a branch circuit almost casually run through an expansion joint in a concrete deck or slab. When this is brought to the attention of the electrical designer, the subject is usually

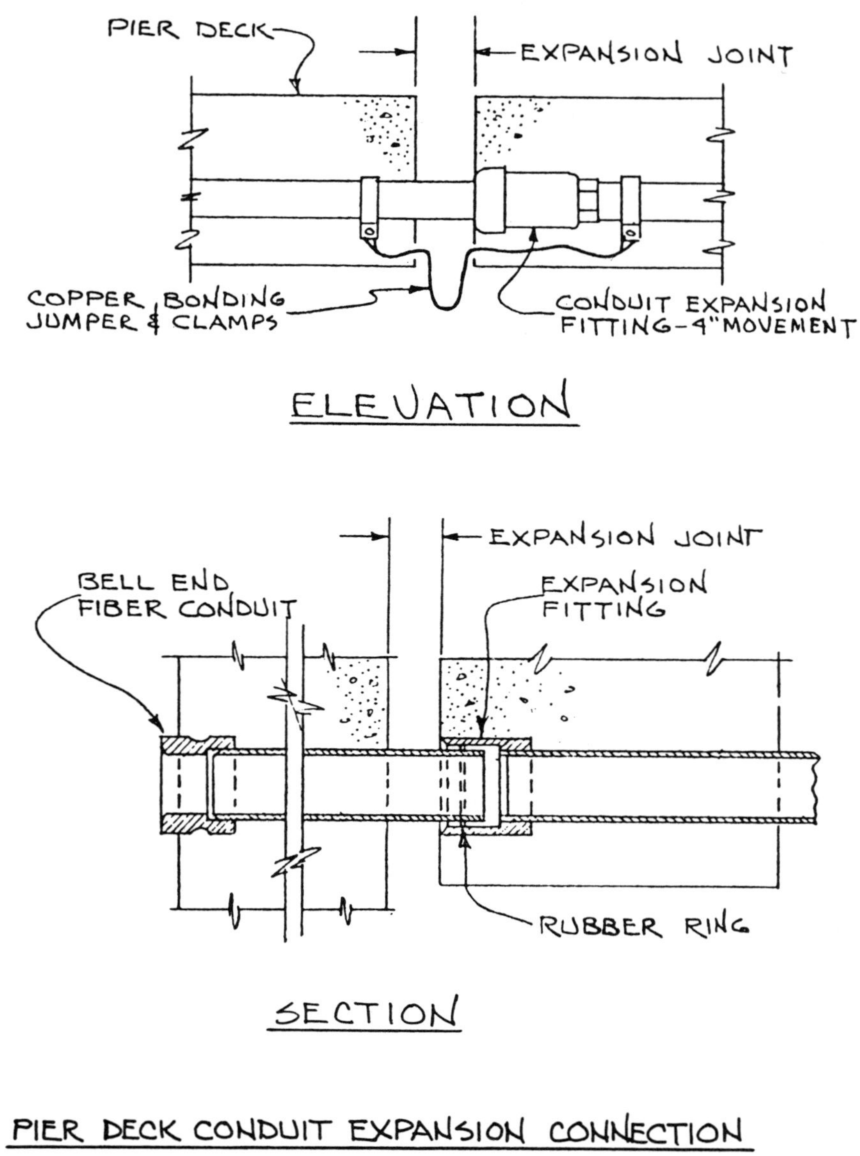

FIGURE 2-6

brushed off with statements such as "We've never had a problem," or "That's covered in the specs," or similar remarks. Maybe one or both statements are true. However, the fact remains that if the slab or deck moves and the conduit is locked in the slab something has to give.

Figure 2-6 shows a proper provision for the expansion and contraction of concrete in a pier deck which, like a bridge or other exposed concrete surface, expands and contracts with the changes in seasons, the action of the sun, and other changes in the weather.

There is nothing unusual about the detail. The important points to be noted are that an expansion fitting be installed in the conduit and equally important, if not more so, that a clamped bonding jumper be installed to maintain conduit ground continuity. Even if this detail is placed on your drawing where the situation requires it, your task is really not fully complete.

It is impossible to merely look at the detail and be able to estimate how many of these expansion connections are required. Ideally you should note each and every case on the plans. However, this is not always possible or necessarily desirable, since it is difficult to spot every situation that might arise. A reasonably fair way of producing a clear-cut design would be to locate all expansion joints, note as many as possible, and advise the contractor that all conduits of all types passing through this joint will be individually treated with expansion joints as per your detail.

THREE

Panels and Branch Circuits

Before going into the detailing of panels and branch circuits, let us pause to define the subject of discussion. In every electrical system design there are generally three basic schemes that, depending on the complexity of the design, are covered in one composite plan or in separate plans. The three schemes, which may end up as one, two, or three separate plans of the same area, can each be described by one word: lighting, power, and systems, respectively.

This chapter covers branch-circuit details of lighting and power systems for panels and branch circuits. Two contrasting examples are a modern industrial plant and a private residence. The industrial plant may have higher voltage and larger amperage branch circuits than the residence's 120-V 20-A circuits, but the detailing fundamentals are similar. In addition, whether simple or complex, the house and the industrial plant have a lighting system that needs to be shown.

The industrial plant frequently has a bus-duct system which could be described either as a feeder or as an extended panel. Bus-duct systems act primarily as extensions of a panel and are directly connected to a power source circuit breaker or connected by a short run of conduit and wire from the power source circuit breaker to the bus duct.

In normal use the bus duct has a series of attached (plug-in) circuit breakers with conduit and wire running from these circuit breakers to machinery mounted adjacent to the bus duct (above, below, or parallel).

The bus duct, in the viewpoint of this book, is really an extended panel, and its branches, however large, are similar to the branch circuits of the residential panel, which is why we have included bus ducts in this chapter.

Since the detailing of lighting fixtures and associated outlet boxes was covered in Chapters 1 and 2, there is little to add here. As is obvious, the electrical circuit proceeds from one fixture to the next until the circuit has reached code capacity. At that point the circuit makes a "home run" to its source of power in the lighting panel. Standard convenience duplex outlets are mounted, recessed, or surfaced on walls and floors and require no particular detailing. For reasons of economy, outlets and lighting fixtures can be combined on the same circuit. However, where possible outlet circuits should be separated from lighting circuits.

Although this book is not on the design of electrical systems, there are areas where design and detailing are so intimately related that it is hard to separate them. In branch-circuit design, for example, there are generally a number of choices of systems, such as 240/120-V single-phase, 208/120-V three-phase, and 480/277 V three-phase systems. The design of those systems does not properly belong in a book concerned with system details. However, when planning the coverage of schematic and other forms of panel detailing for this book, it became obvious that some crossover into the area of design was unavoidable. This situation is noted in this chapter, as well as in Chapter 5, Services and Distribution.

For example, in a residential structure the Code allows only two main disconnects. The allowance noted in the Code for six main disconnects applies only to an *existing* installation if the panel is a lighting and appliance panelboard. Panel main protection is another area of detailing where errors in the panel schedule frequently occur (although not necessarily the detailer's error).

In Section 384-16 of the NEC there are rules on the selection of lighting and appliance branch-circuit panelboard. The Code defines this type of panelboard as one having more than 10 percent of its overcurrent devices rated 30 A or less for which neutral connections are provided.

Since the notation on the detailer's panel schedule for a panel having a main circuit breaker is a simple pencil stroke, the tendency to call for a main circuit breaker in all panels is frequently overdone. These devices add $100 and more to the cost of the panel. What is the detailer to do?

Before completing the finished panel schedule, the detailer should certainly have the electrical designer approve the detail. Under Section 384-14, if all circuits have circuit breakers over 50 A, or if no neutral connections are provided in the panel, no main circuit breaker is required. Generally, it is your panel circuits which create the requirement to install a main panel circuit breaker.

There are other rules that cover such specialized applications as split-bus panels, ground fault protection, and the like. Our point here is not to cast doubt on the details and schematic drawings that are shown in this section and the one that follows, but to alert you to the fact that frequently our detail will almost, but not quite, fit your needs. No amount of detailing in any book can ever cover your requirements exactly.

We are not trying to make excuses for the presentations depicted. Instead, we are pointing out that in the area of branch-circuit wiring at 600 V and less there have been many problems in the past. As a result, the NEC contains some very specific and sometimes complex rules which affect not only your design but, on occasions, your detail. Where there is the slightest concern, a thorough study of Code requirements should be made.

Split-Bus Panels Figures 3-1 and 3-2 show split-bus panels. Both were taken from actual jobs and were successfully installed without contractor complaint.

Figure 3-1 truly contains the barest minimum of detail. The reason for this approach is that there is a separate schedule of all the items contained in and required for the panel. To repeat this information is not only redundant but unwise as problems could be created if any piece of information in one place contradicts the information in another location. Presuming all the above was properly done and itemized and that the circuits shown exactly agree with the detailed account given in the schedule, there is much to be said for this approach. The gym panel detail does show constant-current and remotely controlled sections. However, it is a detail that is open to much criticism and possible misinterpretation by the installing contractor.

Figure 3-2 leaves nothing to a chance interpretation. While it is obvious that the individual circuits are not numbered, this is an item of no great importance. The circuits are described in the accompanying panelboard schedule, to which the detail clearly refers. Since part of this detail refers to operation by remote-control switches, these switches and the panel contactor are both shown and specified right on the detail, leaving no room for doubt or debate on what was intended or required. A point frequently overlooked in many details is the connection of the neutral wires. This, too, is clearly covered.

Why show two details when one, by most any standard, is superior to the other? The answer might very easily be "Who says one is superior to the other?" Figure 3-2 certainly is more detailed than the other, and, as was freely admitted, one is far less open to debate or error than the other. Our reason for doing this is that frequently there is no detailing for this type of plan. Therefore, we decided to display the absolute minimum requirement as well as a more acceptable and usable version.

Like a lot of other details, the split-bus detail may be forgotten because detailers do not know how to show it, nobody tells them how, or no one thinks the detail is necessary. As in many other instances, you only need the detail when a question arises during construction. Then you wish you had it and are busily explaining why what you have not shown but specified is obviously required.

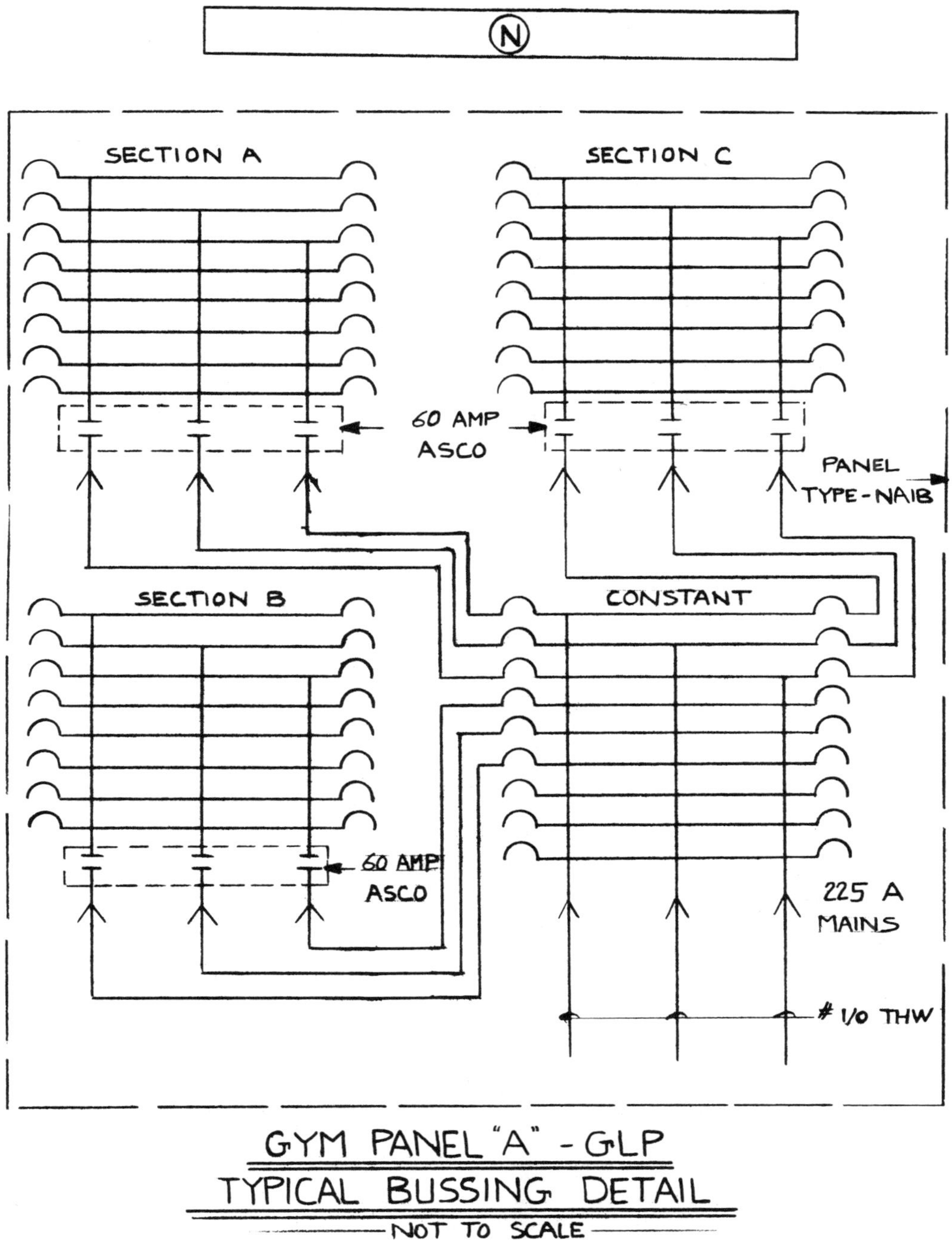
N
SECTION A
SECTION C
60 AMP
ASCO
PANEL
TYPE-NAIB
SECTION B
CONSTANT
60 AMP
ASCO
225 A
MAINS
1/0 THW
GYM PANEL "A" - GLP
TYPICAL BUSSING DETAIL
NOT TO SCALE

FIGURE 3-1

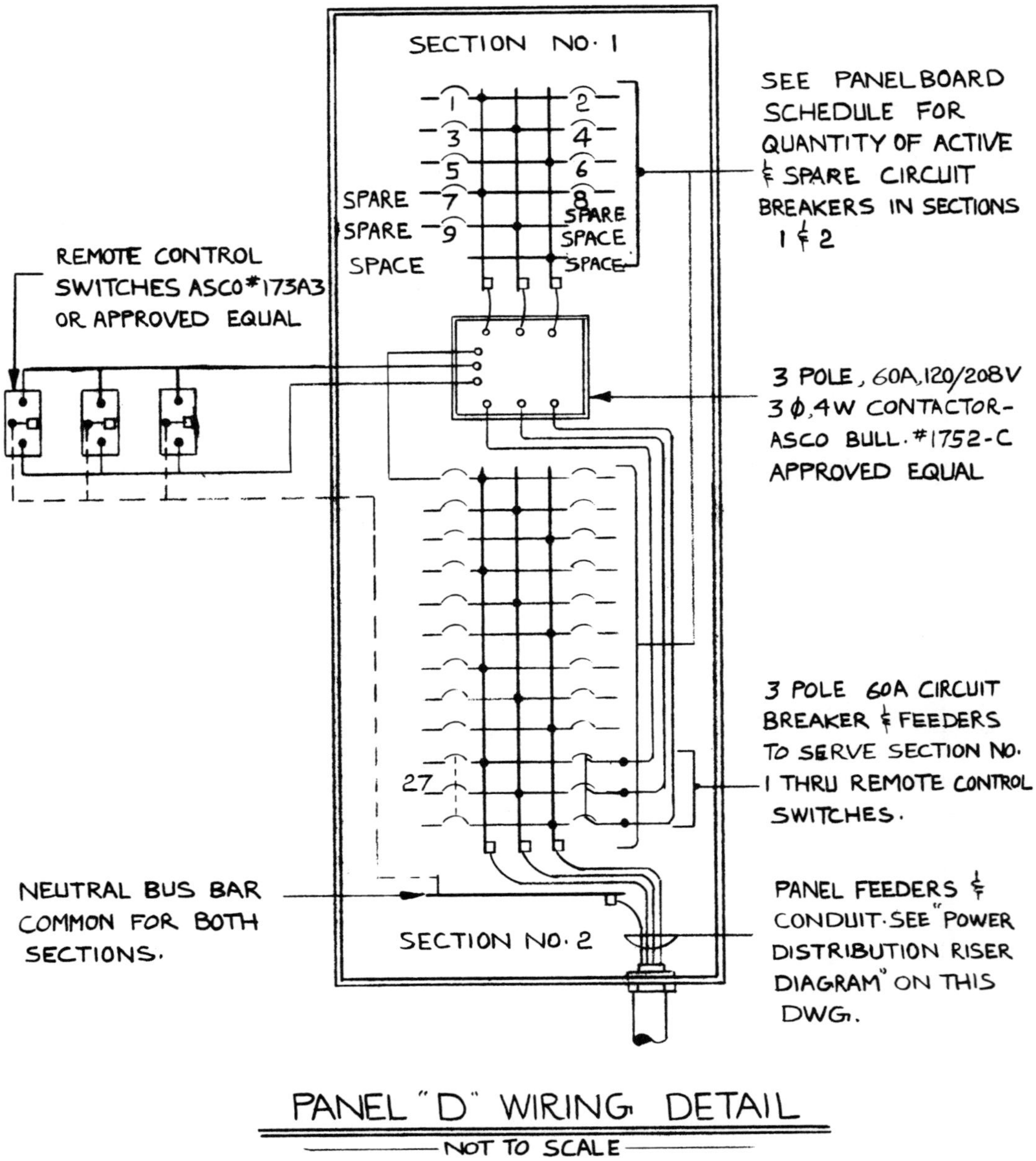

FIGURE 3-2

Schedules In order to complete the blank schedule shown in Figure 3-3 properly, the designer must first be made aware that the schedule must not only relate to but also properly coordinate with the specifications and the completed plans. The written specification commonly states that all the pertinent information is as shown on the drawings. Unfortunately this fact is not fully and carefully conveyed to the designer, and the instructions the detailer receives frequently do not stress the need for accuracy and completeness.

The following is a typical condensed specification for a small project. It might be the type of specification that is usually being typed as the plans are being completed. In many cases this limited version is used when more complex equipment is required, and extreme care should be taken in this event.

Typical Specifications:

1. Furnish and install panelboards as shown on the plans and described herein, equal to General, Westinghouse, or Square D types, as noted on the plans. Panelboards shall be rated for the intended voltage.
2. Boxes shall be galvanized steel, providing ample wir-

PANEL SCHEDULE													
PANELS					BRANCH CIRCUITS							REMARKS
KEY	TYPE	LOCATION	MTG	MAINS	TYPE	FRAME	TRIP	POLES	ACTIVE	SPARE	TOTAL	

FIGURE 3-3

ing gutter space in accordance with the National Electric Code.

3. Fronts shall be sheet steel painted medium-dark gray (ANSI 33) over a rust inhibitor. They shall be equipped with doors, semiconcealed five-knuckle hinges, cylinder tumble-type lock, directory card holder, and quarter-turn adjustable trim clamps. On doors more than 48 in high a combination three-point catch and lock shall be provided. All locks are to be keyed alike.

4. Panel shall consist of a reinforced galvanized sheet steel frame with bus bars and circuit breakers properly supported to prevent vibration, breakage in handling, and to withstand effects of indicated short-circuit current. All terminals shall be solderless type suitable for copper or aluminum cable of the sizes indicated. Bus bars shall be sequenced-phased.

5. Branch circuit breakers shall be fully interchangeable without disturbing adjacent units. Circuit breakers shall be quick-make, quick-break, and trip-indicating. All two- and three-pole circuit breakers shall have common trips.

The first item to note is that the specification is very generalized. Components, sizes, types, and other details are not covered. If the schedule on the plans does not carefully define what is really wanted, needed, and required, then the result will be neither clear, concise, nor usable. The following is an item-by-item discussion of the panelboard schedule components.

Key: This is the identifying number such as LP-I, LP-A, PP-A, or other identifier. Good practice is to have two types of panel, light (LP) or power (PP), numbered or lettered consecutively. The schedule should group the respective types by category, with each category separated by a space. Alternatively, you can create two separate schedules—one for lighting and one for power. Note also the discussion below under Mains.

Type: This refers to the usage of the panel such as LP, PP, or distribution (Dist).

Location: Although self-evident, the location should nevertheless be precise, such as "Storage Room 102." Not infrequently room numbers and space numbers get changed at the last minute. Even though the number or name or both may no longer be correct, it is fairly easy to correct or understand the proper location if a full name and number are shown.

MTG: This is the abbreviation for *mounting,* which is either surface (surf) or recessed (rec). But a competent designer should be carefully checking the building plans to be certain he or she is not trying to put a 6-in recessed panel into a 4-in wall.

Mains: Once upon a time the answer here was simple. The panel had either a main circuit breaker or no main circuit breaker and lugs only. But the NEC has become much more specific. All panelboards meeting the Code definition of Section 384-14 for a lighting and appliance panelboard must have one main protective device, either in the main panel or ahead of it, rated not greater than the panel bus bar. In lieu of this there may be two main protection devices dividing the supply into two circuits, with a combined rating not greater than the bus bar rating. Section 384-14 further states that panels containing more than 10 percent of their devices rated 30 A or less and with neutral connections are lighting and appliance panels. In most common cases the panel falls into the category where a main circuit breaker will be required.

Type: In this instance the word "type," as well as the data that follows, refers to the branch circuits themselves. Here the input of the project engineer is required. Are bolt-in, plug-in, or factory-preassembled and factory-mounted circuit breakers required? NLAB and NLTQ are most commonly used designations.

Frame: Manufacturers generally limit the capacity of frames and the variety of sizes that house the circuit breakers. Commonly they are 50, 100, 225, and 400 A. Thus a circuit breaker rated at 15, 10, 30, or 50 A will fit into a 50-A frame.

Trip: This refers to the amperage you have selected for your specific circuit, such as 15, 20, 30 A.

Poles: This is a simple specification that is frequently the source of error. Panelboards are normally constructed of single 120-V circuits. One 120-V circuit equals one pole. Forty-two poles make up the normal panel. A 120-V circuit equals one pole; a 240-V circuit equals two poles. See the next item, Active or Spare.

Active or Spare: The circuits available in your panel are either being used (active) or not being used (spare). The appropriate column should be checked. In some firms the individual circuits are not listed but only the number of circuits by pole, that is, 20 of one pole, etc. In this case definite values are required. See also discussion under Total that follows.

Total: As previously noted, 42 poles is a limiting value; so is 20 percent spare capacity. Thus, in the maximum-sized panelboard the total available circuits (poles) is 42 minus 20 percent of 42, or 34 poles. More circuits than 34 requires a second panel.

Remarks: This column should be used for brief notes, cautions, or explanations, for example, a special case of a Code exception where a device is used at 100 percent of capacity, or where the circuit breaker is used as a switch.

Having described a panel schedule and what is involved, we show Figure 3-4 as an example of a completed schedule. Because this is a small project, the schedule shows not only the remotely located panels A, B, and C but also the main distribution panel which covers the main service as well as the subfeeders that supply existing panels.

The "Remarks" column, as can be readily seen, can hold only the briefest of notations, and this is as it should be. However, much more information might be required on switches and contactors. If an extended description of any item is required, the "Remarks" column should contain a clear reference, such as "See Note 1 below." Sufficient space therefore should always be allowed below the schedule for such notes.

PANEL SCHEDULE

KEY	TYPE	LOCATION	MTG.	MAINS	TYPE	FRAME	TRIP	POLES	ACTIVE	SPARE	TOTAL	REMARKS
MDP	NAB	BLR. RM.	SURF	400A 3P	TQD	225	200	3	1	0	3	SERVES EXISTING
				TJD	TEB	100	100	3	3	1	12	SERVES 3 NEW PNLS
					TEB	100	70	2	1	1	4	SERVES SAUNA
					TEB	50	50	2	1	1	4	SERVES EXISTING
					TEB	50	45	3	1	1	6	SERVES EXISTING
A	NLAB	POOL OFF	REC.	100A, 3P	THQB	20	20	1	17	7	24	
				THQC	THQB	20	20	3	1	1	6	
B	NLAB	POOL OFF.	REC.	100A, 3P	THQB	20	20	3	8	1	27	
				THQC	THQB	50	40	3	1	0	3	
C	NLAB	BLR. RM.	SURF	100A	THQB	20	15	1	2	0	2	SERVES EXISTING
				LUGS	THQB	20	20	1	13	5	18	BASEMENT
				ONLY	THQB	50	40	2	1	1	4	
					THQB	20	15	3	1	0	3	
					THQB	50	30	3	1	0	3	

FIGURE 3-4

Bus-Duct System Bus-duct systems are commonly thought of as lines of electrical duct, usually in an industrial plant, that are fed by conduit and wire from the main distribution center. These lines of bus duct have cable tap boxes attached to them from which conduit and wire are extended to bring power to motors on the floor below. However, bus ducts have many other uses as well. Figure 3-5 is a sort of composite drawing illustrating the potential of bus duct. Some, all, or none of the figure may apply to your particular situation.

Starting at the utility company or the secondary side of your transformer, the total load to your main secondary disconnect can logically and economically be served by bus duct in lieu of conduit and wire. This could also be accomplished by bus duct if the service were from an overhead source. Obviously you would have one or the other, not both. Note that bus duct, like conduit, also has a weatherhead cable tap box.

From the secondary distribution you could extend a feeder in bus duct to the horizontal bus duct as shown. This would be the typical industrial layout we described in the opening paragraph. What is not shown are the cable tap boxes which would be installed in the horizontal parts of the bus duct and which would feed the motors via conduit and wire.

What if you were involved in a high-rise building instead of an industrial plant? For this case we have illustrated a vertical rise of power to feed panels on a number of floors. Again it could be done in conduit and wire, but a bus-duct riser makes a lot of economic sense where a number of floors and associated panels are to be supplied. As can be seen in the detail, the bus-duct system could begin at the main distribution center and continue up to serve many floors.

In our viewpoint the bus-duct system can serve both as an extended panel system and as a branch circuit. As a distributor of power to other panels it is correctly described as a distribution feeder. Thus the variety of uses to which a bus duct can be applied makes it very hard to classify. This book looks at the bus duct as a feeder device.

Economically it may or may not fit into your design scheme. This choice is up to your firm's engineers and designers. There is no single answer that fits all cases.

Panel Detail The variety and quantity of information shown in individual panel details are among the detailing items most often in error. The plans frequently contain a panel schedule indicating the panel type, type of mains and circuit breakers, and number of operating and spare circuits. There is nothing wrong with this sort of panel description and its usage is widespread.

Since the panel description by schedule is acceptable, then why detail the panel? There are several good reasons. The first arises when circuits are switched from the panel. In this case you or your client may want the switched circuits grouped in a certain position in the panel. While it is true that the circuit number presumably does this, it is also true that the contractor may not necessarily put the numbered circuit on the same circuit breaker number at the panel. Our schedule shows circuit numbers at the panel. To make the point unmistakable, you could add to the schedule, for example, the duty of circuit number 1 is "house lights."

The second reason for the panel schedule is more straightforward. Your client may have had work done by other engineers in the past, and they have provided this detail. If you want to do the work, you must do the same.

The third reason is that in the pressure of completing a project, one or more circuits are added at the last minute and the panel schedule does not get changed. For example, the schedule shows 24 operating single-pole circuits and 6 single-pole spare circuits for a total of 30 single-pole circuits. Someone adds some additional lights or outlets so another circuit or two is quickly entered, but these are not added to the schedule. It is not unusual to find our 24-

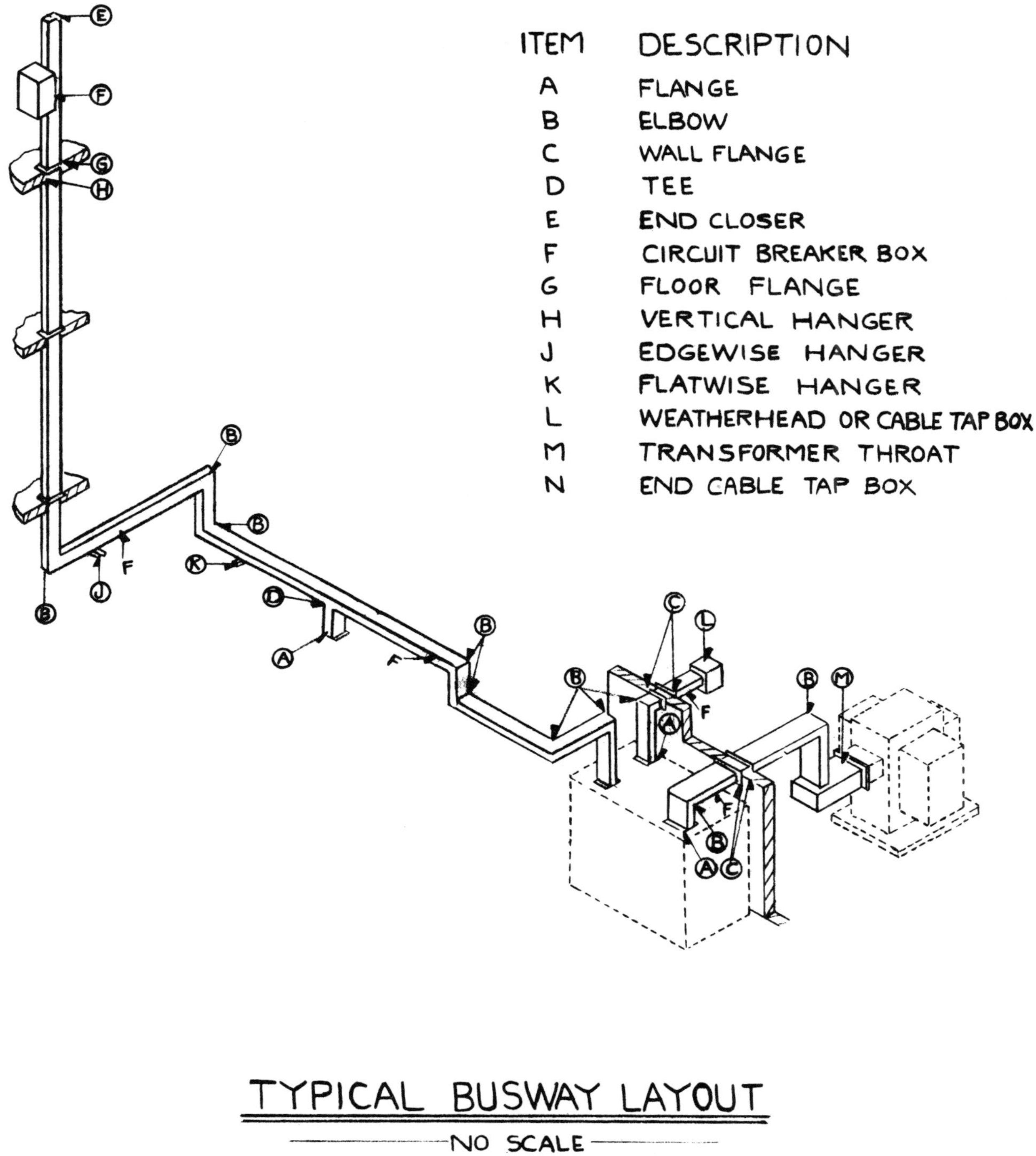
ITEM DESCRIPTION
A FLANGE
B ELBOW
C WALL FLANGE
D TEE
E END CLOSER
F CIRCUIT BREAKER BOX
G FLOOR FLANGE
H VERTICAL HANGER
J EDGEWISE HANGER
K FLATWISE HANGER
L WEATHERHEAD OR CABLE TAP BOX
M TRANSFORMER THROAT
N END CABLE TAP BOX
TYPICAL BUSWAY LAYOUT
NO SCALE

FIGURE 3-5

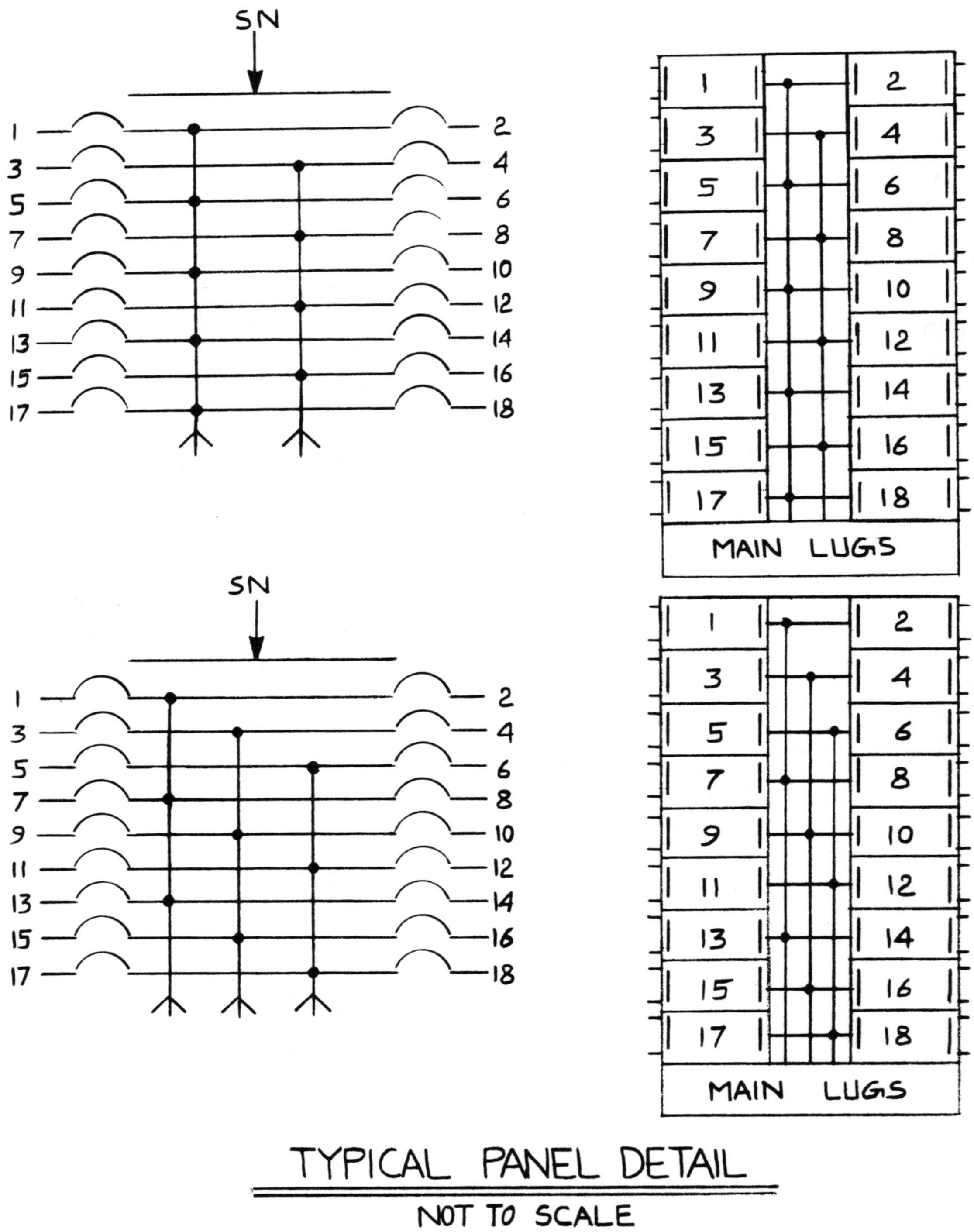

FIGURE 3-6

operating and 6-spare panel turn up in the field as a 27-operating and 3-spare panel. If your system requires you to add these new circuits to your panel detail, they can still be missed but their addition is much more obvious and checkable for all concerned.

Figure 3-6 gives you two choices of presentation—schematic or block style. In either case the ideal system would include the circuit use description associated with the number. Frequently, use is noted and the number deleted. Both number and use not only tie into plan circuit numbering but end all debate as to the intentions of the designer.

FOUR

Ground Fault Protection and Grounding

Ground Fault Protection When panel circuits are overloaded the affected circuit breaker or fuse will trip or blow to protect the circuit. But there are conditions common to grounded systems which create faults not cleared by conventional overcurrent devices. This is called a *phase-to-ground fault,* usually an arcing between a hot conductor and a grounded metal enclosure or conduit, which creates a current value less than that which the circuit breaker or fuse needs to react.

To give you an extreme example, a 200-A fault on a 1000-A circuit breaker will not affect the circuit breaker. Thus an arcing fault of considerable destructive ability can, in effect, be "tolerated" by the breaker. A ground fault interrupter senses this problem and does something about it.

The location and application of ground fault protection is a designer's job and not one that should be left to the detailer. But, as you know, these sorts of omissions and lack of instructions do occur. In the panel schedule you can, and should, call for circuit ground fault protection where required.

There are a number of ways the ground fault requirement could be resolved. Figure 4-1 is a sample diagram of how and where the ground fault could be applied. Article 230-95 of the NEC notes the requirements for ground fault protection, but the information is a little difficult to understand. As previously noted, the NEC handbooks provide explanatory diagrams and text to further clarify the Code.

Finally, it might be asked "How do ground faults really occur?" Uusally the cause is a failure, for whatever reason, in the insulation system. This permits current to return to the source by conduit, building steel, or piping. The problems are created by damaged conductors, loose connections, moisture, and dust. In 208/120-V systems fortunately the characteristic of the voltage drop across the arc and the low level of the fault current usually cause the fault current to be self-extinguishing. But on 277/480-V and larger systems the arc may not be self-extinguishing and considerable or even catastrophic damage to equipment, property, and personnel may occur.

Grounding In countless specifications and on innumerable plans you can see and read the same phrase "Provide all grounding in accordance with Article 250 of the National Electrical Code and all local, state, and federal codes that apply." Sometimes the words "and in accordance with good practice" are added. The users of such phraseology are fortunate that the grounding task is simple in many cases and that the installing electrician knows the Code probably more thoroughly than the designer.

To begin at the beginning, we start with the question "What do we mean by grounding?" There are two separate, clear-cut areas. One is system grounding and the other is equipment grounding. The subject is not treated lightly by the NEC. In fact, Article 250 of the Code has 155 numbered paragraphs on the subject, divided into 13 separate subheadings. In terms of legality and fairness a specification format on grounding would not merely refer to Article 250 but would repeat the many pages of printed material in their entirety. Ridiculous, perhaps, but at least legal, complete, and fair as to content and intent.

We do not intend to repeat the entire Code or detail every possible connection that might be involved. Except for control circuits noted in NEC Section 250-26 and cranes and health care facilities noted in Section 250-7,

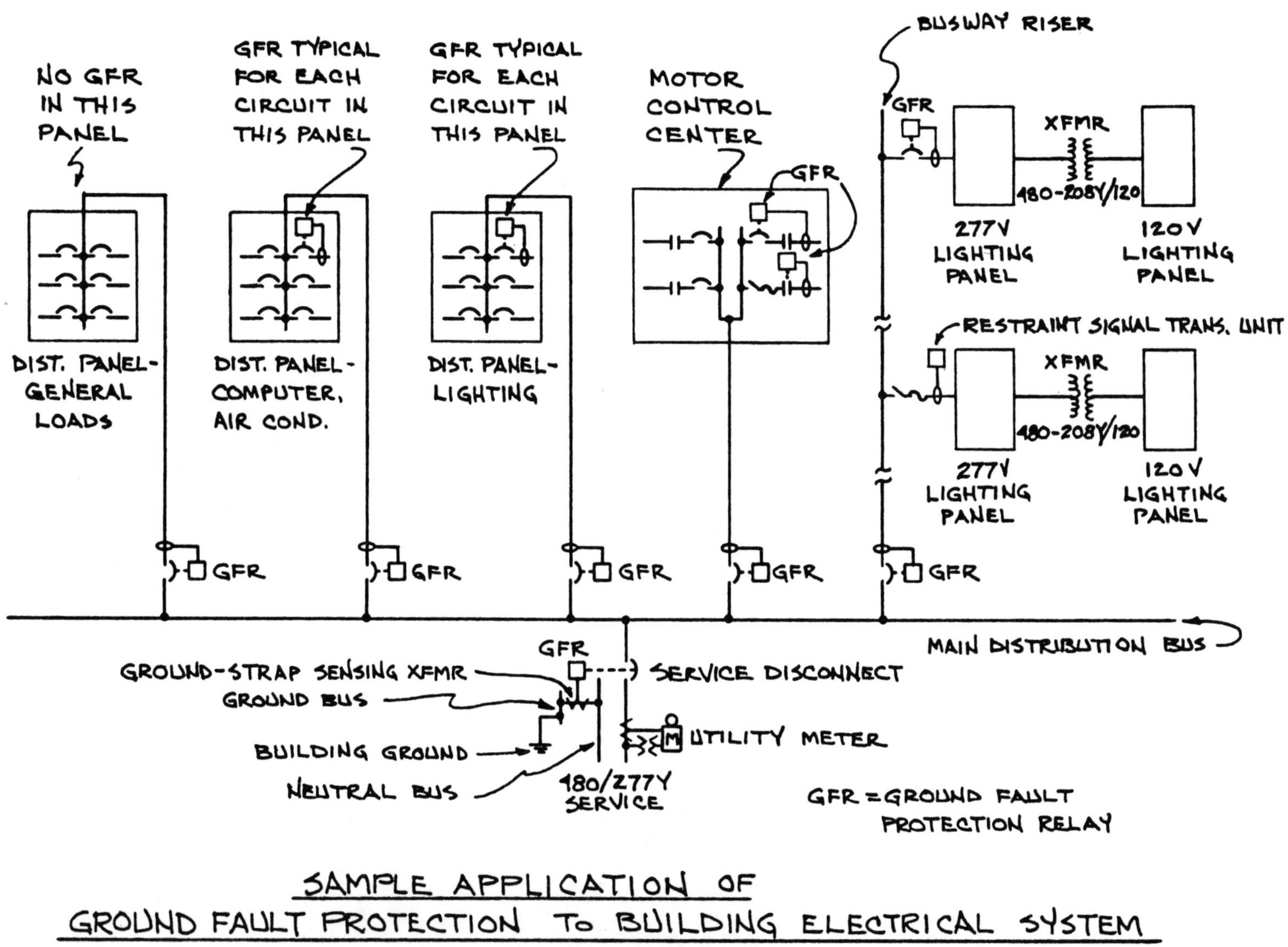
BUSWAY RISER
NO GFR IN THIS PANEL
GFR TYPICAL FOR EACH CIRCUIT IN THIS PANEL
GFR TYPICAL FOR EACH CIRCUIT IN THIS PANEL
MOTOR CONTROL CENTER
GFR
XFMR
480-208Y/120
277V LIGHTING PANEL
120V LIGHTING PANEL
RESTRAINT SIGNAL TRANS. UNIT
XFMR
480-208Y/120
277V LIGHTING PANEL
120V LIGHTING PANEL
DIST. PANEL-GENERAL LOADS
DIST. PANEL-COMPUTER, AIR COND.
DIST. PANEL-LIGHTING
GFR
GFR
GFR
GFR
GFR
MAIN DISTRIBUTION BUS
GFR
GROUND-STRAP SENSING XFMR
GROUND BUS
SERVICE DISCONNECT
BUILDING GROUND
UTILITY METER
NEUTRAL BUS
480/277Y SERVICE
GFR = GROUND FAULT PROTECTION RELAY
SAMPLE APPLICATION OF
GROUND FAULT PROTECTION TO BUILDING ELECTRICAL SYSTEM

FIGURE 4-1

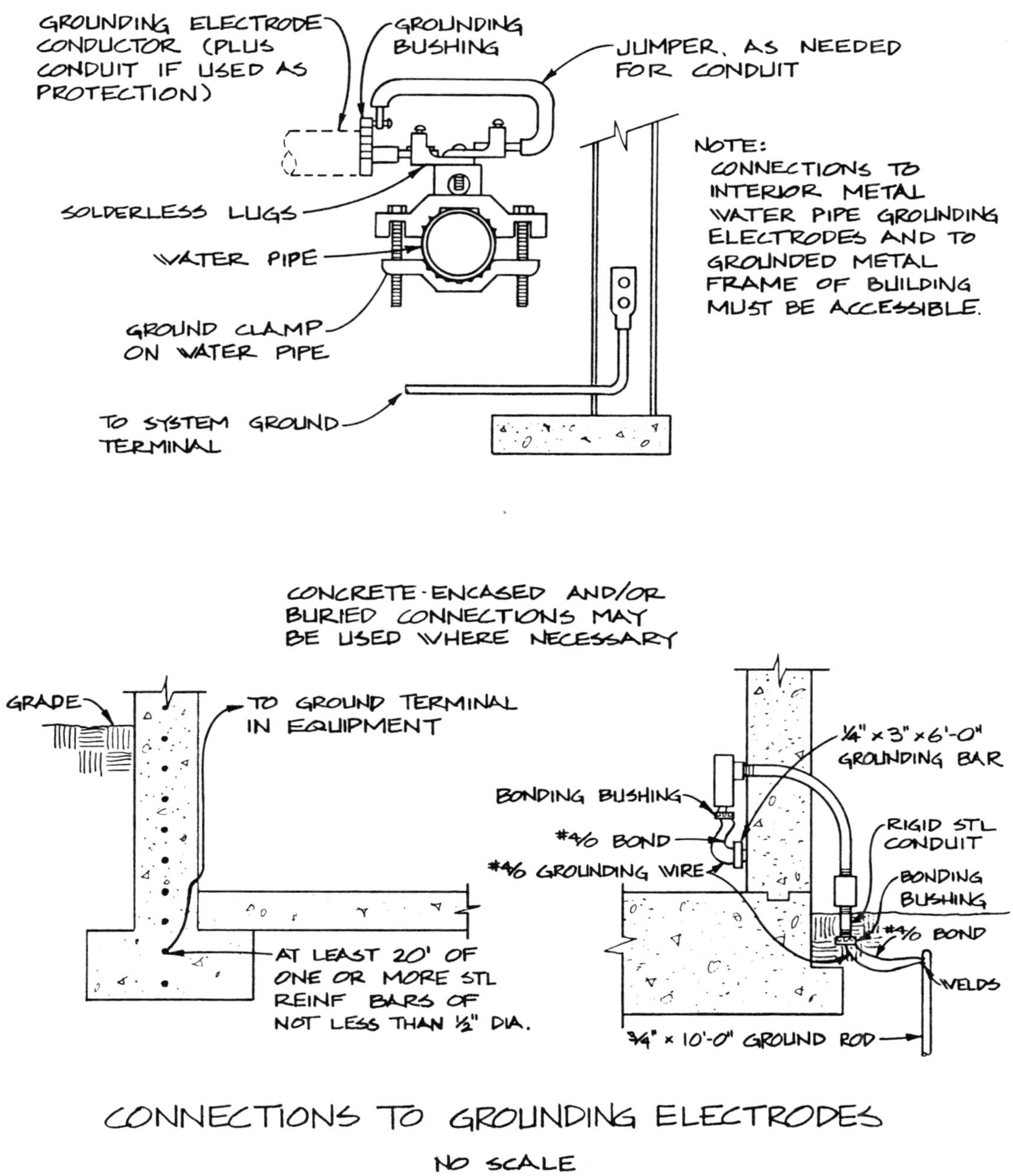

FIGURE 4-2

the basic system requirements state that a 120-V two-wire system or circuit must have one of its wires grounded, a 240/120-V three-wire single-phase system must have its neutral conductor grounded, and a 208/120-V four-wire three-phase wye-connected system or circuit must be operated with the neutral conductor grounded. In each case the maximum voltage to ground does not exceed 150 V from any other conductor when the neutral conductor is grounded. In effect, any 120-V circuit must have one of its conductors grounded. For control circuits there are exceptions which will not be repeated here in detail.

An equipment ground is the permanent and continuous connecting together of all non-current-carrying metal parts of equipment enclosures, such as conduit, boxes, cabinets, housings, frames of motors, and lighting fixtures, to a system grounding electrode located at the service equipment or on the secondary side of the system transformer. Grounding essentially provides a low-impedance path for current. The grounding device is frequently a service entrance water pipe, building steel, or a driven rod.

While knowledge of the Code is mandatory for anyone involved in electrical design, everything said so far does not solve the detailer's problem of how to show the actual connections. To this end we have included Figures 4-2 to

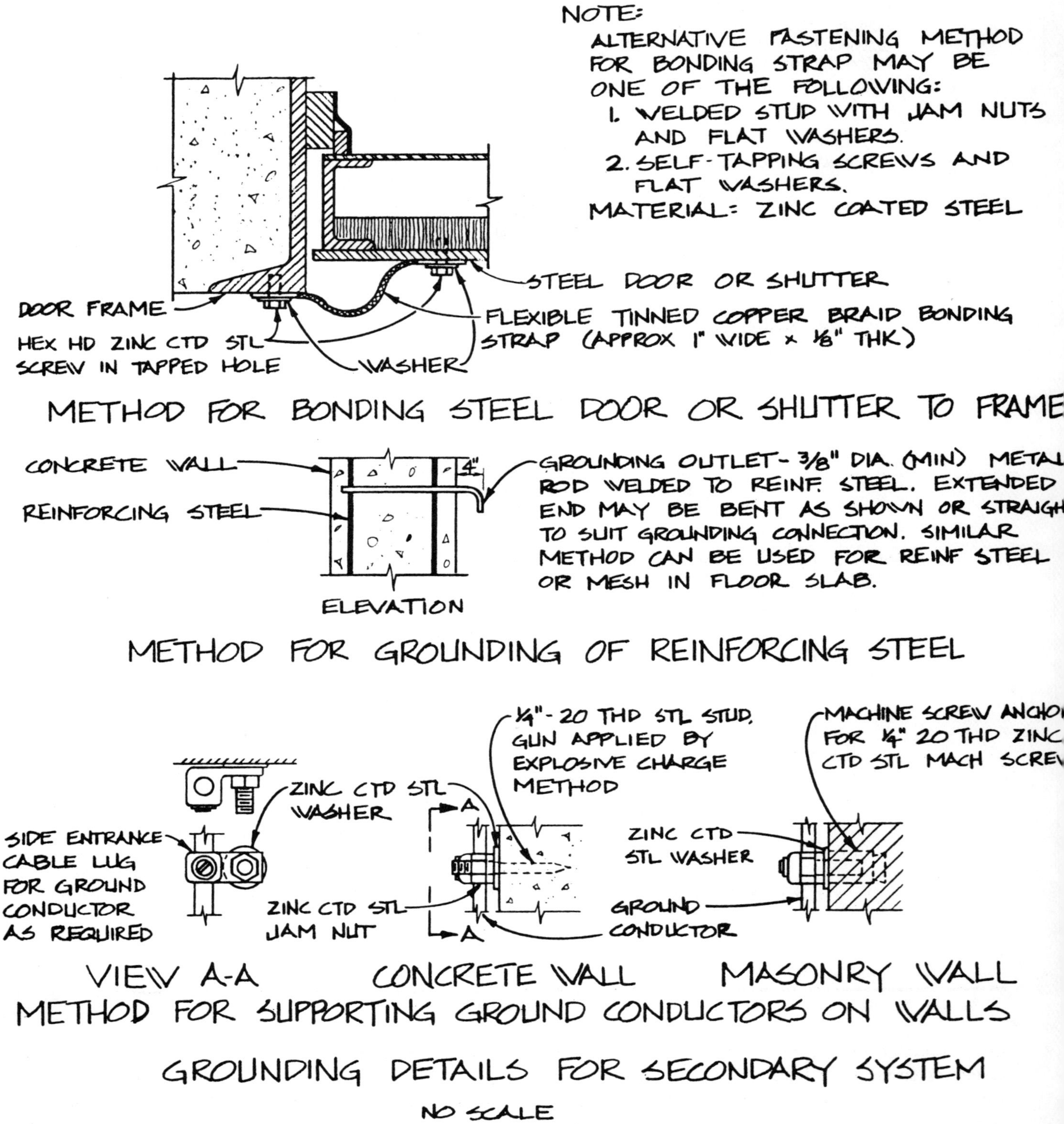

FIGURE 4-3

4-8, which present seven separate pages of illustrated details. We begin with Figure 4-2, which illustrates how to detail grounding electrode connections to a water service pipe or to the building steel. As in all our details there are other ways. If you use this detail, use it as is; do not be an inventor.

Figures 4-3 and 4-4 cover methods of bonding and grounding for secondary systems. These include bonding a steel door or shutter to a frame, supporting ground conductors on walls, grounding steel columns, and universal ground connections. What is attempted here is to cover the problems you will most likely have in your design.

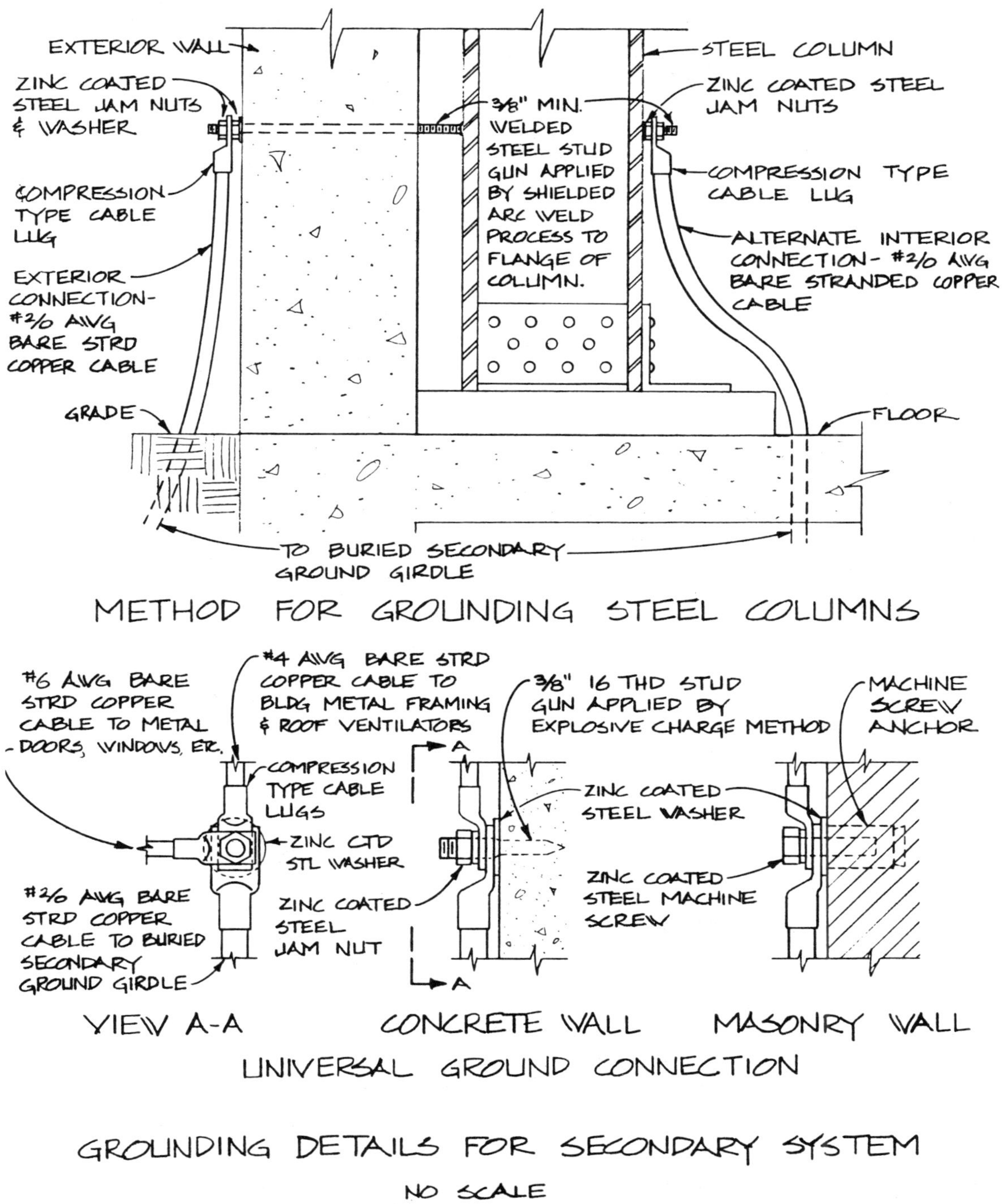
EXTERIOR WALL
STEEL COLUMN
ZINC COATED STEEL JAM NUTS & WASHER
3/8" MIN. WELDED STEEL STUD GUN APPLIED BY SHIELDED ARC WELD PROCESS TO FLANGE OF COLUMN.
ZINC COATED STEEL JAM NUTS
COMPRESSION TYPE CABLE LUG
COMPRESSION TYPE CABLE LUG
ALTERNATE INTERIOR CONNECTION- #2/0 AWG BARE STRANDED COPPER CABLE
EXTERIOR CONNECTION- #2/0 AWG BARE STRD COPPER CABLE
GRADE
FLOOR
TO BURIED SECONDARY GROUND GIRDLE
METHOD FOR GROUNDING STEEL COLUMNS
#6 AWG BARE STRD COPPER CABLE TO METAL DOORS, WINDOWS, ETC.
#4 AWG BARE STRD COPPER CABLE TO BLDG METAL FRAMING & ROOF VENTILATORS
3/8" 16 THD STUD GUN APPLIED BY EXPLOSIVE CHARGE METHOD
MACHINE SCREW ANCHOR
A
COMPRESSION TYPE CABLE LUGS
ZINC CTD STL WASHER
ZINC COATED STEEL WASHER
#2/0 AWG BARE STRD COPPER CABLE TO BURIED SECONDARY GROUND GIRDLE
ZINC COATED STEEL JAM NUT
ZINC COATED STEEL MACHINE SCREW
A
VIEW A-A
CONCRETE WALL
MASONRY WALL
UNIVERSAL GROUND CONNECTION
GROUNDING DETAILS FOR SECONDARY SYSTEM
NO SCALE

FIGURE 4-4

On occasion you will have tanks, stacks, and structures requiring the installation of lightning rods, which have to be grounded to be effective. Figure 4-5 describes the proper grounding procedure.

One of the newer (new in that little has been done in this area for years) types of projects has to do with renovation and retrofit of railroad stations. In your design criteria there may be a requirement for track grounding. Figure 4-6 covers this subject.

Of all the projects that require proper grounding, certainly a swimming pool has to be one of the most important. There are potentials for stray current distribution in

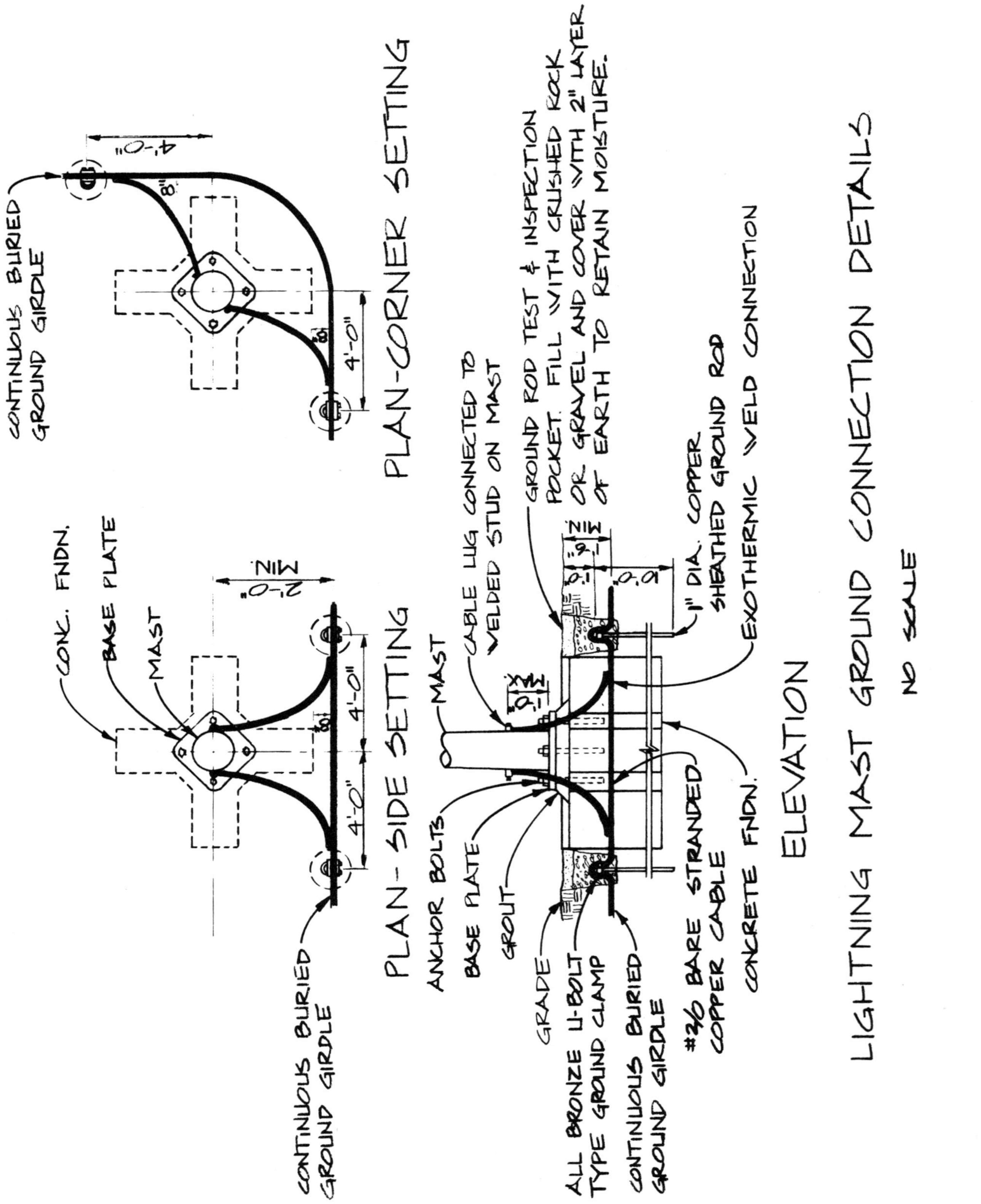

FIGURE 4-5

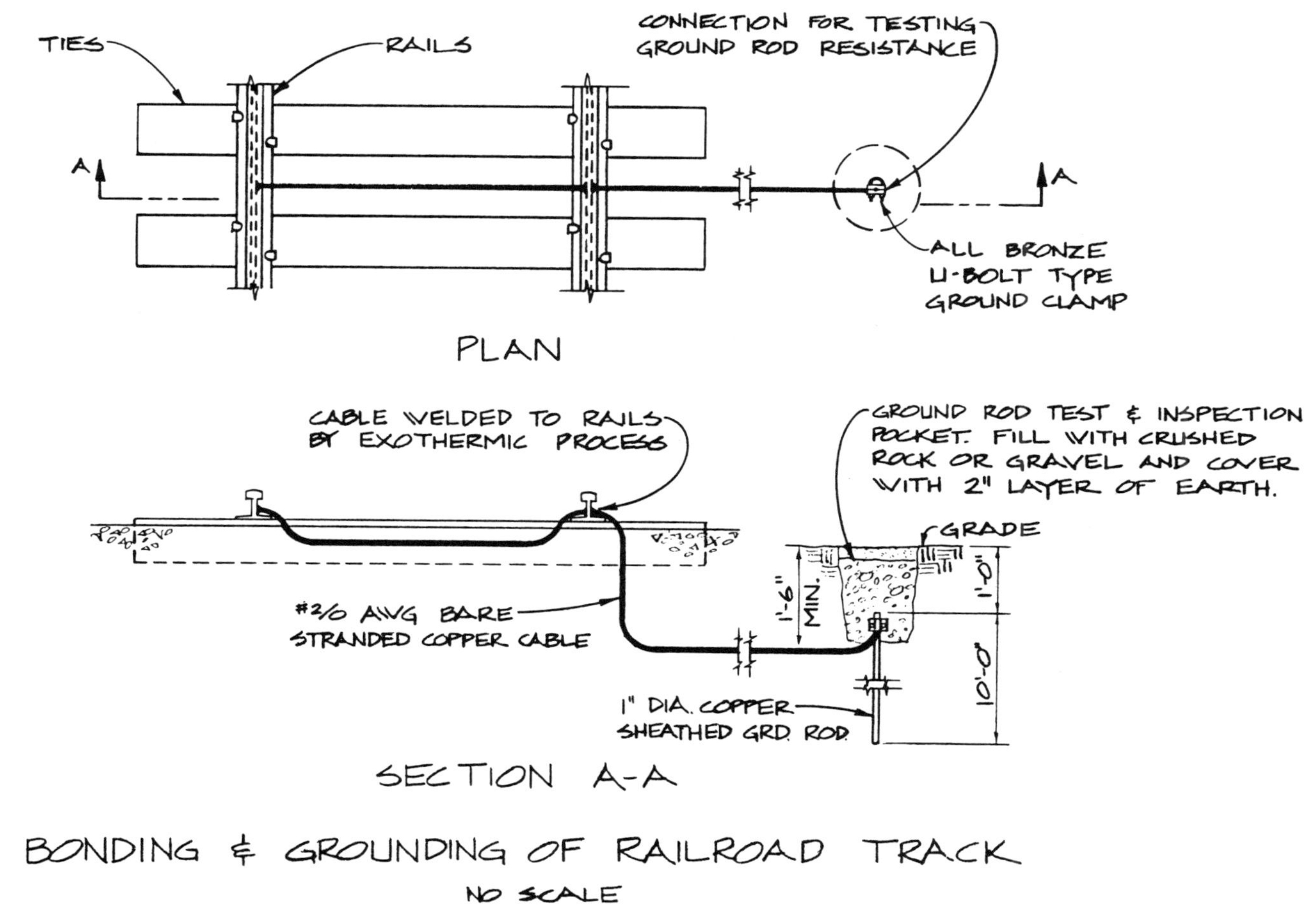

FIGURE 4-6

literally every piece of steel. Thus it is not merely the electrical system but the steel itself and the piping, drains, drinking fountains, motor frames, and the like which constitute equipment to be grounded. We show a pool bonding and grounding system in Figures 4-7 and 4-8.

Where driven ground rods are noted, they are to be ¾- or 1-in copper rods 8 ft 0 in long and driven to that depth in the soil. Where rock or other obstacles preclude direct driving of the rods, the Code allows for a rod to be driven at a 45° angle or to be totally buried horizontally at least 2 ft 6 in below the surface.

While this concludes our discussion of grounding in this chapter, grounding for services, transformers, and high-voltage switchgear will be discussed in later chapters.

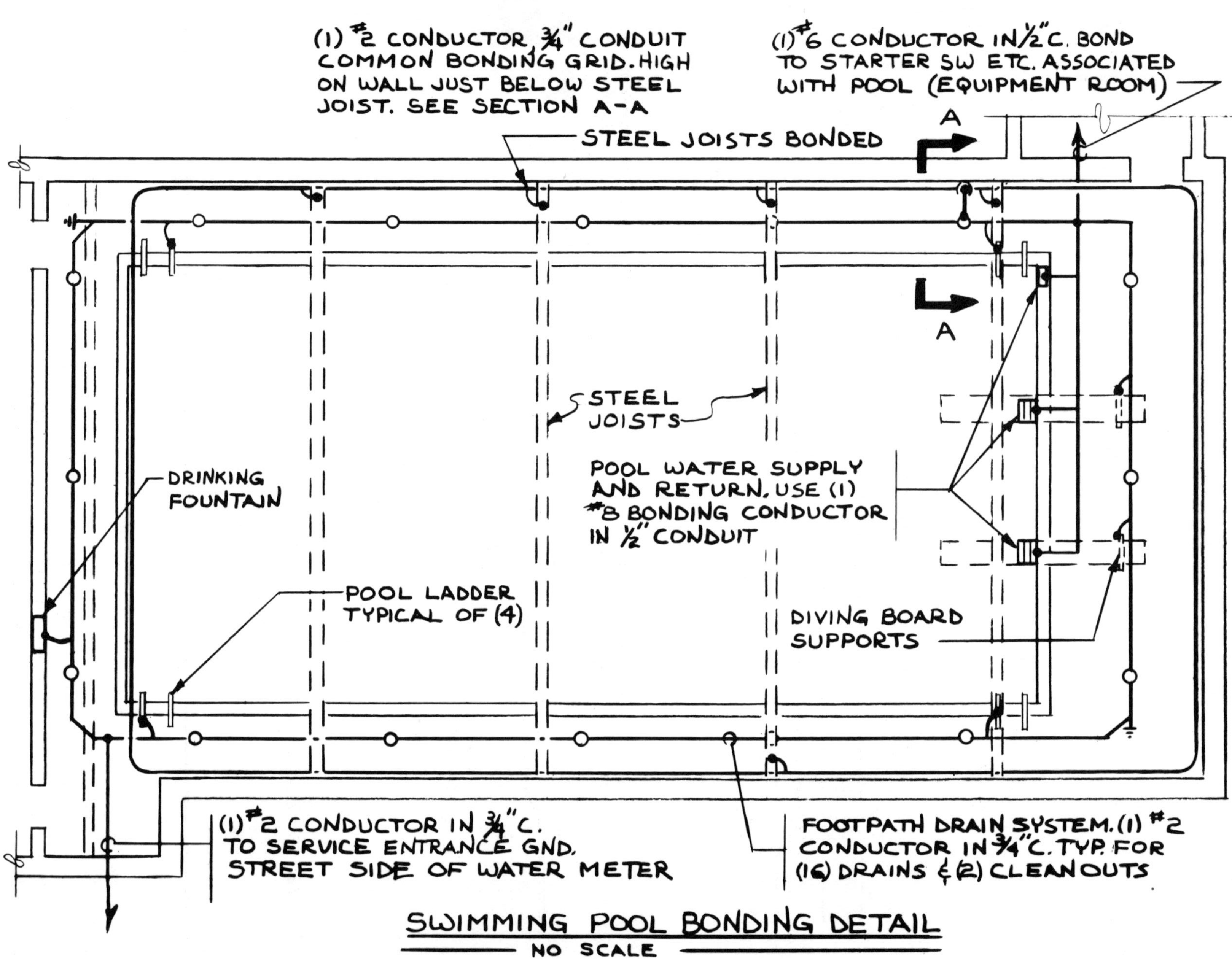
(1) #2 CONDUCTOR, 3/4" CONDUIT
COMMON BONDING GRID. HIGH
ON WALL JUST BELOW STEEL
JOIST. SEE SECTION A-A
(1) #6 CONDUCTOR IN 1/2" C. BOND
TO STARTER SW ETC. ASSOCIATED
WITH POOL (EQUIPMENT ROOM)
A
STEEL JOISTS BONDED
A
STEEL
JOISTS
POOL WATER SUPPLY
AND RETURN, USE (1)
#8 BONDING CONDUCTOR
IN 1/2" CONDUIT
DRINKING
FOUNTAIN
POOL LADDER
TYPICAL OF (4)
DIVING BOARD
SUPPORTS
(1) #2 CONDUCTOR IN 3/4" C.
TO SERVICE ENTRANCE GND.
STREET SIDE OF WATER METER
FOOTPATH DRAIN SYSTEM. (1) #2
CONDUCTOR IN 3/4" C. TYP. FOR
(16) DRAINS & (2) CLEANOUTS
SWIMMING POOL BONDING DETAIL
NO SCALE

FIGURE 4-7

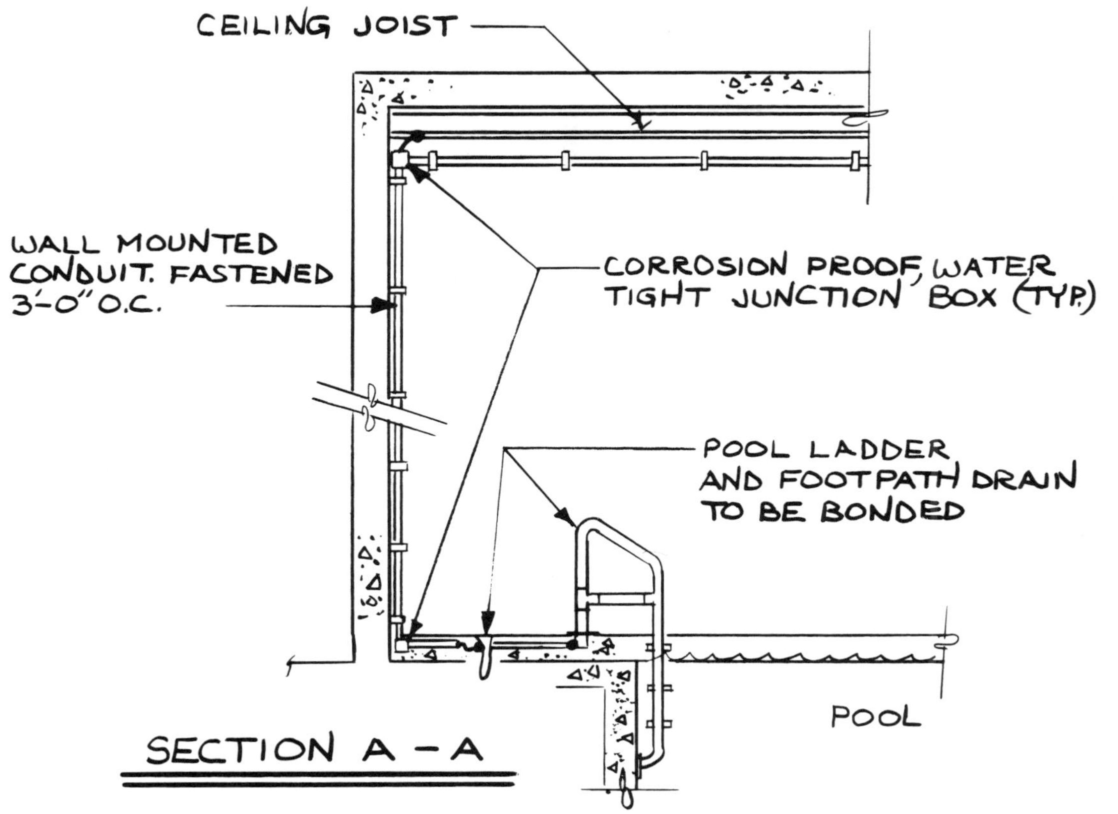

NOTE:

THE FOLLOWING PARTS SHALL BE BONDED TOGETHER: ALL METAL FITTINGS WITHIN OR ATTACHED TO THE POOL STRUCTURE; METAL PARTS OF ELECTRIC EQUIPMENT ASSOCIATED WITH THE POOL CIRCULATING SYSTEM INCLUDING PUMP MOTORS; METAL CONDUITS AND PIPING; DUCTWORK WITHIN (5) FEET OF THE INSIDE WALLS OF THE POOL AND ARE NOT SEPARATED FROM THE POOL AREA BY A PERMANENT BARRIER; ALL METAL PARTS THAT ARE WITHIN (5) FEET OF THE INSIDE WALL OF THE POOL AND IS NOT SEPARATED BY A PERMANENT BARRIER; METAL REINFORCING OF THE POOL STRUCTURE. ALL CONNECTIONS SHALL BE OF THE EXOTHERMIC TYPE. SEE ARTICLE 680 OF THE N.E.C. FOR ADDITIONAL BONDING INFORMATION.

BONDING CONDUCTOR INSTALLATION DETAIL

NOT TO SCALE

FIGURE 4-8

FIVE

Services and Distribution

In this chapter we are concerned with services from the incoming line to the building, to the distribution point, and to panels within the building. This includes the line from the pole, the main service entry and distribution equipment, and the lines from this equipment to the panels.

The line that comes from the utility company source to the building can be installed overhead or underground. If it is underground, the situation is generally simplified and little or no detailing is needed, since the connections shown on the drawing merely indicate a line from the utility transformer or point of connection to the building service. However, if the line runs overhead, then detailing is required to show how the connections are to be made at the utility service point, at the building, and at the service entrance equipment.

If the incoming service line is a long overhead run, one or more poles may be required. It is general practice that the distance between the poles be some 250 ft or less. Overhead lines are normally built in accordance with the local utility company standards and the requirements of the National Electrical Safety Code (NESC), also known as ANSI C2. It is important to note that many states have requirements that exceed those of the NESC, and these requirements should be carefully investigated.

The installation of open wires over buildings should be avoided if at all possible, as they interfere with the activities of fire fighters and the functions of building maintenance and security personnel. They also present a safety hazard. If it is impossible to avoid passing close to or over a building, the incoming service wires should be enclosed in a grounded metallic conduit or encased in concrete if the voltage is above 300 V. Grounded metallic aerial cable can be used where the line passes over a building. The minimum clearance should be sufficient so that all personnel can pass freely under the wires.

When service conductors enter the building to connect to the service equipment, a safety hazard exists regardless of the voltage. Where the distance from the point of building penetration to the connection is 10 ft or less, the hazard is considered very small. Circuits over 10 ft long from the point of entrance to the building to the service equipment should be installed in a metallic raceway and encased in concrete at least 2 in thick.

Safety and property life are always enhanced by encasing the incoming line in concrete inside the building. The concrete-encased raceway can be installed along the ceilings, under the basement floor, or on the roof. There are specific minimum requirements contained in the NEC, and these should be adhered to and even exceeded in many cases.

A checklist of items that should be evaluated in any service system includes the following:

1. Loads to be served
2. The kilovoltampere demand
3. Continuity requirements
4. Voltage requirements and limitation to voltage variations
5. Special loads such as x-ray machines and computers
6. Special requirements of electrical systems such as signals, clocks, and communications
7. Size of largest motor involved

A second area of investigation should be the characteristics of all types of service available from the utility and the various rate structures. These investigations include voltages available, billing demand clauses, rates and rate clauses, space requirements, and transformer or other special requirements.

A third area of investigation is the physical and mechanical requirements of the service. This should cover the type of service, where the service will be, how the meters are to be located, the accessibility of the meters, the space requirements for the service, and the fault capabilities and conditions, both present and future. The future is usually defined as 10 to 20 years from now. Finally, the requirements for coordination of service protection systems should be investigated.

The 480/277-V three-phase four-wire grounded wye system is rapidly replacing the well-known 208/120-V system currently in use in many buildings. However, the mere fact that this change is occurring does not mean that this voltage is the proper one for your project. There are many practical and technical factors affecting the choice of voltage within the building, and even in a large, complex system, as we will show later in our details, there may be reasons for using a 208/120-V system.

Generally the determining factor between the 208-V system and the 480-V system is the amount of lighting that must be placed on a 120-V system as well as outlets, small motors, and other special requirements in the building, which, if connected to a 480-V system, would need dry-type transformers in many locations to accommodate them.

In any form of distribution system there are various methods that can be used to choose the type of design. One of the ways is to divide the building systems, or the buildings themselves, into various classifications. A typical classification system is as follows:

1. Large vertical buildings under 12 stories
2. Tall office buildings and commercial buildings
3. Large horizontal building structures
4. Groups of small buildings
5. Industrial plants

While this is one way of categorizing the types of services required, the additions to the above five areas could be expanded considerably into such categories as hospitals, research laboratories, computer facilities, and the like.

The primary consideration in all distribution systems from the point of the main service entrance to the panels and areas of use is *distance*. Whether vertical or horizontal, distance at the higher voltages obviously reduces the wiring costs. By using a higher voltage the voltage-drop factor is also considerably assisted.

Since this book is primarily concerned with details, we will not continue with this discussion and, instead, refer you to the publications of the Institute of Electrical and Electronics Engineers (IEEE) and to the publications of the major equipment manufacturers, all of which are excellent sources of reference material and guidance for equipment and system selection.

Utility Connections Generally the utility company requires that a sufficient length of slack wire, usually 10 ft, be left at the top of the incoming pole or in the vicinity of their transformer for the final connections to their equipment, which they customarily make. However, there are occasions where an engineer may be faced with a situation requiring connections to an owner's primary distribution system. Thus we have included Figures 5-1 to 5-5, which present details that are relatively standard for this type of installation.

Figure 5-1 illustrates the connection of an aerial-to-underground system at a normal primary pole line distribution system. The pole has both lightning arresters and primary fuse cutouts. These primary fuse cutouts accept the primary distribution system and act as the means of disconnecting the incoming service. They are normally operated by a pole with a hook at its end.

From the primary service, Figure 5-1 shows a pothead for each of the incoming wires. These wires are tied to a multiple-wire spreader head on the pole, and the conductors then normally run down the pole in conduit and from there underground to the building. The specification descriptions of each of the items shown, such as lightning arresters, primary fuse cutouts, single-conductor potheads, and spreader head, as well as the wire involved, are contained in the catalog of a number of well-known manufacturers. The materials can be specified on the detail. If space in the area surrounding the detail is available, we would recommend using this area for specific data and description.

This particular detail could have been presented in the primary distribution discussion in Chapter 6. It seemed more logical to show it here and the grounding required with this type of installation in Chapter 6.

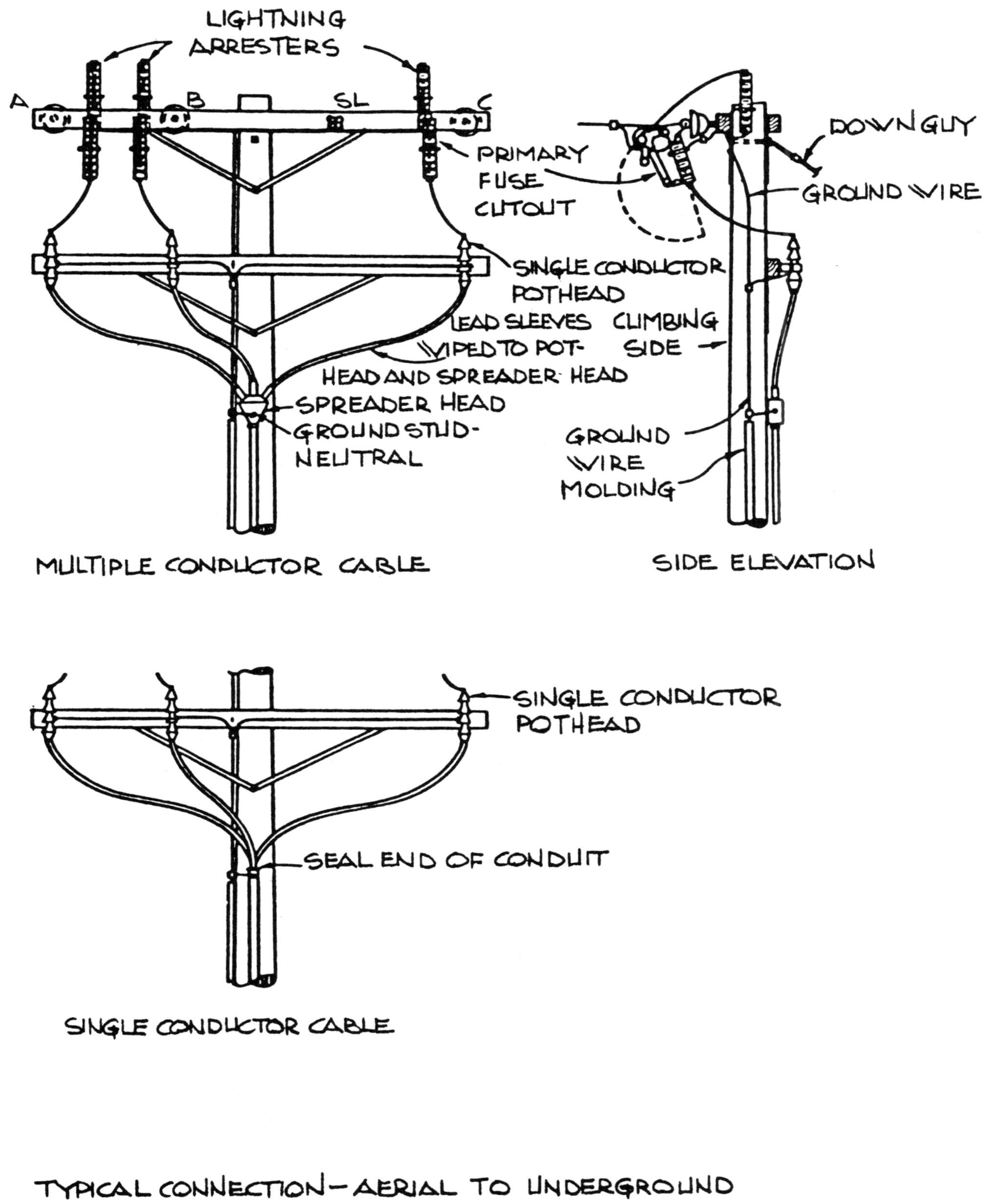
LIGHTNING ARRESTERS
A
B
SL
C
PRIMARY FUSE CUTOUT
DOWN GUY
GROUND WIRE
SINGLE CONDUCTOR POTHEAD
LEAD SLEEVES WIPED TO POT-HEAD AND SPREADER HEAD
CLIMBING SIDE
SPREADER HEAD
GROUND STUD-NEUTRAL
GROUND WIRE MOLDING
MULTIPLE CONDUCTOR CABLE
SIDE ELEVATION
SINGLE CONDUCTOR POTHEAD
SEAL END OF CONDUIT
SINGLE CONDUCTOR CABLE
TYPICAL CONNECTION—AERIAL TO UNDERGROUND

FIGURE 5-1

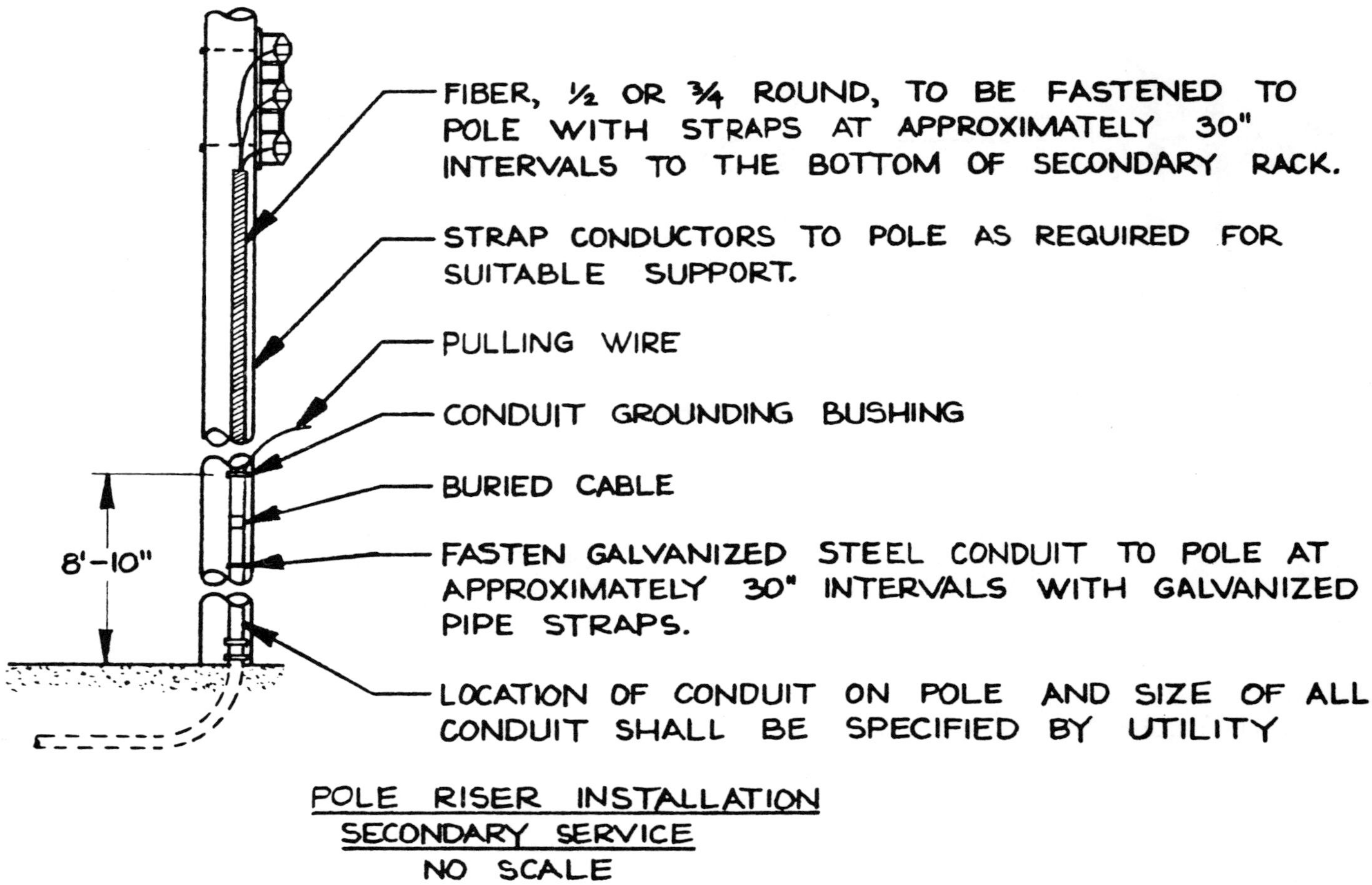

FIGURE 5-2

Figure 5-2 shows another version of a service to a pole and indicates what happens to the line coming down from the spreader head and going to the building. While the details are not directly related, they do go together. The conduit shown would terminate at a given distance above the ground, depending on the situation and utility requirements (or your own requirements). In the detail shown, the conduit terminates at some 8 ft 10 in above the ground with a weather proof cap. From the cap the wires can extend in a fiber duct or other means of protection up to the pole connection point.

Frequently, details are required to cover a situation where the service comes from a pole (as shown in Figure 5-1) to the building and then continues down the side of the building into the service entrance equipment. One of the things frequently missed in the installation detail of overhead wires from a pole to a building is the often misunderstood fact that you generally cannot simply tie the wires to the pole on one end and simply tie them to the building on the other. You most often must provide a supporting cable called a *messenger cable* from the pole to the building. You cannot directly fasten the wire itself.

There are situations where large buildings are, for economic reasons, supplied by overhead rather than underground services. These systems are typically arranged as shown in Figures 5-3 and 5-4. In viewing these figures, the most important parts to be emphasized are the messenger splice and the service messenger cable itself. This is a steel cable which may or may not be joined in the middle. In addition, the service wires which it supports must be secured to the cable in some fashion so that the cable can truly act as a support.

In these two details we have shown several different and important items. Each detail shows an arrangement of the service wire and the messenger cable to hold the entire distribution system together. Both figures also indicate how the messenger cable is attached to the building. Figure 5-4 shows the connection to the pole. Finally, the loose service cable that freely connects into the point of service

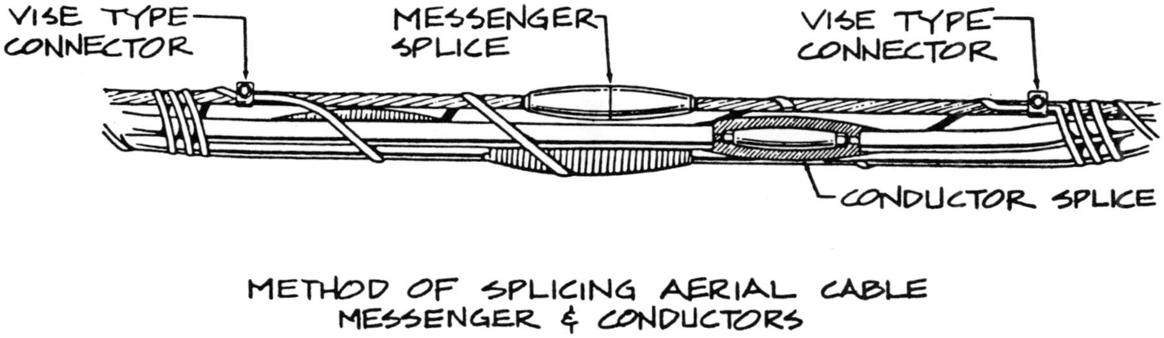

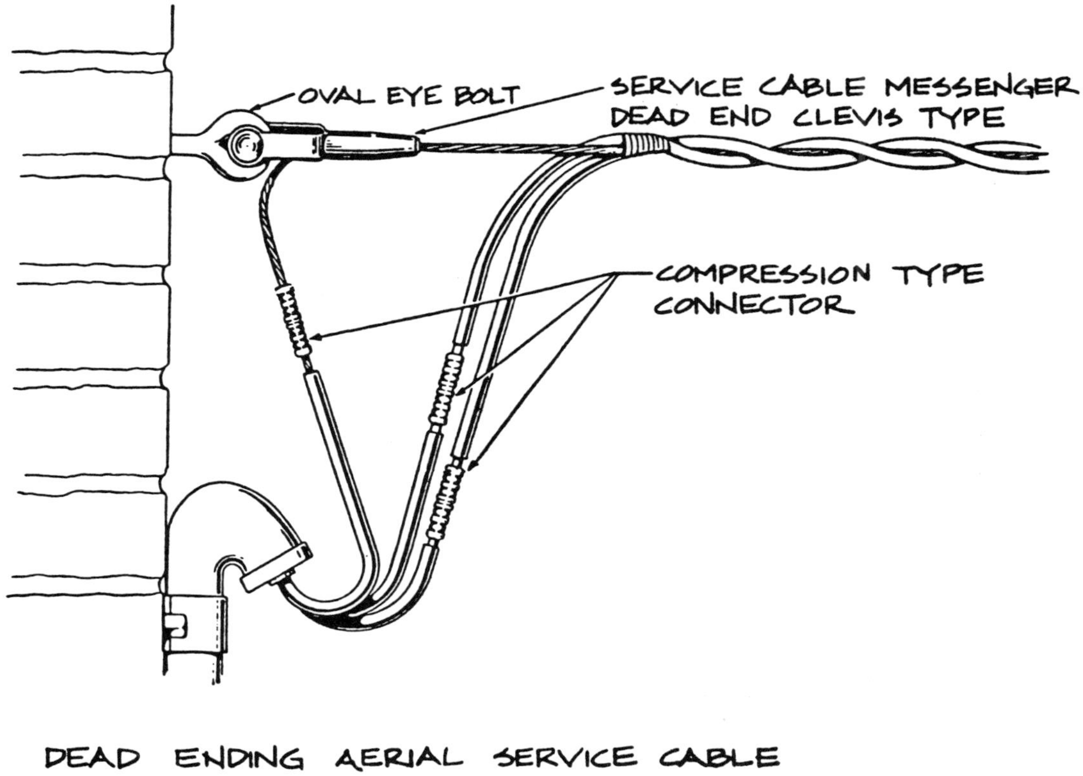

AERIAL CABLE SPLICE & SERVICE ENTRY

FIGURE 5-3

entry on each end can be seen in both figures. The line terminating at the pole has a special dead-end connecting device. This allows the wires to be easily connected to the primary service. On the building there is an oval eye bolt screwed into the building to support the messenger cable and the service cables it is supporting. The service entry cable enters into a conduit with a weatherhead connection as shown on the building wall in the detail. This sort of detailing may seem very simple and almost unnecessary. It is recommended that every overhead service to a building have this sort of detailing to define the installation requirements clearly.

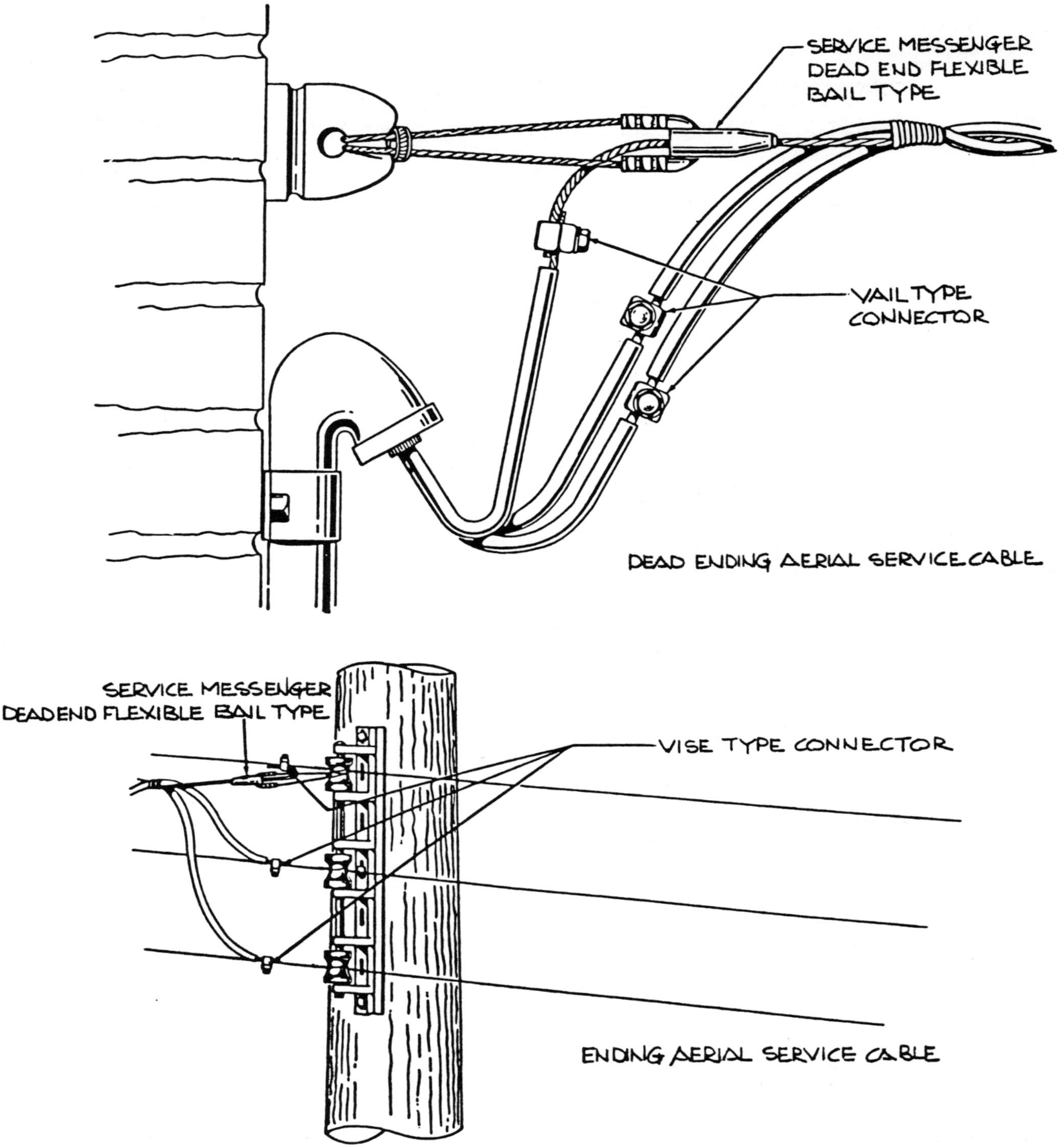

ENDING AERIAL SERVICE CABLE

FIGURE 5-4

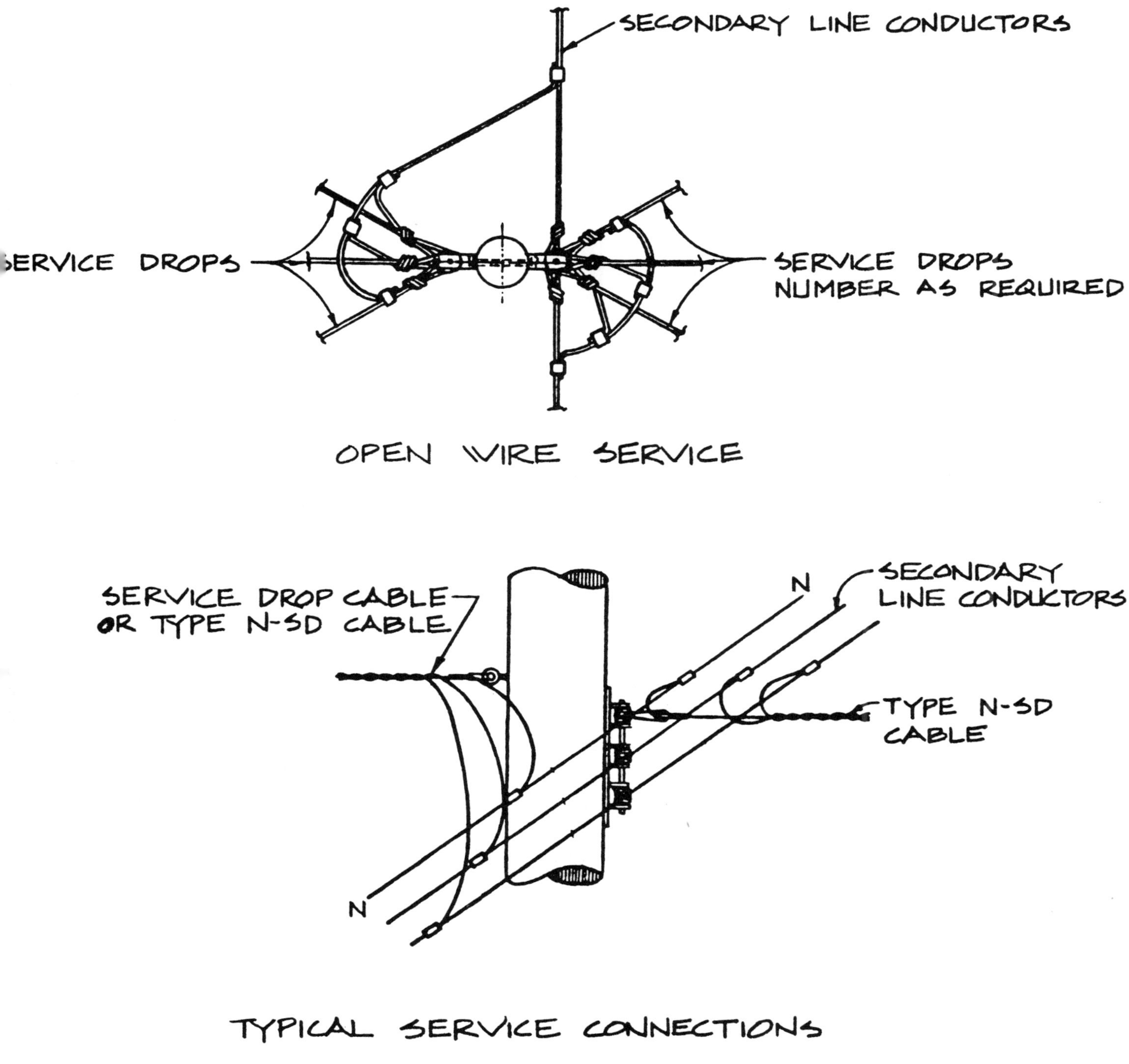

FIGURE 5-5

The overhead service is common to fairly small installations, as illustrated in Figure 5-5. These systems are normally overhead and would usually have the primary type of connection shown. These connections are more thoroughly detailed in the following chapter.

Electrical Distribution Riser Diagram Figure 5-6 illustrates a problem typical in many engineering designs, that is, installing a new service simply and economically to replace an existing service and distribution system in a building of 20,000 square feet.

In the detail the incoming service entrance equipment includes provisions for the utility company meters. The new main distribution system will supply power to the new

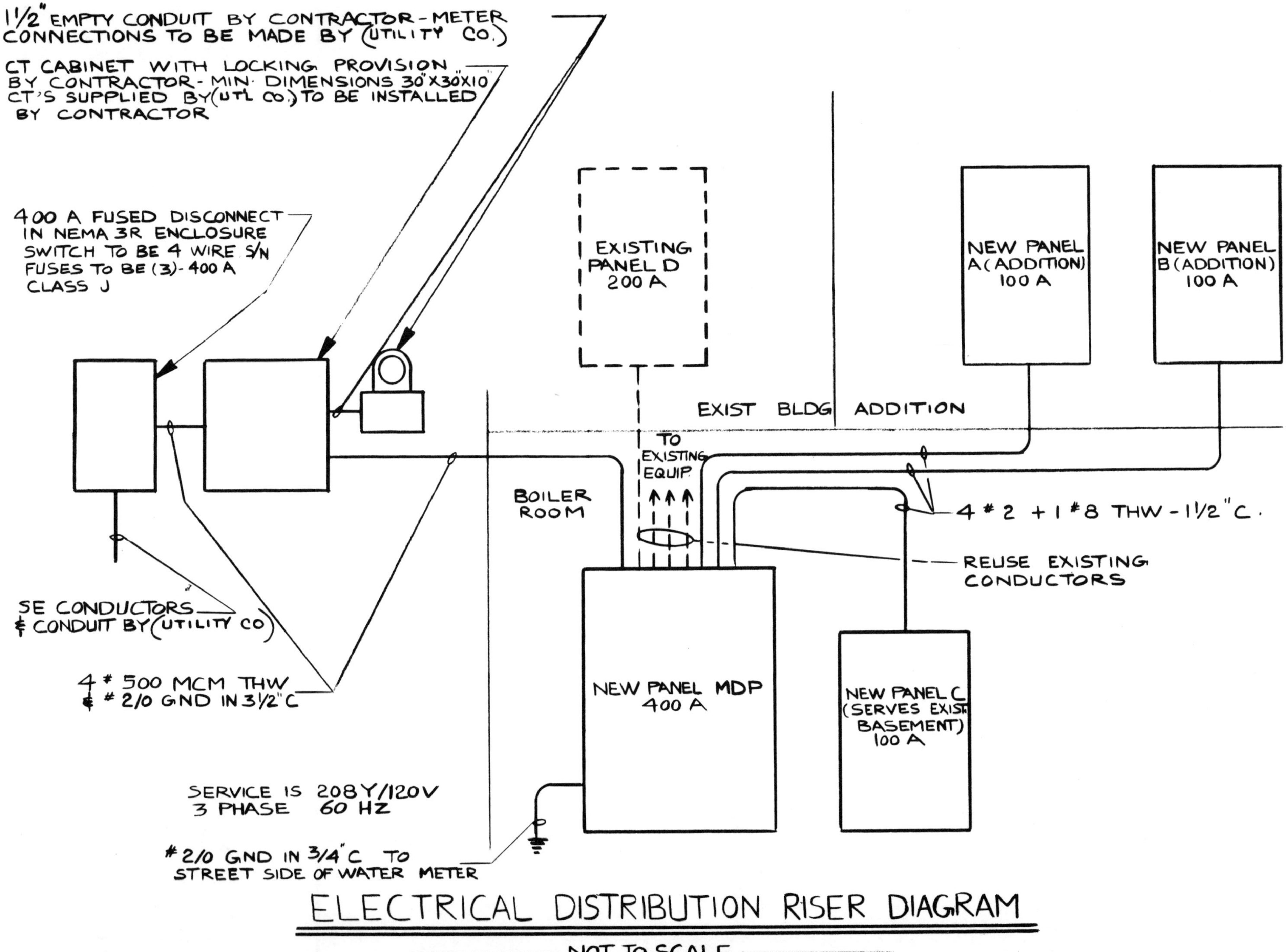

ELECTRICAL DISTRIBUTION RISER DIAGRAM

NOT TO SCALE

F[illegible]URE 5.6

panels as well as the panels serving the existing distribution system. As is obvious from the size of the equipment shown, the electrical load involved was not particularly large.

The first item to note is that the detail very clearly shows what work will be done by the contractor and what work will be done by the utility company at the point of service entry. Note also that both the original distribution panel and the new panels specify main capacity and that the conduit and wire to each new panel are sized.

The distribution system shown also illustrates a point frequently overlooked, that is, the grounding of the distribution panel. Many engineers consider this requirement obvious. But many contractors consider this, when not shown, to be an extra. While a detail such as this obviously cannot be used directly, it can form a basis of design for your particular situation.

Another point of importance frequently overlooked is the clear definition of the reuse of the existing conductors and existing panels when they are a part of the new system. These existing panels are shown as dashed lines. In essence what this detail portrays is not how to design your particular project but those elements essential to the successful detailing of any similar project.

Power Distribution Riser Diagram The title of Figure 5-7 could have just as easily been "Electrical Distribution Riser Diagram," but the designer insisted that the panels served only power requirements and therefore it was more appropriate to use the word "power."

This detail is similar to Figure 5-6 except for one additional area. The owner in this case has an exterior primary distribution system. As a result, the owner must supply the pad-mounted transformer and the connections from the primary feeder to the transformer as well as the secondary connections to the main power distribution panel, entitled "DP."

In the following chapter we will cover the connections to an exterior transformer in more detail, where, similarly, the designer has done certain things which create slight but significant differences in detailing. The first difference to note is that the conduits which run underground from the primary source to the building are shown as dotted lines to distinguish underground from aboveground work. Second, the panel DP is presumed to be a service entrance panel, which is described in an accompanying schedule not shown on this detail.

In addition, the design shows lines from the distribution panel to an existing panel. In this case these lines are not dashed because this is new wiring to an existing panel and is not the reuse of the existing wiring.

Again, because this was a small job, the size of the pad, the fact that the pad was grounded, and the size of the other items of equipment are given in a brief description on the detail itself. This sort of abbreviated detailing must be accompanied by another detail or by considerably more description in the specification, since, by itself, detailing of the primary service is not sufficient to cover the requirements of the total installation.

Load Centers Pursuing our discussion of exterior primary systems that are a part of the owner's service, the most logical areas of investigation are load centers and distribution systems, which were first shown in small scale in the previous detail. These can grow to indoor or outdoor substations of considerable size. Where the primary voltage is 15 kV or less, the primary system and the substation can be indoors. The higher-voltage systems normally require outdoor substations.

Figure 5-8 is very nearly the ultimate in this type of installation. First of all it is double-ended. Generally this degree of sophistication is not required. In the more common situation there frequently is a dual source of primary service to a single dry-type transformer and secondary switchgear. While the secondary voltage may be 208/120 V, it is much more commonly 480/277 V.

Any properly selected substation is commonly metal enclosed, with factory-coordinated built-in primary disconnects and a transformer. The secondary protective devices can easily be coordinated and the secondary system should have provisions for cost-effective expansion.

Clearances become very important. While slight variations may be allowed by the Code inspecting authority, the minimum clearance should be 5 ft in front of the equipment, 3 ft on the sides and top of the equipment, and 7 ft between facing pieces.

There are certain other criteria to be noted. The room or area of the substation should have emergency lighting. Ventilation of a positive mechanical type should also be included. The noise level of the transformer specified should be checked. There is frequently a need for sound insulation in the room.

In Figure 5-8 the primary disconnect switches are not part of the composite load center, simply because the room was too small to allow for the 3 ft 0 in side clearance if the disconnects were mounted next to the transformers. Also the wire partition was arranged so that the secondary side of the equipment was readily accessible but the primary side could be locked off from the casual visitor.

In the incoming service line there is always a pull box to handle the service cables involved. This box should be sized by the designer (it usually is of considerable size). It probably will be an approximate 3-ft cube. In addition spare ducts for future primary cables equal to the ducts for present cables should be provided. Ideally all the cables for the primary service should be below the floor slab and encased in concrete. If they are above the slab, they should still be encased in concrete.

Generally the selection of all of the primary disconnects, transformers, and secondary distribution breakers as well as all the interior cabling, connectors, isolators, and associated hardware can be made from the catalogs of any of several major equipment manufacturers.

Usually when a detail like Figure 5-8 is shown, in whole or in part, on the floor-plan drawings, it is shown to proper scale. To avoid argument and conflict, our detail is noted to be to "no scale." Actually it should be drawn to one-eighth or one-quarter scale to assure that it really fits, but the scale does not need to be labeled on the detail.

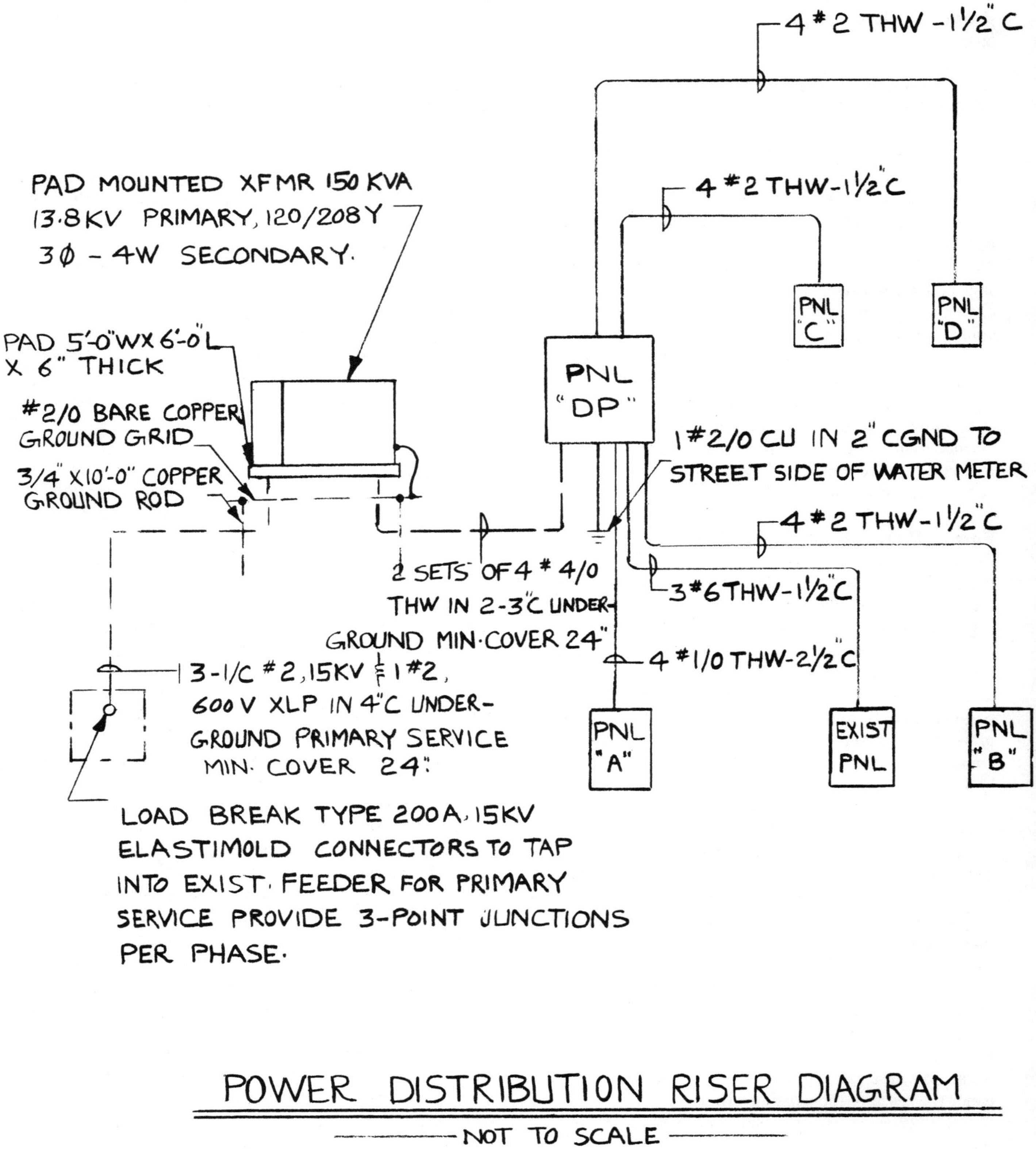
4 # 2 THW - 1½" C
4 # 2 THW-1½" C
PAD MOUNTED XFMR 150 KVA
13.8KV PRIMARY, 120/208Y
3Ø - 4W SECONDARY.
PNL "C"
PNL "D"
PAD 5'-0"W X 6'-0"L
X 6" THICK
PNL "DP"
#2/0 BARE COPPER
GROUND GRID
1 #2/0 CU IN 2" C GND TO
STREET SIDE OF WATER METER
3/4" X 10'-0" COPPER
GROUND ROD
4 # 2 THW-1½" C
2 SETS OF 4 # 4/0
THW IN 2-3"C UNDER-
GROUND MIN. COVER 24"
3 #6 THW-1½" C
4 #1/0 THW-2½" C
3-1/C #2, 15KV & 1 #2,
600 V XLP IN 4"C UNDER-
GROUND PRIMARY SERVICE
MIN. COVER 24".
PNL "A"
EXIST PNL
PNL "B"
LOAD BREAK TYPE 200A, 15KV
ELASTIMOLD CONNECTORS TO TAP
INTO EXIST. FEEDER FOR PRIMARY
SERVICE PROVIDE 3-POINT JUNCTIONS
PER PHASE.
POWER DISTRIBUTION RISER DIAGRAM
NOT TO SCALE

FIGURE 5-7

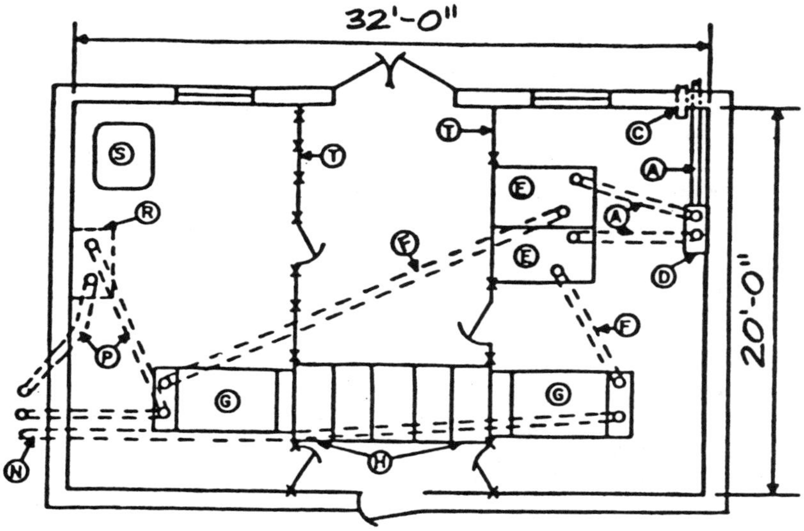

TRANSFORMER ROOM FLOOR PLAN
NO SCALE

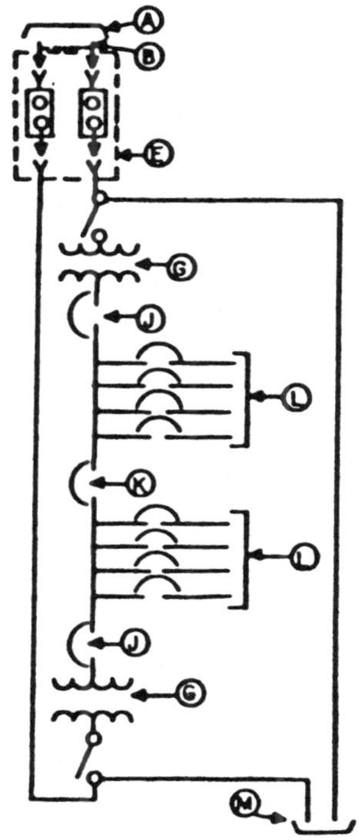

ONE LINE
WIRING DIAGRAM
NO SCALE

A INCOMING SERVICE - TWO FEEDERS PREFERRED
B TIE BUS - INSTALL WHEN ONLY ONE FEEDER IS AVAILABLE
C SPARE DUCTS
D PULL BOX
E MAIN CIRCUIT BREAKERS
F PRIMARY CABLE IN DUCT
G THREE PHASE TRANSFORMER
H SECONDARY SWITCHGEAR SECTION
J SECONDARY BREAKERS WITH AUXILIARY ALARM CONTACTS
K TIE BREAKER INTERLOCKED WITH MAIN BREAKERS
L SECONDARY FEEDERS
M CABLES TO OTHER LOAD CENTERS, IF REQUIRED
N DUCTS FOR CABLES TO OTHER LOAD CENTERS, IF REQUIRED
P ALTERNATE ROUTE OF ONE PRIMARY FEEDER FOR TAPS TO OTHER BUILDINGS, IF REQUIRED
R JUNCTION BOX AND OIL FUSE CUTOUTS OR LOAD-BREAK FUSED DISCONNECTS, IF REQUIRED FOR TAP
S STREET LIGHTING TRANSFORMER
T REMOVABLE WIRE PARTITION

TYPICAL DOUBLE-ENDED TRANSFORMER LOAD CENTER- PRIMARY CIRCUIT BREAKERS

NO SCALE

FIGURE 5-8

A schematic diagram is an important requirement since it delineates electrically how the various components interrelate and, most especially, the location and relationship of the tie circuit breaker between the two sections.

Usually a system, as shown in Figure 5-8, is arranged to provide a backup in case of failure or loss of power in one of the two incoming cables. The designer may also have in mind a future load growth that could most economically be served by a double-ended system. Generally the assumption is that either of the transformers can accept the total load.

Where the future expansion is possibly greater than normal or is unpredictable, this arrangement has obvious limitations for any additional secondary circuit breakers. In this case the lineups would have to be separated with a connecting tie bus and the tie circuit breaker located in one or the other set of circuit breakers. Another possibility would be to locate the transformers and primary disconnects separately.

Grounding is not shown in this particular detail but is absolutely required. Not only must the switchgear be grounded but the wire enclosure as well. For entry gates provide bonding straps across hinges. Lightning protection is usually required when supply or load circuits are overhead and when equipment is located in elevated structures. Simply because it was convenient for this particular project, there is a site-lighting transformer located in the room. There is no other reason for its location.

In this detail any form of specification data would most likely do more harm than good. The description of the various components will usually fill a number of specification pages, and any additional data on the detail would be of little or no help and could very well be misleading.

Power Distribution Centers Much of the discussion in the preceding description of the load center detail applies to the schematic power distribution detail in Figure 5-9. The schematic represents more than $200,000 worth of equipment spread over two substations, which are part of an overall power distribution system at a large industrial plant. The two substations noted as no. 7 and no. 8 are a portion of 14 similar substations. If a detailed block plan similar to that shown in Figure 5-8 were drawn, the design would show the primary disconnect of dual feeders, the 15-kV to 480-V transformer, and a series of large secondary circuit breakers.

Standard schematic legend detailing is used to show many of the items in Figure 5-9. The first thing to note is the fused and unfused 600-A 15-kV primary voltage disconnects. These are manual changeover switches which represent a client's preference.

Note that the 2000-kVA dry-type transformers are schematically grounded. The 3000-A 480-V secondary main circuit breakers contain both voltmeters and ammeters along with the necessary instrument transformers. The 3000-A bus indicates the ampere rating of the bus copper within the secondary switchgear circuit-breaker compartments. This busing is internal, not external.

The schematic symbols for the circuit breakers indicate they are the draw-out type and that they are fused. Each secondary section consists of 11 circuit breakers, of which 3 are 1600-A and 8 are 800-A. In equipment of this size both the circuit-breaker frame size (which denotes the largest circuit breaker that can be accepted) and the actual circuit-breaker size to be installed are noted. For example, in Figure 5-9 the circuit-breaker frame size is 1200 A, and below the 1200 A there is noted 800. Where the circuit breaker has a trip setting different from the circuit-breaker size, that also is noted below the breaker; for example, 2000 A—1600 (1200).

This detail is one of several used to portray the overall substation. Obviously there would be a scaled floor plan and a no-scale block detail similar to that shown in Figure 5-8. The schematic detail shown in Figure 5-9 is most closely related to the specification covering the equipment. The numbers and letters in the symbol enclosures next to various items of equipment are related to a complete numbered equipment list which accompanied the detail. Primary fuses are also listed. They are noted as 125E. Note that either primary circuit breaker can feed through this fuse. The LA item is a lightning arrester. The dotted connection between primary and secondary disconnects is a key interlock (EO), which requires both circuit breakers to be in the off position before work can be done on the transformer.

The purpose of showing only a portion of the total plan is to emphasize the value and importance of the schematic detail in, of, and by itself. Obviously the plan must show the total equipment space, its enclosure, and the proper clearances. In this case the detailing of the very large pull box requires proper dimensioning of the box.

While all of the accompanying schedules, plans, and elevations of the switchgear are important, the schematic really is the depiction that best shows the electrical functions. In the proper completion of this plan and specification, a considerable amount of installation description must be either on the plans or in the specification. The following paragraphs have been added to illustrate the basic equipment specification. As you can readily see, the specification of equipment is rarely, if ever, a complete picture, even when covered by the schematic and other plan details. Guidance concerning installation and component parts not shown or described in the details must be carefully specified.

Equipment Specification Furnish and deliver to job site where shown on the plans two substations as detailed, connected to a looped 13.2-kV primary system. Each substation consists of two 13.2-kV primary circuit breakers, a 13,200/480-V dry-type transformer, and a 480-V secondary circuit-breaker section joined together to form an enclosed, continuous, and connected substation.

Each incoming line section shall consist of an indoor metal-enclosed cubicle providing a duplex selector switch

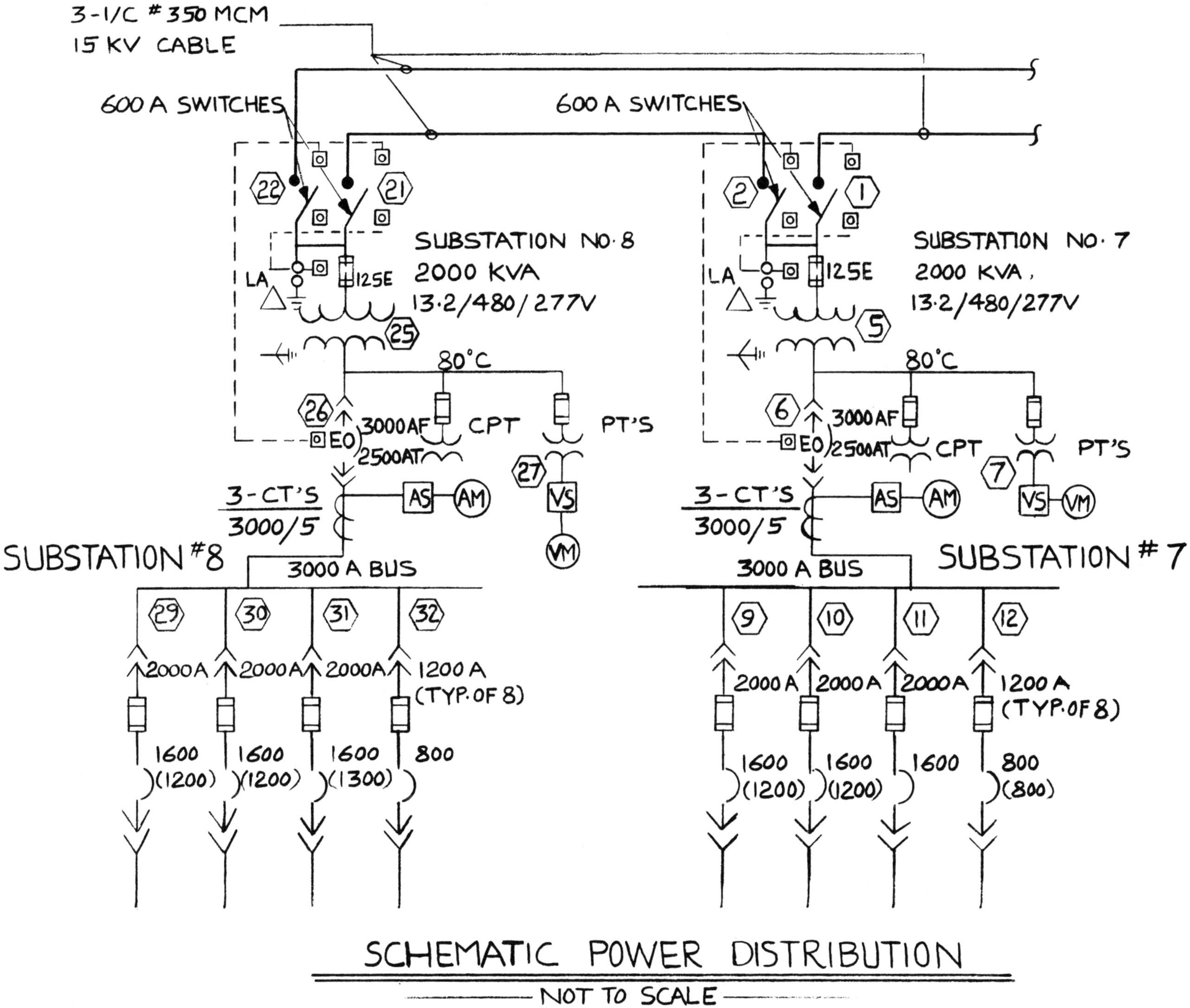
3-1/C #350 MCM
15 KV CABLE
600 A SWITCHES
600 A SWITCHES
22
21
2
1
SUBSTATION NO·8
2000 KVA
13·2/480/277V
SUBSTATION NO·7
2000 KVA,
13·2/480/277V
LA
125E
LA
125E
25
5
80°C
80°C
26
3000AF
2500AT
EO
CPT
PT'S
6
3000AF
2500AT
EO
CPT
PT'S
27
7
3-CT'S
3000/5
AS
AM
VS
VM
3-CT'S
3000/5
AS
AM
VS
VM
SUBSTATION #8
3000 A BUS
3000 A BUS
SUBSTATION #7
29
30
31
32
9
10
11
12
2000A
2000A
2000A
1200 A
(TYP. OF 8)
2000 A
2000 A
2000 A
1200 A
(TYP. OF 8)
1600
(1200)
1600
(1200)
1600
(1300)
800
1600
(1200)
1600
(1200)
1600
800
(800)

FIGURE 5-9

of 2 three-pole two-position air-interrupter switches. The two switches will provide three positions (line 1, open, line 2). The switches will be key interlocked to prevent both from being closed at the same time. Each interrupter switch shall be quick-make, stored-energy, and manually operated, with the operator located on the front of the unit. Each operator will have indicators to show position of the switch blades (open-closed). Spring charging and switch closing or opening will be accomplished in a single 180° rotation, upward or downward, of the operating handle. Two windows will be located in the upper front panel for visual inspection of the switch blades. The lower interlock handle will have padlock provision. Each interrupter switch will be rated at 13.2 kV, voltage class 13.8 kV, basic impulse level (BIL) 95 kV, continuous current 600 A, interrupter rating 600 A, momentary rating 40,000 A, and fault-losing 61,000 A.

The incoming line section shall include three sets of current-limiting fuses mounted in the lower compartment of the unit between the switch and outgoing sections. The fuse access panel will be hinged and key interlocked with the operating mechanism to prevent opening when the switch is closed or the operator springs charged, and to prevent operating the switch when the panel is open. The upper front panel shall be bolted.

Incoming loop-feed lines will enter the top of the unit sized as per plan. Provision shall be made for connection to the adjacent transformer section with key interlock so that operation is possible only when the associated transformer secondary circuit breaker is open. An auxiliary unit with a hinged front panel containing voltmeter, voltmeter transfer switch, ammeter, and ammeter transfer switch as well as current and potential transformers for future primary metering shall be provided.

Each secondary unit substation transformer shall be of the ventilated dry type for load-center application. Each indoor three-phase transformer shall be rated at 2000 kVA, 60 Hz, 8.7 percent impedance, BIL of 50 kV, 220°C insulation at 80° rise, 13.2-kV delta primary, 480-V wye secondary. Provide four 2½ percent full-capacity taps—two above and two below rated voltage. Test and Construct unit per American National Standards Institute (ANSI) and National Electrical Manufacturers Association (NEMA) codes. Enclosure shall be heavy-guage steel, ventilated, and built per NEMA and NEC standards. Cabinet enclosure finish shall be light-gray (ANSI 61) enamel. High- and low-voltage terminals shall be provided for direct connection to lines and secondary switchgear sections.

Each secondary circuit-breaker unit shall consist of the number of 800-A and 1600-A 480-V three-phase 60-HA grounded wye breakers shown on the plans and/or described herein. Switchgear section shall be of indoor type, welded steel with prime and finish coats of ANSI 61 light-gray paint. Guide rails are to be racked to the connected, test, and disconnected position with the doors closed. An interlock will prevent racking the circuit breaker when it is closed. Insulation shall be polyglass polyester.

Circuit breakers will be the low-voltage power type, 600-V class, three-pole, single-throw, drawout-mounted, electrical and mechanical trip free with stored-energy operator. Each will have arc quenches, main and arcing contract structure, Static Trip II LimTrip solid-state overcurrent trip device, a contact-position indicator (open-closed), stored-energy mechanical indicator (charged-discharged), primary disconnection devices, and a mechanical interlock to prevent making or breaking contact on the primary disconnects. Breakers shall be electrically operated with a spring release coil, secondary disconnecting devices, and with a four-stage auxiliary switch.

A ground bus will extend the length of the lineup. A one-half full-capacity neutral bus will be carried with the phase buses in addition to a ground bus. All buses will be tin-plated aluminum with bolted joints.

Main and feeder circuit breakers shall be as sized on the plans with control switches, each with red and green indicating lights. Breakers shall be rated 65,000 A.I.C. Each breaker shall have suitable clamp terminals for outgoing feeders. The main breaker only shall have voltmeter, ammeters, and associated switches and transformers for main bus metering.

208/120-V Three-Phase Distribution In the opening discussion of this chapter we made the point that there are still many situations where 208/120-V distribution is preferable to 480/277-V distribution, even though the loads may be fairly heavy. Figure 5-10 covers this sort of situation. The fairly small three-story building of some 70,000 square feet has a large number of tenants. Some of the air conditioning is from a central system and some is from smaller individual units. There are extensive lighting and outlet systems and many small motors for a variety of purposes. In addition, many of the tenants pay their own electric bills and have separate meters for this purpose. Figure 5-10 shows all these problems.

As you can see, the 108/120-V secondary load center at 3000 A represents slightly more than 1000 kVA. Because we are getting into some fairly large loads and are supplied by a utility company network, the available short-circuit capacity of the utility company at the service entrance must affect the equipment short-circuit capacity and coordination calculations.

Although there are a number of panels and feeders, this fact by itself does not break any new ground as far as comparing this detail to the previously presented smaller details of secondary services. What is new and different in this detail is the subfeeders from the main distribution panel to the tenant meters. In addition, in previous details we have alluded to panel schedules and other information

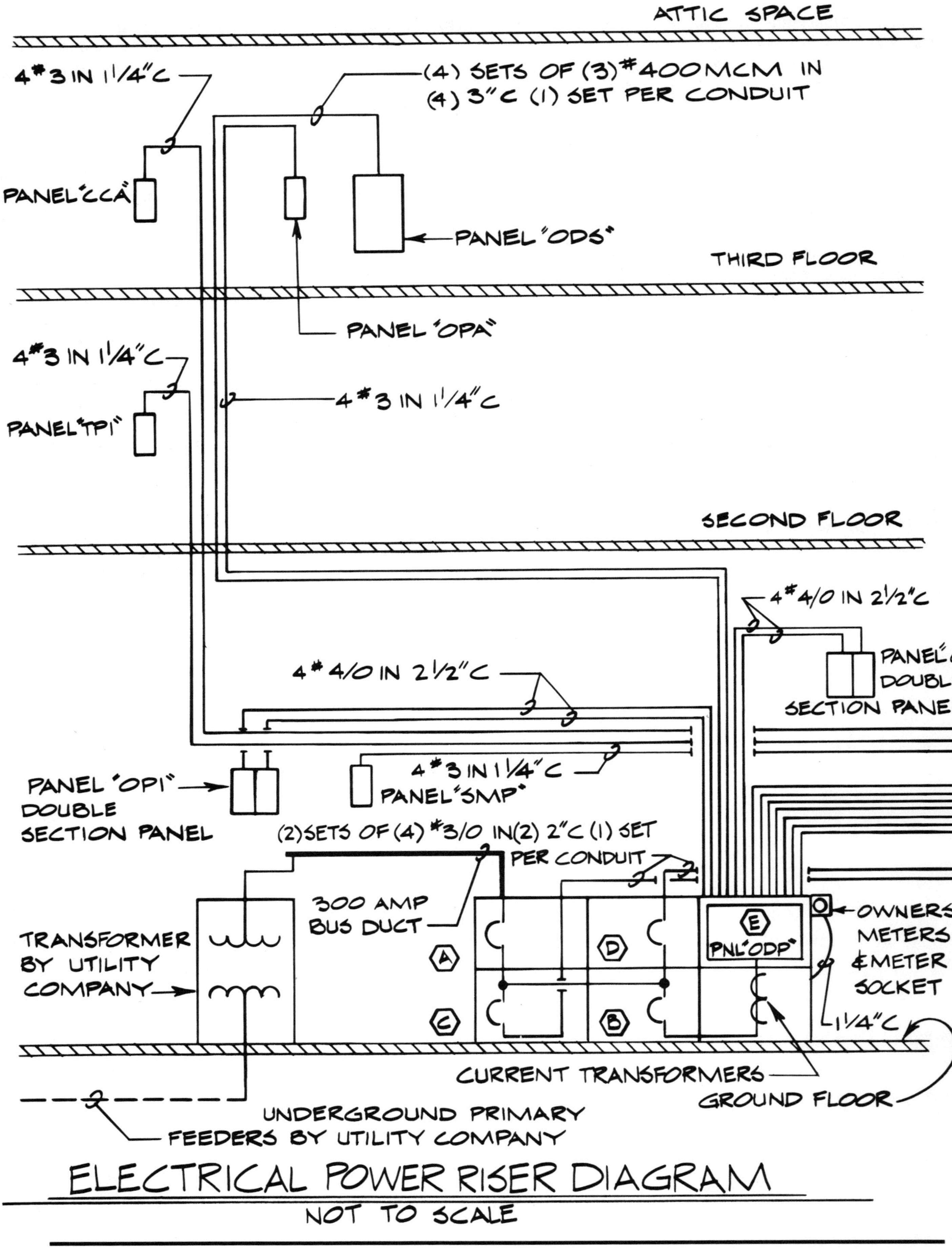
ATTIC SPACE
4#3 IN 1 1/4"C
(4) SETS OF (3)#400MCM IN
(4) 3"C (1) SET PER CONDUIT
PANEL "CCA"
PANEL "ODS"
THIRD FLOOR
PANEL "OPA"
4#3 IN 1 1/4"C
PANEL "TP1"
4#3 IN 1 1/4"C
SECOND FLOOR
4#4/0 IN 2 1/2"C
PANEL "OP2" DOUBLE SECTION PANEL
4#4/0 IN 2 1/2"C
4#3 IN 1 1/4"C
PANEL "OP1" DOUBLE SECTION PANEL
PANEL "SMP"
(2) SETS OF (4) #3/0 IN (2) 2"C (1) SET PER CONDUIT
300 AMP BUS DUCT
TRANSFORMER BY UTILITY COMPANY
A
C
D
B
E
PNL "ODP"
OWNERS METERS & METER SOCKET
1 1/4"C
CURRENT TRANSFORMERS
GROUND FLOOR
UNDERGROUND PRIMARY FEEDERS BY UTILITY COMPANY
ELECTRICAL POWER RISER DIAGRAM
NOT TO SCALE

FIGURE 5-10

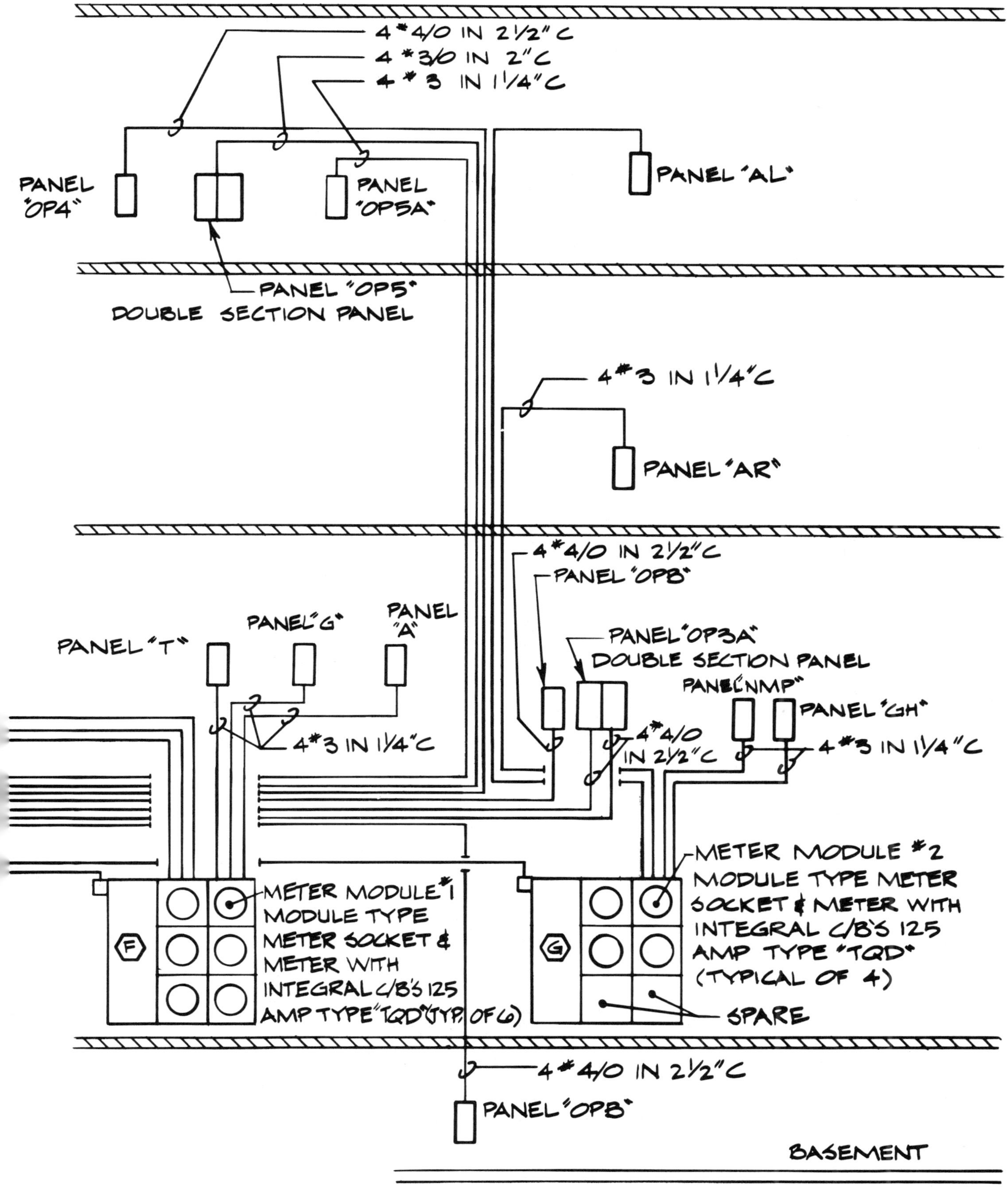
4 #4/0 IN 2 1/2" C
4 #3/0 IN 2" C
4 #3 IN 1 1/4" C
PANEL "OP4"
PANEL "OP5A"
PANEL "AL"
PANEL "OP5"
DOUBLE SECTION PANEL
4 #3 IN 1 1/4" C
PANEL "AR"
4 #4/0 IN 2 1/2" C
PANEL "OP8"
PANEL "T"
PANEL "G"
PANEL "A"
PANEL "OP3A"
DOUBLE SECTION PANEL
PANEL "NMP"
PANEL "GH"
4 #3 IN 1 1/4" C
4 #4/0 IN 2 1/2" C
4 #3 IN 1 1/4" C
METER MODULE #1
MODULE TYPE METER SOCKET & METER WITH INTEGRAL C/B'S 125 AMP TYPE "TQD" (TYP. OF 6)
F
G
METER MODULE #2
MODULE TYPE METER SOCKET & METER WITH INTEGRAL C/B'S 125 AMP TYPE "TQD" (TYPICAL OF 4)
SPARE
4 #4/0 IN 2 1/2" C
PANEL "OP8"
BASEMENT

that is required in a given detail. In Figure 5-11 we have included a secondary switchgear schedule and in Figures 5-12 and 5-13 two pages of panelboard schedules.

A word of caution. The mere showing of the panelboard and switchgear schedules plus the overall distribution detail and the equipment specifications is still not the total presentation. There are many notes, explanations, and other details plus the total plans and specifications that are required to complete this design. With this caveat we present typical specifications in the paragraphs that follow.

Equipment Specification The estimated maximum root-mean-square (RMS) symmetrical short-circuit amperes available at this location are as follows:

	Transformer and Street-Tie Capacity, kVA	*Volts*	*Amperes*
Present	1000	216	53,500
Ultimate	1500	216	80,250

Consideration shall be given to additional short-circuit amperes which may be contributed by other connected equipment. Provide appropriately rated secondary distribution assembly and protective devices.

Space is to be provided in the secondary switchgear for the utility company's instrument transformers. Adequate space shall be provided for the billing meter conveniently located in the vicinity of the service entrance, as specified by the utility company representative. The metering panel, including the utility company's instrument transformers and the required meter socket, is to be completely installed and wired by the electrical contractor.

The switchboard shall consist of the required number of vertical sections bolted together to form one metal-enclosed vertical switchboard, including all protective devices and equipment listed on the plans and/or specified herein, together with necessary interconnections, instrumentation, and control wiring conforming to NEMA # PB2 and UL-891 standards.

Buses shall be tin-plated aluminum, sized not to exceed 705 A/in^2 current density, mounted on high-impact non-tracking insulating material and braced to withstand mechanical forces exerted under short-circuit conditions.

Small wiring and necessary fuse and terminal blocks are to be included. Control indicated on plans and ground fault protection will be provided by a solid-state relay adjustable from 100 to 1200 A.

The main protective device shall be individually mounted with only utility company current transformer or

SECONDARY SWITCHGEAR SCHEDULE						
DESG'N	TYPE	FRAME	TRIP	QUAN	POLE	SERVES
A	TPRR	3000	3000	1	3	MAIN SERVICE
B	TPRR	2500	2500	1	3	PANEL "ODP"
C	TPRR	400	400	1	3	METER MODULE # 1
D	TPRR	400	400	1	3	METER MODULE #2
E (PANEL "ODP")	THUK	400	400	1	3	NORTH CHILLER ON GRADE LEVEL
	THUK	225	225	9	3	PANELS "OP1", "OP2", "OP3A", "OP3" & "OP8" SOUTH CHILLER ON GRADE LEVEL
	THUK	225	150	2	3	PANELS "OP5", "OP4"
	THED	100	100	2	3	PANEL "OPA", "OP5A"
	THKM	1200	1200	1	3	PANEL "ODS"
F	TUKG	400	400	1	3	METER MODULE # 1
G	TUKG	400	400	1	3	METER MODULE #2

FIGURE 5-11

PANELBOARD SCHEDULE									
PANEL		MAINS	BRANCH CIRCUITS						SERVES
DESIG'N	TYPE	LUG'S AMP'S	BK'R	TRIP	POLE	ACTIVE	SPARES	TOTAL	
"OPA"	NLAB	100	THQB	20	1	18	6	24	EXHF #'S 1 TO 9 & LIGHTS
			TEB	100	3	1	0	1	MAIN C/B
"OPB"	NLAB	225	THQB	20	1	7	5	12	BOILER RM. LIGHTS
			THQB	20	3	6	1	7	BOILER RM. PUMPS
			THQB	40	3	1	0	1	" " PUMP
			THQB	50	3	1	0	1	" " "
			THQB	100	3	1	0	1	EXIST. NORTH ELEVATOR
			TQD	225	3	1	0	1	MAIN C/B
DOUBLE "OP1" SECTION	NLAB	225	THQB	20	1	42	10	52	MAIN WAITING RM. LIGHTS
			THQB	20	3	1	1	2	SO. BASEMENT CIRC. PUMP
			TEB	100	3	1	0	1	EXIST. BLDG. NEW ELEVATOR
			TEB	150	3	2	0	2	MAIN C/B
DOUBLE "OP2" SECTION	NLAB	225	THQB	20	1	46	5	51	PLATFORM LIGHTS
			THQB	20	3	7	3	10	NEW ADDITION AC UNITS
			TEB	70	3	1	0	1	NEW ADDITION ESCALATOR
			TEB	225	3	2	0	2	MAIN C/B
"OP3"	NLAB	225	THQB	20	1	40	8	48	NEW ADDITION & PLATFORM LIGHTS
			TEB	225	3	1	0	1	MAIN C/B
DOUBLE "OP3A" SECTION	NLAB	225	THQB	20	1	19	5	24	PLATFORM & STAIR LIGHTS
			THQB	20	3	6	3	10	NEW ADDITION AC UNITS
			THQB	50	3	2	0	2	CONDENSER UNITS
			TEB	150	3	1	1	2	NEW ADDITION - NEW ELEVATOR
			TEB	225	3	2	0	2	MAIN C/B
"OP4"	NLAB	225	THQB	20	1	26	7	33	MAIN WAITING RM. LIGHTS
			TEB	150	3	1	0	1	MAIN C/B
DOUBLE "OP3" SECTION	NLAB	100	THQB	20	1	40	12	52	3RD FLOOR OFFICE LIGHTS
			TEB	70	3	2	0	2	MAIN C/B

FIGURE 5-12

main metering or instrumentation located in this compartment. No feeder or branch-circuit protection device shall be located in this section.

The switchboard shall be arranged for 208/120-V service at 60 Hz with a full-capacity neutral. It is to be fused and rated at 100,000 RMS symmetrical short-circuit amperes at the line terminals.

The board shall be arranged for cable in conduit feeders with a full-length ground bus and Underwriters Laboratories (UL) service entrance label.

The incoming line main circuit breakers shall be molded-case with adjustable trips, manually operated, three-pole, and rated at 65,000 RMS amperes symmetrical.

The individual distribution circuit breakers shall be molded-case, current-limiting fused circuit breakers. Each device shall have a standard thermal magnetic trip as well as current limiters. All circuit breakers shall be three-pole with separately mounted current-limiting fuses to provide a range of capacity form 15 to 1200 A. All circuit breakers shall be rated at 100,000 RMS amperes symmetrical.

Sizes, types, and wiring arrangements of all items of equipment are as shown on the plans.

Furnish and install panelboards as shown on the plans

PANELBOARD SCHEDULE - CONTINUED									
PANEL		MAINS	BRANCH CIRCUITS						SERVES
DESIG'N	TYPE	LUG'S AMP'S	BK'R	TRIP	POLE	ACTIVE	SPARE	TOTAL	
"OPSA"	NLAB	100	THQB	20	1	25	5	30	OFFICE RECEPT'S
			TEB	100	3	1	0	1	MAIN C/B
"ODS"	CCB	1200	TJD	400	3	3	0	3	(3) ROOF CHILLER UNITS
			THQC	20	3	11	3	14	(8) AHU'S & (3) CW PUMPS
			TKM12	1200	3	1	0	1	MAIN C/B
"A"	NLAB	100	THQB	20	1	1	8	9	AMTRAK TICKET COUNTER
			TEB	70	3	1	0	1	MAIN C/B
"AL"	NLAB	100	THQB	20	1	17	10	27	LOFT LIGHT'G & RECEPT'G
			TEB	100	3	1	0	1	MAIN C/B
"CCA"	NLAB	100	THQB	20	1	15	9	24	LOFT LIGHTS & RECEPT'S
			TEB	100	3	1	0	1	MAIN C/B
"G"	NLAB	100	THQB	20	1	2	7	9	GREYHOUND TICKET COUNTER
			TEB	70	3	1	0	1	MAIN C/B
"GH"	NLAB	100	THQB	20	1	15	3	18	GLASS ENCLOSURE LT'S & RECEPT'S
			TEB	50	3	1	0	1	MAIN C/B
"NMP"	NLAB	100	THQB	20	1	1	19	20	FUTURE USE
			TEB	70	3	1	0	1	MAIN C/B
"SMP"	NLAB	100	THQB	20	1	1	19	20	FUTURE USE
			TEB	70	3	1	0	1	MAIN C/B
"T"	NLAB	100	THQB	20	1	1	8	9	TRAILWAYS TICKET COUNTER
			TEB	70	3	1	0	1	MAIN C/B

PANELBOARD SCHEDULE - CONTINUED									
PANEL		MAINS	BRANCH CIRCUITS						SERVES
DESIG'N	TYPE	LUG'S AMP'S	BK'R	TRIP	POLE	ACTIVE	SPARE	TOTAL	
"TP1"	NLAB	100	THQB	20	1	1	19	20	FUTURE USE
			TEB	70	3	1	0	1	MAIN C/B
"AR"	NLAB	100	THQB	20	1	11	4	15	EXIST. CIRCUITS
			THQB	20	2	1	1	2	" "
			TEB	100	3	1	0	1	MAIN C/B

FIGURE 5-13

and described herein, types as noted on plans. Panelboard shall be listed by UL, and so labeled. All panelboards shall be rated for the intended voltage.

Boxes shall be galvanized sheet metal, providing ample wiring gutter space in accordance with the NEC.

Fronts shall be of sheet steel painted medium dark gray (ANSI 33) over a rust inhibitor. They shall be equipped with doors, semiconcealed five-knuckle hinges, cylinder tumbler-type lock, directory card holder, and quarter-turn adjustable trim clamps. On doors more than 48 in high a combination three-point catch and lock shall be provided. All locks are to be keyed alike.

Panels shall consist of a reinforced galvanized sheet steel frame with bus bars and circuit breakers properly supported to prevent vibration, breakage in handling, and to withstand effects of indicated short-circuit current. All terminals shall be the solderless type, suitable for copper or aluminum cable of sizes indicated. Bus bars shall be sequence-phased.

Branch-circuit breakers shall be fully interchangeable without disturbing adjacent units. Circuit breakers shall be quick-make, quick-break, and trip-indicating. All two- and three-pole circuit breakers shall have common trips.

The wireway shall consist of an enclosed wiring trough, or auxiliary wiring gutter system. All equipment shall be listed by UL and be so-labeled. The housing shall be die-formed from sheet steel. The arrangement shall be such that with the covers open one side of the complete system is open for laying in wires without threading through couplings or the like. The top piece of connectors between units shall snap into place, thus holding the wires within the trough.

Ample knockouts shall be provided as standard on both sides of the wireway. The wireway shall be finished in ANSI 33 dark-gray enamel over a rust inhibitor.

Each length of wireway shall be provided with knockout holes in the sides of the trough to permit bolting to bracket supports or attaching directly to a wall.

Furnish and install fused or unfused disconnect switches sized as shown on the plans with type R fuses where fuses are used. Indoor enclosures shall be NEMA 1 and outdoor enclosures shall be NEMA 3R.

Furnish and install grouped socket metering modules where shown on the plans. Each socket shall contain its own bolt-on circuit breaker rated 150 to 200 A positioned directly under the associated meter.

Include a main device circuit breaker and cable terminal box rated at 1200 A. The total package shall include self-contained bus joints, wiring gutters, and aluminum bus rated at 1200 A. The utility company will furnish the meters.

Pack or crate the equipment and materials so they will not be damaged during shipping, storing, and handling. Protect the painted surfaces with removable heavy kraft paper, or equal, installed at the factory. Store the equipment and materials where they will have adequate dryness and heat prior to installation. Remove the protective coverings prior to the final inspection. Repaint damaged paint on the equipment and materials with spray-painting equipment and with the same quality of paint and workmanship used at the factory so the repainted areas will not be noticeable.

For each item which will not be readily identifiable without an identification sign, install such a sign to indicate clearly the information required by people who will use and maintain the items. Install signs on such equipment as panelboards, cabinets, motor controllers (starters), safety switches, separately enclosed circuit breakers, individual circuit breakers and controllers in switchgear and motor control assemblies, control devices, and other significant items.

The signs are to be laminated black prenolic resin with a white core, or equal, and lettering not less than $\frac{3}{16}$ in high. Attach the signs suitably and permanently to the items which they identify.

SIX

Transformers and Primary Systems

Before going into the detailing of transformers and primary systems, there are several items to consider. In the broad sense the use of the words "transformers and primary systems" implies any primary system going back to the utility company's generating plant. Consultants primarily involved in high-voltage transmission are a very special breed. This book does not explore primary systems to that depth and extent.

While the details we present do show some of the general arrangement of outdoor service entrance equipment in the 35- to 70-kV range, the detailing shown is far from complete. For systems in that range there is, first of all, a very considerable design requirement for the various supporting towers and structures. Second, there is a considerable amount of detailing required of the component parts. There are a number of sources of information on this high-voltage-distribution subject. One of the better ones is a three-volume set published by the Westinghouse Corporation which covers generation, system analysis, and transmission and distribution. It can be purchased from their Utility Engineering Department in Pittsburgh, Pennsylvania.

This book and its details, especially this chapter, are concerned with utility-supplied energy in the 5- to 25-kV range. This range covers the vast majority of situations faced by consultants. The primary voltage system could be primary-metered, and from that point the system would effectively belong to the owner. There can be an exterior or interior transformation system at one or many locations. The owner can utilize this distribution at a variety of voltages from either an overhead or underground system. And it can be installed inside or outside of various structures. As we noted in our discussion on switchgear in Chapter 5, there is a 15-kV limitation on interior high-voltage distribution.

Two of the most common systems are the 13.8-kV three-phase system to a series of interior substations and the 23.8-kV three-phase system to a series of exterior substations. These are both common and popular because many standard electrical equipment components are rated with either a 15- or 25-kV limitation. Higher voltages commonly result in a need for special-order components.

Most of the details shown in this chapter relate to outdoor and underground installation and distribution systems. Packaged interior systems, which were covered in the previous chapter, can be found in the catalogs of many major electrical equipment manufacturers. This is not true of many outdoor installations. In these cases one must research the components, frequently using several different catalogs, and then figure out how to detail the assembly of the parts properly.

In all of our details a serious attempt was made to describe pertinent parts. But these are not totally detailed where the actual connection to a transformer or, for example, a standoff isolator on a pole or support frame is noted. The reasoning was that such depiction is relatively self-evident, and application and intent are obvious. Such compromises were necessary at times to avoid cluttering the details with notes and arrows.

Basic to any distribution system is a high-voltage distribution schematic. Anytime a system delivering power at 5- to 25-kV is planned, there are always a number of ways the power distribution can be arranged. Seven of the more common methods are shown. Each has advantages and disadvantages, but one of the seven will usually provide the best, most *cost-effective* results for your particular situation. There are variations to each of these schematics that could have been shown, but we felt this would be both confusing and redundant.

In all exterior and interior installations grounding is an

important element of design. This is as true for a simple one- or three-unit pole installation as it is for a far more complex fenced-in transformer grouping.

In the distribution system there are always poles, raceways, and manholes to be detailed. Far too frequently these items are merely noted or casually alluded to when a more thorough detailing would solve a lot of on-the-job claims for "extras." In this same vein we have shown such things as transformer pads, shrubbery around a transformer installation, and the like.

Much of the need for this kind of detailing has been reduced by modern interior and exterior pad-mounted transformer packages. However, in large industrial and military installations many of the situations detailed still occur, especially in partial replacement of existing systems.

It would not be wise to look at a given situation and decide either it is old-fashioned or that it is ideal. Neither or both opinions may be correct. In any design of a primary system, both reliability and cost are important factors and frequently contradict each other. Least cost frequently provides least reliability, but greatest cost does not guarantee greatest reliability. This is one area of electrical design that is truly dependent on intelligent compromises of various factors, and, very frequently, two similar situations have two very different answers as to which is the best overall arrangement.

Splices and Stress Relief Two details are shown in Figure 6-1. The upper detail delineates a taped splice in a shielded cable, and the lower detail covers a stress-relief cone. Referring first to the upper splice detail, we begin by noting that splices are subjected to a set of voltage gradients and dielectric stresses different from those of the cable. The highest stress is around the conductor and connector area. The splice must contain these stresses within the working limits of the material used to make it. On shielded cables the splice is also in the direct line of the cable system and must handle any ground currents or faults that pass through the cable shielding. The splice must also be capable of carrying full-rated load, emergency overload, and fault currents without overheating or separating. The cover must provide adequate protection to the splice under all sorts of environmental conditions.

Taped splices for shielded cables have been successfully used for a long time. The tapes must have the same dielectric properties as the cable insulation and be compatible with it. Semiconducting tapes are recommended for covering the connector and exposed conductor. Connector surfaces must be free of sharp edges and have tapered ends to provide a gradual transition between the conductor connector and the cable insulation. This is required to hold voltage gradients and stress within the working limits of the insulating materials. To avoid restricting heat dissipation, the splice should not be overinsulated.

Preferably, a tinned copper braid is used to provide shielding and is referred to as shielding tape or braid on the detail. As shown, a grounding strap is provided on one end of the splice and a heavy braid jumper is applied across the splice to carry ground fault current. Cover tapes are built up over the splice to seal against moisture.

In the lower half of Figure 6-1 there are two versions of a stress-relief cone illustrated. Electrically speaking there are a number of types of terminations such as taped, armor, pothead, and preassembled. In terminating a shielded cable, cutting back the shielding provides proper creepage distances between conductor and shielding and introduces a longitudinal stress over the surface of the exposed cable insulation. To control the resultant radial and longitudinal electrical stress at that point, insulating tapes are commonly used to form a cone of gradually increasing thickness. The cable shielding is brought up to the cone surface and ended about ⅛ in behind the largest diameter of the cone. This is commonly referred to as a *stress-relief cone.*

Because some cable systems are more resistant to high levels of electrical stress than others, it is advisable to consult the cable manufacturer for a recommendation on termination and splicing of the shield cable.

Potheads and Taped Terminators Two more very common terminators, potheads and taped, are shown in Figure 6-2. We refer first to the pothead detail, which is a common method of enclosing and protecting a cable end. Basically it is a metallic body with one or more porcelain insulators arranged to accept a variety of cable entrance sealing fittings. The porcelain ends will accept a variety of cable conduit or aerial connections.

Potheads accept 5 kV or more in indoor, outdoor, or liquid-immersed situations. They can be bracket-, plate-, or flange-mounted. They are also generally superior in protective ability to taped connections and are preferred in tougher applications. Three-conductor potheads frequently have an exterior bolted neutral connection as noted in some of the following details.

Cable construction usually controls the selection of proper entrance sealing fittings, stress-relief-cone materials, and filling compound. Application is the prime point to consider in choosing the type of pothead, mounting, and acceptable aerial connections. Cable systems are generally broken down into pressurized and nonpressurized conditions. Most normally encountered power cable systems are nonpressurized with solid dielectric cables.

For solid cables the cap nut or the solder-seal pothead is commonly used. The pothead is, by design, filled with a high-strength dielectric material and is electrically stronger internally than externally. Dielectrics used include asphaltic-based materials, resins, and oils.

The lower half of Figure 6-2 depicts a typical taped termination, indoor or outdoor, for shielded or unshielded cables. This sort of terminator is normally used at 15 kV and lower voltages. It can, however, be made to work at voltages up to 69 kV. Where the cables are unshielded the device consists only of a lug and tape—indoors or outdoors.

Where cables are shielded, the situation generally re-

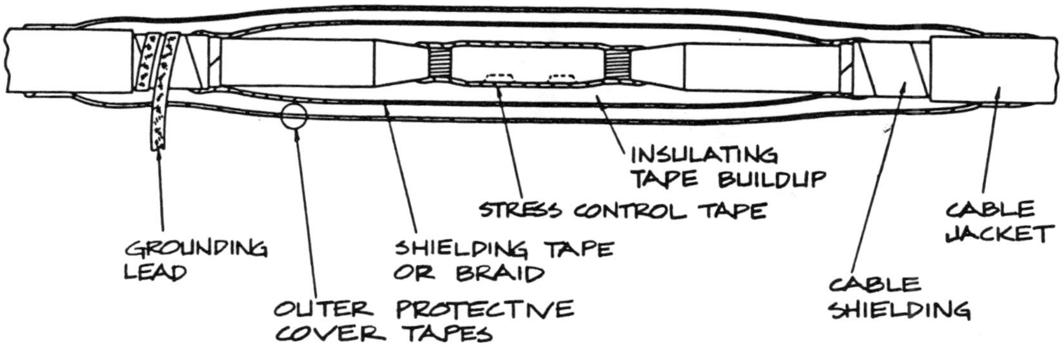

NOTE: HEAVY BRAID JUMPER SHOULD BE USED ACROSS SPLICE TO CARRY POSSIBLE GROUND-FAULT CURRENT. STRESS CONTROL TAPE SHOULD COVER STRANDS COMPLETELY, LAPPING SLIGHTLY ONTO INSULATION TAPER.

TYPICAL TAPED SPLICE IN SHIELDED CABLE

NO SCALE

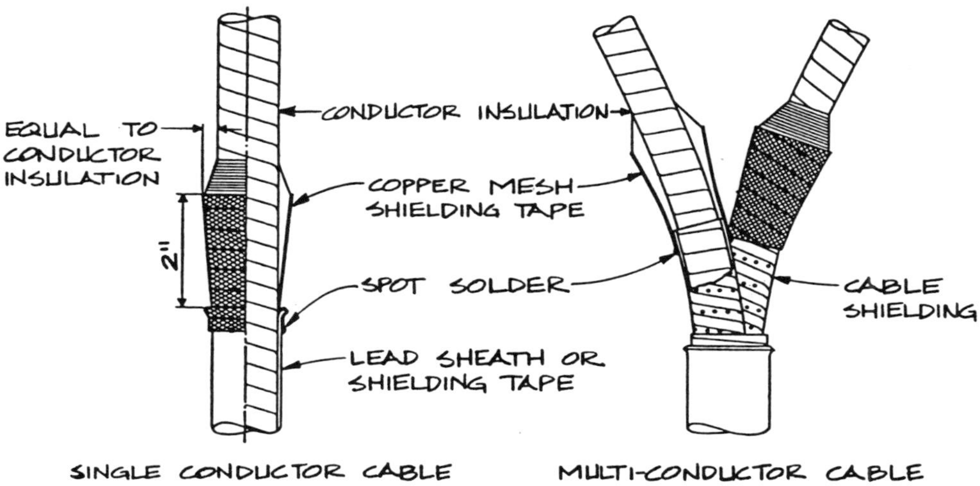

TYPICAL STRESS RELIEF CONES

NO SCALE

FIGURE 6-1

quires use of a stress-relief cone and cover tapes in addition to the lug. As before, the size of the stress-relief cone is controlled by voltage value and whether the cable is indoors or outdoors. Creepage must be allowed for in outdoor installations, and additional creepage may be secured by installing a rain hood made of neoprene, rubber, plastic, or porcelain, as depicted in the detail.

As shown, cover tapes are applied over the stress-relief cone, cable insulation, and on up to the connecting lug. This tape can be of an anhydrous substance painted with a weatherproofing compound, vinyl, silicone, rubber, or other taping materials such as a laminated build up of coarse mesh tape saturated with epoxy.

Used on outdoor installations, rain hoods are in the creepage path between the lug and the stress-relief cone. If the installation is vertical, the rain hood goes right over the cone to keep at least some part of the cable insulation along the creepage path dry. Again, where unusual environmental conditions are expected or encountered, the pothead should be used.

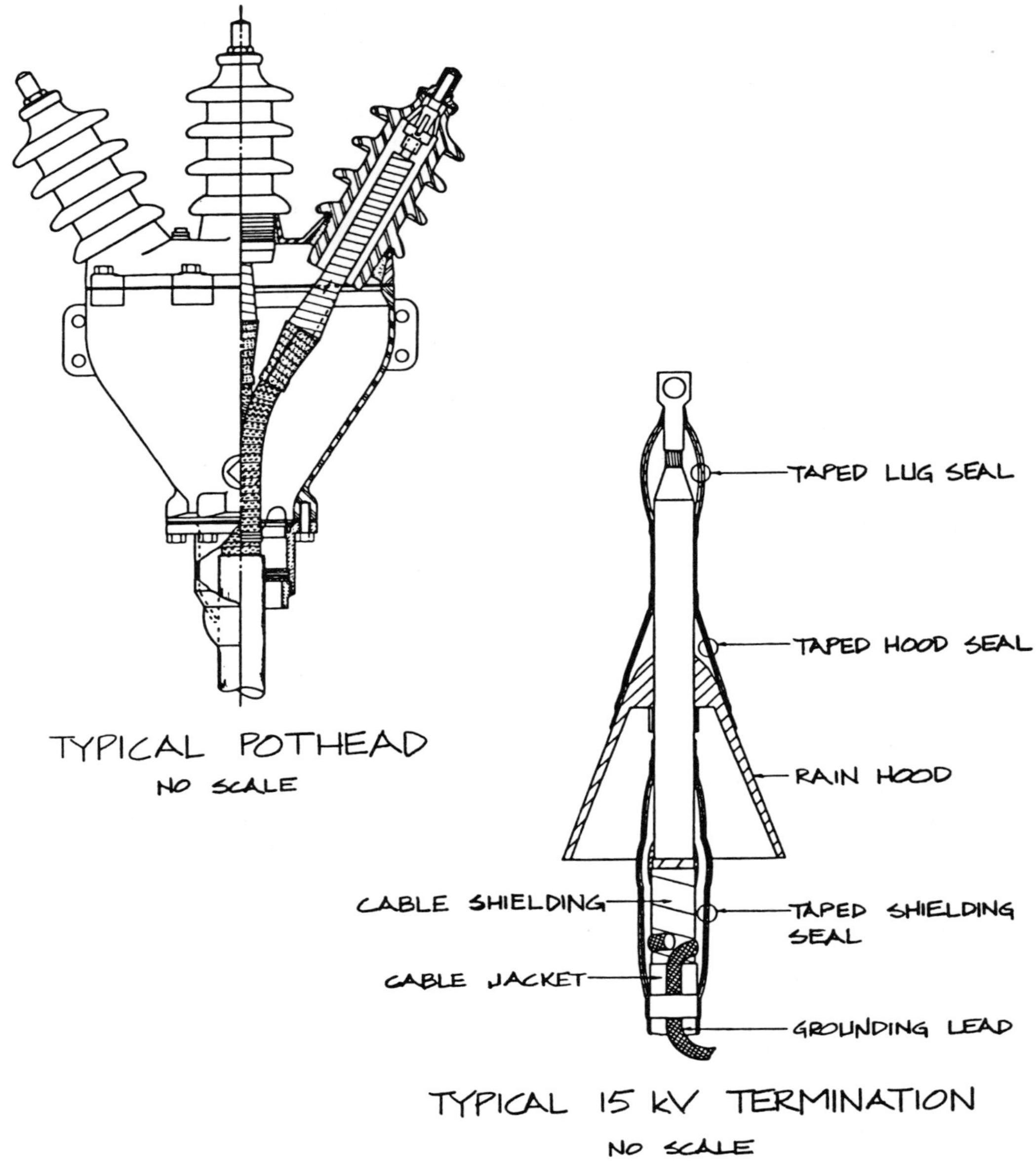

FIGURE 6-2

Aerial Cable, Pole Guy A current copy of the NEC is mandatory for anyone involved in designing exterior overhead electrical wiring. Also needed is a current copy of the National Electrical Safety Code ANSI-C2. This document is published by the IEEE and can be obtained at IEEE headquarters 345 East 47th Street, New York, NY 10017. The text describes the rules and regulations in a clear, logical, and readily understandable manner. Figures 6-3 and 6-4 illustrate pole and guy connections. They should help clarify the statements in your specifications and the general layout on your plans.

Figures 6-3 and 6-4 can be modified depending on your particular requirements. Commonly the messenger cable carrying the electrical conductors terminates with an eye bolt on the pole.

In theory the force on the pole can be balanced with an anchored guy pulling in the opposite direction.

As is obvious in Figure 6-3, a triangular balance of forces not only guarantees a zero reaction at the pole but assures that the pole will remain vertical. However, be-

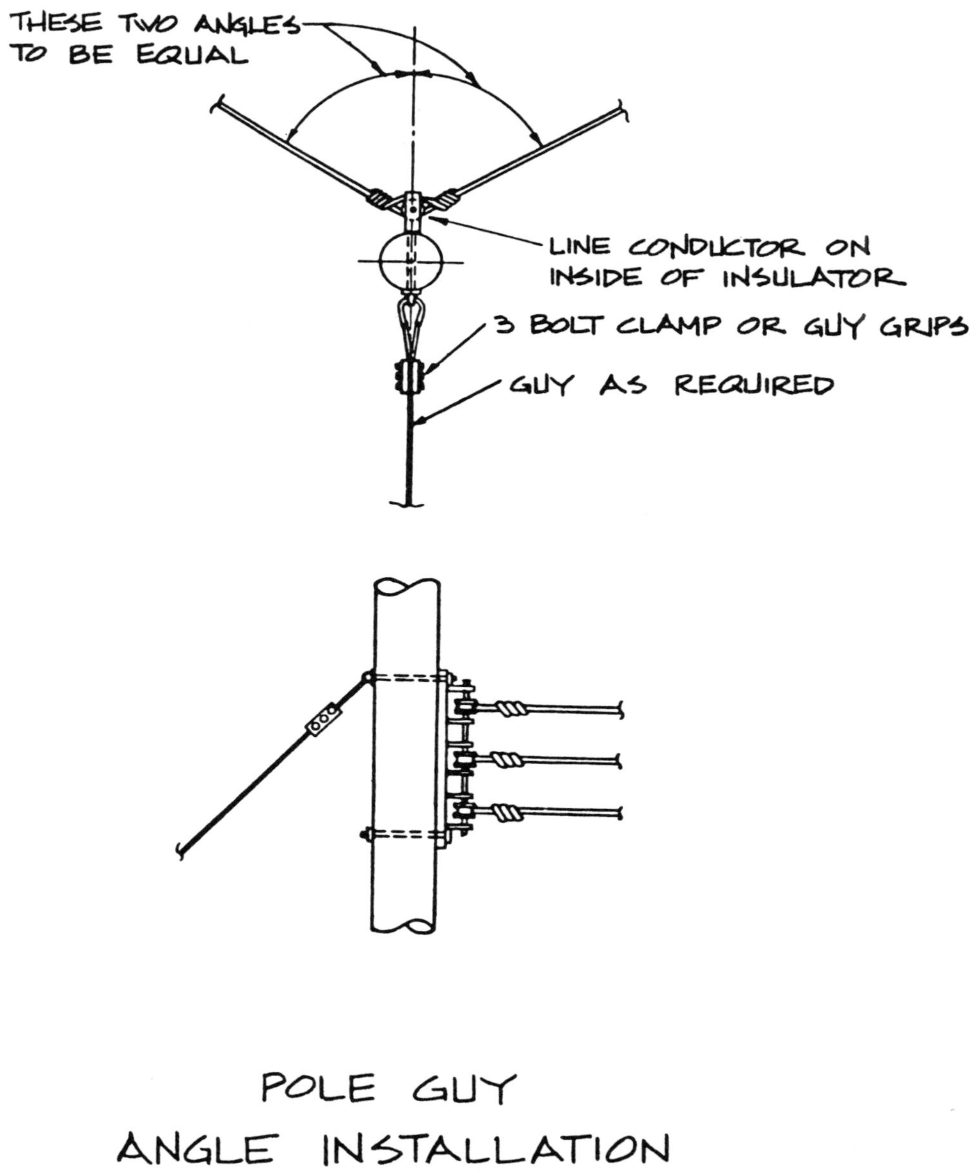

FIGURE 6-3

cause of space limitations it is not always possible to get this ideal arrangement and the detail does not say the angle between forces is, or must be, exactly 120°. Nor is this absolutely necessary. The detail does say that whatever the angle between the centerline of the opposing force and the two guys is, it should be equal for each.

Figure 6-4 shows messenger cables at right angles, and guys exactly opposite the pull of the messenger cables. The obvious question would be "Why not one guy bisecting and exactly opposite the force line bisecting the two cables?" That could be done, but frequently it is difficult to be certain of the forces under all weather conditions. This detail merely guarantees peace of mind.

While no new ground is broken or covered in this detail, a serious attempt was made to show the cable, conductor, and pole hardware in order to preclude any debate by the contractor over your intent.

Aerial Pole Grounding Grounding aerial cable installations is, at best, an inexact science, and our presentation

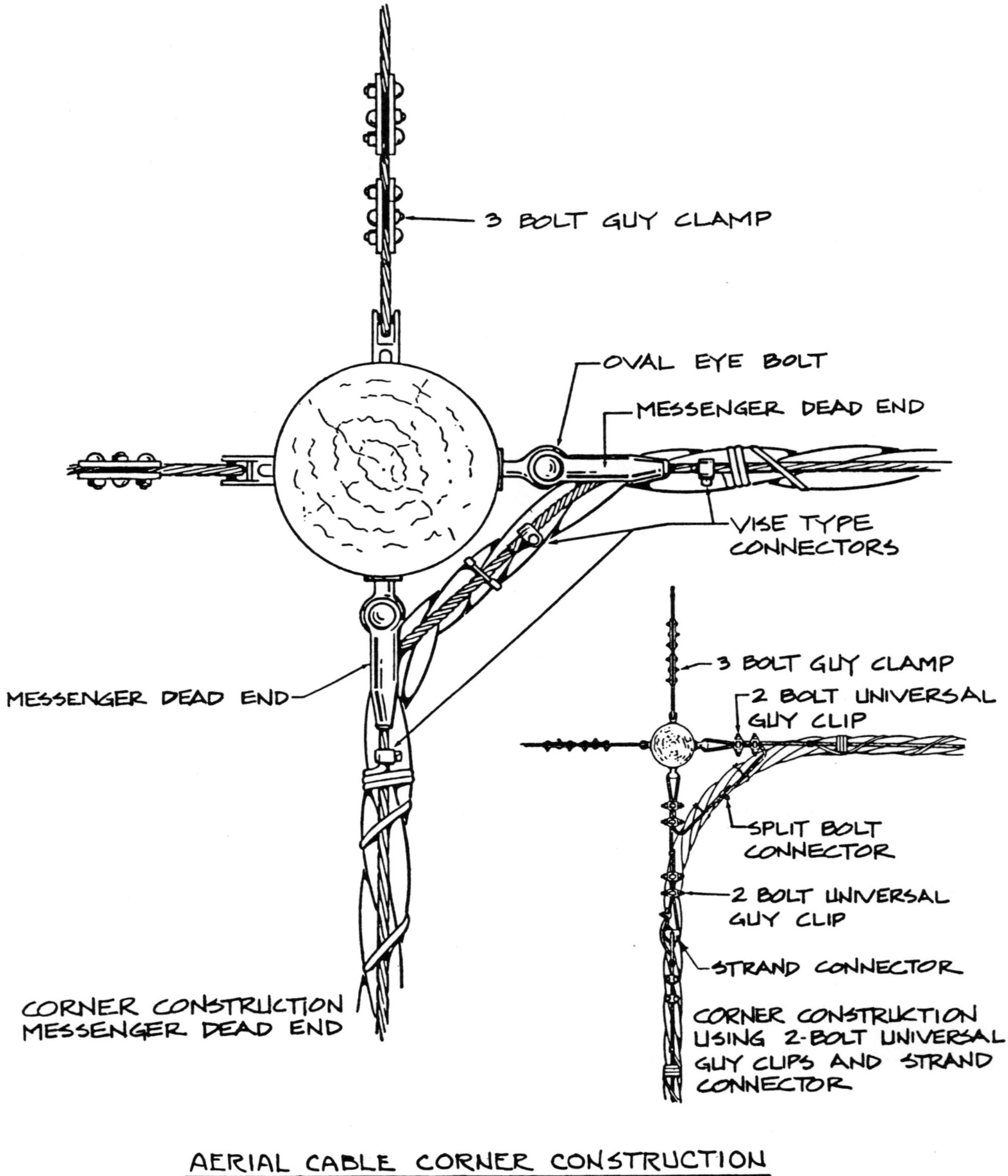
3 BOLT GUY CLAMP
OVAL EYE BOLT
MESSENGER DEAD END
VISE TYPE CONNECTORS
MESSENGER DEAD END
CORNER CONSTRUCTION MESSENGER DEAD END
3 BOLT GUY CLAMP
2 BOLT UNIVERSAL GUY CLIP
SPLIT BOLT CONNECTOR
2 BOLT UNIVERSAL GUY CLIP
STRAND CONNECTOR
CORNER CONSTRUCTION USING 2-BOLT UNIVERSAL GUY CLIPS AND STRAND CONNECTOR
AERIAL CABLE CORNER CONSTRUCTION

FIGURE 6-4

of ways to ground a line at a pole will undoubtedly not satisfy everyone. In order to establish proper grounding, a knowledge of the currents involved and the conductivity of the ground itself must be known. Figure 6-5 really does not address those questions nor is it the intent of the detail. Certainly there are other ways of grounding.

In our detail two ways of grounding are illustrated. One is the typical coil for type around the pole with the coil base below the surface. The actual length of coil on the butt end of the pole still requires some calculation. The other common way to ground a pole is by using a driven ground rod or rods as required. As we mentioned before,

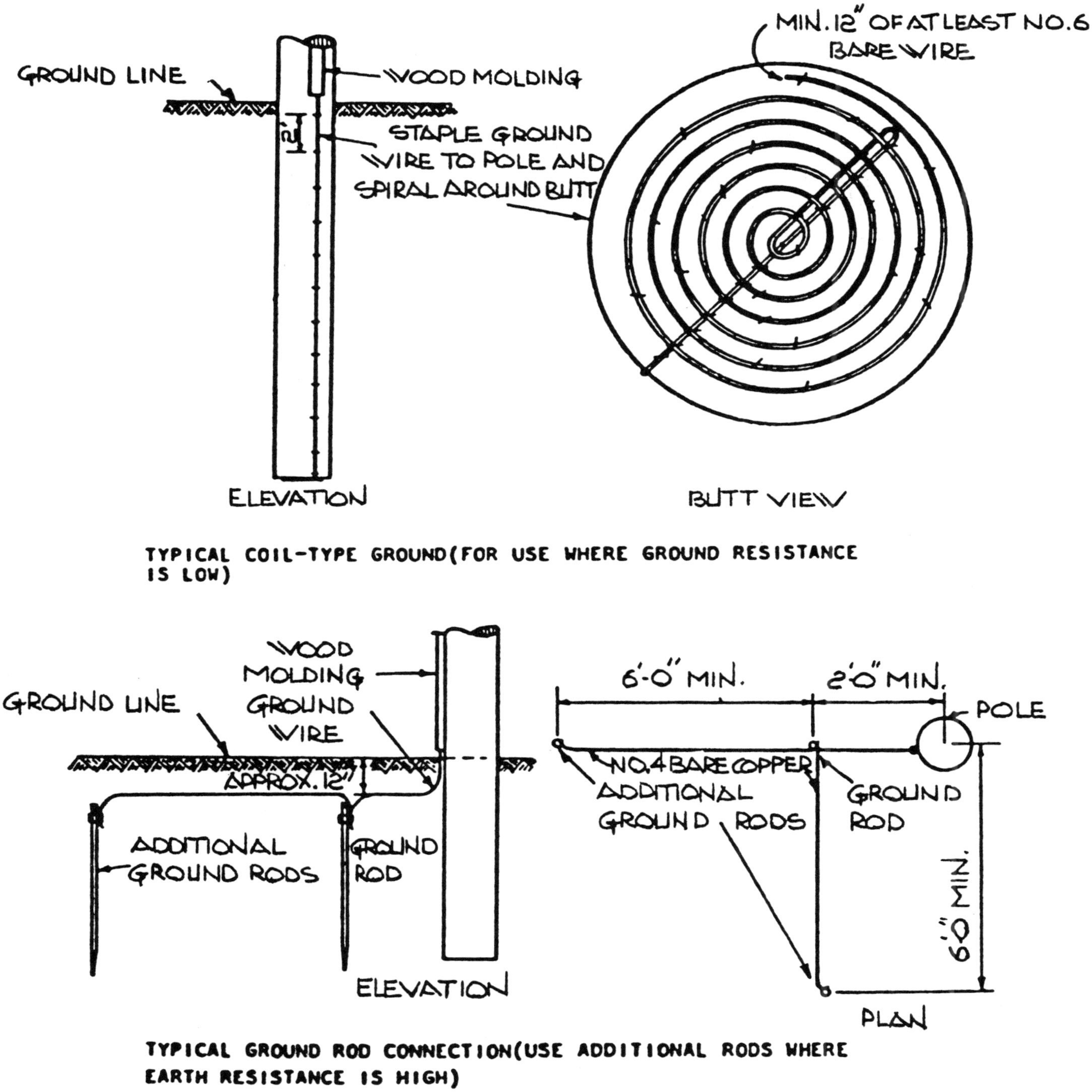

FIGURE 6-5

the NESC, ANSI-C2, covers this subject in 18 pages of text, which is why we are so brief in our description.

There are basically three types of ground rods.

1. Copper-encased steel ground rods, where the copper normally has a minimum wall thickness of 0.013 in or greater.
2. Solid copper rods made from rolled or drawn electrolytic copper.
3. Stainless steel rods with a diameter of not less than ⅝ in.

Underground Raceways Figure 6-6, detailing duct raceways for electrical conduits, seems so obvious that we almost left it out. Perhaps our almost too casual treatment of this detail is the best reason for including it.

Many electrical plans requiring only a few underground conductors contain a casual statement allowing the contractor to put the conductors in separate conduits "or a suitable raceway." Actually any buried system should have sufficient spare raceways to provide an alternate path for most if not all conductors in case there is a problem with any or all of them.

There are many potential problems. As we all know, the chemical reactions that occur underground can attack and destroy metal conduits. Conduits chemically treated to survive under these conditions can be expensive. Voltage requirements frequently mandate conduits encased in concrete. The net result frequently is a duct system. Figure 6-6 shows various types of duct materials, construction, and spacing. All possible requirements and situations are obviously not covered.

The most important point, from our experience, is not shown on the detail and should be carefully covered under excavation, fill, structural fill, and backfill. The underground ducts are invariably cast in a series of sections and sealed at the section joints. *The* most important item is to require that the entire duct installation be either level or, preferably, pitched at a continuous grade (for drainage purposes) from manhole to manhole (or building or pole to manhole or manholes). Nothing is more serious than to try to pull a conductor through an uneven duct bank and have the rough edges abraid the insulation. Usually, but not always, a conductor insulation test will bring out this problem. However, an incipient problem may not show up for years. But, at some point, it most likely will show up as a cable failure—with considerable aggravation and public furor.

Manholes Our manhole detail, shown as Figure 6-7, is a rather fancy set of twin manholes with well-located 45° angles, corners, and reinforcing that, with proper type and thickness of concrete and reinforcing, could be located in any roadway.

We have not totally described the concrete or the reinforcing for two reasons. First, frequently in a detail describing concrete and reinforcing, the detail supplier uses descriptions that would cover all sorts of conditions and may even be an acceptable fallout shelter. Second, people who construct prefabricated concrete septic tanks have for some time been actively promoting prefabricated manholes built to a variety of wheel-loading and utility company requirements. We have shown our somewhat old-fashioned manhole so that if you use a specified prefab manhole you can ensure that it has all the necessary parts.

As can be seen, our manhole is reinforced on all sides and top and bottom. It has a roadway-type cast-iron frame and cover for each unit. Each unit also has a moisture-collection sump. Since ours is custom-built, we have a neat fit that two prefab manholes may not have. We show ideally arranged conduit openings. Your prefab manhole will undoubtedly have a slot or slots, and you will have to specify a filler for the void areas between incoming conduits. Whether you copy a standard prefab detail or use ours, you definitely should show the manhole.

What is frequently missed in manhole design is the pulling iron connection and the racks for the cables. These are standard items that are shown in the catalogs of a number of manufacturers. Our manhole does not show a fixed ladder or ladder rings built into the wall of the manhole. This point is open to debate as the manhole access could have been located to one side and the cables arranged in such a fashion as to allow the ladder to be installed. Having checked out old cable installations with some current leakage through the old insulation, we would much prefer to descend slowly on a wooden ladder, even if the built-in metal ladder were available.

One final note about the sump shown in the manhole. Some engineers and authorities will require this drainage sump to be piped away from the manhole to a dry well around the sump itself. Some authorities and utility companies will ask that the sump be deleted from the work. It would be best to check local utility company requirements and standard practices before doing anything about the sump and its drainage. Where a high water table exists the sump obviously is useless and should be omitted.

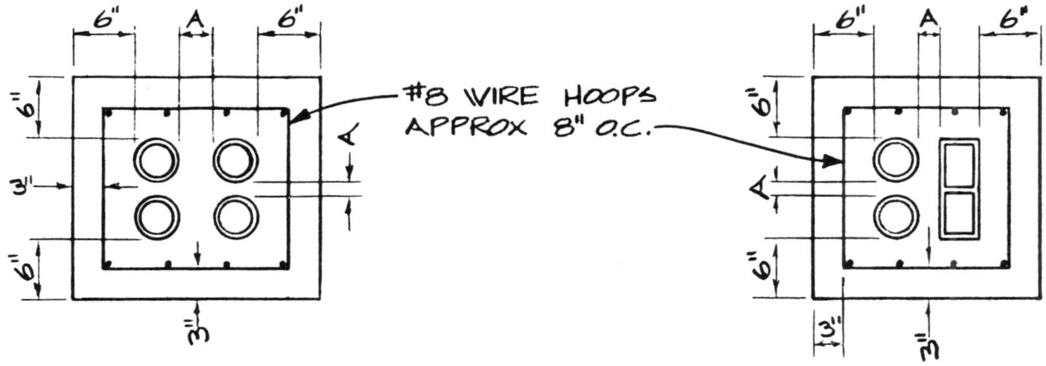

TYPICAL SECTIONS OF REINFORCED DUCT LINE ENCASEMENT. USE UNDER ALL RAILROAD TRACKS AND IN OTHER AREAS AS NEEDED. REINFORCED ENCASEMENT TO EXTEND 15' MIN. ON EACH SIDE OF CENTERLINE OF RR TRACKS.

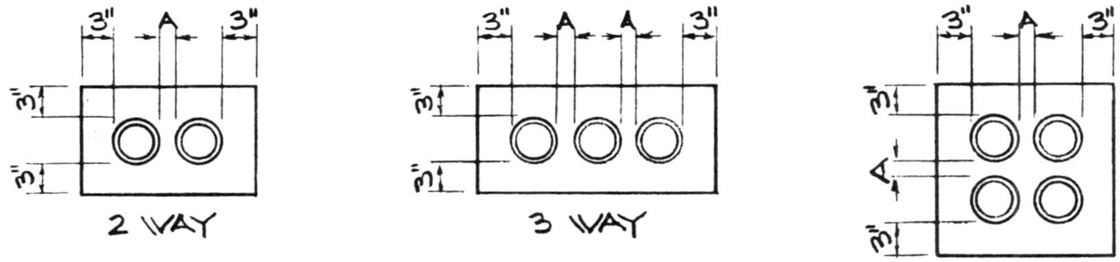

4 WAY

INSTALL EDGEWISE BETWEEN MANHOLE AND SPLAY OF JOINT-USE DUCT LINE

NOTES:

1. DIMENSION "A" IS 1½" FOR CLAY OR SOAPSTONE DUCT AND 2" FOR FIBER OR ASBESTOS CEMENT DUCT.
2. ALL DIMENSIONS ARE MINIMUM FROM OUTSIDE SURFACE OF DUCT.
3. CONCRETE NOT REQUIRED BETWEEN HORIZONTAL FACES OF CLAY DUCT.

TYPICAL SECTIONS OF DUCT LINES

NO SCALE

FIGURE 6-6

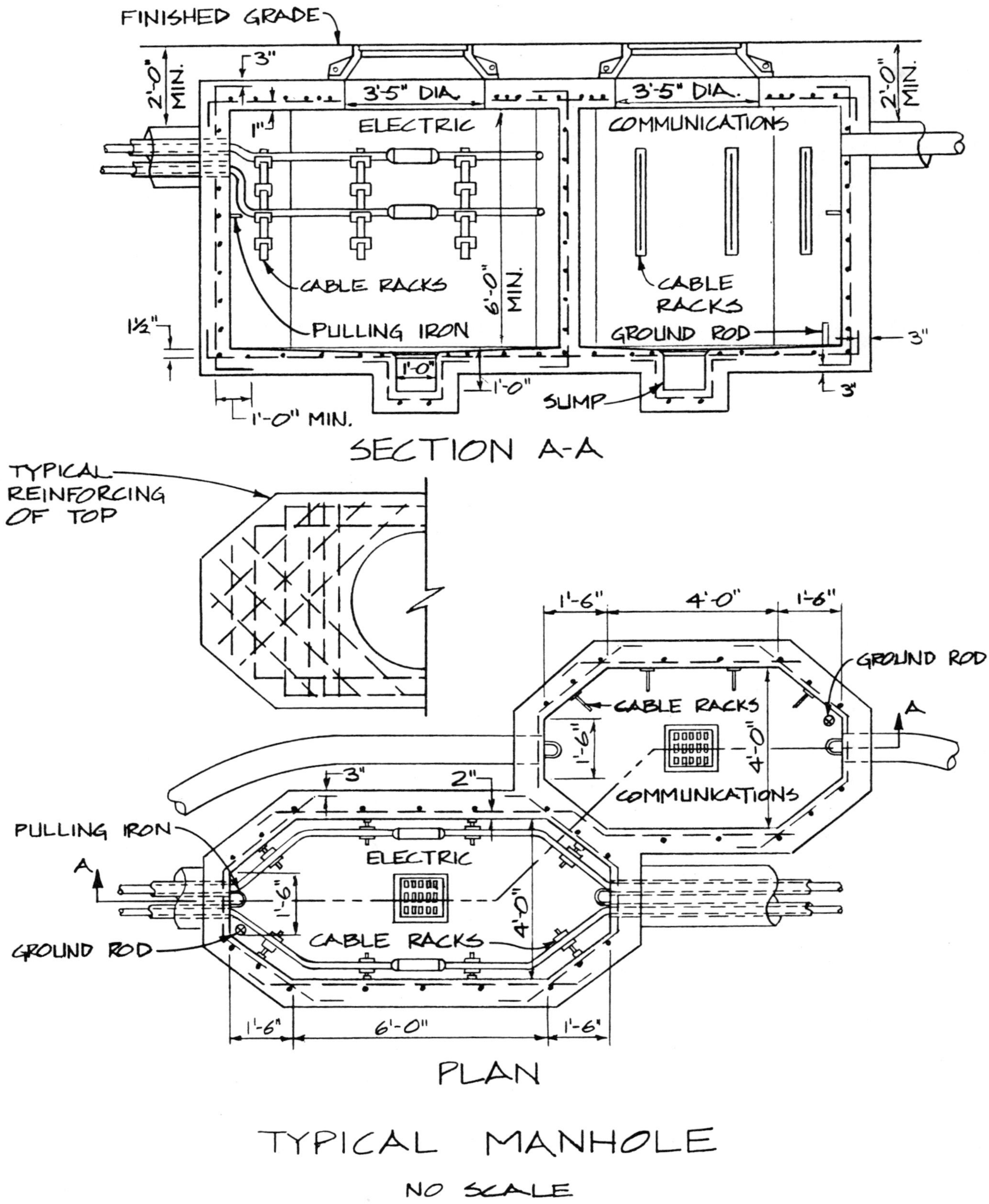
FINISHED GRADE
3"
2'-0" MIN.
3'5" DIA.
3'5" DIA.
2'-0" MIN.
1"
ELECTRIC
COMMUNICATIONS
CABLE RACKS
6'-0" MIN.
CABLE RACKS
1½"
PULLING IRON
GROUND ROD
3"
1'-0"
1'-0"
SUMP
3"
1'-0" MIN.
SECTION A-A
TYPICAL REINFORCING OF TOP
1'-6"
4'-0"
1'-6"
GROUND ROD
CABLE RACKS
A
1'-6"
4'-0"
3"
2"
COMMUNICATIONS
PULLING IRON
ELECTRIC
A
1'-6"
4'-0"
CABLE RACKS
GROUND ROD
1'-6"
6'-0"
1'-6"
PLAN
TYPICAL MANHOLE
NO SCALE

FIGURE 6-7

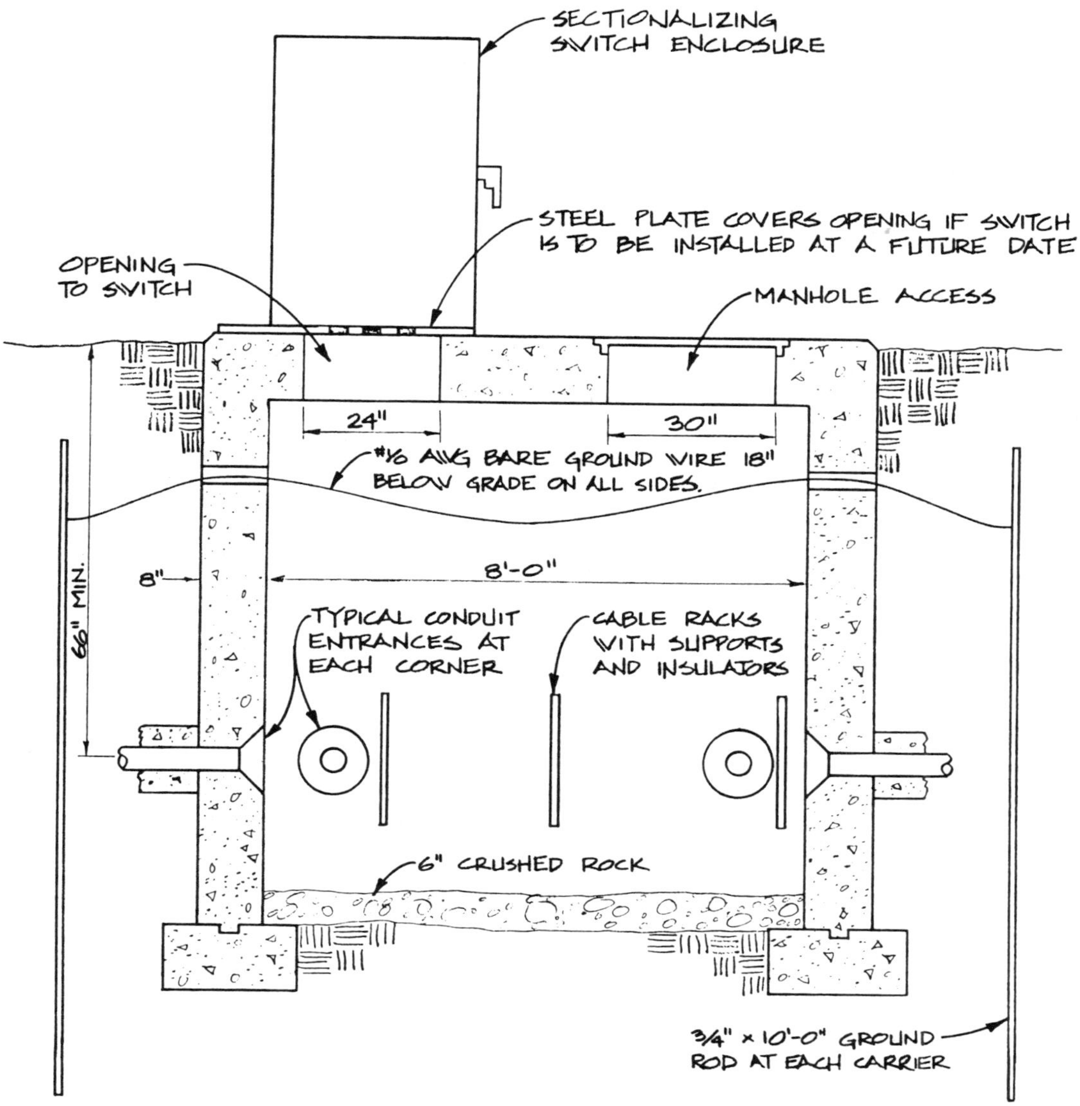

FIGURE 6-8

Switch Manhole Somewhat different is the manhole used to feed services to an exterior sectionalizing switch as shown in Figure 6-8. The switch manhole shown could have had a concrete bottom and the drainage previously described. To vary the possibilities, it was decided to show a manhole in a dry location with a simple crushed-rock bottom.

This manhole is supporting equipment and, lacking a bottom, requires a footing to support the walls. Because of the uncertainty of just what the future cabling and connections will be, the egress and entry to the manhole is located fairly tight in the corners. In addition a spare center cable rack is noted, which your project may or may not require.

The manhole access location with respect to the 8 × 8 square it serves is not nailed down either. What we are attempting to show here is approximately how this manhole should be built. We leave the exact position of various items to the design engineer.

Even the switch itself may not be a part of the present design. Our detail notes the opening should be covered with a steel plate if the switch is a future item. Normally this opening is 24 × 24 in, while the manhole has a circumference 30 in, and the conduit entry openings have a 6-in circumference.

Most important is that whether or not the probable use of this manhole is for a higher-voltage device, the grounding requirement is a serious matter. To cover ourselves (and you) we have called for ground rods at all four corners (preferably copper or copper clad) connected by a # 1/0 bare ground wire.

As we said, this is not a complete detail. Your specific requirements will have to be spelled out, but the major parts of this type of manhole are shown.

Exterior Transformers and Hardware Following this discussion there are details that describe exterior uncovered transformers and related equipment, followed by one detail of similar equipment located indoors. Because certain basic items appear in each detail the following descriptions are given to avoid constant repetition.

Grounding: Every detail contains pole grounding, sometimes simply pointed out as molding. In all cases the grounding is similar to the pole grounding previously presented.

Insulators: Every line connected to a pole is held in position by an insulator. If it must take the pulling force of a line from another pole it is called a *strain* insulator. A strain insulator is a porcelain device that is securely fastened to the pole. The size varies according to voltage in the line attached. Some of these are noted as well as depicted, and others, to maintain detail clarity, are merely depicted.

Disconnects: Pole-mounted transformers have fused primary disconnects which may be accompanied by surge arresters. The disconnects usually have an eye arrangement so that they can be opened or closed using a long wooden pole with a hook on the end to push or pull the eye connection. These disconnects can also be equipped with fused cutouts. The surge arrester controls stresses beyond the insulation strength of the connected equipment.

Bushings: All connections to transformers are made through low- and high-voltage bushings on the transformer that are usually a part of the transformer bill of material.

Lightning Arresters: The usual pole has lines connected to its high-voltage cables that extend to pole-mounted lightning arresters. From these arresters a separate line leads down the pole to its own ground rod.

Cable Clamps: To attach both high-voltage lines and low-voltage lines other than service secondaries, connections are made by simple clamping devices referred to as *hot line clamps* or *solderless connectors*. These are similar to a clamp on a pipe.

All of the above items are a part of normal pole line hardware, along with messenger cables, guy rods, and the like. A number of manufacturers have full and complete descriptions of all of these items in their catalogs, and you should refer to them for more complete and detailed description and specification.

Depth of Buried Pole: There are various tables that tell you how deep to bury a pole relative to its height. The mere fact that you have specified the depth of burial does not eliminate the need for guying of end poles, corner poles, or poles in any situation where unequal forces are created. The following table is of fairly conservative depth-of-pole values based on relatively firm surrounding earth conditions.

Height of Pole; ft	*Depth of Burial*
20	5 ft 0 in
25	5 ft 6 in
30	5 ft 6 in
35	6 ft 0 in
40	6 ft 6 in
45	7 ft 0 in
50	7 ft 6 in
55	8 ft 0 in
60	8 ft 6 in

Single-Phase Transformer To begin our depiction of pole-mounted equipment, we present Figure 6-9, which is a typical single-phase transformer installation. As described in our list of standard items, the voltage line has pole-mounted insulators and feeds the transformer through hot-line clamps to primary disconnects, and thence to the primary bushings of the single-phase transformer. Three lines of primary are shown because this is a three-phase delta-connected primary. We need only two of the lines because we are in a single-phase situation. If you check the normal residential neighborhood, you commonly see two primary lines because the utility company is delivering 2400-V (sometimes 4800-V) single-phase current transformed down to the usual 240/120-single-phase house service.

To illustrate street lighting, we have previously shown luminaires. In this detail we show a photo-electric-cell relay controller for a series of street lights and its cable and feeder just below the transformer. This device, activated by lack of natural light, automatically energizes the street lights, which are not shown in this detail.

From the secondary side of the transformer, lines are extended to a vertical rack installation of the secondary distribution system and connected to the secondary distribution system by solderless connectors. The primary has no neutral connection. The transformer provides a neutral connection for the secondary system.

The pole, as with all good pole-mounted transformer installations, has lightning arresters which are grounded through a molding usually made of wood. Finally we show the control wires of the photoelectric relay tied to a manual switch 5 ft 0 in above ground.

Not all of these items are necessary on every detail. Our detail is a composite to show the possibilities. If telephone lines were strung on the same poles, they would be below the secondary distribution.

For proper heights of all wires above ground you must be guided by the NESC, local codes and ordinances, and utility company requirements.

Three-Phase Transformer Bank, 3-50 kVA or Less Units The manner by which transformers are supported on a pole and the number of poles, cross arms, and other supporting arrangements depend on the weight

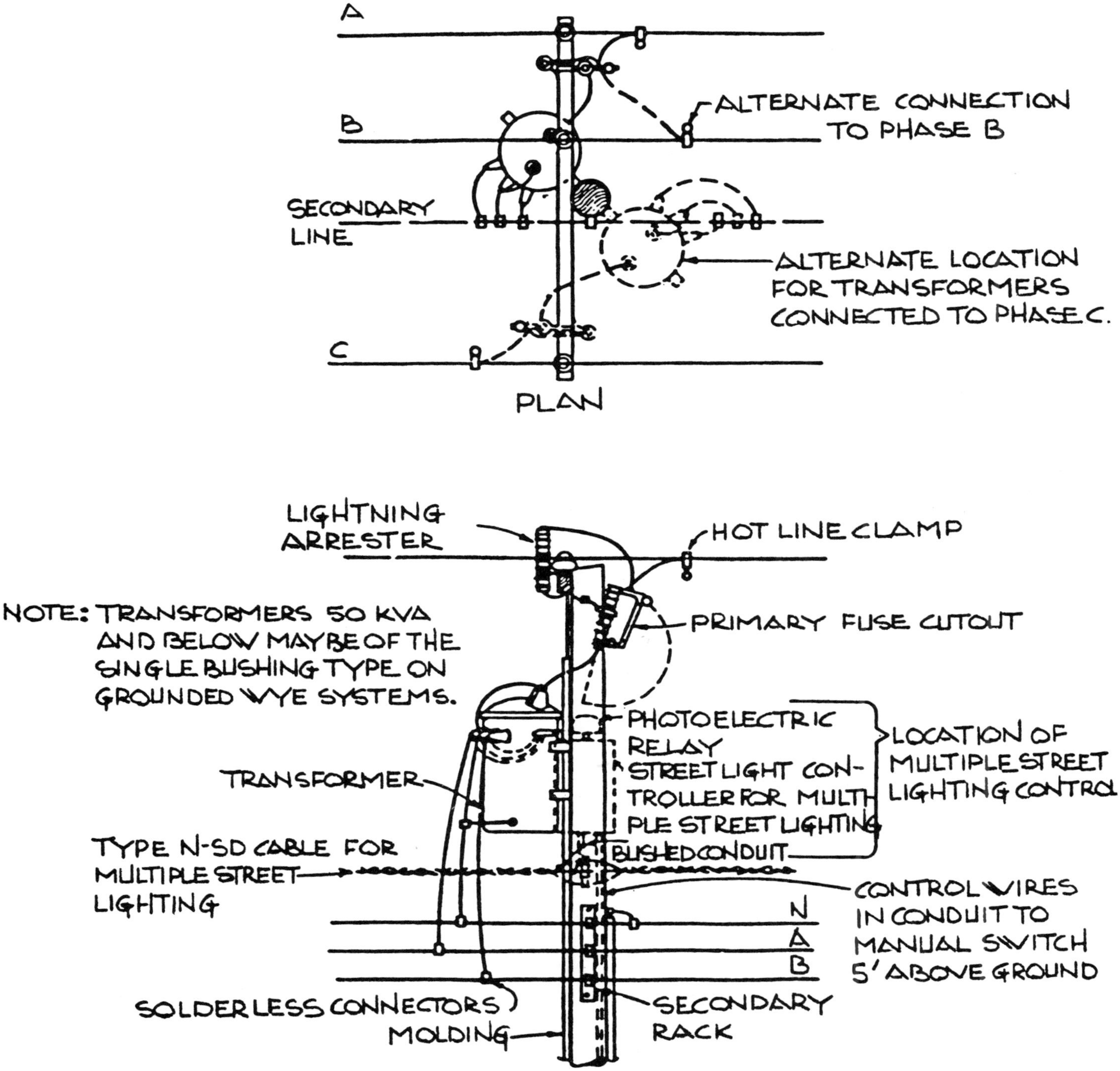

TYPICAL SINGLE-PHASE TRANSFORMER INSTALLATION

FIGURE 6-9

and size of the transformers. Figures 6-10 and 6-11 show two different ways a single three-phase pole-mounted transformer installation can be made.

Besides the basic hardware and pole covered in our general discussion, there are a number of items to be noted. As indicated in Figure 6-11, 50-kVA and lower-rated transformers may be of the single-bushing type. This is allowable at the 480, 208, and 240 voltage levels common on the secondary side of this type of installation.

The primary lines on top of the pole have no neutral connection. The neutral connection is provided by the transformer. Note also the location of primary fused cutouts, the primary line lightning arresters and the ground, and the primary and secondary line connections from the transformers.

The primary line passes along the top cross arm and is held in place by primary insulators. The secondary lines are mounted vertically and run below the transformer, held in place by a standard manufactured rack holding four insulators.

The two details give you a choice of mounting the transformers around the pole in a three-sided arrangement called a cluster, or adding two cross arms and lining up the transformers across one side of the pole.

One final note: Keep the phasing the same on both the primary and secondary sides of the transformer.

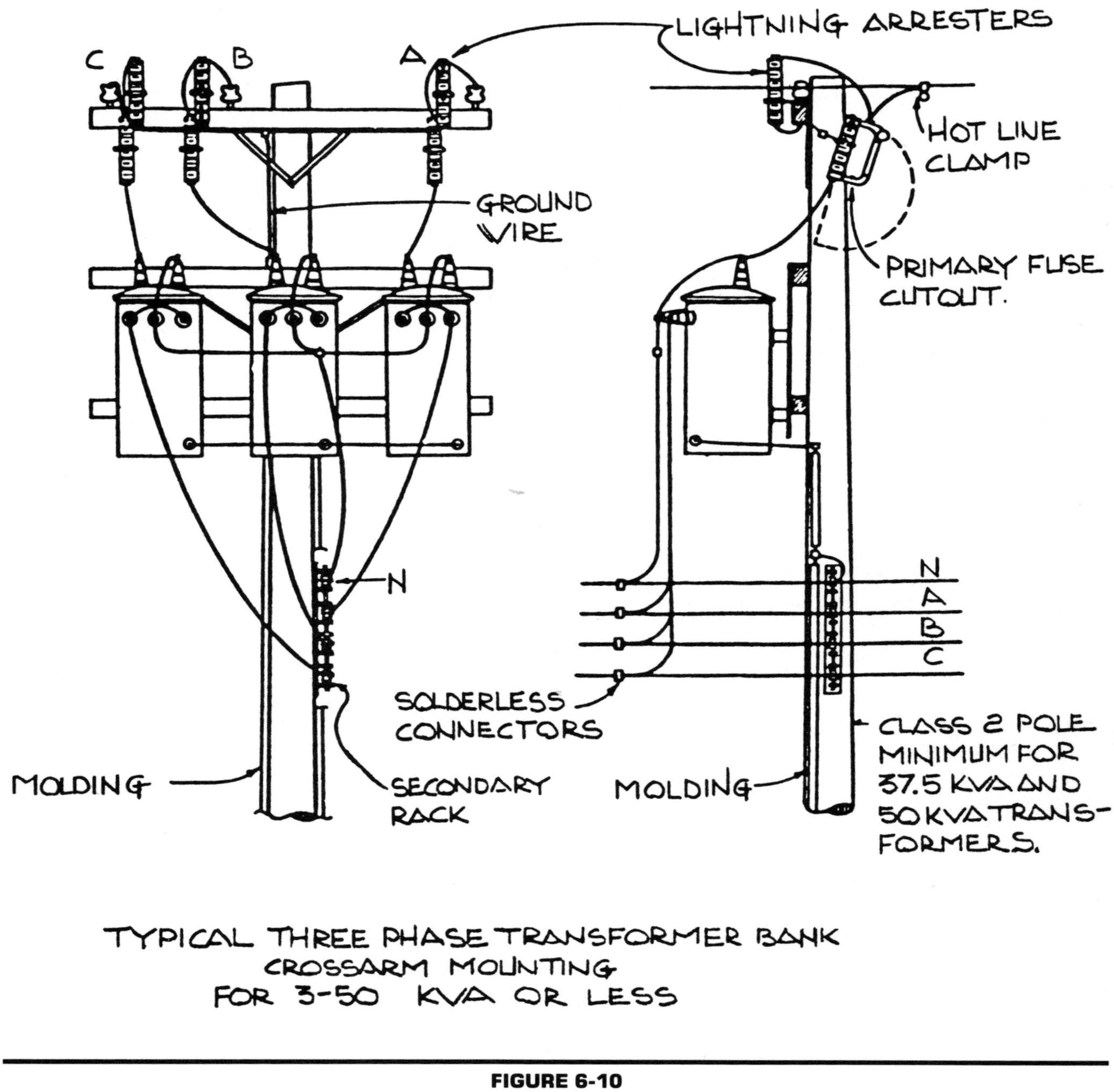

FIGURE 6-10

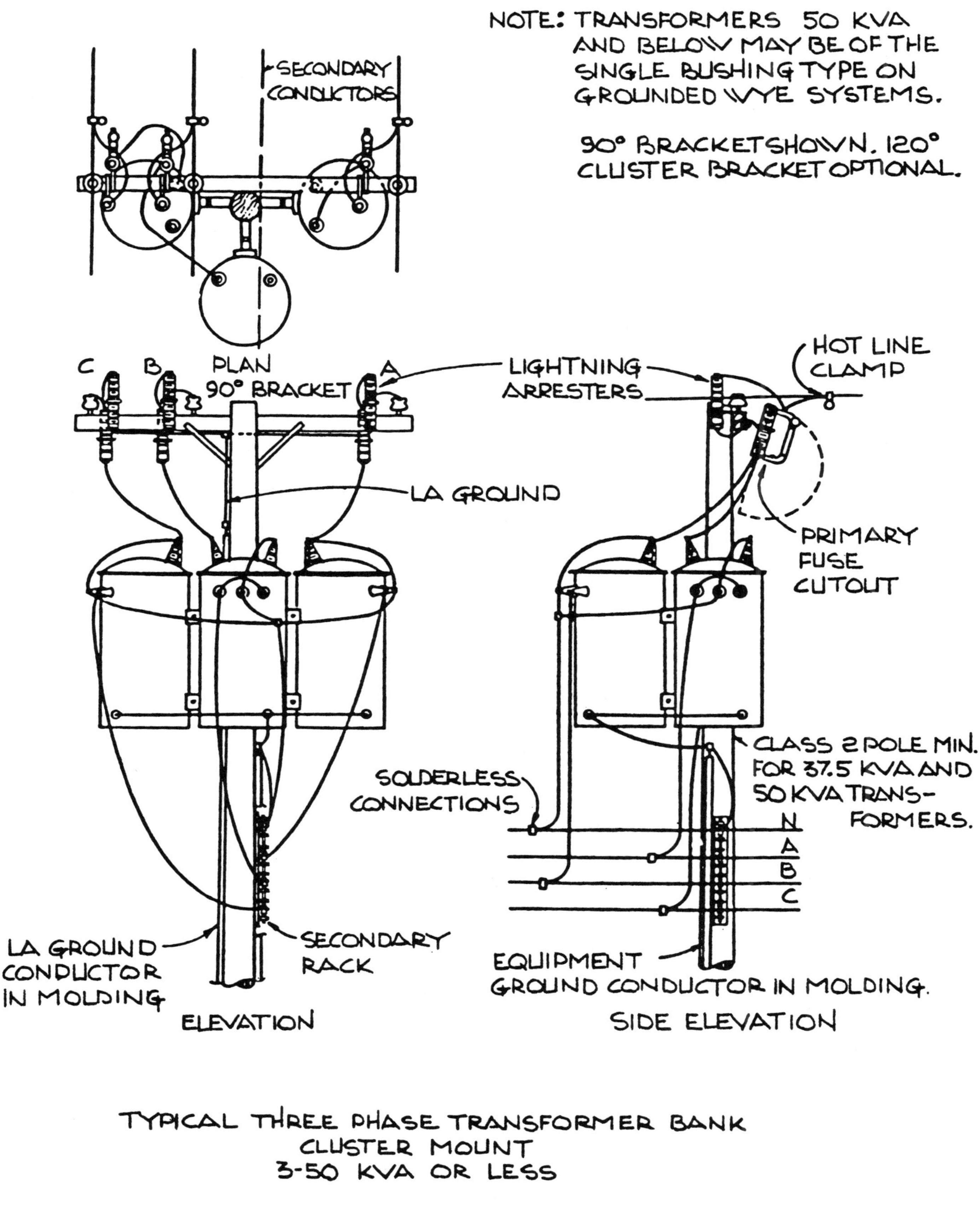
NOTE: TRANSFORMERS 50 KVA AND BELOW MAY BE OF THE SINGLE BUSHING TYPE ON GROUNDED WYE SYSTEMS.
90° BRACKET SHOWN. 120° CLUSTER BRACKET OPTIONAL.
SECONDARY CONDUCTORS
PLAN 90° BRACKET
C
B
A
LIGHTNING ARRESTERS
HOT LINE CLAMP
LA GROUND
PRIMARY FUSE CUTOUT
CLASS 2 POLE MIN. FOR 37.5 KVA AND 50 KVA TRANSFORMERS.
SOLDERLESS CONNECTIONS
N
A
B
C
LA GROUND CONDUCTOR IN MOLDING
SECONDARY RACK
EQUIPMENT GROUND CONDUCTOR IN MOLDING.
ELEVATION
SIDE ELEVATION
TYPICAL THREE PHASE TRANSFORMER BANK
CLUSTER MOUNT
3-50 KVA OR LESS

FIGURE 6-11

Three-Phase Transformer Bank Using 75- to 167-kVA Units As can be seen in Figure 6-12, the over 50-kVA transformer is simply too large and too heavy to mount on a single pole. Our detail ends when the transformer gets to 167 kVA because, at that size, it is getting too large and too heavy for any reasonable type of pole-mounting installation. Of course, if you are really determined and do not mind looking ridiculous, you could construct some kind of aboveground system that would support almost any size transformer.

While this detail looks much more complex than Figures 6-10 and 6-11, the complexity is primarily due to the dimensioning of the supporting frame and to the fact that we wanted to show secondary aerial feeders as well as a typical aerial secondary service conduit to some user.

Our primary transformers are now sitting on a wooden base some 12 ft 0 in above the ground. The base is supported between two poles with proper clearances for all items involved. There is nothing very unusual about the frame, and we believe the various notes make the pole (standard size) and the frame installation fairly obvious. These poles are buried 6 ft 0 in in the ground.

The primary lines have their own crosspiece and insulators to maintain their positions as well as to provide a location for the primary fused cutout. From these they proceed to the primary transformer bushings.

To show all we wanted, we installed a second cross arm with a support to carry 2-in bare copper feeder conductors on pin insulators. To these copper pipes we connect the secondary connections of the transformer. On the left-hand side of the side elevation, a typical building service conduit rises up the pole and terminates in a weatherhead connection. From the weatherhead our service lines are connected to our bare copper secondary pipe conductors.

On the right-hand side of our side elevation, secondary lines connect from bare copper pipes to an aerial secondary distribution system like the one in Figures 6-10 and 6-11. As before, keep primary and secondary phase relationships identical.

In note 3 in the detail there is a final bit of advice. Anything larger than 501 kVA from three transformers shall be mounted on a concrete pad on the ground. The general hardware associated with this installation is the same as it was for the 50-kVA detail.

Three-Phase Circuit Recloser A not very common part of electrical design work is the installation or application of a circuit recloser. It is analogous to the thermal overload device used to protect secondary equipment. As you know, the thermal overload device is usually arranged so that a problem in any or all of the phases will cause current to be disconnected from the device. The problem then has to be investigated and/or corrected. In a primary distribution system this manual reenergizing could take a lot of time and create endless complaints and misunderstandings.

The circuit recloser (three phase or single phase as required) is a self-contained device installed in series in the primary distribution line as shown in Figure 6-13. Pole line hardware is similar to that previously presented, and there are properly grounded lightning arresters.

The recloser has devices built into it which sense overcurrents, time and interrupt the overcurrent, and reclose automatically to reenergize the primary line. If the fault is sensed as "permanent," the recloser will lock in the open position after a preset number of attempts to reclose, normally three or four tries. This isolates the faulted section from the main part of the distribution system. What may surprise you is the fact that almost 90 percent of all primary system problems are temporary, lasting only a few seconds. The recloser solves these annoying blips in the system.

The three-phase recloser illustrated in Figure 6-13 is used where lockout of all three phases is required for any permanent fault to prevent single phasing of three-phase loads in, for example, large three-phase motor feeders. They usually have two methods of functioning: single-phase trip three-phase lockout and three-phase trip three-phase lockout.

Not shown but sometimes installed are single-phase reclosers used to protect branches of a three-phase feeder. They can also be used on three-phase circuits where loads are predominantly single phase. In such a case a failure on one phase would not eliminate current on the other two phases.

Single-Phase Pad-Mounted Transformer In Figure 6-14 a single-phase transformer of fairly large capacity is fed from an underground system. There are a number of things to be noted before describing this installation.

First, there is a concrete pad, usually 8 in thick, with some footing to maintain stability. This pad is at least 8 in larger than the tranaformer on all four sides. Second, we need a frame anchored to the pad about 8 ft 0 in high to support the primary fused cutout and lightning arrester. Third, we need a sturdy wire fence some 8 ft 0 in high with barbed wire at the top and a locked gate to control entrance to the transformer installation. Clearance between the fence and the pad is controlled by codes, rules, and ordinances, but the 4 ft 0 in is a minimum value for any installation. The fence, the frame. and the lightning arrester all must be grounded.

Actually our electrical work is no different from the single-phase pole-mounted transformer. To make the primary connection from an underground primary system, we put a lower level support cross member on our frame. We rise out of the underground system with primary cable in a rigid conduit, and this time we terminate in a pothead which we have previously described. From the pothead, in open wire, we proceed to our primary fuse cutout. Again, in open wire, we proceed to the primary transformer bushing.

From the secondary lines we go underground via a weatherproof junction box to a secondary fused switch or

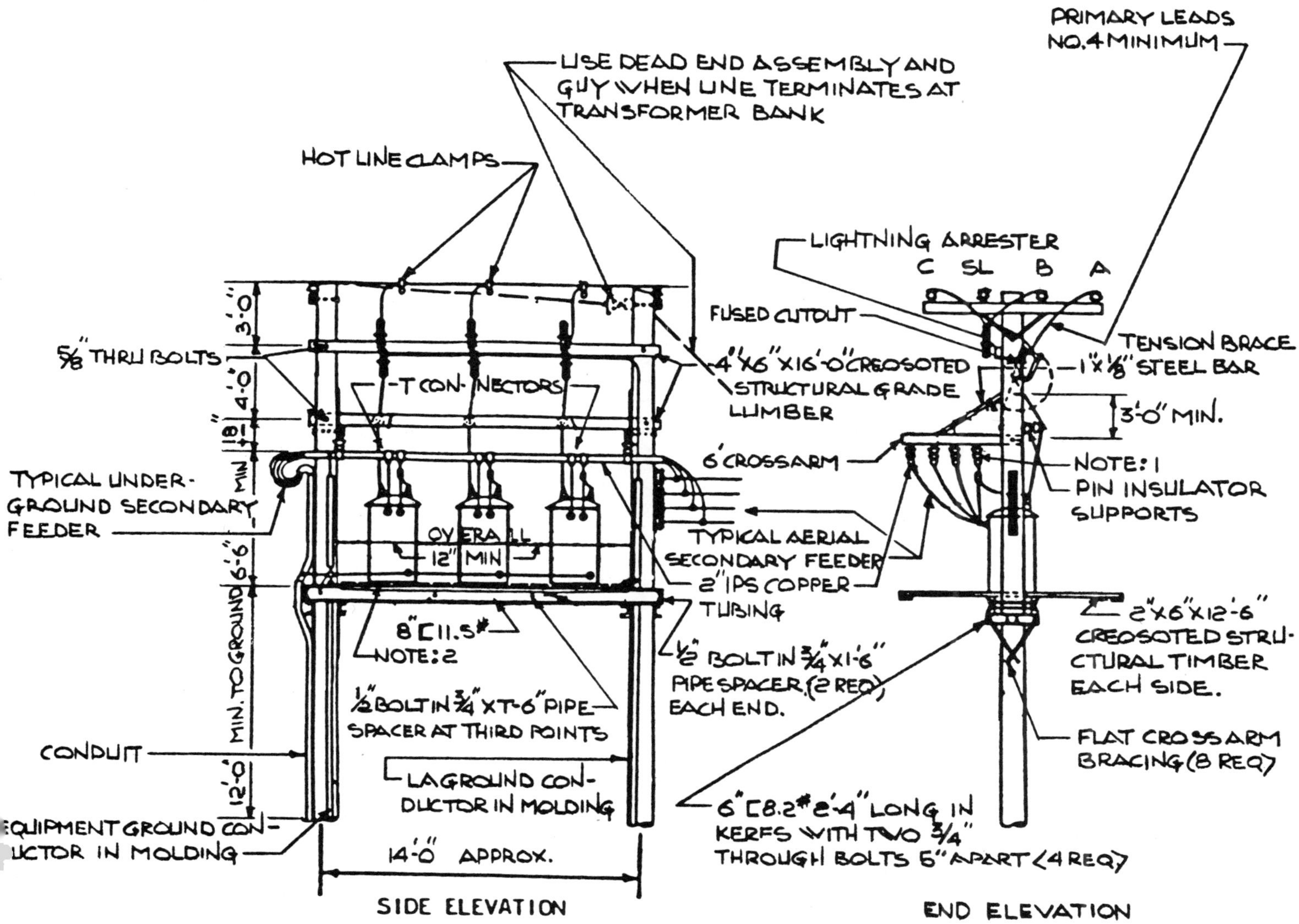

NOTES: 1. INSULATOR BUS SUPPORTS SHALL BE NEMA TYPE TR-1 APPARATUS CAP AND PIN INSULATOR WITH TOP CLAMP FOR COPPER TUBBING BUS. BOLT TO CROSSARM WITH TWO 1/2" THROUGH BOLTS.

2. SIXTEEN 2"X8"X8'-0" CREOSOTED STRUCTURAL GRADE TIMBERS ON 10"± CENTERS HELD WITH A 3/8" X 3 1/2" CARRIAGE BOLT IN EACH CHANNEL.

3. TRANSFORMERS OR TRANSFORMER BANK HAVING A CAPACITY IN EXCESS OF 501 KVA SHALL BE MOUNTED ON A FENCE ENCLOSED CONCRETE MAT. CONNECTIONS TO BE DELTA OR WYE AS REQUIRED. PRIMARY AND SECONDARY BUS NEUTRALS MAY BE COMBINED.

TYPICAL THREE PHASE TRANSFORMER BANK

3-75 KVA TO 3-167 KVA, INCL.

FIGURE 6-12

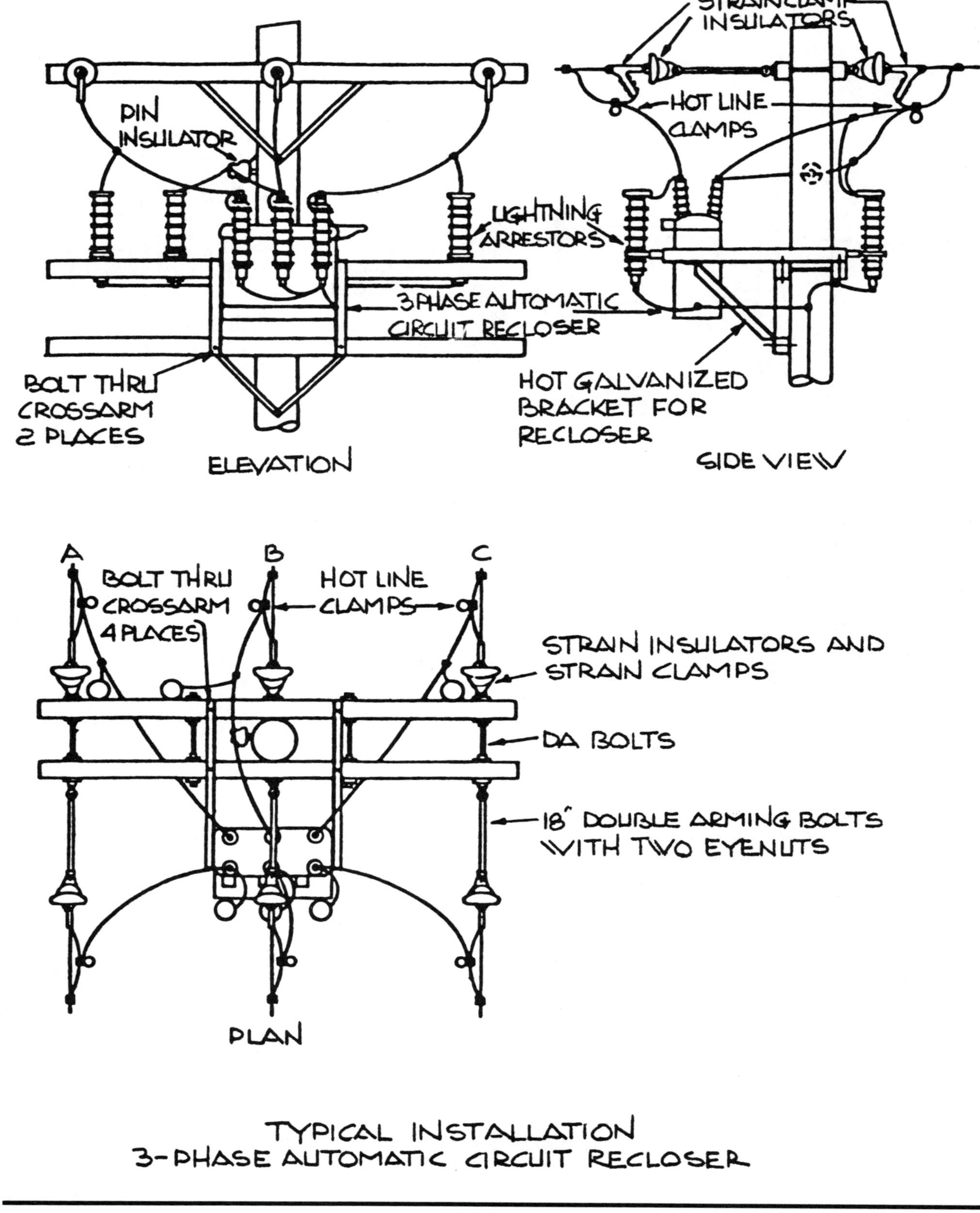
STRAIN CLAMP
INSULATORS
PIN
INSULATOR
HOT LINE
CLAMPS
LIGHTNING
ARRESTORS
3 PHASE AUTOMATIC
CIRCUIT RECLOSER
BOLT THRU
CROSSARM
2 PLACES
HOT GALVANIZED
BRACKET FOR
RECLOSER
ELEVATION
SIDE VIEW
A
B
C
BOLT THRU
CROSSARM
4 PLACES
HOT LINE
CLAMPS
STRAIN INSULATORS AND
STRAIN CLAMPS
DA BOLTS
18" DOUBLE ARMING BOLTS
WITH TWO EYENUTS
PLAN
TYPICAL INSTALLATION
3-PHASE AUTOMATIC CIRCUIT RECLOSER

FIGURE 6-13

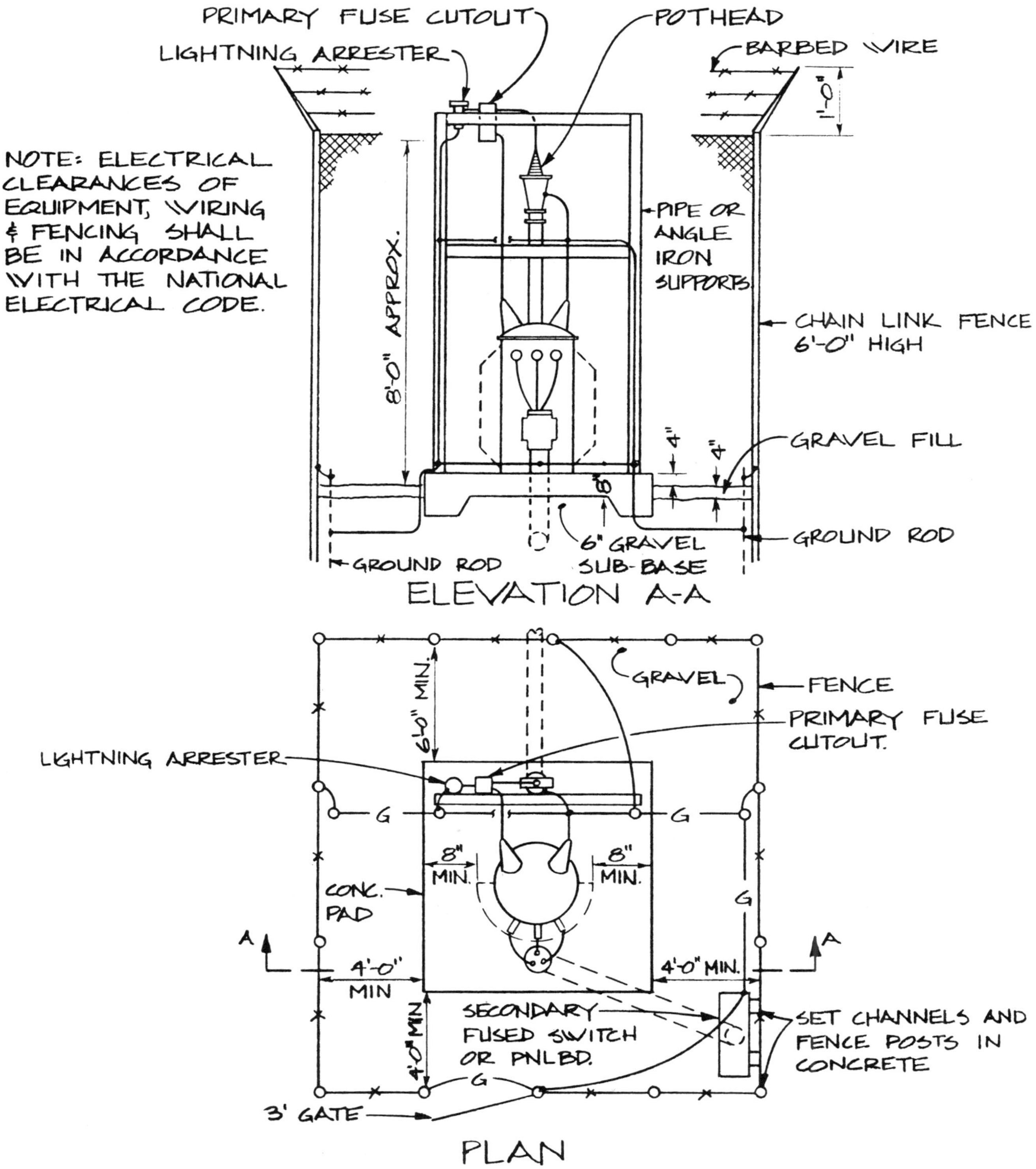

TYPICAL SINGLE PHASE OUTDOOR CONVENTIONAL TRANSFORMER INSTALLATION

NO SCALE

FIGURE 6-14

panel circuit breaker. From these we can go overhead or underground, as conditions warrant, to our building.

Note that the transformer has line and neutral connections, and that the transformer neutral connection is tied to the primary neutral connection at the pothead.

Three-Phase Pad-Mounted Transformer If you have been reading our outdoor transformer details consecutively, Figure 6-15, although it appears complex, is actually very clear and there is little to add to our previous discussion. For those of you who immediately went to this detail because it applied to your problem, we suggest you first read our general introduction to this set of details for all the standard hardware and then refer to the previous single-phase detail for the concrete base, fence, grounding, pothead, and primary fused cutouts.

Since there are three transformers instead of one, the concrete base is larger and we have noted the space between transformers. Our supporting frame is larger than that used for the single-phase installation, and our open wiring from the pothead is beginning to dangle. As on the pole installation, we installed fixed insulators and fastened our primary cables to them to control their position and relationship with each other. We have primary fused cutouts—one per phase. We also have lightning arresters—one per primary line.

On both the primary and secondary sides of the transformer the neutral connections are tied together and continued as one to their primary or secondary source.

As in the single-phase detail, we come up from the underground primary, terminate in a pothead, and go to the primary disconnects. On the secondary side we come together in a junction box, extend our cables underground in conduit to a switch or panelboard, and then can go overhead or underground to our building. If you carefully read all our descriptions before getting to this detail, including all the pole-mounted description, you will find this is a relatively simple system that just looks complex.

Three-Phase Indoor Transformer Installation In Figures 6-10 to 6-15 we showed typical outdoor installations. What we now illustrate in Figure 6-16 is the same three-phase outdoor installation moved into a vault in the building. If your work begins at the secondary side of the utility-supplied transformers, this detail has little or no value. The premise here, as before, is that your client has primary metering and all transformation is your client's—and your—problem.

What is handy in this system is that we have a room with walls, ceiling, and floor to attach our equipment and set our transformers. The connection details are the same as they were outdoors. We still come from our primary underground line to our primary fuse cutouts and then proceed to our transformers, and we still have our lightning arresters. Now that we have a wall to work with, the various pieces of hardware can be neatly mounted on the wall. We have added one more dimension showing a clearance from the transformers to the wall. The grounding is similar to what it was outdoors.

Electrically there is one difference. Vaults are generally kept locked with keys furnished only to authorized personnel. This is as it should be. To work on the secondary distribution system, all that needs to be done is to shut off the power. This can be accomplished as shown, by a secondary main switch in a fused or circuit-breaker panelboard located outside the vault.

The vault itself requires some comment. The detail has a note that the vault is to be constructed as per the NEC. The Code, as well as all other local and state codes and ordinances, should be carefully reviewed and all recommendations followed. Generally the vaults are constructed of poured concrete with fireproof doors.

Rules vary somewhat on the relationship of vault doors and curbs. Our detail shows a 4-in curb to contain spills of transformer liquid and a drain which normally goes to a special tank with a pump-out connection so that the spilled liquid can be recovered and properly disposed of without contamination to any drainage or soil system. Some utility companies would recommend that no drain be installed and the liquid held within the vault for removal by emergency suction spill-removal equipment.

As described in the Code, ventilation is obtained by natural circulation and, as we show, ventilation louvers are required both high and low on the sidewall of the vault. The high louvers may be on the roof of the vault rather than on the sidewall. Section 450 of the Code is very specific and your attention is directed to that section.

Finally, we show the transformers mounted on a 3-in-high pad. We think this is both a safety feature and a good working arrangement. It is not mandatory.

Specialized Exterior Transformers In our details to this point we have covered the usual indoor primary and/or secondary substation and the common exterior primary installations. On occasion you can become involved in large site-lighting installations such as college campuses or a very large industrial park. With the requirement of inconspicuous equipment, underground lines, and long runs, higher-voltage distribution is usually selected for site-lighting substations.

Needless to say, the equipment is going to be totally enclosed. Your detailing would consist of an electrical schematic imposed on the enclosures as we show in Figure 6-17. This detail shows a series of enclosures fed from underground on both the primary and secondary side. We indicate a metal enclosure called a *throat enclosure* on the primary side of the transformer between it and the primary disconnect. Do not forget to cover it in your equipment specification along with all the other devices. We also show the transformer as one section attached to the secondary distribution. This may or may not be true in your case. Be careful to check whether or not you need a throat piece on the secondary side of the transformer enclosure you select.

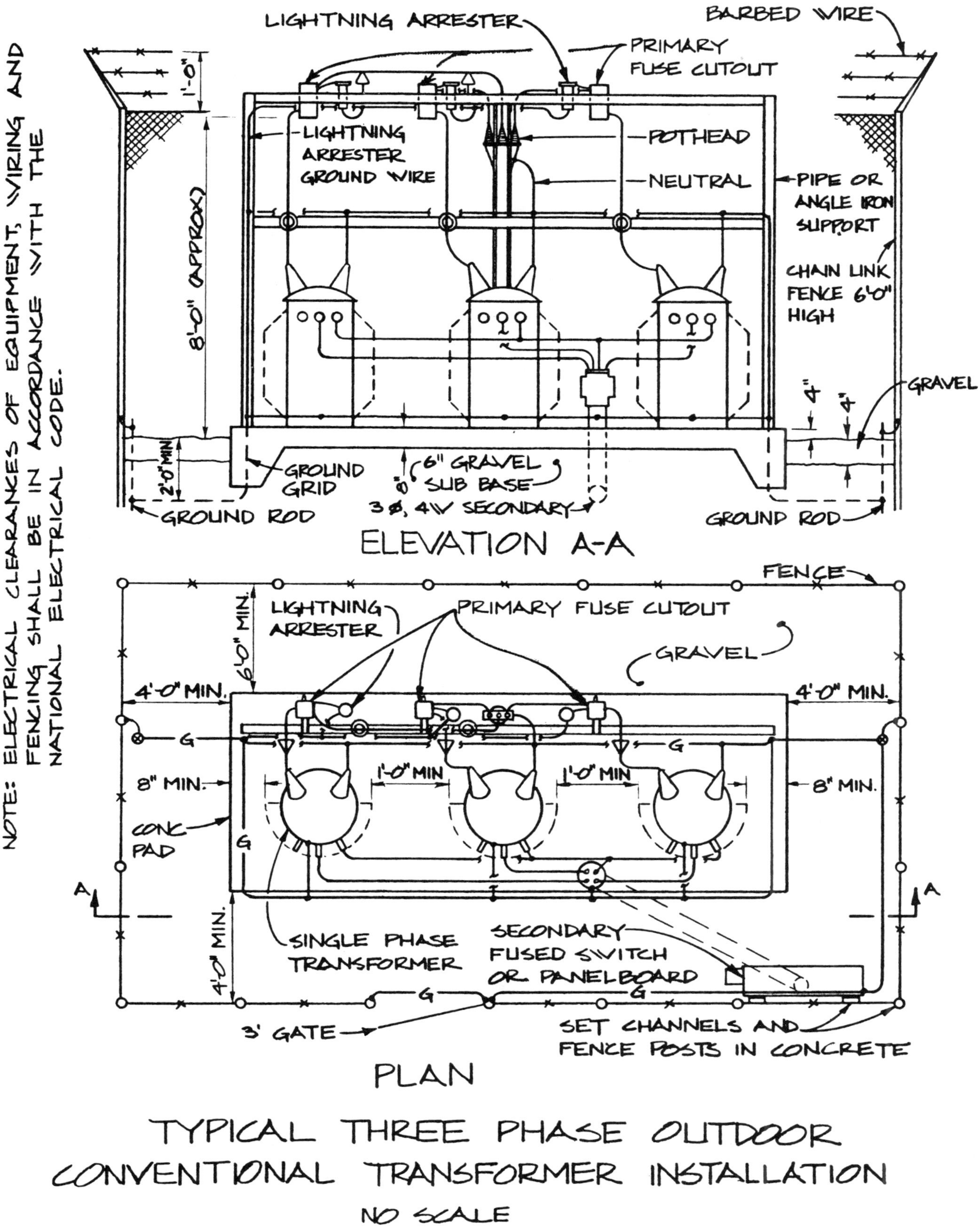
NOTE: ELECTRICAL CLEARANCES OF EQUIPMENT, WIRING AND FENCING SHALL BE IN ACCORDANCE WITH THE NATIONAL ELECTRICAL CODE.
LIGHTNING ARRESTER
BARBED WIRE
PRIMARY FUSE CUTOUT
1'-0"
LIGHTNING ARRESTER GROUND WIRE
POTHEAD
NEUTRAL
PIPE OR ANGLE IRON SUPPORT
8'-0" (APPROX)
CHAIN LINK FENCE 6'-0" HIGH
GRAVEL
4"
4"
2'-0" MIN
GROUND GRID
8"
6" GRAVEL SUB BASE
3 Ø, 4W SECONDARY
GROUND ROD
GROUND ROD
ELEVATION A-A
FENCE
6'-0" MIN.
LIGHTNING ARRESTER
PRIMARY FUSE CUTOUT
GRAVEL
4'-0" MIN.
4'-0" MIN.
G
G
8" MIN.
1'-0" MIN
1'-0" MIN
8" MIN.
CONC PAD
G
A
A
4'-0" MIN.
SINGLE PHASE TRANSFORMER
SECONDARY FUSED SWITCH OR PANELBOARD
G
G
3' GATE
SET CHANNELS AND FENCE POSTS IN CONCRETE
PLAN
TYPICAL THREE PHASE OUTDOOR CONVENTIONAL TRANSFORMER INSTALLATION
NO SCALE

FIGURE 6-15

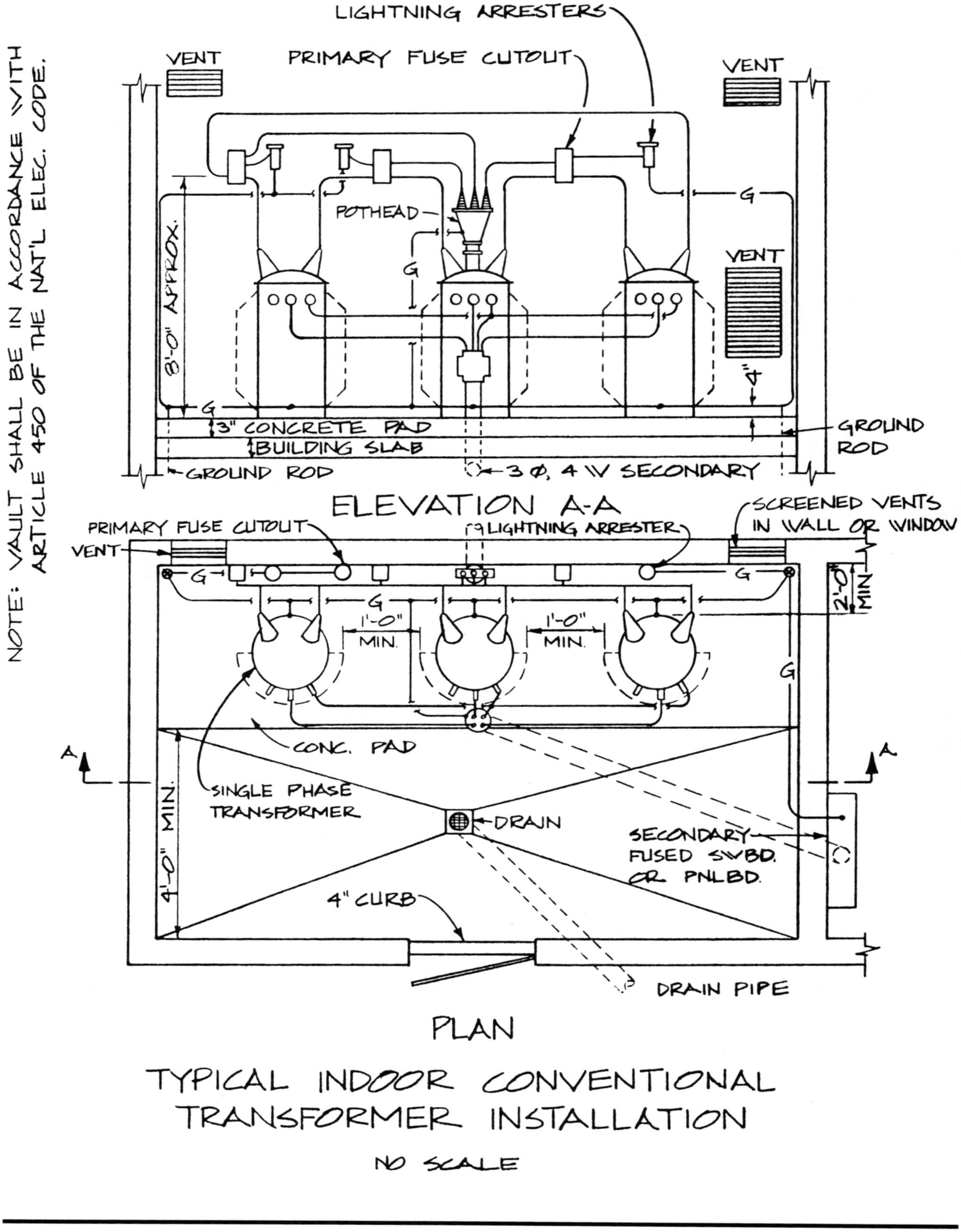

LIGHTNING ARRESTERS
VENT
PRIMARY FUSE CUTOUT
VENT
POTHEAD
G
8'-0" APPROX.
VENT
4"
3" CONCRETE PAD
BUILDING SLAB
GROUND ROD
GROUND ROD
3 Φ, 4 W SECONDARY
ELEVATION A-A
NOTE: VAULT SHALL BE IN ACCORDANCE WITH ARTICLE 450 OF THE NAT'L ELEC. CODE.
PRIMARY FUSE CUTOUT
LIGHTNING ARRESTER
SCREENED VENTS IN WALL OR WINDOW
VENT
1'-0" MIN.
1'-0" MIN.
2'-0" MIN
CONC. PAD
A
A
SINGLE PHASE TRANSFORMER
4'-0" MIN.
DRAIN
SECONDARY FUSED SWBD. OR PNLBD.
4" CURB
DRAIN PIPE
PLAN
TYPICAL INDOOR CONVENTIONAL TRANSFORMER INSTALLATION
NO SCALE

FIGURE 6-16

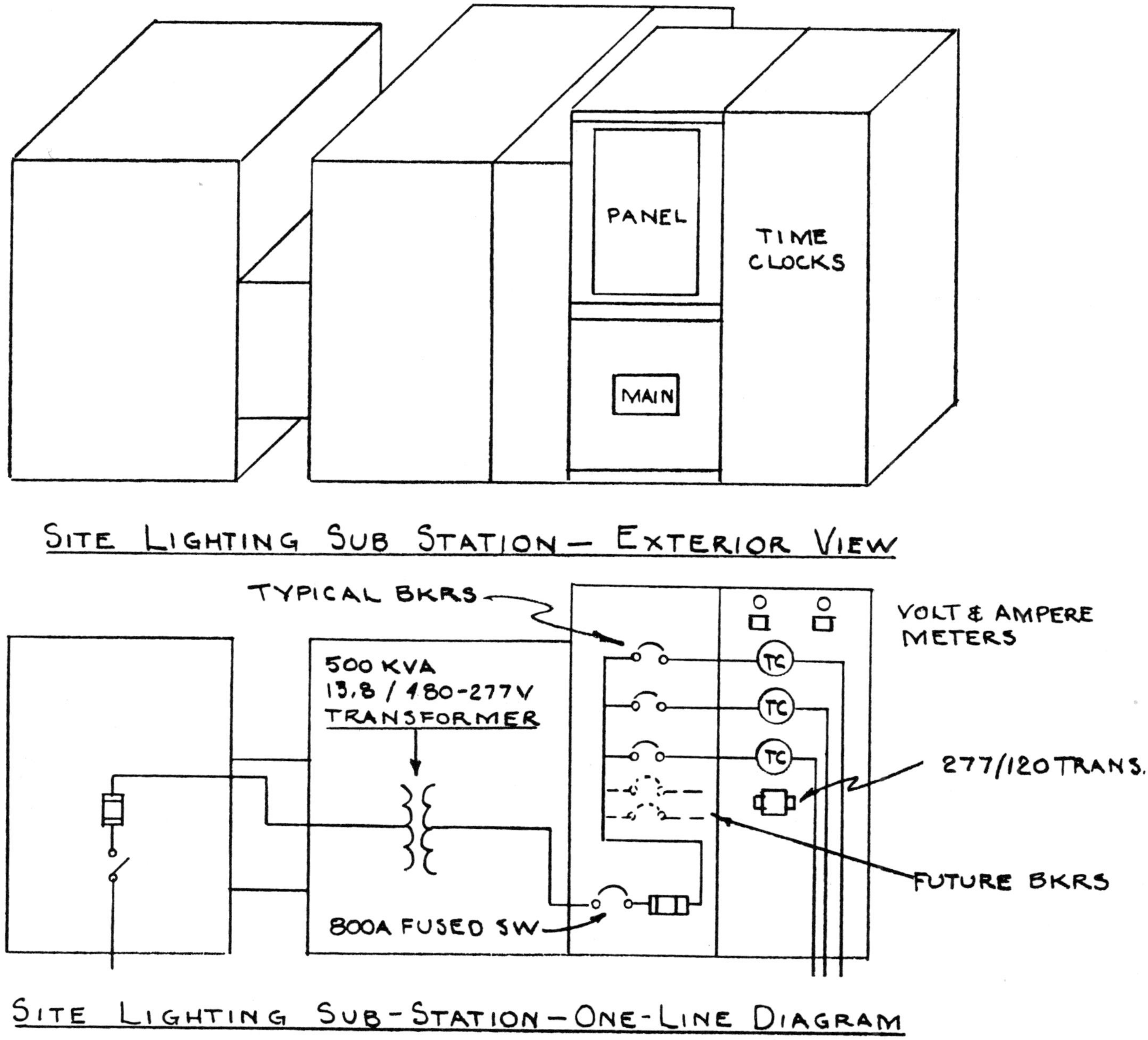

FIGURE 6-17

Figure 6-18 shows the concrete pad on which the site-lighting transformer is mounted. We have shown some details such as primary openings, secondary openings, primary and secondary conduits, and a secondary junction box. These were taken from an actual job where this detail was used. We show a lot of pad and equipment dimensions which may not apply since we do not know the size of the equipment you may select. When you complete your selection be sure to dimension your pad about 2 ft 0 in larger than your equipment on all four sides.

As in every other transformer installation, grounding of the primary and secondary systems is a must. We have a 500-kVA installation and a grounding system for a particular set of site conditions. This will have to be altered to suit your particular project.

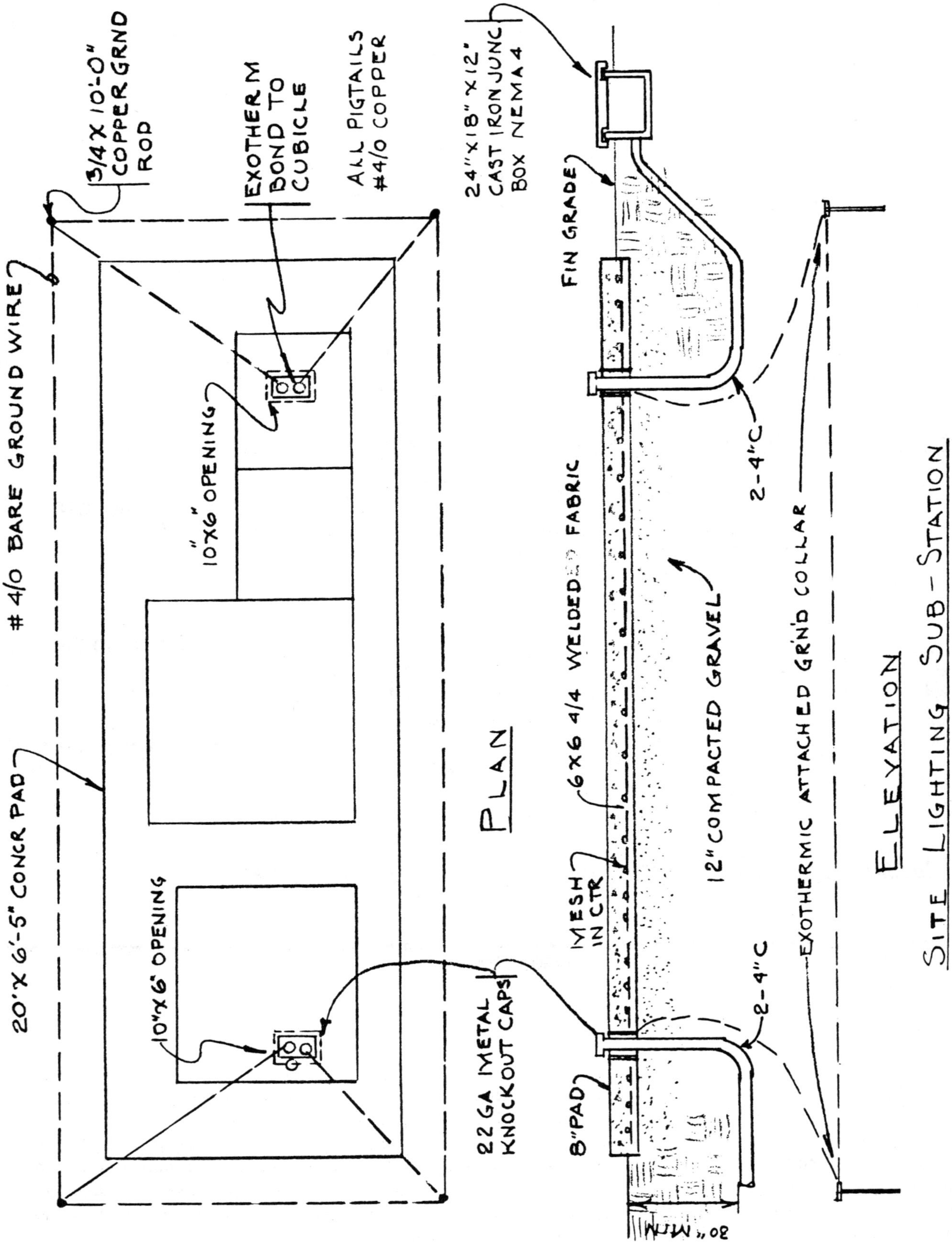

20'X 6'-5" CONCR PAD
4/0 BARE GROUND WIRE
3/4 X 10'-0" COPPER GRND ROD
EXOTHERM BOND TO CUBICLE
ALL PIGTAILS #4/0 COPPER
10"X6' OPENING
" " 10X6 OPENING
PLAN
22 GA METAL KNOCKOUT CAPS
8"PAD
MESH IN CTR
6X6 4/4 WELDED FABRIC
12" COMPACTED GRAVEL
FIN GRADE
24"X18"X12" CAST IRON JUNC BOX NEMA 4
2-4"C
2-4"C
EXOTHERMIC ATTACHED GR'ND COLLAR
30" MIN
ELEVATION
SITE LIGHTING SUB-STATION

FIGURE 6-18

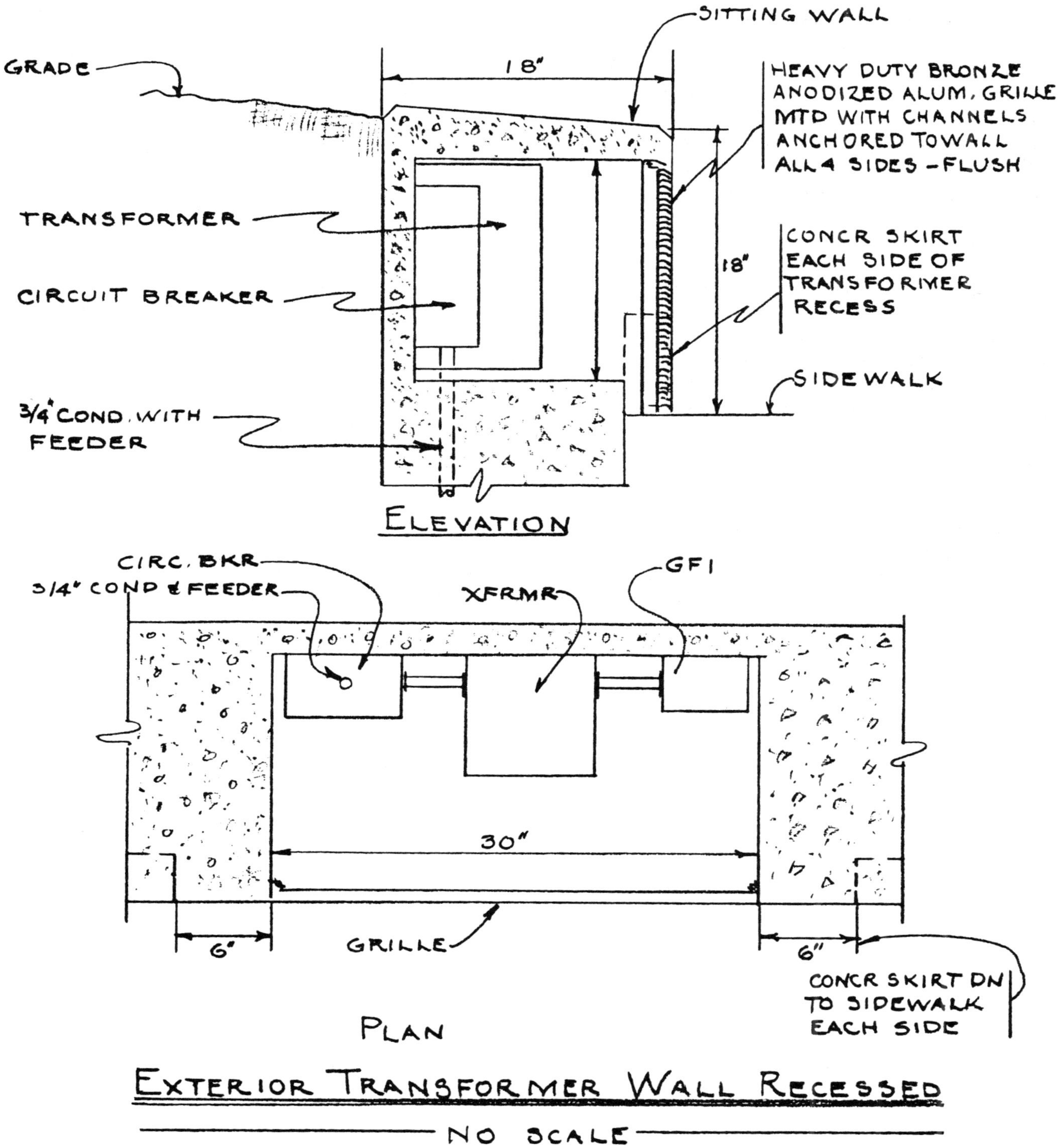

FIGURE 6-19

In Figure 6-19 we show the installation of a little transformer buried in a wall seat. We managed to hide this small transformer in a space 2 ft 6 in long by less than 1 ft 6 in high. In the elevation on Figure 6-19 you can see that the seat is 18 in high and we have failed to set the transformer enclosure interior height. This is because we were still arguing with the landscape architect who wanted the seat 18 in and the enclosure 14 in, while we were holding out for 18 in. What we will tell you that we would not tell him was that we could have made it in 14 in. But it was unbelievably tight for the transformer, and we wanted a little more clearance.

Transformer Pad Details In our details of outdoor, pad-mounted transformers we have not stressed the pad construction itself because we were showing individual transformers in more of an industrial, economical-type arrangement which better illustrate electrical connection features. Since most modern transformers are provided with complete enclosures by the manufacturer, one frequently forgets the requirements of all the parts and connections. However, the modern transformer installation does require pad detailing, and frequently landscaping, to make the installation aesthetically acceptable.

Figures 6-20 to 6-23 show four transformer pad details. In Figures 6-21 and 6-22 there are two typical pads for two typical outdoor enclosed transformers, fed from underground on both the primary and secondary side.

Figure 6-20 shows a bumper detail to be used when the transformer installation can possibly be damaged by vehicular traffic: The objective of the bumpers around the transformer pad shown in Figure 6-20 is to really stop traffic—at the bumper not at the transformer. Our detail gives you a choice of bumper material and a warning that even stiffer bumpers may be required. Presumably a car could not slip between the 66-in clearance between bumpers. There are engineers who insist on 54-in clearance just to be sure. If there is any doubt about soil density you might add a requirement to encase the buried portion of the bumper with 12 in of concrete.

The dimensions shown in Figures 6-21 and 6-22 for primary and secondary conduit openings are not necessarily going to fit your particular transformer selection, and you will have to adjust them. Conduit openings can also vary the size of the pad to some degree. We do not really dimension the thickness of the pad since that can vary a little, but we imply it is 6-in-thick concrete, which is a reasonable average.

The grounding is normally 6 in inside the outer edge of the pad with ground rods in opposite corners, as shown in Figures 6-21 and 6-22. We also call your attention to the oil sump trench which gets both wider and deeper for a larger installation. Finally, the large installation and sometimes the smaller installation require anchor bolts, as may be noted.

While we do not profess to be landscape architects or site-planning consultants, we have stuck our neck out to show in Figure 6-23 a possible shrubbery plan to conceal the transformer. In many industrial parks the transformer, for any number of reasons, is not located next to the building but perhaps some 10 to 20 ft away and on a lovely rolling, grassy lawn. Obviously something must be done to conceal it or it will stick out as the sorest of sore thumbs.

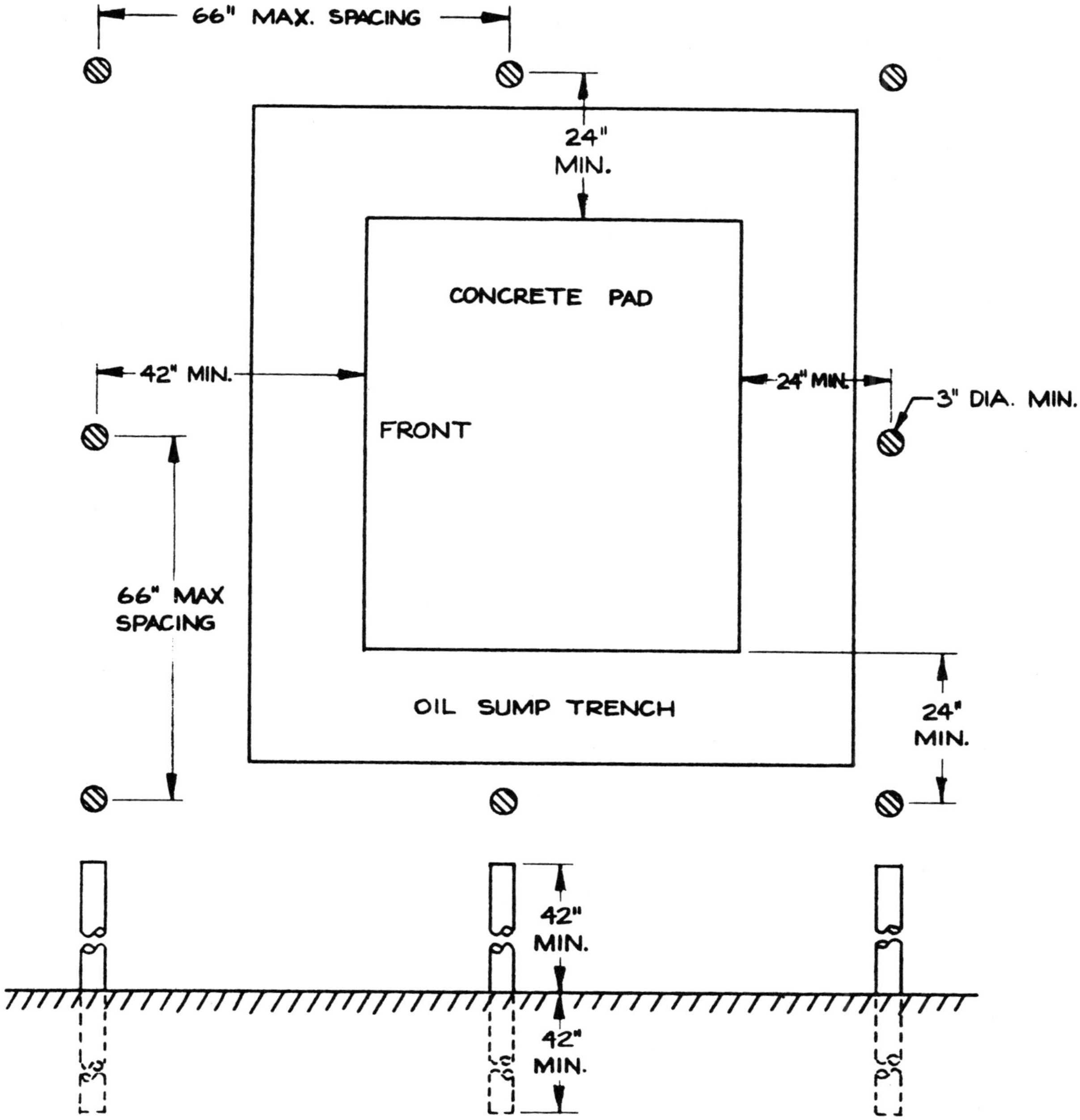

NOTES:

1. PROTECTIVE BUMPERS ARE TO BE INSTALLED ON SIDES EXPOSED TO VEHICULAR TRAFFIC, ONLY.
2. STEEL PIPES 3" MINIMUM FILLED WITH CONCRETE, I-BEAMS 5" MINIMUM OR OTHER SUITABLE MEANS OF PROTECTION MAY BE USED. HEAVIER BUMPERS SET DEEPER SHOULD BE CONSIDERED WHERE EXPOSED TO HEAVY TRUCKS.

PAD MOUNTED TRANSFORMER OR TRANSCLOSURE
PROTECTIVE BUMPER INSTALLATION
NO SCALE

FIGURE 6-20

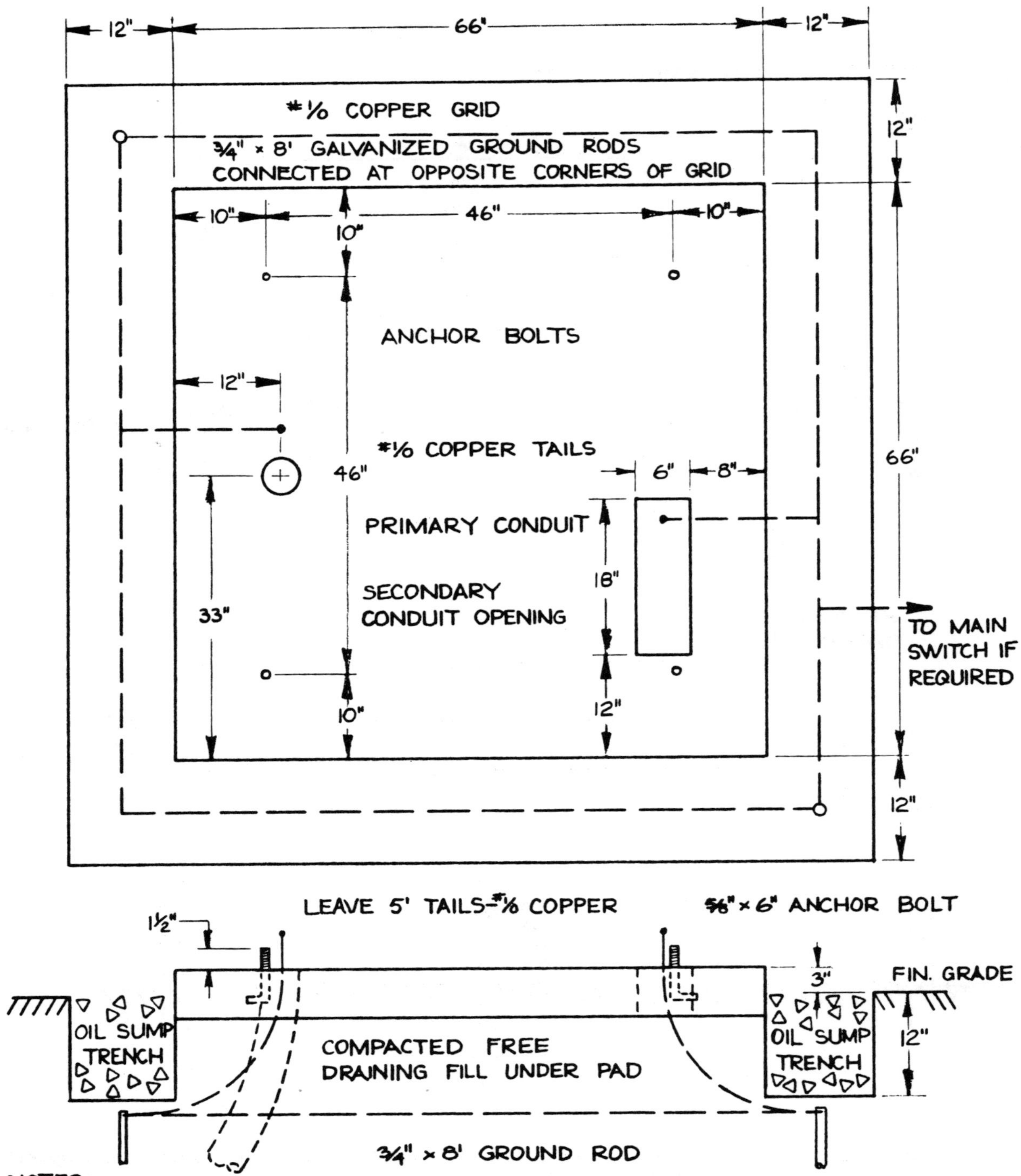

NOTES:

1. CONCRETE PAD TO BE 36" MINIMUM FROM WALL, FENCE OR OTHER OBSTRUCTION.
2. EXACT LOCATION OF ANCHOR BOLTS AND PRIMARY & SECONDARY CONDUIT OPENINGS TO BE DETERMINED FROM FINAL APPROVED SHOP DRAWINGS

<u>TRANSCLOSURE PAD INSTALLATION</u>

<u>4.16 KV & 4.8 KV 3φ 150 KVA MAX.</u>

NO SCALE

FIGURE 6-21

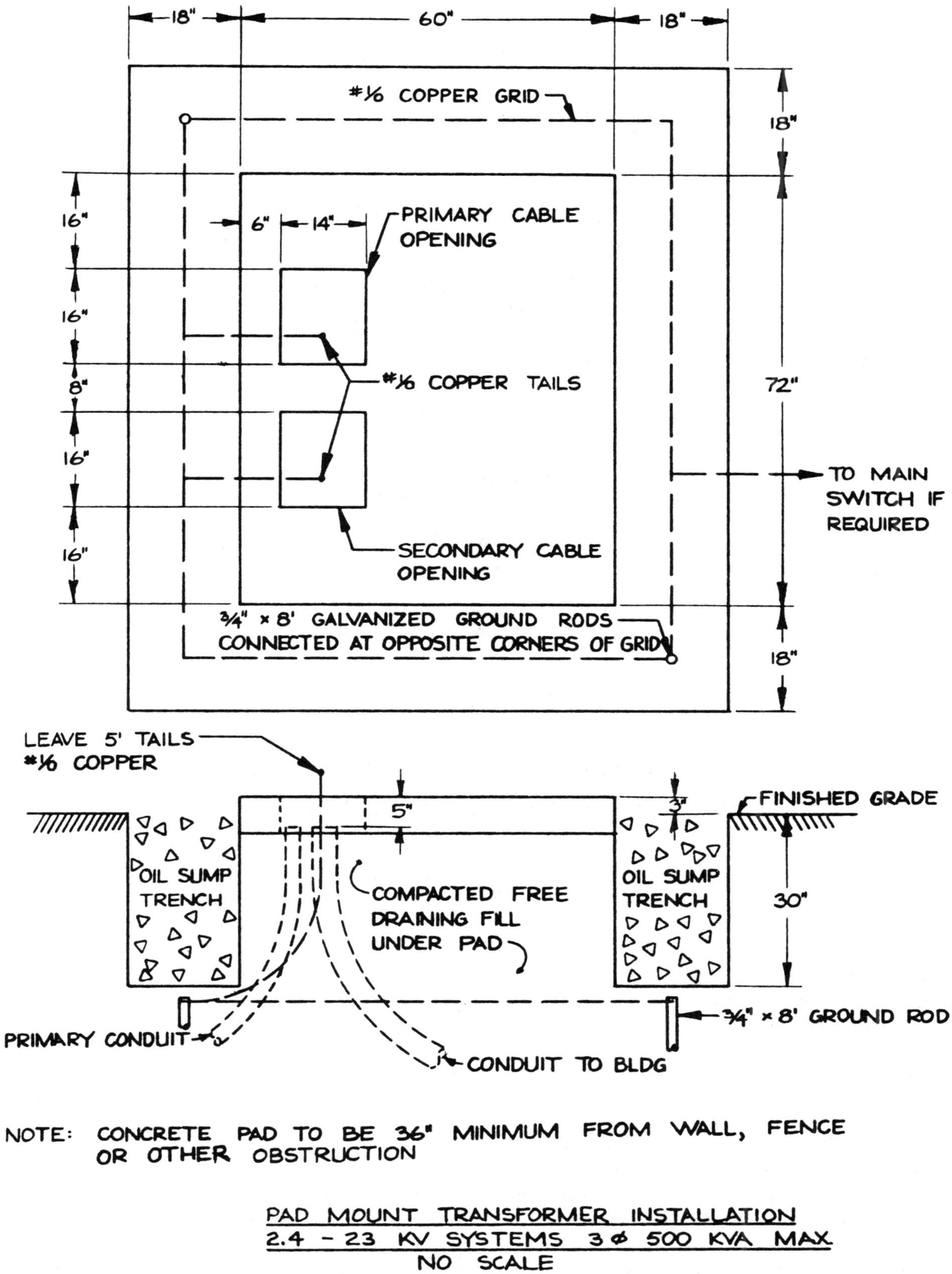
18"
60"
18"
#1/0 COPPER GRID
18"
16"
6"
14"
PRIMARY CABLE OPENING
16"
#1/0 COPPER TAILS
8"
72"
16"
TO MAIN SWITCH IF REQUIRED
16"
SECONDARY CABLE OPENING
3/4" × 8' GALVANIZED GROUND RODS CONNECTED AT OPPOSITE CORNERS OF GRID
18"
LEAVE 5' TAILS #1/0 COPPER
5"
3"
FINISHED GRADE
OIL SUMP TRENCH
COMPACTED FREE DRAINING FILL UNDER PAD
OIL SUMP TRENCH
30"
3/4" × 8' GROUND ROD
PRIMARY CONDUIT
CONDUIT TO BLDG
NOTE: CONCRETE PAD TO BE 36" MINIMUM FROM WALL, FENCE OR OTHER OBSTRUCTION
PAD MOUNT TRANSFORMER INSTALLATION
2.4 - 23 KV SYSTEMS 3 ϕ 500 KVA MAX
NO SCALE

FIGURE 6-22

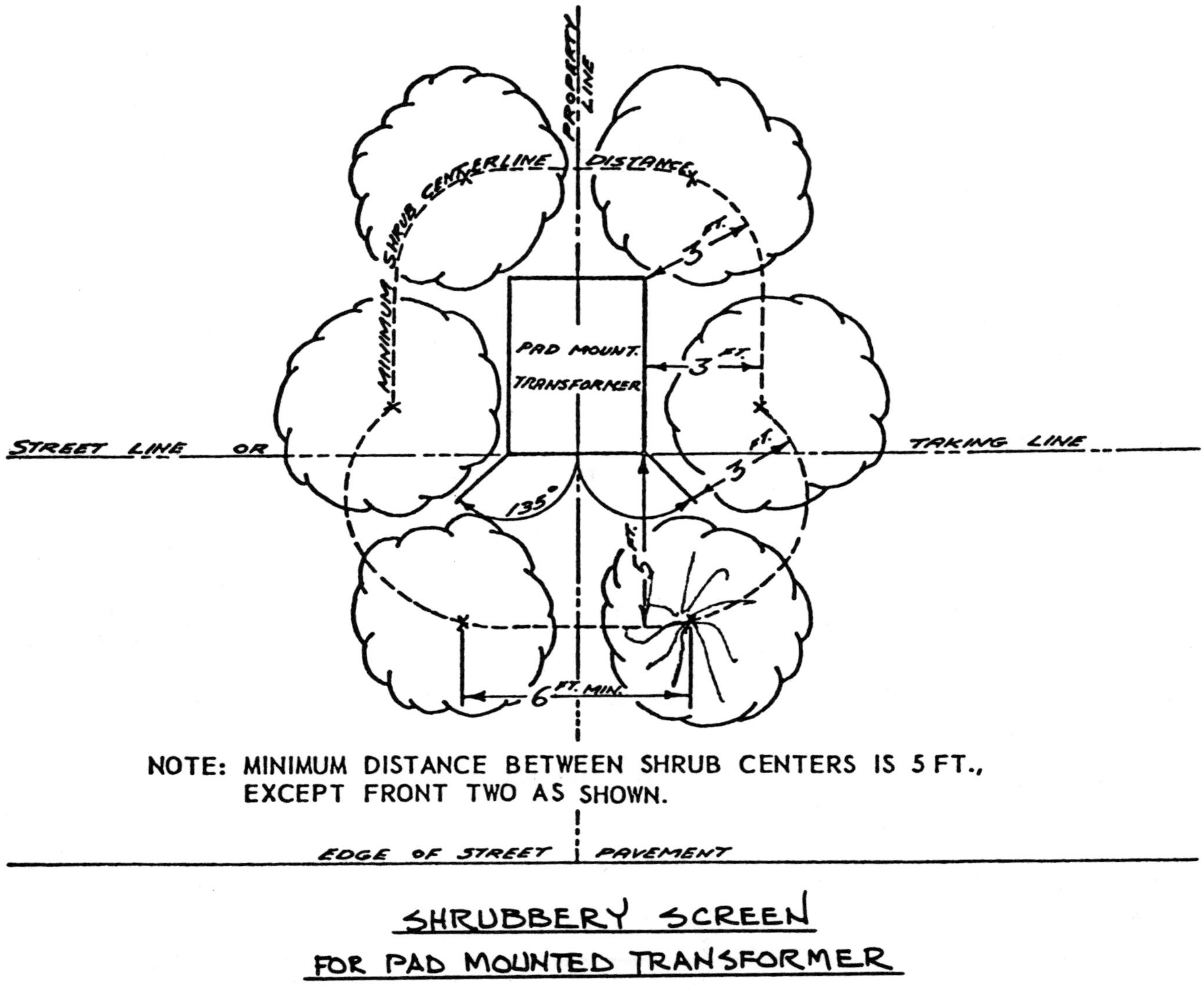

FIGURE 6-23

Very High Voltage Distribution In this chapter we have previously mentioned the expertise of those firms specializing in high-voltage distribution. Generally this is a special area of practice of firms who do work directly for utility companies and are familiar with requirements at voltages 69 kV and higher. On rare occasions this may become a part of your work.

Figures 6-24 to 6-29 are by no means the total collection of the elements of this design. As you will quickly note, there is no description of the structures that support the electrical equipment. This omission is intentional. The design of the supporting structures does not come out of a catalog, handbook, or from this book. The serious investigation and design of structure and foundation must originate with a competent, experienced, qualified, and licensed structural engineer.

For your guidance we have attempted to give you an overview of the design and its related elements. Our first detail, Figure 6-24, is a plan view of a typical multicircuit structure. Depending on the problem, the incoming primary lines can come from one, two, or three directions. At the top of the detail is an associated schematic of the typical horn-gap switch and a circuit breaker or voltage regulator, which is one of the first items to be considered.

In Figure 6-25 there are two sectional views of some of the equipment which may be involved. We have shown two more or less typical arrangements of incoming lines and how the circuiting progresses. Again these are not complete details but merely major elements in the design. The structure is described as a typical single-circuit primary or secondary structure, and that is all it is—typical. So, too, are the related schematic details.

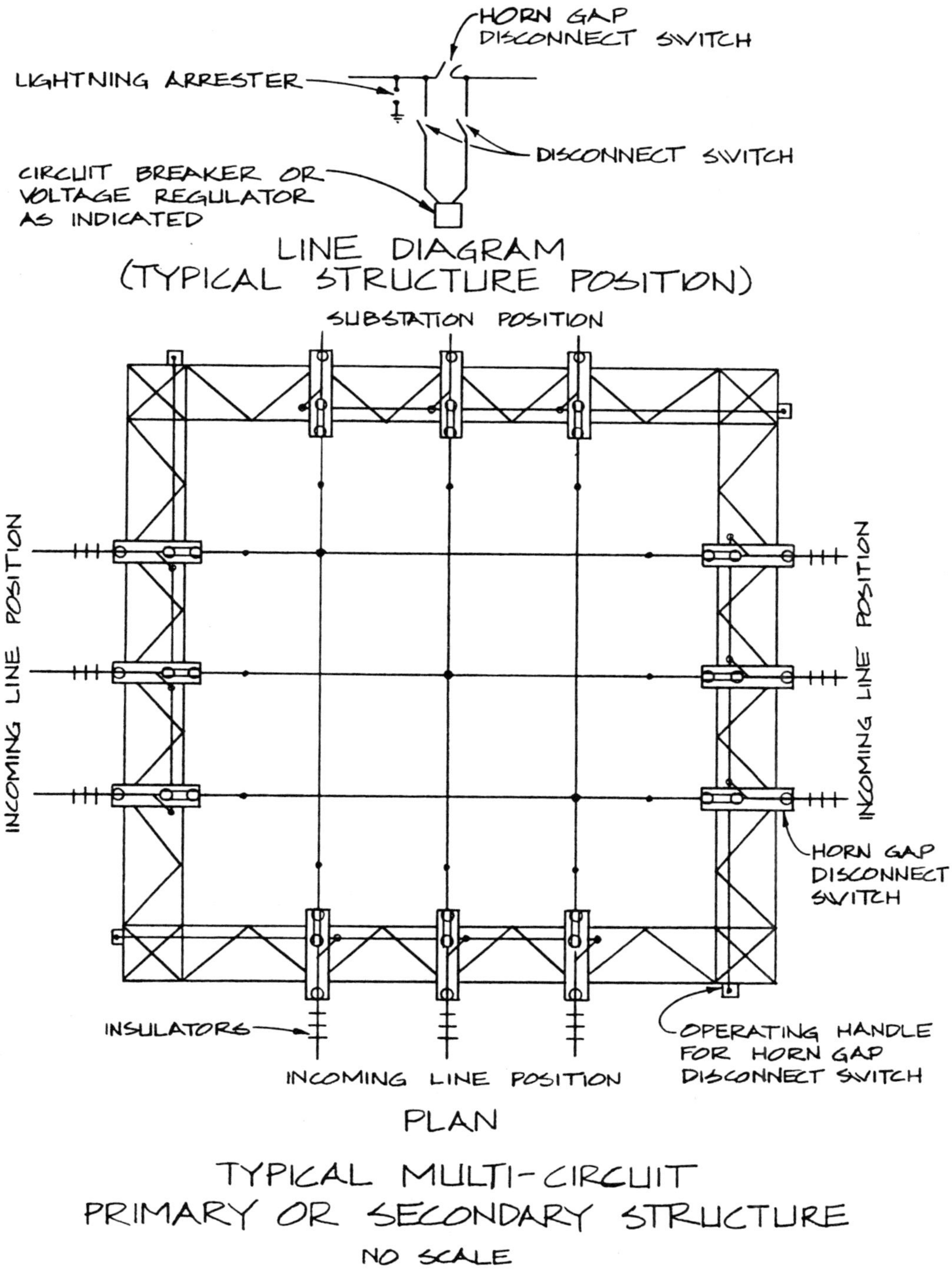
HORN GAP
DISCONNECT SWITCH
LIGHTNING ARRESTER
DISCONNECT SWITCH
CIRCUIT BREAKER OR
VOLTAGE REGULATOR
AS INDICATED
LINE DIAGRAM
(TYPICAL STRUCTURE POSITION)
SUBSTATION POSITION
INCOMING LINE POSITION
INCOMING LINE POSITION
HORN GAP
DISCONNECT
SWITCH
INSULATORS
INCOMING LINE POSITION
OPERATING HANDLE
FOR HORN GAP
DISCONNECT SWITCH
PLAN
TYPICAL MULTI-CIRCUIT
PRIMARY OR SECONDARY STRUCTURE
NO SCALE

FIGURE 6-24

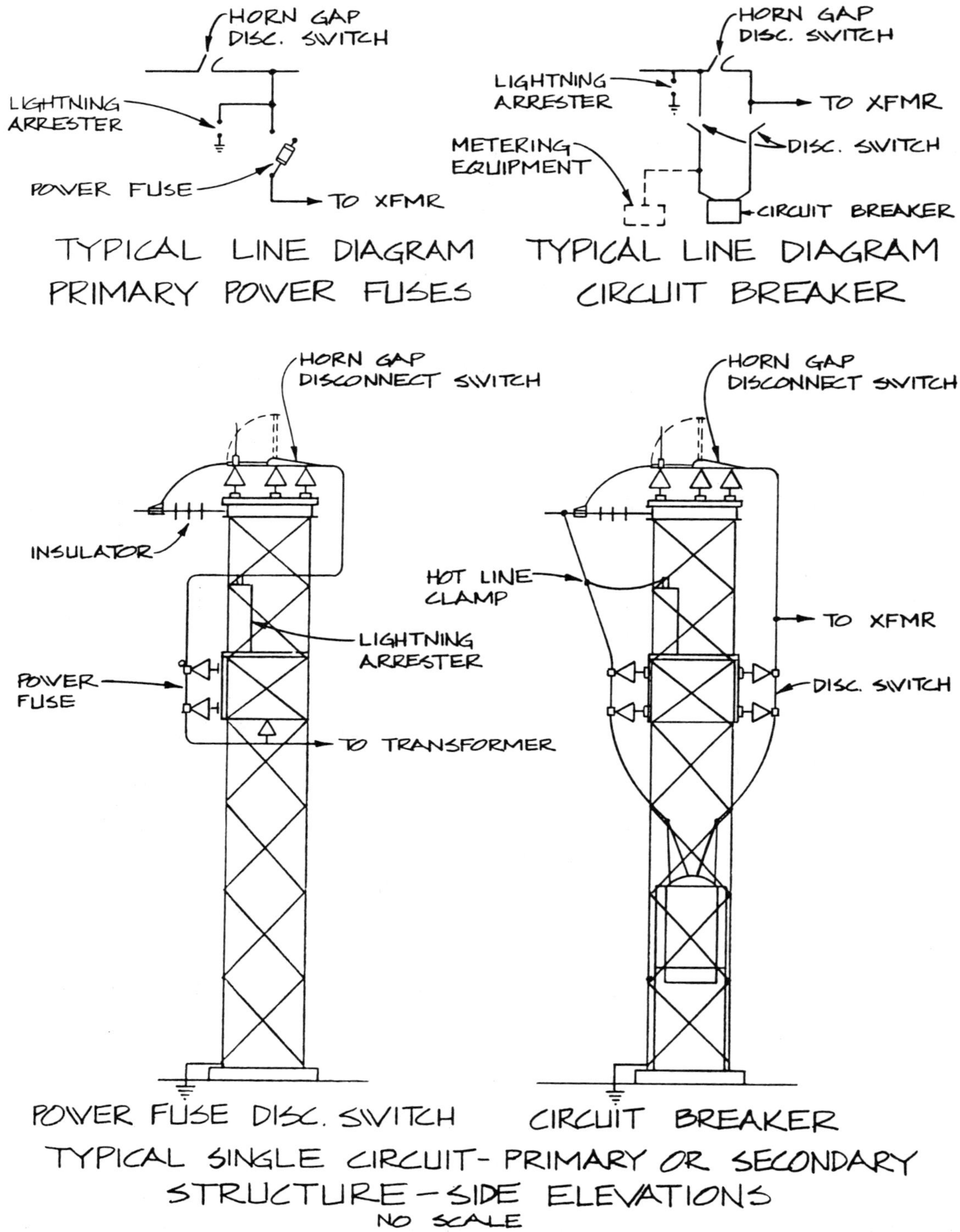
HORN GAP
DISC. SWITCH
LIGHTNING
ARRESTER
POWER FUSE
TO XFMR
TYPICAL LINE DIAGRAM
PRIMARY POWER FUSES
HORN GAP
DISC. SWITCH
LIGHTNING
ARRESTER
TO XFMR
METERING
EQUIPMENT
DISC. SWITCH
CIRCUIT BREAKER
TYPICAL LINE DIAGRAM
CIRCUIT BREAKER
HORN GAP
DISCONNECT SWITCH
INSULATOR
LIGHTNING
ARRESTER
POWER
FUSE
TO TRANSFORMER
HORN GAP
DISCONNECT SWITCH
HOT LINE
CLAMP
TO XFMR
DISC. SWITCH
POWER FUSE DISC. SWITCH
CIRCUIT BREAKER
TYPICAL SINGLE CIRCUIT - PRIMARY OR SECONDARY
STRUCTURE - SIDE ELEVATIONS
NO SCALE

FIGURE 6-25

In Figure 6-26 we show what could be one side of the plan view from the first detail. Even though voltages and equipment are much larger, some elements are very similar to previously presented pad-mounted outdoor transformer details. We have a support carrying a bus on insulators. A switch is a switch, even if it is a horn-gap switch. These usually are separated from a locked mechanism at ground level as we have indicated. Big or little, a lightning arrester is a lightning arrester, and we still connect primary lines with hot-line clamps.

Figure 6-27 is a different view and continuation of Figure 6-26. It emphasizes the primary disconnects and the

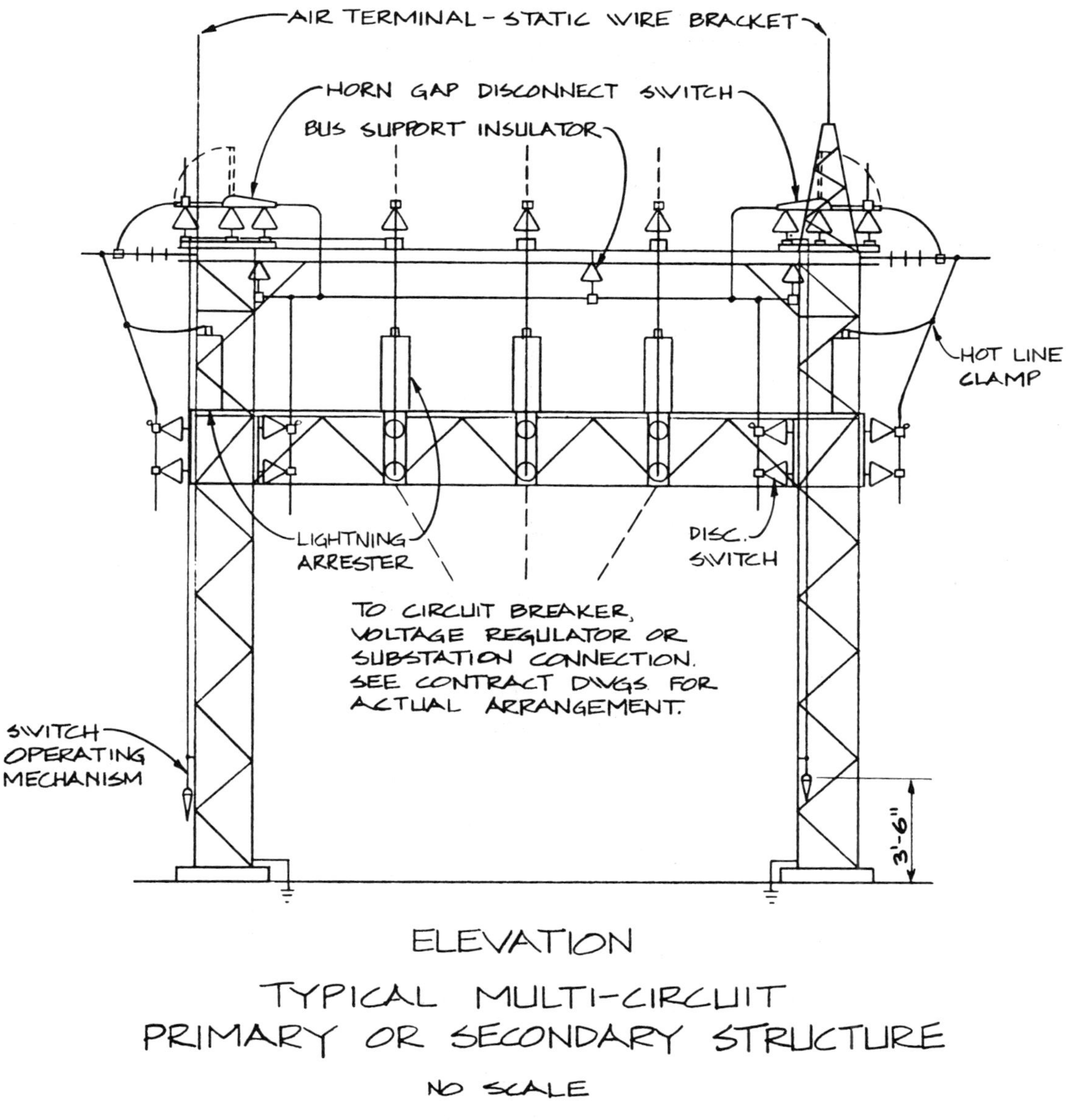

FIGURE 6-26

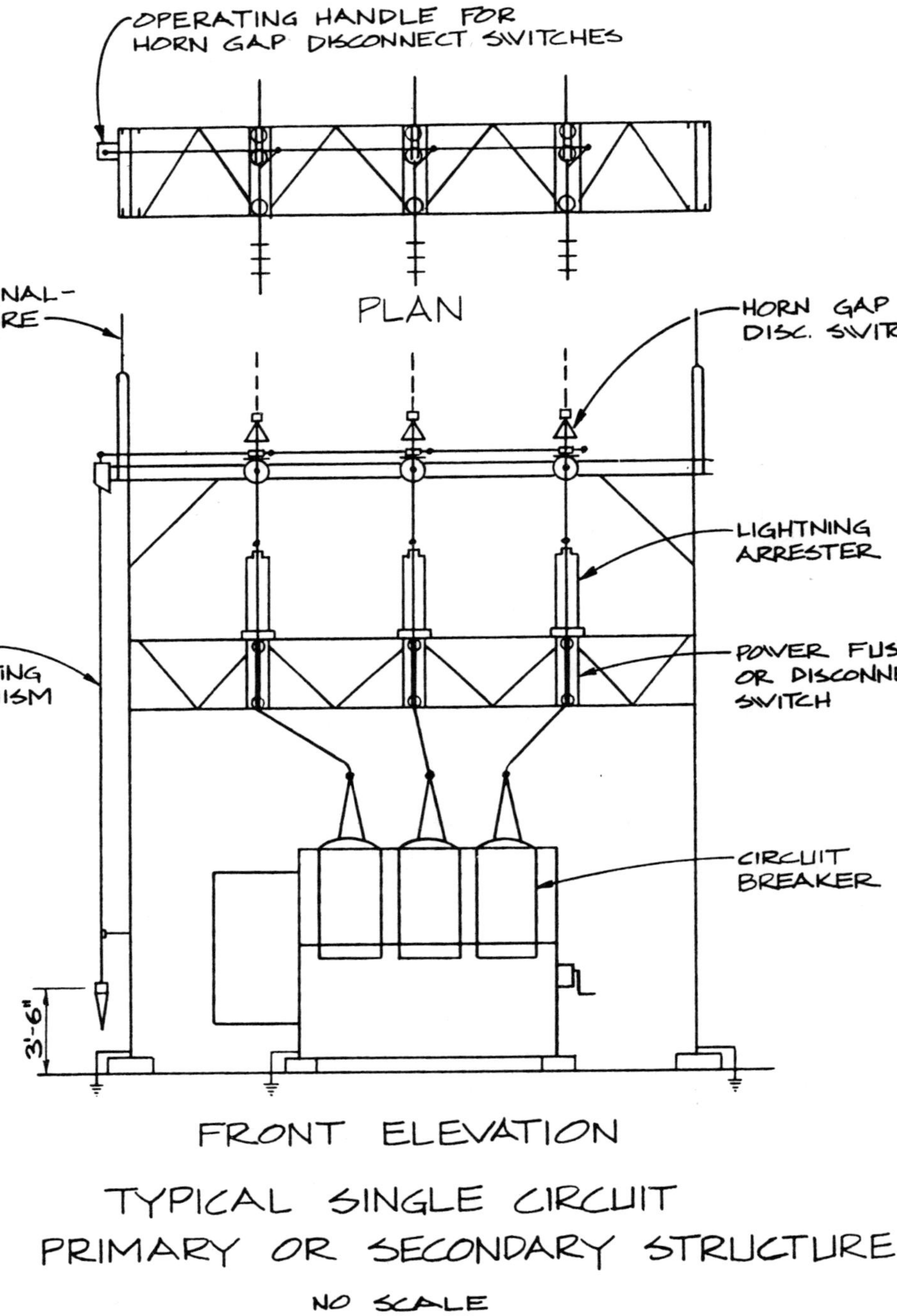
OPERATING HANDLE FOR
HORN GAP DISCONNECT SWITCHES
PLAN
AIR TERMINAL-
STATIC WIRE
BRACKET
HORN GAP
DISC. SWITCH
LIGHTNING
ARRESTER
SWITCH
OPERATING
MECHANISM
POWER FUSE
OR DISCONNECT
SWITCH
CIRCUIT
BREAKER
3'-6"
FRONT ELEVATION
TYPICAL SINGLE CIRCUIT
PRIMARY OR SECONDARY STRUCTURE
NO SCALE

FIGURE 6-27

lines that continue to a transformer and circuit breaker.

Figure 6-28 is a design element that does not necessarily directly relate to the previous four details and is by no means complete. The primary metering is a special part of the structure as well as a special part of the electrical system. Actually this detail merely locates the usual position of the current transformers and the metering cabinet.

Finally, in Figure 6-29 the usual elements of a primary unit substation are shown along with the grounding of equipment, structure, and fenced enclosure. One new element suddenly appears, the capacitors that usually are present to improve the secondary distribution power factor. This is not a very involved hookup and as such is not detailed.

It might be asked "Why show a bunch of partially complete, not directly usable details?" Our answer is that a completed detail could create some serious false impressions. This area of design is exceedingly complex and is always tailored to a given situation. Considerable engineering input is required from the manufacturer to complete the design, and there are many details that are unique to a given situation. We chose to highlight the equipment arrangement on the premise that this would provide a sort of design guide as to what should be thoroughly detailed. Also, we did not want to lead any firm not thoroughly grounded in the fundamentals of this type of work to come to the very mistaken conclusion that some worked out set of details for a high-voltage substation could readily be modified to fit another, different situation. This is very difficult to do even if the situation is very similar, and casual copying can lead to disastrous results where real differences occur.

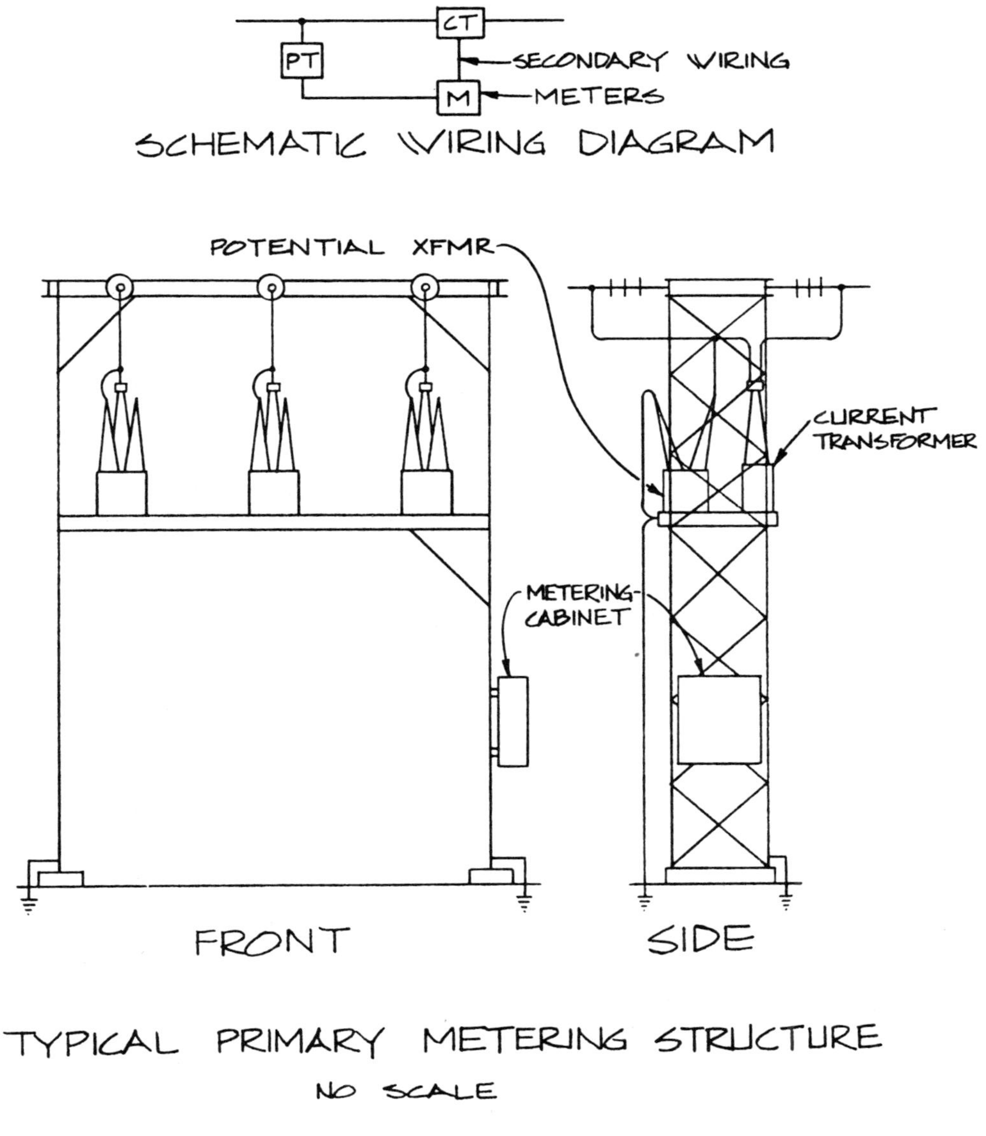

FIGURE 6-28

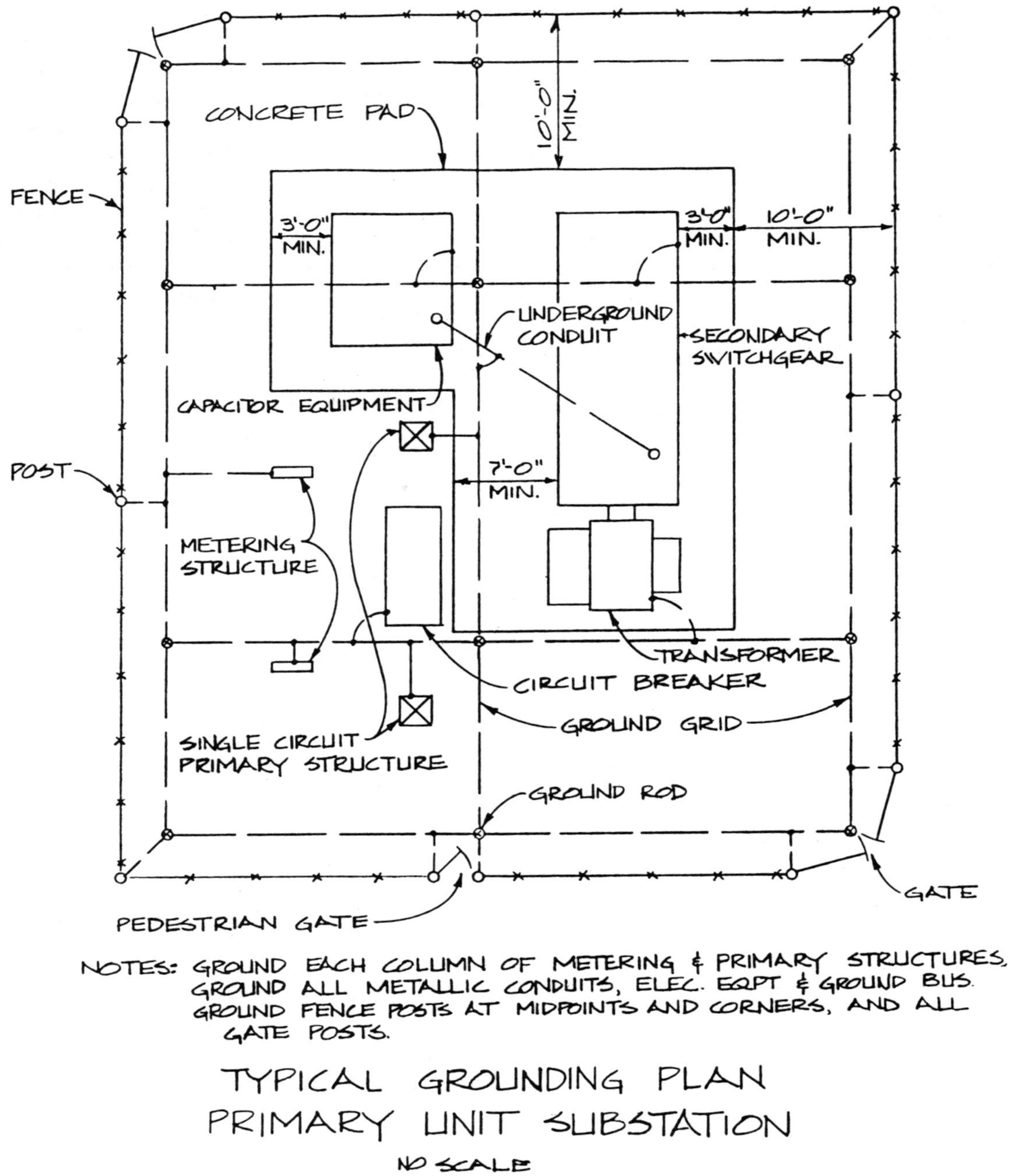

FIGURE 6-29

Transformer Load Center Getting back to a more common consultant problem, we would like to close our discussion of transformer detailing with the problem of an economical way to set up a small (approximately 500 kVA) transformer for an owner who has an underground primary and wants to avoid the problem of constructing indoor vaults.

The arrangement in Figure 6-30 could be used inside or outside. The transformer has a primary disconnect and a limited amount of secondary distribution. The equipment is usually totally enclosed. For the more common types of primary distribution, we have included some guidelines on the most cost-effective and logical type of primary-disconnect devices, which usually can be specified to fit neatly into the primary side of the transformer.

On the secondary side you usually need a throat section on the transformer to fit it to the enclosed secondary switchgear. The secondary switchgear consists of fused or infused circuit breakers depending on your short-circuit calculations.

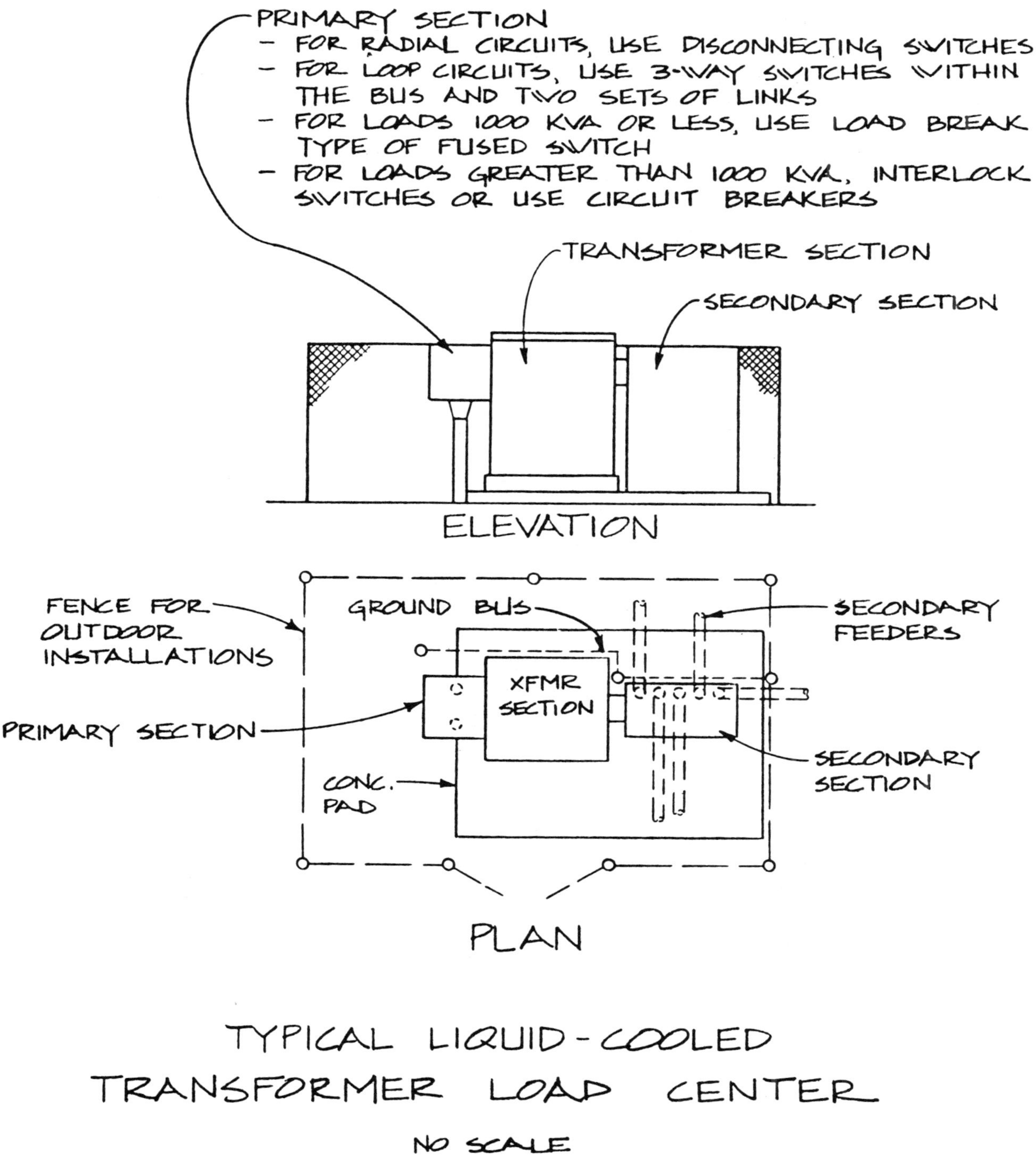

FIGURE 6-30

All of the equipment is housed in weatherproof enclosures which usually are *not* the walk-in type. If you set up your space dimensions and select appropriate pad, fence, and shrubbery details from our previously presented details, you can readily assemble a workable, pleasing, economical installation.

Power Distribution Systems In various preceding details we have frequently alluded to schematic power system presentations and their use on higher-voltage distribution systems. Because this is a somewhat vague type of reference, especially in this particular area, we have included seven schematics of fairly commonly used distribution systems. In many design problems you have the situation where power is brought to the building and transformed down to its utilization voltage. In these cases the details have no related value. But higher-voltage distribution systems are becoming more common, and we feel the future use of these systems will expand fairly rapidly.

Thus to relate our designs to your problems and illustrate how growth occurs, we begin in Figure 6-31 with the simple radial distribution system which receives power at the utility-supplied voltage and steps it down to building use level. The entire load is fed from a single company- or utility-owned source. Utilizing diversity to its fullest extent, transformer capacity can be minimized. While low-voltage design and distribution through circuit breakers is straightforward, the voltage regulation and efficiency of this system is poor due to the low-voltage feeder and single

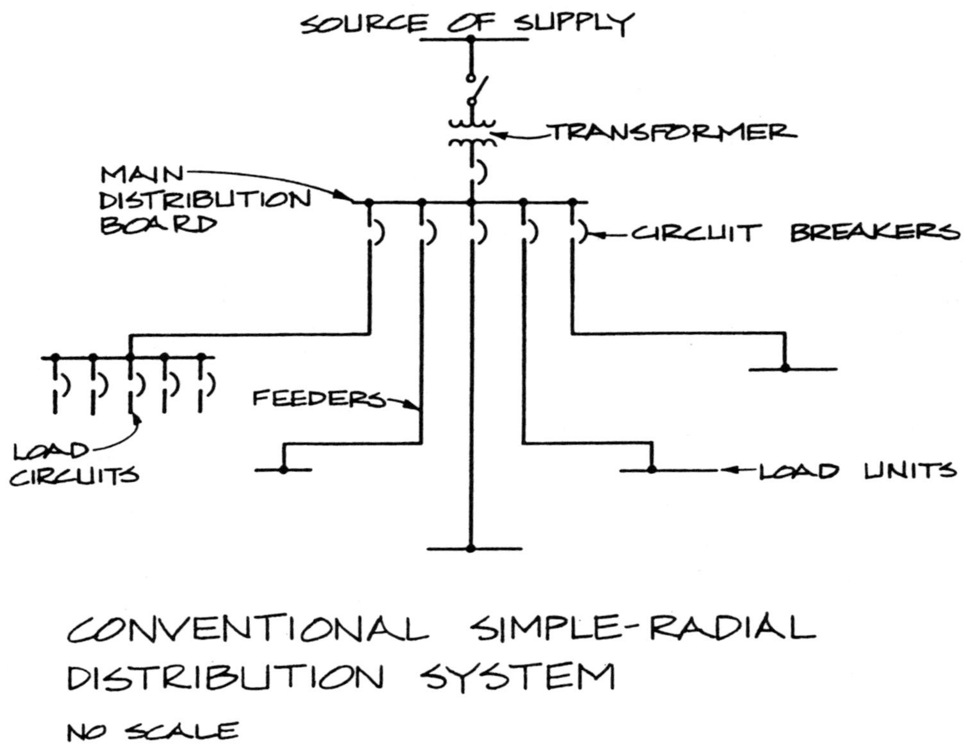

FIGURE 6-31

source. Also, costs can be surprisingly high for long low-voltage feeders and related circuit breakers, especially when peak demand is above 1000 kVA. A fault on the secondary bus shuts everything down.

Figure 6-32 is a loop primary radial distribution system. This is one of the simplest and most economical ways of distributing power for an owner-operated primary system. The sectionalized primary feeder circuit breaker and two manually operated load break switches sectionalize the loop intersection. Disconnect devices at the individual transformers are arranged to isolate load fault problems. When a transformer or primary feeder fault occurs, the system shuts down and the fault is located. Service can then be manually restored and the faulted area bypassed. Adding loop feeders would help solve that problem but increase the cost.

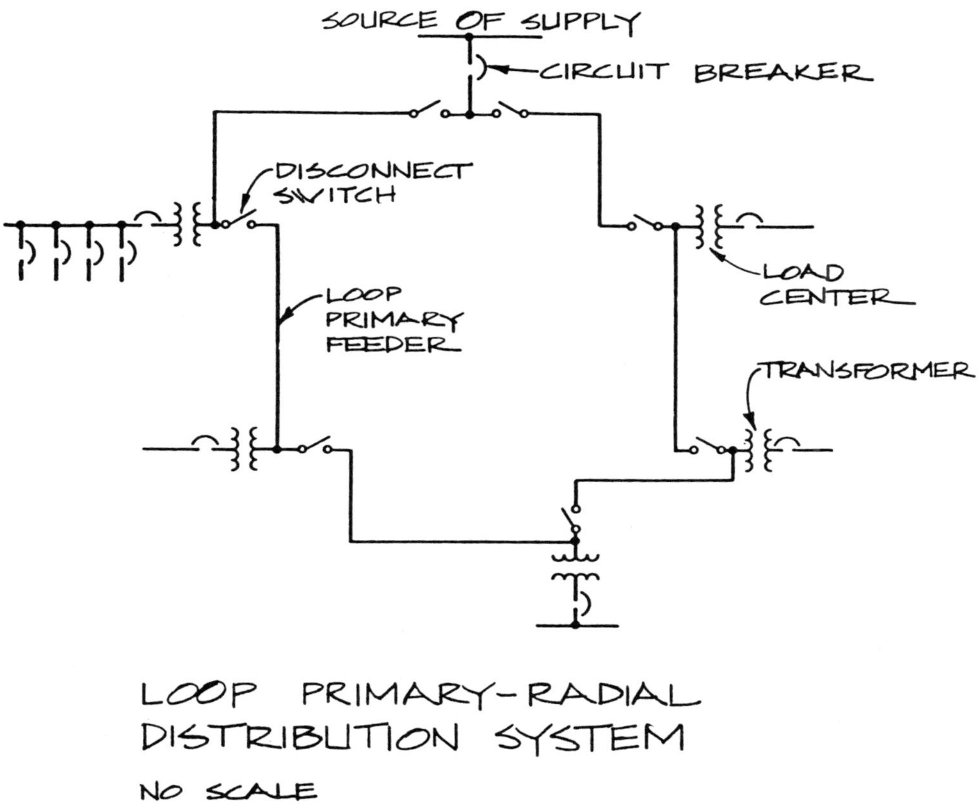

FIGURE 6-32

Having raised a problem with our economical loop design and implied that better solutions also cost more money, we present two solutions. Figure 6-33 employs two primary feeders to each load center. It is designed so that either feeder can accept the loads imposed. Double-throw or primary-selector switches are utilized at all transformer locations. When a fault occurs in a primary feeder, the transformer can be switched to the good feeder and service restored. For considerably more money this can be done automatically.

However, the problem may be on the other side of the transformer—the secondary side. Figure 6-34 carries the duplicate-feed solution one step further. It also has two sources on the secondary switchgear side. Each system has

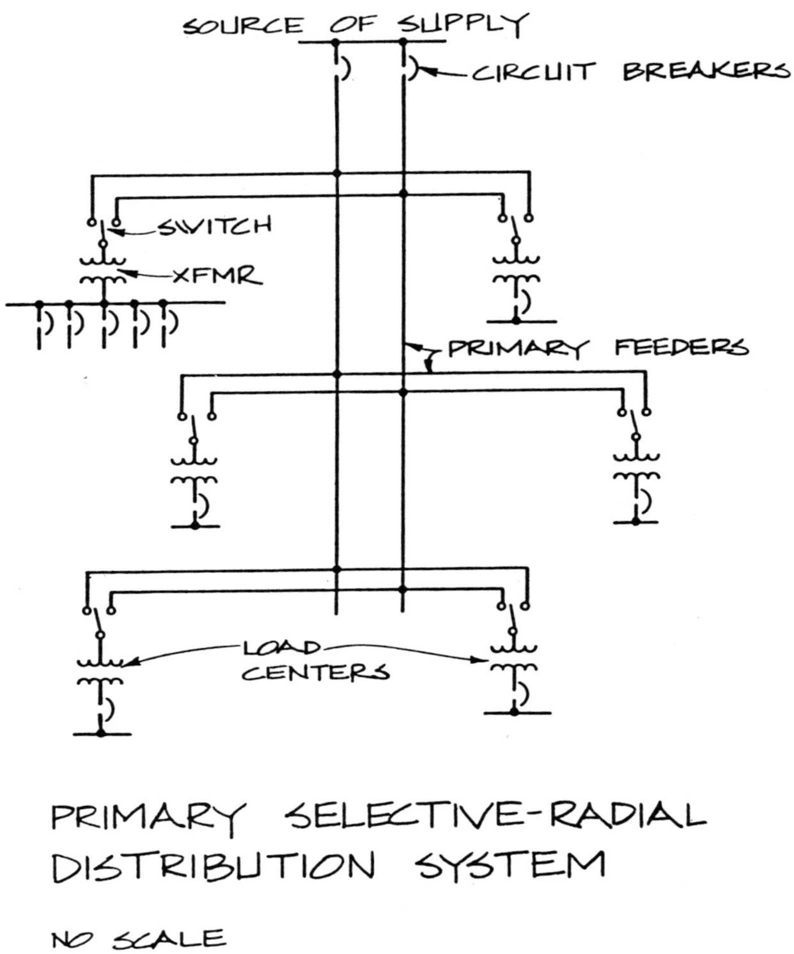

FIGURE 6-33

two transformers, and these are connected to the secondary bus. A bus tie breaker is interlocked with the two transformer circuit breakers so that it cannot be closed unless one of the transformer circuit breakers is open. A fault in a transformer (normally the system operates as two separate halves) causes the associated primary circuit breaker to open and half the area load is lost. Service can be quickly restored by opening the disconnect switch on the primary and the circuit breakers on the secondary and closing the tie breaker switch. The basic advantage of this system is that the usual fault only breaks half the normal load.

The simple network system detail has been used by utility companies for years to service densely loaded midtown

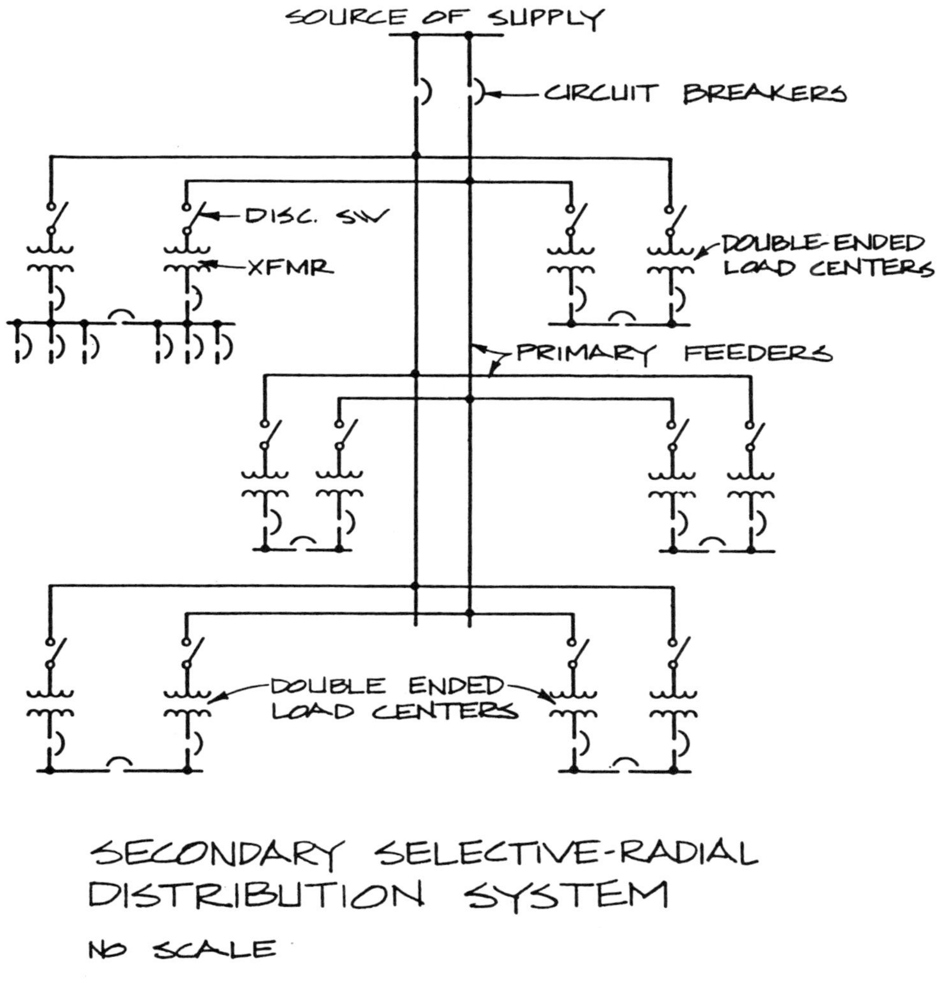

FIGURE 6-34

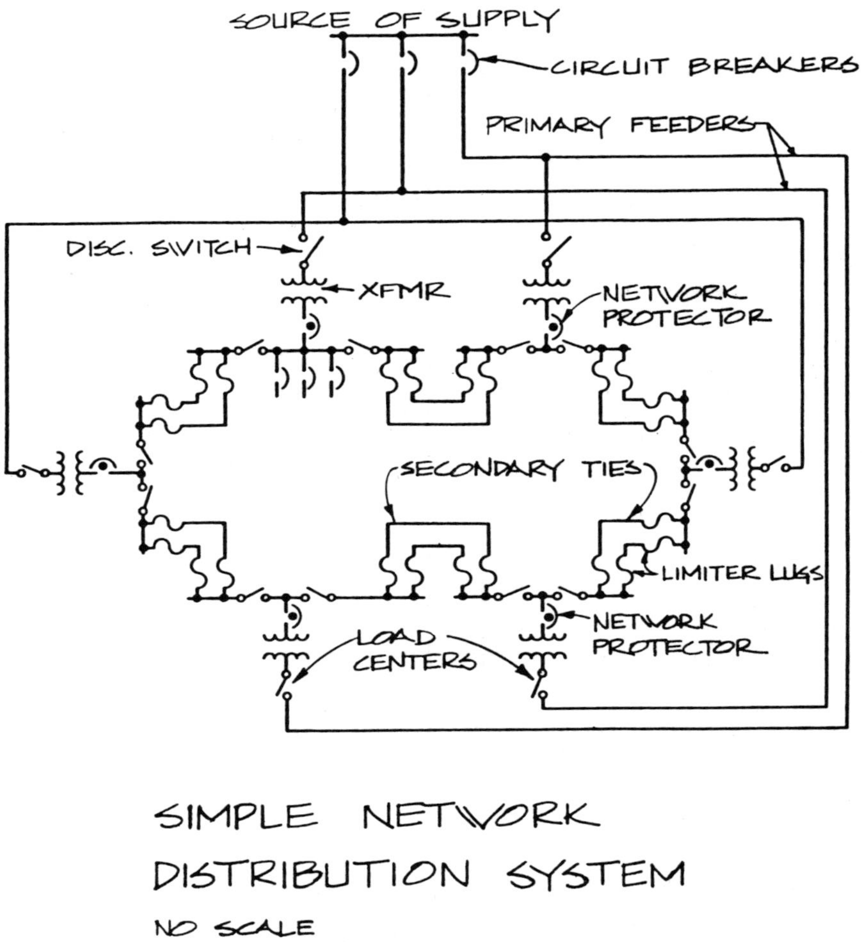

FIGURE 6-35

areas. Figure 6-35 is a modification of this system for building use. Its major advantage is continuity of service. No single fault anywhere will affect more than a small part of the load. The chief purpose of the secondary ties is to provide an alternative supply to a load bus when its primary source fails. The network protector is a specially designed motor-operated shunt-tripped air circuit breaker controlled by network relays. The relays automatically close the circuit breaker only when voltage conditions indicate the associated transformer can supply the load to the secondary loop, and they automatically open when power flows from the secondary loop back to the transformer.

Figure 6-36 is a primary selective spot-network system,

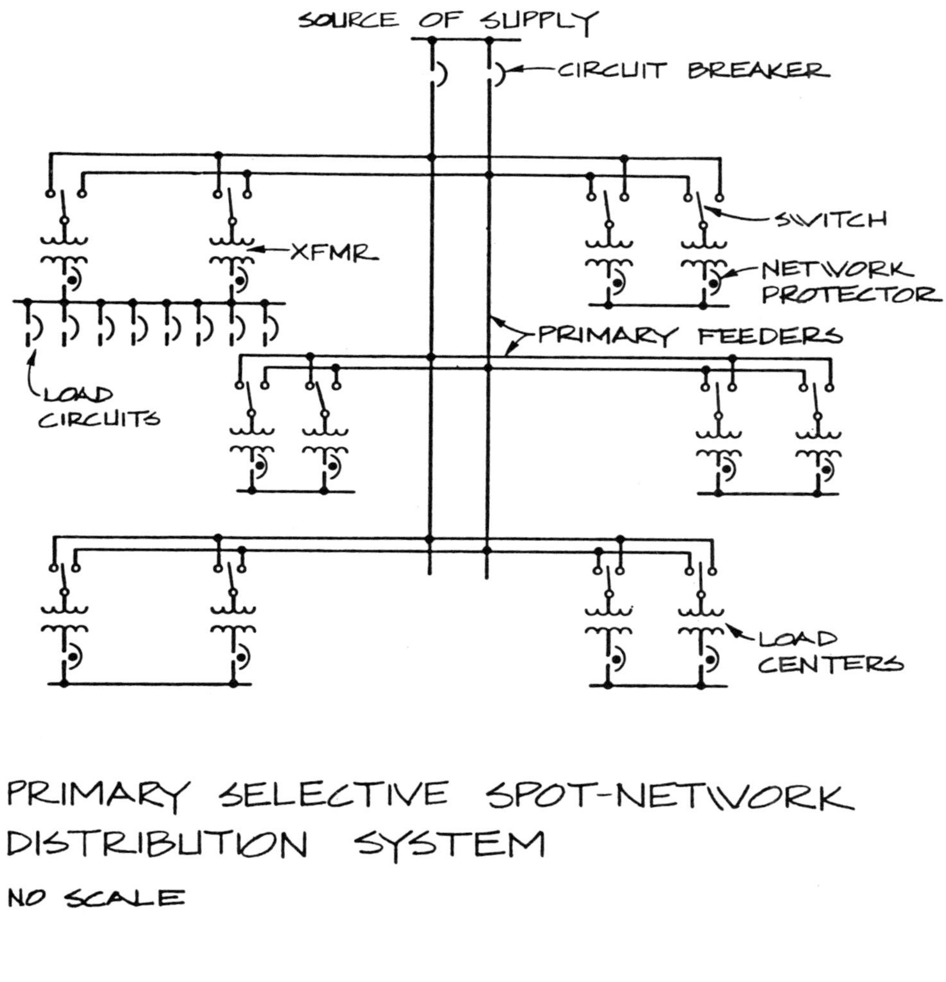

FIGURE 6-36

which is a sort of evolutionary growth of the simple network detail. The arrangement resembles our detail for a secondary selective radial distribution system. The transformers are connected through feeders to a single bus. Since the transformers are in parallel, a primary fault does not cause circuit interruption. The transformers normally carry equal loads—a difficult-to-obtain situation in the secondary selective scheme. This spot network system works best in buildings where there are concentrated electrical loads widely separated with very little load in between. The chief disadvantage of the system is inflexibility compared to other network systems when the loads shift or relocate.

Finally, in Figure 6-37 we show the primary network

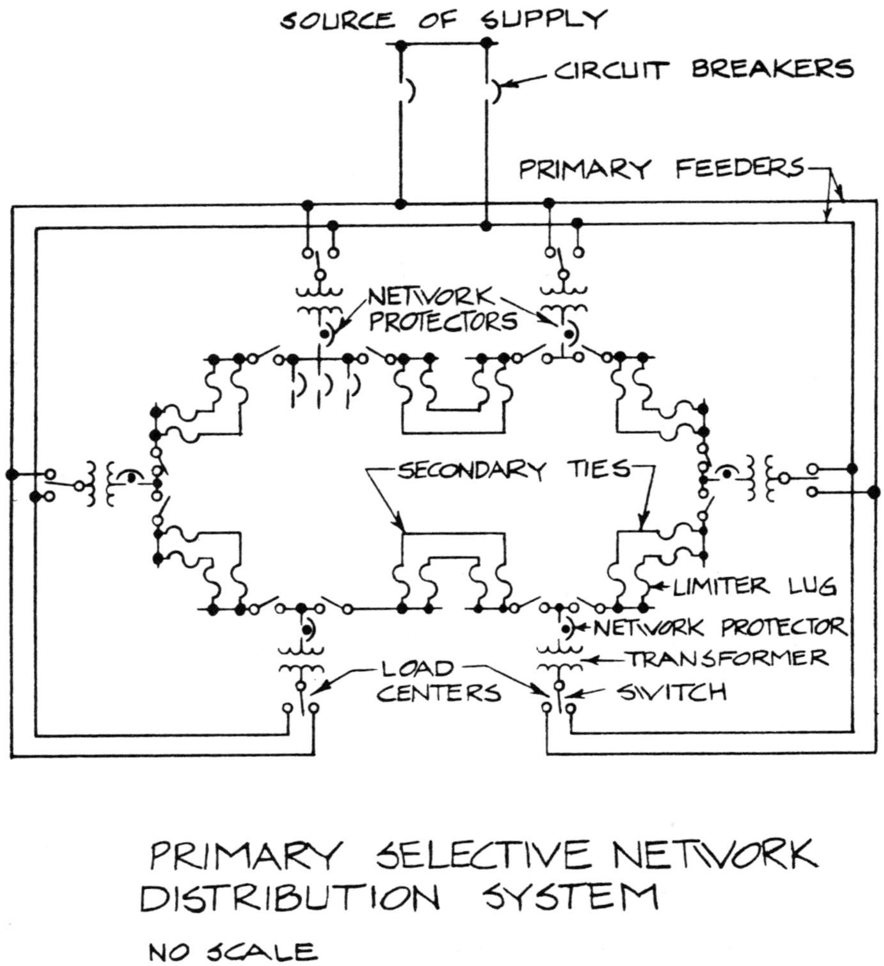

FIGURE 6-37

which is widely used for many industrial plant systems. It is similar to the simple network and low in cost where only two or three primary feeders are required. Savings are accomplished by duplicate feed or alternate feed to each transformer, which eliminates the need for spare transformers. And also, like the simple network system, it can take the form of a spot network system. In that case a transformer failure requires a reduction in the load served from its loading bus to prevent serious overloading of the remaining transformer connected to that bus. But these infrequent and remote faults rarely require and usually do not justify the expense of providing spare capacity throughout the system.

In the above descriptions we have only provided an abbreviated outline of each system, since we are concentrating on detailing and not design. For a full elaboration of each system, its details, and application, we refer you to the IEEE power distribution manuals and to the excellent printed material and descriptions available from a number of major electrical equipment manufacturers.

SEVEN
Sound Systems

In presenting detailing for sound systems, it should be clearly understood what the detailing does and does not do. Sound system detailing is generally schematic and depicts the major parts of the system and how they connect and relate to one another. Coupled with a set of plans that shows precisely where each of the parts is located, such detailing will usually adequately describe the installation of a sound system. Unless a large amount of wiring is carried in a specific manner and in a single conduit, not much thought is required to size the conduit system. Also, unless certain specific conduit paths are required by the building construction, even the physical location of conduits is not an item of importance.

The above paragraph describes what the detailing of a sound system may accomplish. Its tone is casual or even careless in describing system design. The above description unfortunately fits the approach taken far too often. What this approach *cannot* do is guarantee a good usable, workable sound system. A good sound system requires good design. While this book is not intended as a design manual, some of the basics of design will be covered in the paragraphs that follow.

In any well-designed (and this definitely means thoroughly specified) sound system the first question to be answered by the designer is "What purpose or result are you trying to achieve?" The following items are considered reasonable objectives.

1. A sound system in a hall or church that creates an illusion in the listener's ear that the speaker is talking directly to him or her
2. A series of microphones on a large conference table tied to a system that ensures everyone will hear what is said and that enables one to make a clear tape recording of the meeting
3. A public announcement set up in a sports arena, bus station, or airport in such a manner as is to make certain everyone gets the message
4. A sound system that plays music or projects it at an outdoor concert

In each of the above cases we have microphones, amplifiers, tape or record players, and speakers, and the detailing would look much the same for all cases. There are also a number of elements and degrees of quality required. The following is a list of some of the problems to be resolved in the design.

1. Because of complexity, some systems may require several separate conduit systems, that is, one for microphones, one for line level, and one for loudspeakers. Put lines together having level differences of more than 30 db and you are asking for trouble.
2. A basic requirement is low *acoustic distortion,* which includes both loudspeaker and amplifier overloading. Pushed beyond their limits, sound-class loudspeakers and amplifiers, or even first-class items, are simply *built-in* problems.
3. Not only must the system be loud enough it also must be designed to cover all of the area involved.

Without going into the technicalities of design, a few facts might be of interest. For the average ear, half as loud or twice as loud equal about 10 db down or up. Outdoor sound will drop about 6 db in 100 ft. Since sound varies inversely as the square of the distance, the usual calculation will show a 10-db drop in 216 ft.

Indoors the system is quite another matter. Investigation must be made into how "live" or "dead" the room is. That is, how much sound can be bounced off the walls, ceilings, and floors (live) or, conversely, how much will be absorbed by these surfaces (dead)? Again calculation must show not only how much bounce back is required but also how long it takes to bounce back (reverberate).

All of the foregoing discussion should serve to alert you to the fact that careful selection of the system to meet the situation is very important. Generally the system design to a large degree revolves around the amplification system, which, in oversimplified terms, involves the type and number of speakers and their location. There are three types of sound amplification system.

1. A high-level single source. In an auditorium the preferred location is usually over the proscenium and as near the ceiling as possible. In a sports arena the position is usually high up in the middle of the area. In a baseball park it is at the center-field fence.
2. A low-level system, wherein a collection of speakers is mounted on the ceiling, projects straight down. A variation of this may be noted in a grandstand of a racetrack where a collection of spaced horns aims sound at the crowd.
3. A combination of high and low speakers are commonly used in an auditorium with a balcony. Speakers are not only located in the proscenium but also mounted on the ceiling of the balcony.

In all of the above cases the proper kinds of loudspeakers must be selected. For high locations the best speakers are either multicellular or sectoral horns. The multicell horn is a group of cells, each covering an angle of about 20°, with a common throat. Multicells have been used in motion picture theater sound systems for years, but for some reason they are not in widespread use in auditoriums. Where music is projected, the multicell must be supplemented by one or more low-frequency speakers called *woofers*. The sectoral horn is a small horn with an experimental taper and a mouth arranged as a narrow rectangle. The coverage pattern is not as good as a multicell and the high-frequency response is frequently poor in quality.

The ceiling speakers are usually a direct radiator cone loudspeaker with a diameter of about 8 in. The usual metal box commonly seen in the ceiling installation frequently is smaller than the 2-ft^3 enclosure recommended by the speaker manufacturers. As a result, these speakers have a bass deficiency because of this smaller-than-recommended enclosure.

The preceding paragraphs described the sound system, its components, and its design problems in abbreviated fashion. In the details that follow, the schematic design detailing is shown with the caution but, under no circumstances, is it *the* solution. A proper system selection with complete, thorough, and detailed specifications coupled with plan and detailing is required. And finally, a rigid enforcement of the specifications by the shop-drawing reviewers is a must.

Sound System Riser Diagram Figure 7-1 is a typical diagram seen on many plans. Basically it is a public address system, which is one of the common types of sound systems. As can be seen, there are speakers in lobbies, waiting rooms, mezzanines, a passenger loading dock, and exterior platforms. There are no microphones as such. The system will be connected to telephone lines at each ticket counter and announcements will be made, as you have often seen, by agents picking up the phone, dialing a number, and then speaking into the phone in a normal tone of voice. The phones and their connections were not shown because, at the time of the bidding, the client was uncertain of how the phone equipment and its system would be provided and who would own what.

This detail is an excellent illustration of the cautions mentioned in our opening discussion. Certainly this detail, combined with specific locations noted on the floor plans, will aid in the proper connection of equipment. But the system as shown describes nothing and specifies nothing in a real sense. All you can deduce from the detail and from the plans is that certain components are required. Thus the specification that must accompany this detail becomes extremely important. The specifications should, and literally must, cover every component of the system in exact detail. Quality, function, size, finish, performance, and much more must be carefully and thoroughly described in considerable detail.

As we also stated in our opening discussion, one of the classic problems of ceiling-mounted speaker systems is that of too few speakers. Thus, to answer the question of how many is enough, we first look at the speaker itself. The coverage angle of the ceiling-mounted speaker is 60°. What you are first of all concerned with is the distance from the speaker to the *ear* of the listener and *not to the floor* of the space. Thus, the distance from a speaker in a 15-ft-high ceiling to the user's ear is approximately 10 ft (ear height being 5 ft 0 in on average), and not the 15 ft floor-to-ceiling height. The average of 5 ft 0 in is based on the average height of 5 ft 6 in for standing and 4 ft 6 in for sitting.

The basic coverage area formulas are $R = H \tan 30°$ or $R = 0.577H$. The area is $A = 3.14R^2$. For our example, using 10 ft from speaker to ear, the coverage area is, in round figures, 105 ft^2. Speaker locations should be staggered, and they should overlap. The general rule is that there should be a 50 percent overlap.

Presuming in our sample problem the area involved is 1000 ft^2, dividing 1000 ft^2 by 105 ft^2 per speaker gives us an answer of 10 speakers. Allowing for 50 percent overlap would add 0.50×10 or 5 more speakers. Thus the correct number of speakers for our problem room with a 15-ft ceiling height and 1000 ft^2 of area is 15 speakers.

Sound System Schematic Figure 7-2 is a fairly simple schematic diagram that actually covers many much more seemingly complex installations. As we said in our opening discussion, there are only three types of sound systems—a high source, a collection of ceiling speakers, or a combination of high and ceiling sources. If we chose to

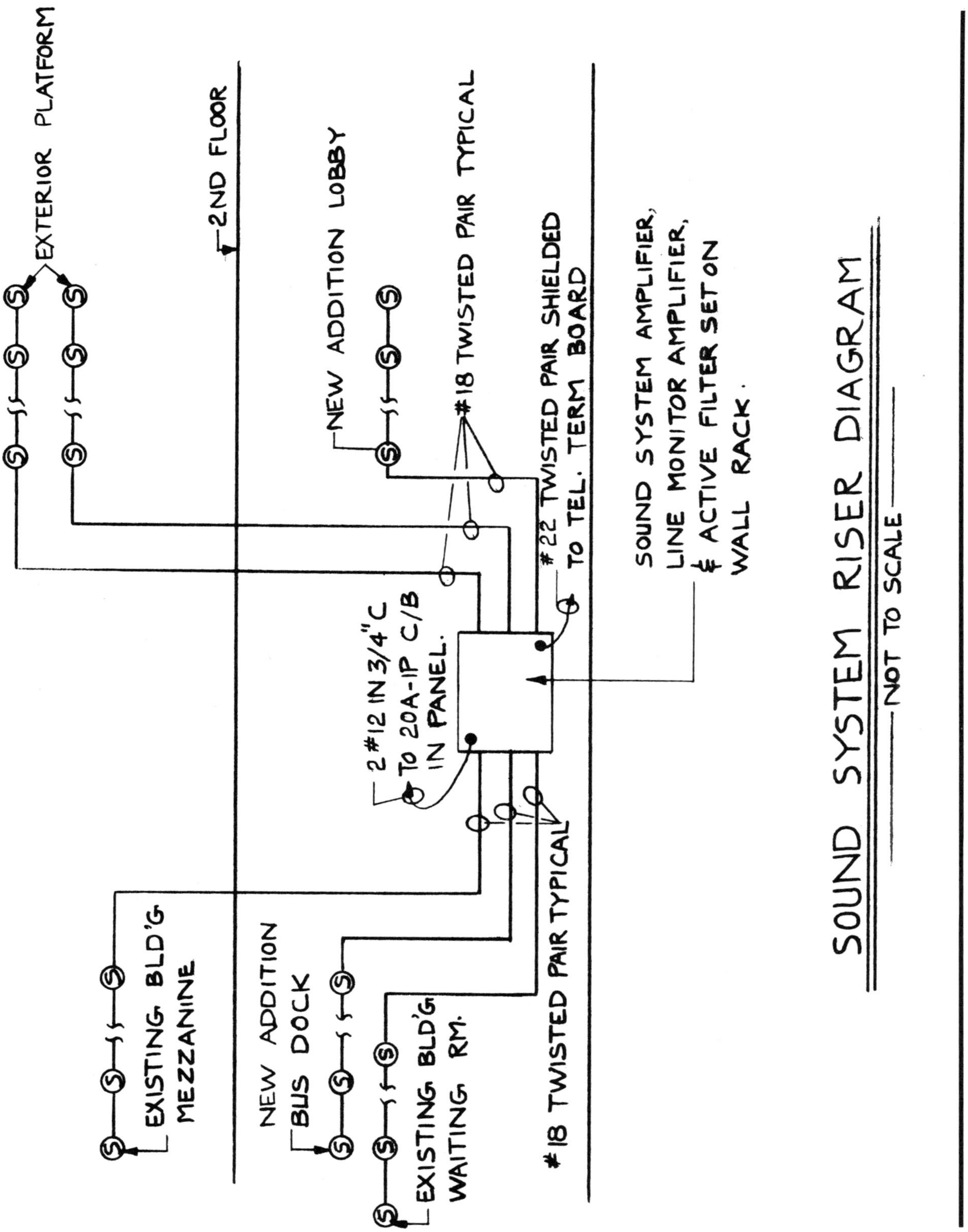
EXTERIOR PLATFORM
2ND FLOOR
NEW ADDITION LOBBY
#18 TWISTED PAIR TYPICAL
#22 TWISTED PAIR SHIELDED TO TEL. TERM BOARD
SOUND SYSTEM AMPLIFIER, LINE MONITOR AMPLIFIER, & ACTIVE FILTER SET ON WALL RACK.
2 #12 IN 3/4"C TO 20A-1P C/B IN PANEL.
EXISTING BLD'G MEZZANINE
NEW ADDITION BUS DOCK
EXISTING BLD'G WAITING RM.
#18 TWISTED PAIR TYPICAL
SOUND SYSTEM RISER DIAGRAM
NOT TO SCALE

FIGURE 7-1

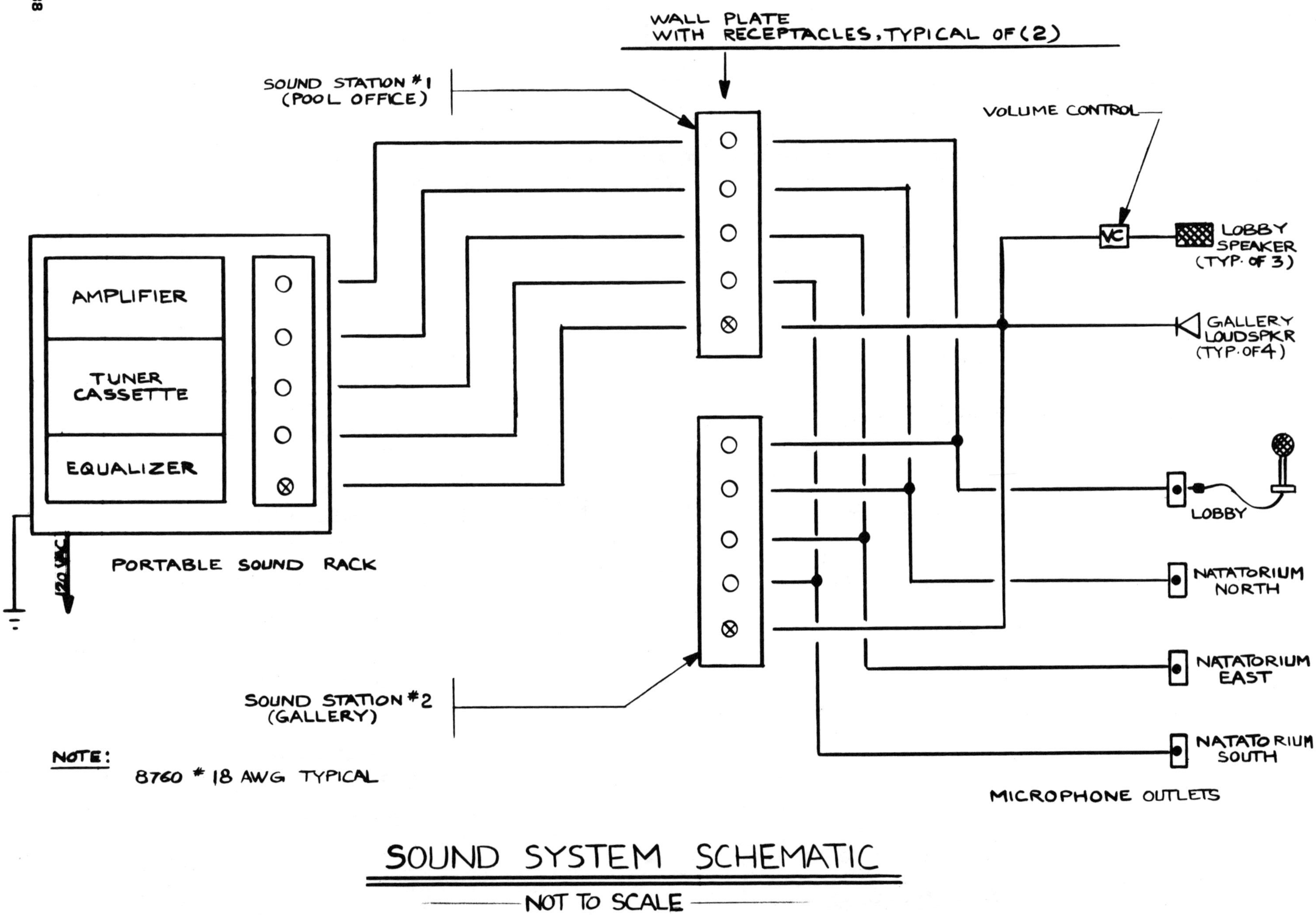
WALL PLATE
WITH RECEPTACLES, TYPICAL OF (2)
SOUND STATION #1
(POOL OFFICE)
VOLUME CONTROL
VC
LOBBY
SPEAKER
(TYP. OF 3)
GALLERY
LOUDSPKR
(TYP. OF 4)
AMPLIFIER
TUNER
CASSETTE
EQUALIZER
PORTABLE SOUND RACK
120 VAC
LOBBY
NATATORIUM
NORTH
NATATORIUM
EAST
NATATORIUM
SOUTH
MICROPHONE OUTLETS
SOUND STATION #2
(GALLERY)
NOTE:
8760 #18 AWG TYPICAL
SOUND SYSTEM SCHEMATIC
NOT TO SCALE

FIGURE 7-2

leave out the lobby speakers shown, there would be a single high source. Conversely, we could have only the low-source lobby speakers.

Customarily the system has an amplifier which, in this case, can also become a record-player outlet. The gallery speakers were selected as a group of four to better distribute the sound in a noisy, highly reverberating, indoor pool atmosphere. This was a fairly difficult choice and considerable investigation was made into the number of speakers and type. A case was made and can still be made for other numbers and types of speakers. Another problem was microphone location. Were the microphones all actually needed, and how many microphones might be used at one time?

Acoustic gain is a factor to be considered in sound design. Two open microphones cause a gain loss of 3 db, five microphones lose 7 db, and so forth. Do not forget that a loss of 10 db is perceived by the human ear as a 50 percent loss in sound. Each time another microphone is in use or "open," the potential acoustic gain is *decreased* and the requirement for needed gain is *increased*.

You will also note that we have a volume control on the lobby speakers. We were concerned about a possible reverberation or echo problem, for which there was no easy answer. One of the solutions in a case like ours with speakers far ahead of the microphone is a time-delay device which corrects mixing of two sounds from two sources far apart. Sometimes making the added delay between speakers 10 to 15 ms greater than the acoustic delay caused by the path length difference has the effect of causing the separate speakers not to be noticed as separate sources of sound. We covered that as well as the volume control in the specifications, because the lobby noise level varies the speaker sound level demand.

Sound System Installation Usually the installation of a sound system does not require much in the way of detailing of support for the individual parts because these items are both light in weight and self-contained, and they are either readily attachable to wall surfaces or readily recessed in walls or ceiling.

The two details shown in Figures 7-3 and 7-4 illustrate

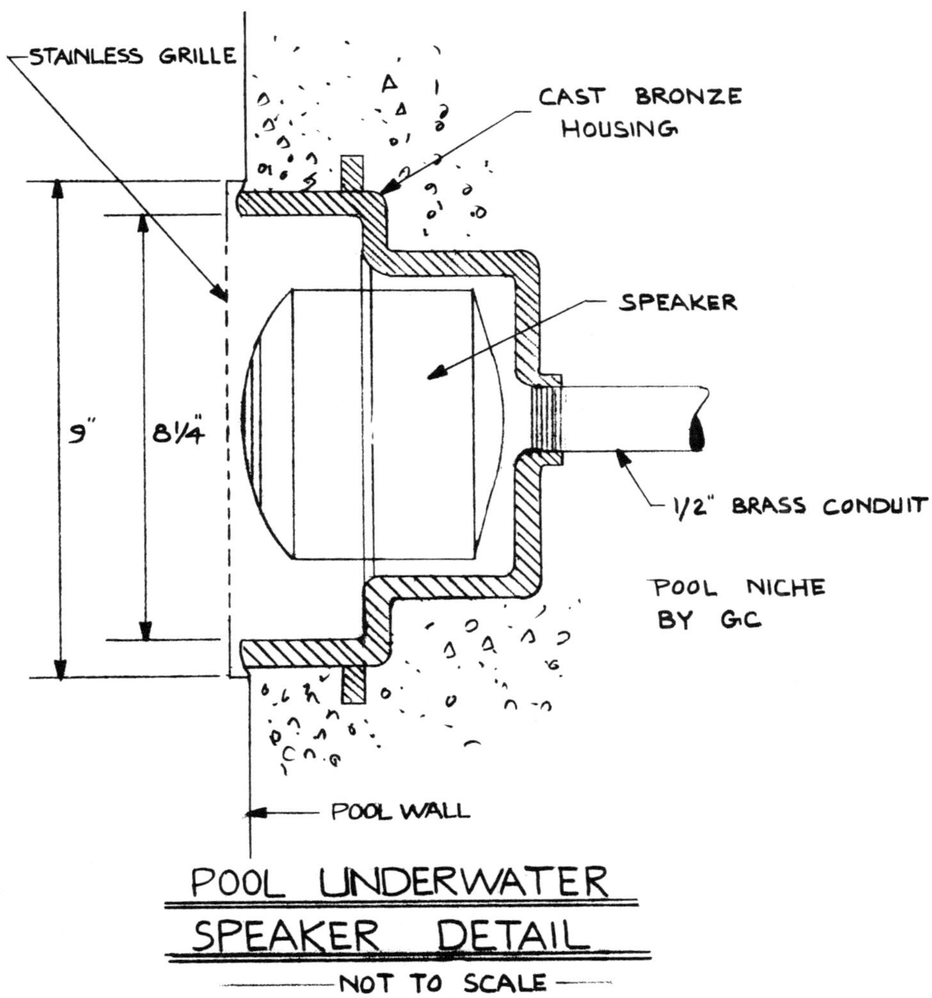

FIGURE 7-3

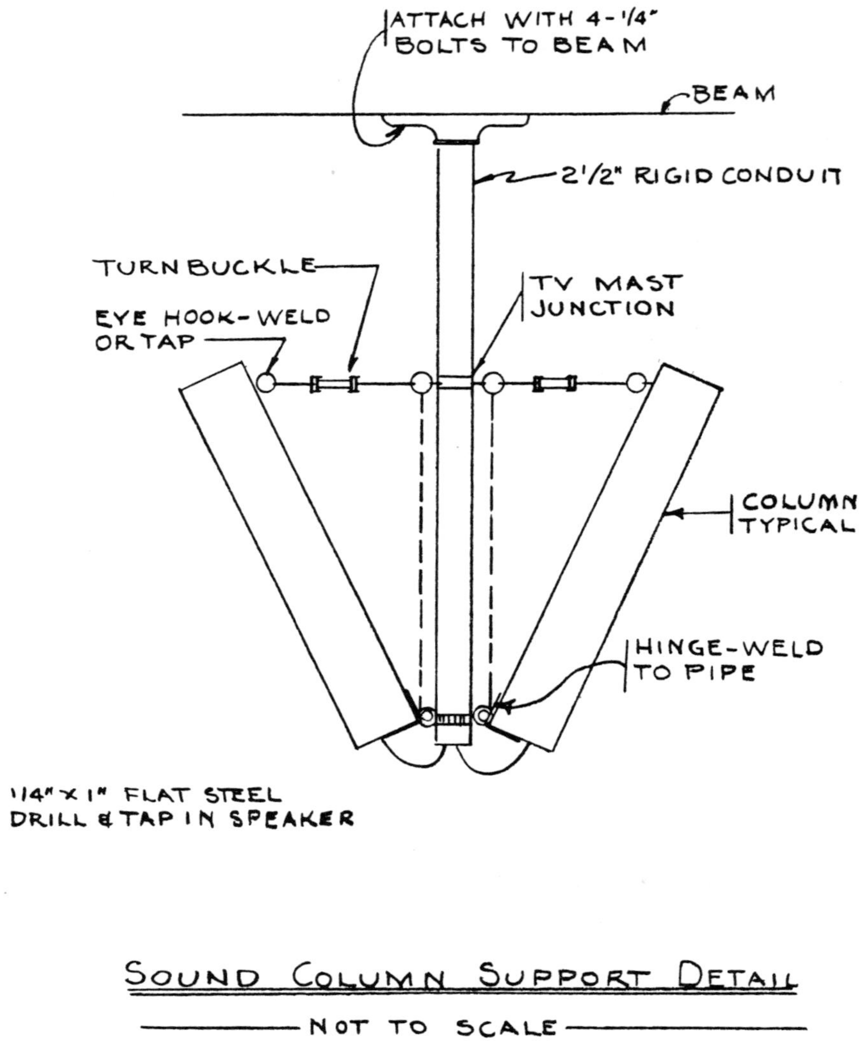

FIGURE 7-4

some of the very few situations where detailing may be required. Before presenting any discussion, perhaps some additional explanation may be in order. The physical space taken up by speakers, consoles, and related equipment requires careful notation on the floor plans and elevations. Appearance, clearance, and arrangement are very important. But this is not a detail. This is an integral part of plans, elevations, and space layout.

Figure 7-3 shows an underwater speaker, which is a somewhat unusual application. The underwater speaker itself is of a special type. What is important to your plans is that something should be shown concerning the installation of this speaker in a pool niche. There may be many nonexpert people involved in the construction of the pool who find an installation such as this hard to believe. We have not fully dimensioned the detail as there are variations which would make further dimensioning useless. Your design chief should provide complete dimension and installation details for your particular detail.

Figure 7-4 illustrates a collection of four speakers mounted in what would normally be a gym or other high-ceilinged area of public assembly. This detail might be considered a "source of information if and when needed." Commonly your specification would describe a central multidimensional speaker that is usually part of a complete manufacturer's package. It merely has to be attached to the ceiling structure and wired. No detail is required. Our detail was created for a client who wanted to reuse some speakers instead of buying new units. What is shown is a rather obvious (at least no contractor complained) method we devised to create a suspension system for four existing speakers. While not a very eye appealing arrangement, it was installed high enough so that no one really noticed, and it did solve the problem.

EIGHT
Telephone Systems

Figure 8-1 is the only telephone system detail we show, since there is normally little detailing required in this area. The normal set of contract plans will show telephone outlets on the floor plans and a schematic riser diagram indicating the relationship of conduit runs between the various phone outlets, the central switchboard (if used), and the telephone panel or backboard.

Your client may own or lease the system, but, in either case, the selection of equipment and the location of individual phones and their interrelationship is all a part of a separate contract with some phone company. In light of the AT&T breakup this is changing very rapidly. But whether purchased by the client or owned by the phone company, the equipment selection is not a part of the detailer's task. Not infrequently a building is prewired. In this event the phone cables are installed with, and frequently without, the use of conduit raceways.

There is little debate over which is the better way to install a phone system. A properly sized conduit raceway is decidedly superior. The ability to add, subtract, or alter lines without taking the building apart is certainly better. The argument that a raceway system cannot cover every possible future requirement applies even more forcefully to the prewired cable system. If an unusual change is required that is not covered by the raceway system, it is, of course, a problem. Frequently, however, the existing raceway system can solve at least part of the problem. The prewired cable system cannot help at all in this case.

The telephone outlet is normally a standard wall box similar to the outlet box. Underfloor systems such as Walker duct are not shown here since they are standard and would normally be shown on the plans for power and communications. This system would be fed, at some point, from a conduit raceway system.

There is one point in this seemingly simple detail that frequently gets away from both the designer and detailer. As is common in most details, the detailing is schematic and has no dimensions. Thus the casual treatment of the telephone backboard frequently becomes a habit. The specification can also become too casual and thoughtlessly repeat a size from an old job of 4 × 4 ft or 4 × 8 ft. Most of the time this sort of sizing will suffice. But, there are cases where a telephone system's board-mounted equipment not only would not fit on one 4- × 8-ft board, it also would not fit on two 4- × 8-ft boards. Even more of a problem occurs when the phone equipment will not fit in the carelessly sized telephone equipment room.

Thus it pays to check carefully with both the owner and the phone company to discover what sort of backboard and equipment space is really required.

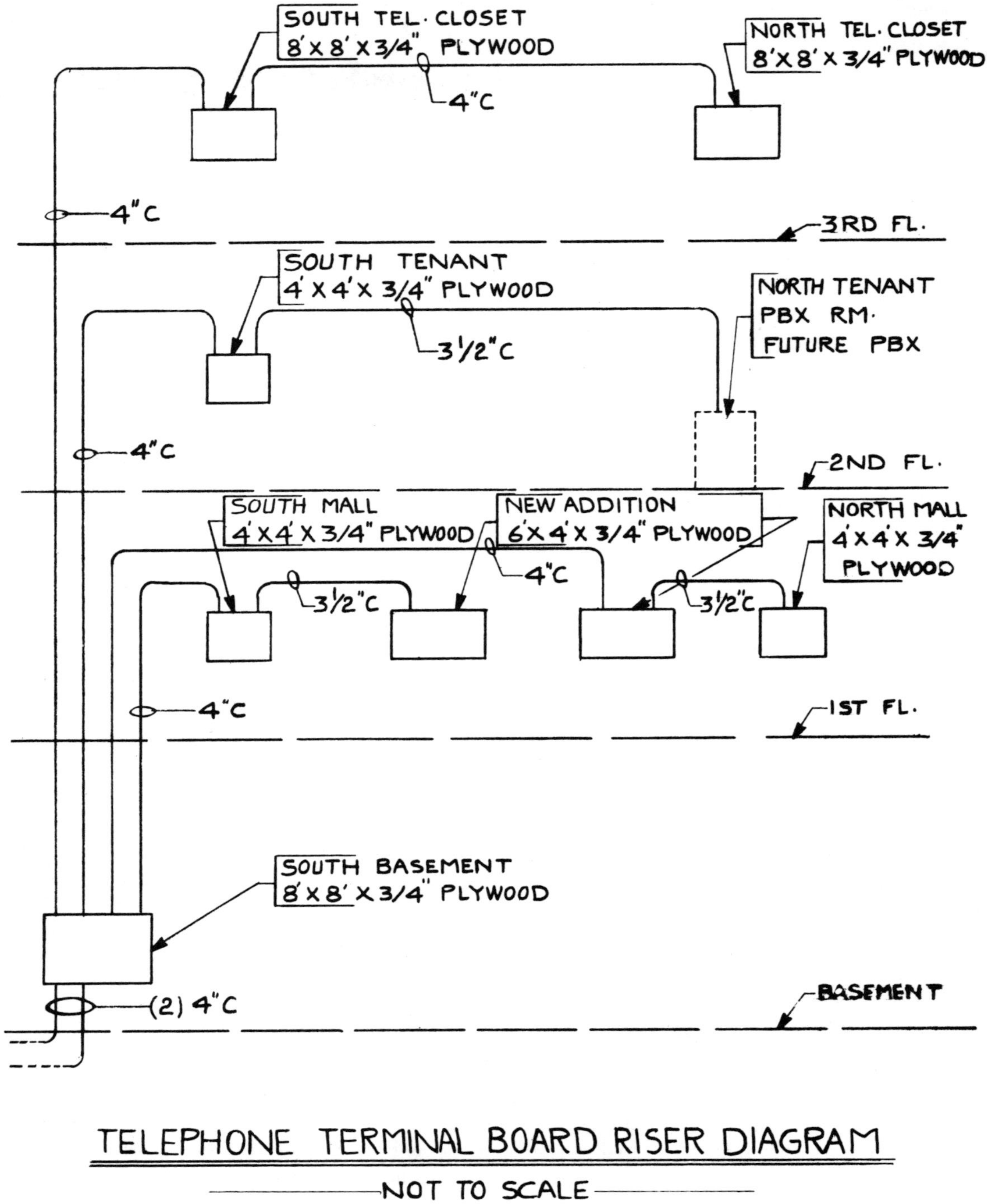
SOUTH TEL. CLOSET
8'X8'X3/4" PLYWOOD
NORTH TEL. CLOSET
8'X8'X3/4" PLYWOOD
4"C
4"C
3RD FL.
SOUTH TENANT
4'X4'X3/4" PLYWOOD
NORTH TENANT
PBX RM.
FUTURE PBX
3 1/2"C
4"C
2ND FL.
SOUTH MALL
4'X4'X3/4" PLYWOOD
NEW ADDITION
6'X4'X3/4" PLYWOOD
NORTH MALL
4'X4'X3/4"
PLYWOOD
4"C
3 1/2"C
3 1/2"C
4"C
1ST FL.
SOUTH BASEMENT
8'X8'X3/4" PLYWOOD
(2) 4"C
BASEMENT
TELEPHONE TERMINAL BOARD RISER DIAGRAM
NOT TO SCALE

FIGURE 8-1

NINE
Emergency Power Systems

In the event the normal electrical supply to any electrical installation fails, the purpose of the emergency system is to supply the necessary electrical energy for lighting or power or both. Lighting has precedence over power, since safety depends on the ability to see one's way out of a situation.

The NEC recognizes the following five items as alternative, or emergency, sources of energy.

1. Storage batteries to supply and maintain 87½ percent of system voltage of the emergency light and power circuits for a period of 1½ hours.
2. An emergency generator driven by an internal-combustion engine with an on-site fuel supply sufficient to operate the unit at full demand for 2 hours.
3. An independent service widely separated from the normal service to minimize physically the simultaneous disruption of the supply source.
4. Similar to item 3 above, a connection on the line side of the main service disconnect sufficiently separated to avoid simultaneous interruption.
5. Unit equipment for emergency lighting consisting of a rechargeable battery, battery-charging means, one or more lamps, and a relay device to start the system when normal power fails.

In addition to the above description, the Code, under Article 517, provides a very complex and complete description of emergency systems for hospitals, which is further modified for health care facilities based on the type of patient care provided. Article 517 carefully details the rules to cover both the emergency system and the equipment system. The emergency system is further broken down into life safety and critical areas. Essentially, power and lighting must be provided as required for life safety and effective operation while normal service is interrupted. Thus for a hospital the source of energy is an emergency generator plus frequently a second separate source of power from the utility company.

In all emergency sources of energy there must be a way of automatically shifting the load from normal to standby power. Whether this is done with a simple transfer relay in a small self-contained battery unit or by increasingly elaborate transfer switches, self-starters, and other devices, the transfer function is always a part of the emergency system design.

The vast majority of emergency system installations are for lighting requirements only. Because of this fact most emergency lighting is done by battery-supplied local units or from a central battery cabinet supplying a number of remote heads. A modified version of this scheme uses an inverter to change dc to ac and use existing or separate 120-V luminaires.

Our detailing does not cover the simple individual unit, because its application is obvious. Nor does it become involved too deeply in battery-supplied systems, which essentially are all very similar. A large amount of the detailing concerns what seem to be mechanical details of emergency generator installation.

The reasoning behind the presentation of mechanical details of generator installation in a book on electrical details is that the emergency system is a hybrid sort of system. Far too often the system and its installation are separated in the planning into mechanical electrical areas. The usual result is that a lot of very pertinent items get lost between the two departments.

Electrical code handbooks are becoming more and more aware of this situation and currently present at least one composite detail so that the detailer and designer can become more aware of the problem. Regardless of where shown, the details of the generator and its parts simply take the problem one step further and show what should be shown somewhere, at least, on the total set of drawings.

Before we go to our details there is one final note. The Code requires that emergency-circuit wiring be kept entirely independent of all other wiring and must not enter the same raceway, cable, box, or cabinet with any other wiring.

Battery-Powered System There are several values of low voltage that may be chosen for any particular project. The detail shown in Figure 9-1 depicts a 12-Volt dc system. The essential parts of any of these systems consist of a battery cabinet, a monitoring contactor, a terminal for all the circuits, associated lighting, and exit lighting fixtures.

Figure 9-1 shows a limited system applied to an indoor pool and associated corridors, exits, offices, and reviewing gallery. Because of size limitations of the battery cabinet, sources of energy, and the associated low voltage and voltage drop, a larger structure would require a number of these systems and a number of separately located battery cabinets. This is a common result created by the above limitations. In terms of pure cost the individual battery units mentioned in our opening discussion are frequently the least expensive solution. They are also frequently considered neither attractive nor desirable by architects.

There are several items to be noted in Figure 9-1. First the circuits are fused as denoted by the symbol S, which should be in your symbol list. The same line of reasoning could have precluded our adding a description to the lighting symbols. We really did not need to note that certain symbols were for exit signs and lights, but we felt it was better to be certain no one confused items on the plans, since the exit signs and emergency heads are both shown there. The solid square for the pool office indicates a square recessed head which is similar in appearance to any normal recessed square incandescent fixture.

In the detail the monitoring contactor monitors the feeders to panel A, which feeds the pool lighting and power systems. Failure of an individual circuit will not activate the emergency lighting system. While the system could, at greater cost, have been arranged with separate monitors for specific lighting circuits, it was felt that individual circuit failure would not be a problem, and individual lighting circuit failure would be a highly unlikely occurrence. The normal failure is panel-feeder failure or loss of utility power. Thus the solution shown is economically realistic. All battery cabinets should have a built-in charger.

The detail will not solve many special requirements and, as noted, is only good for 1½ hours. Sometimes misunderstood is the function of this type of emergency power. Its only purpose is to allow personnel within the building to exit safely and the building owners to shut down equipment properly and lock up. In an extended power failure the building will be dark and unusable. Generally the best of these systems supplies some light energy for 6 hours. They are not sources of energy for any extended power use.

Finally, there are various types of batteries that may be installed in the battery cabinet. Because of their overall superiority of performance and the acceptable cost-benefit ratios, the nickel-cadmium battery has become the choice for most situations.

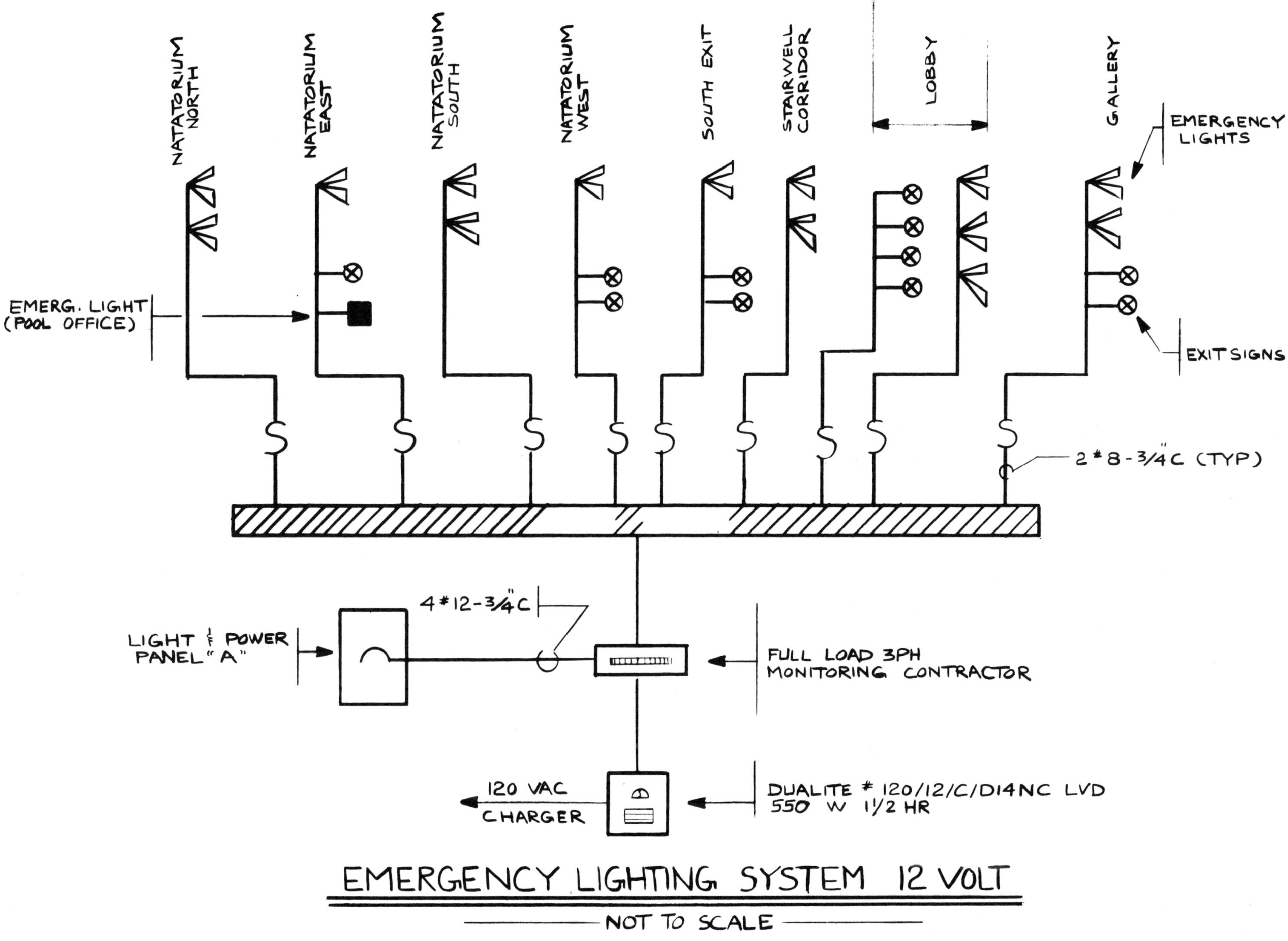
NATATORIUM NORTH
NATATORIUM EAST
NATATORIUM SOUTH
NATATORIUM WEST
SOUTH EXIT
STAIRWELL CORRIDOR
LOBBY
GALLERY
EMERGENCY LIGHTS
EXIT SIGNS
EMERG. LIGHT (POOL OFFICE)
2 # 8 - 3/4" C (TYP)
4 # 12 - 3/4" C
LIGHT & POWER PANEL "A"
FULL LOAD 3PH MONITORING CONTRACTOR
120 VAC CHARGER
DUALITE # 120/12/C/D14NC LVD 550 W 1 1/2 HR
EMERGENCY LIGHTING SYSTEM 12 VOLT
NOT TO SCALE

FIGURE 9-1

Inverter Systems As noted in our discussion of battery-operated emergency lighting systems, power is supplied from batteries at a variety of dc voltages that include 6, 12, 24, 32, 48, and 110 V. Battery systems can also supply current at 120-V ac. Figure 9-2 depicts such a system. This more expensive system is used primarily where long runs, few control sources, and ac operation are desired.

The inverter system is an alternative to the emergency generator and is being selected by consultants in increasing numbers. This system can accommodate normally on or normally off loads or split loads. Items that can be powered include fluorescent and incandescent lamps, small shaded-pole motors, universal motors, public address and fire alarm systems, and the like, so long as the total load does not exceed the system rating plus the starting inrush.

When there is a power failure the transfer relay will drop out, and its contacts will energize the inverter and transfer any normally on loads from the utility line to the inverter. Normally off loads are always tied directly to the inverter and thus will become energized only when the inverter is on. In a lengthy power failure which depletes the batteries, a built-in low-voltage disconnect will turn off the inverter, precluding deep discharge. Upon a return to normal power the transfer relay pulls in and disconnects the inverter. The charger then recharges the batteries.

The heart of the system depicted in Figure 9-2 is the inverter, which is a patented solid-state device which converts dc into sinusoidal ac, which is comparable to line power rather than to square-wave output. This provides for efficient operation of electronic loads and eliminates the overheating of inductive and motor loads commonly associated with square-wave inverters.

Figure 9-2 also illustrates some but not all, of the connections. Not shown is the line from the battery units to the utility service line. Also not shown is the transfer relay built into the battery units. These items are fairly self-evident. From the inverters, circuits feed both normally off and normally on loads through fused distribution panels.

As can be seen in the detail, the emergency load is much larger and more widespread than is normally encountered. The sizes shown are taken from an actual situation and cannot be directly used to fit your particular design. All lines should be carefully sized for your particular requirements.

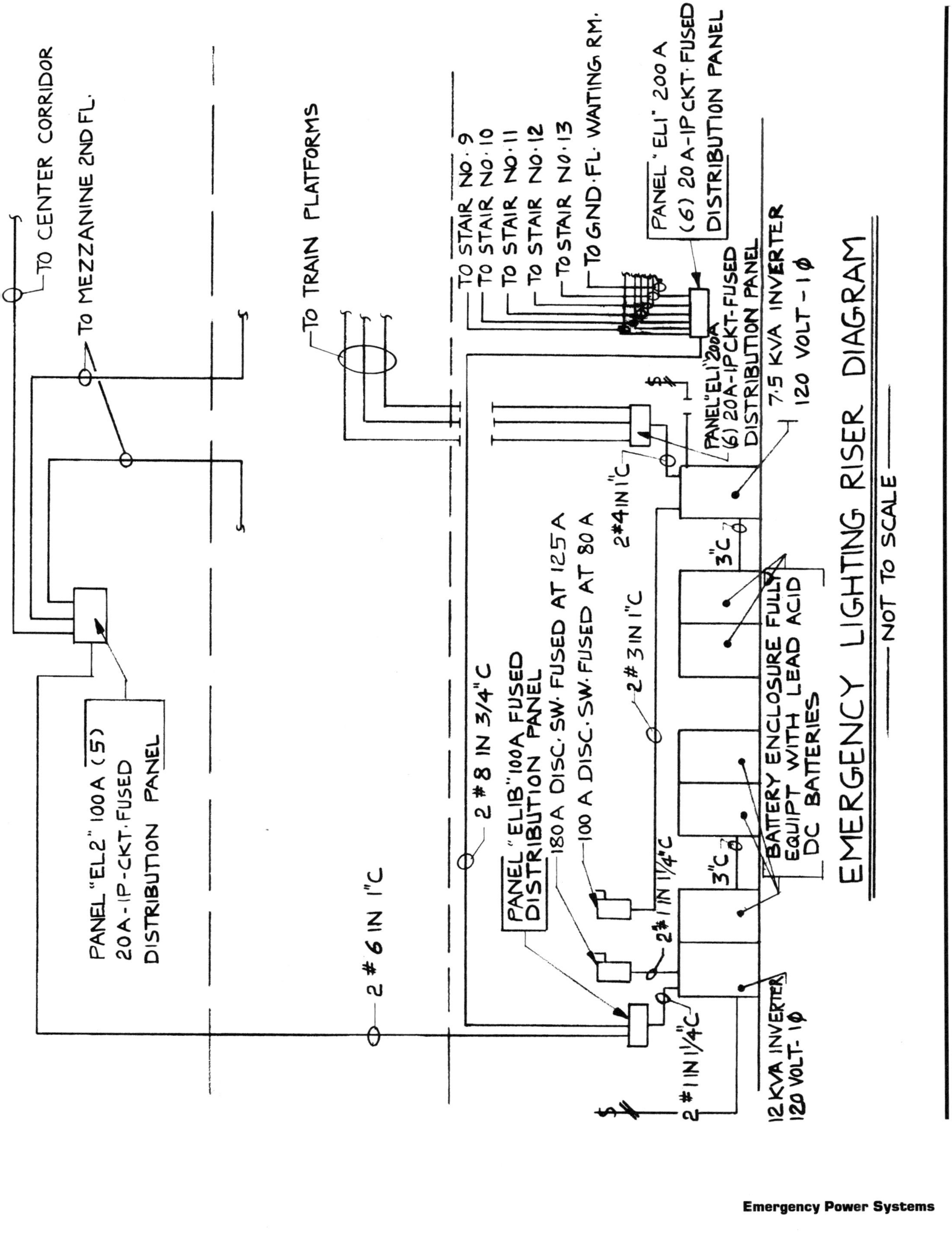
TO CENTER CORRIDOR
TO MEZZANINE 2ND FL.
TO TRAIN PLATFORMS
TO STAIR NO. 9
TO STAIR NO. 10
TO STAIR NO. 11
TO STAIR NO. 12
TO STAIR NO. 13
TO GND. FL. WAITING RM.
PANEL "ELI" 200 A
(6) 20 A-1P CKT. FUSED
DISTRIBUTION PANEL
PANEL "ELI" 200A
(6) 20A-1P CKT-FUSED
DISTRIBUTION PANEL
7.5 KVA INVERTER
120 VOLT - 1Φ
PANEL "EL2" 100A (5)
20A-1P-CKT. FUSED
DISTRIBUTION PANEL
2 #6 IN 1"C
2 #8 IN 3/4"C
PANEL "ELIB" 100A FUSED
DISTRIBUTION PANEL
180 A DISC. SW. FUSED AT 125 A
100 A DISC. SW. FUSED AT 80 A
2 # 3 IN 1"C
2 #4 IN 1"C
2 # 1 IN 1 1/4"C
2 #1 IN 1 1/4"C
3"C
3"C
BATTERY ENCLOSURE FULLY
EQUIPT WITH LEAD ACID
DC BATTERIES
12 KVA INVERTER
120 VOLT - 1Φ
EMERGENCY LIGHTING RISER DIAGRAM
NOT TO SCALE

FIGURE 9-2

Water-Cooled Emergency Generator An emergency generator is usually powered by a liquid- or gas-fueled internal-combustion engine. As in any power plant the work performed creates heat which must be dissipated. Figures 9-3 and 9-4 illustrate how the piping is arranged for a water-cooled remote radiator system.

Commonly the radiator is an integral part of the unit, and if that is your situation, it would have to be modified. Where versatility, less power-room ventilation, and low-noise fans are part of the design criteria, the details shown will apply provided you understand that they are more expensive.

In Figure 9-3 we show a high remote radiator installation which is similar to the problem of a nearby remote radiator installation. In this detail, when the radiator top is more than 15 ft above the engine centerline, a hot-well tank and auxiliary pump are required. A storage tank reduces the water head pressure on the engine to a value that the engine can accept. It is a two-section tank with a partial baffle to separate the hot outlet water from the cold inlet water.

The engine pump circulates water between the engine and the hot-well tank. The auxiliary pump circulates water between the hot-well tank and the remote radiator. The auxiliary pump must supply the same flow as the engine pump. If not, a larger radiator is required.

Because the radiator drains into the cold side of the hot-well tank after the generator is shut down, the baffle between the hot and the cold side must have an opening large enough to allow free passage of water up to the flow rate of the engine or auxiliary pump, whichever is larger. In Figure 9-4 we show a typical hot-well tank which illustrates what we have been saying in greater detail.

To further reduce airflow ventilation requirements, water-cooled exhaust can be used. Models cooled by city water need sufficient airflow to remove heat radiated from the engine generator, exhaust, and plumbing. Figure 9-4 also shows a standpipe installation. In this case, the engine pump pushes cooling water from the standpipe, through the engine, and back to the standpipe. The city water supply forces water into the standpipe and out an overflow line to drain. Heated water from the engine and cooler water

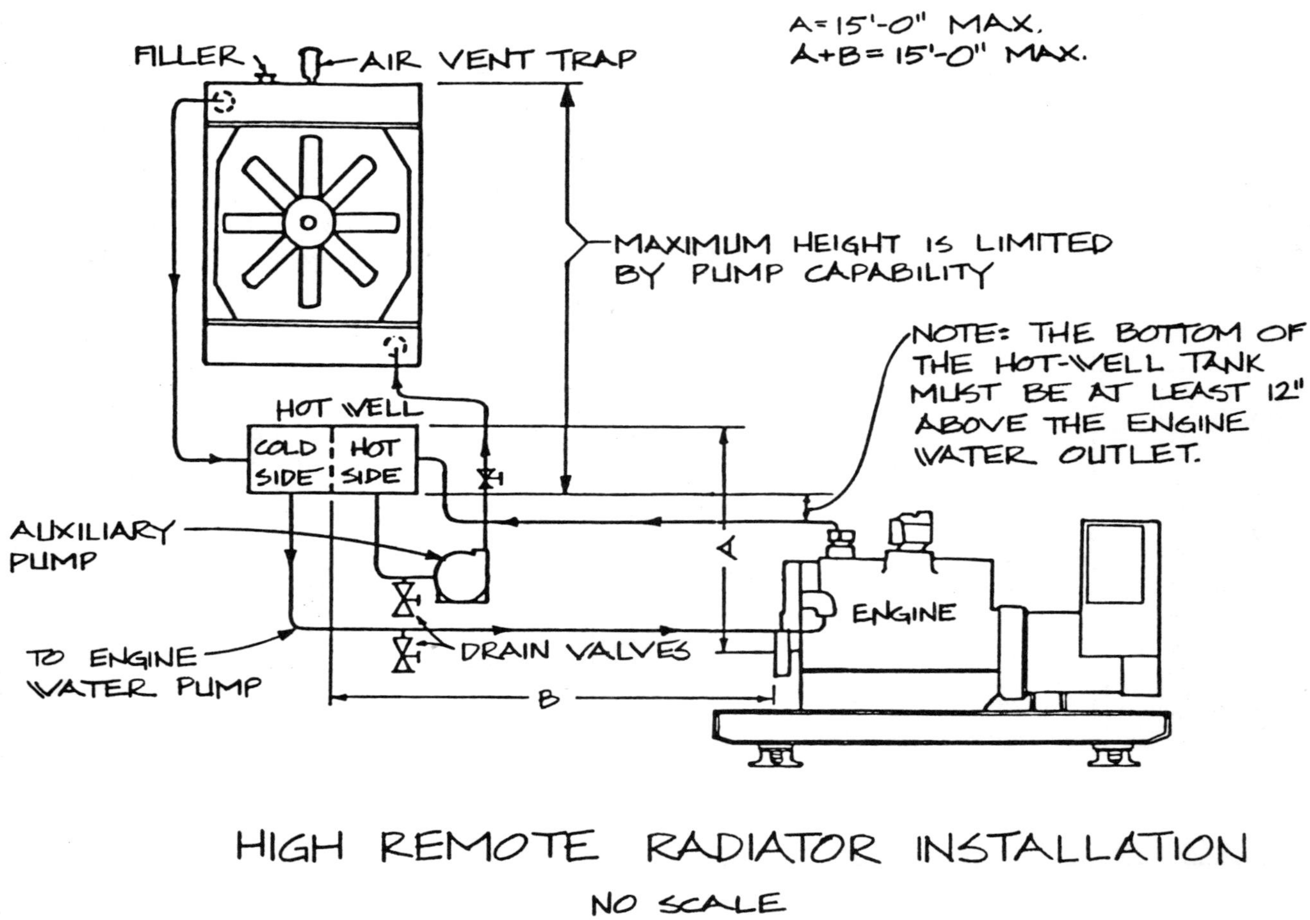

FIGURE 9-3

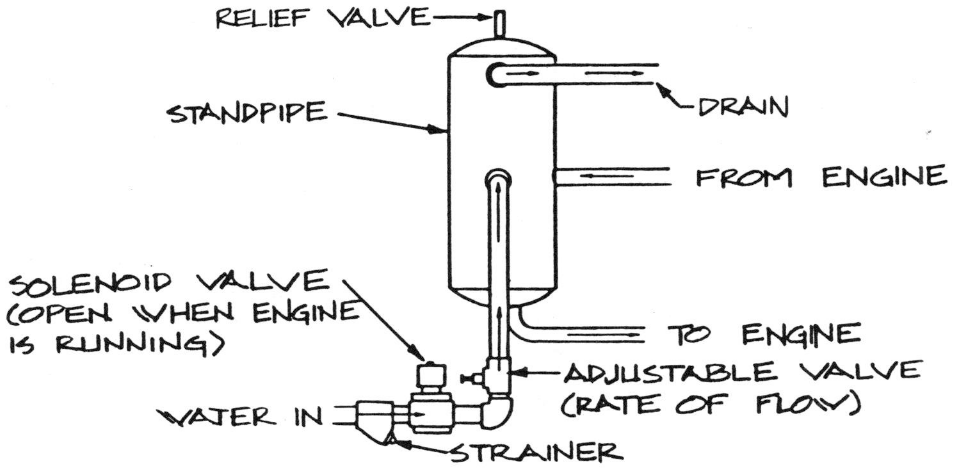

STANDPIPE COOLING SYSTEM

NO SCALE

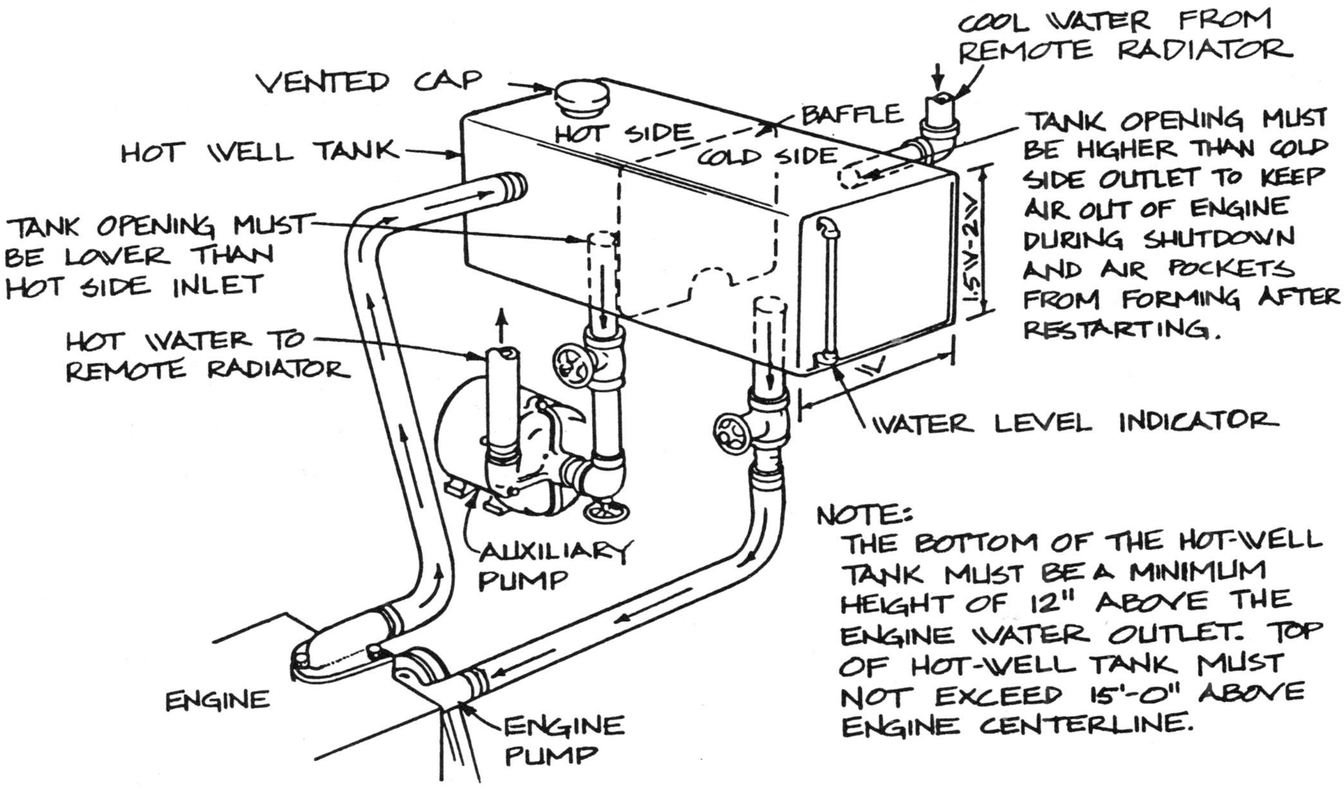

TYPICAL HOT WELL TANK

NO SCALE

FIGURE 9-4

mix in the standpipe to provide a proper-temperature supply for engine cooling. The vacuum-relief valve in the standpipe eliminates a siphoning effect caused by a long drain discharge line.

Long Remote Radiator—Emergency Generator As described in our discussion of water-cooled emergency generators, the radiator used in the cooling system can be remotely located. If it is on the same level and nearby it is referred to as a "short remote radiator," and, if it is some distance away, it is described as a "long remote radiator." As in the vertical system previously shown, 15 ft 0 in is the dividing line. Over 15 ft is long and under 15 ft is short.

Figure 9-5 shows the long system. To be very specific, "When the sum of the vertical distance taken from the engine centerline to the radiator top (A) and the horizontal distance from engine front to the radiator centerline (B)

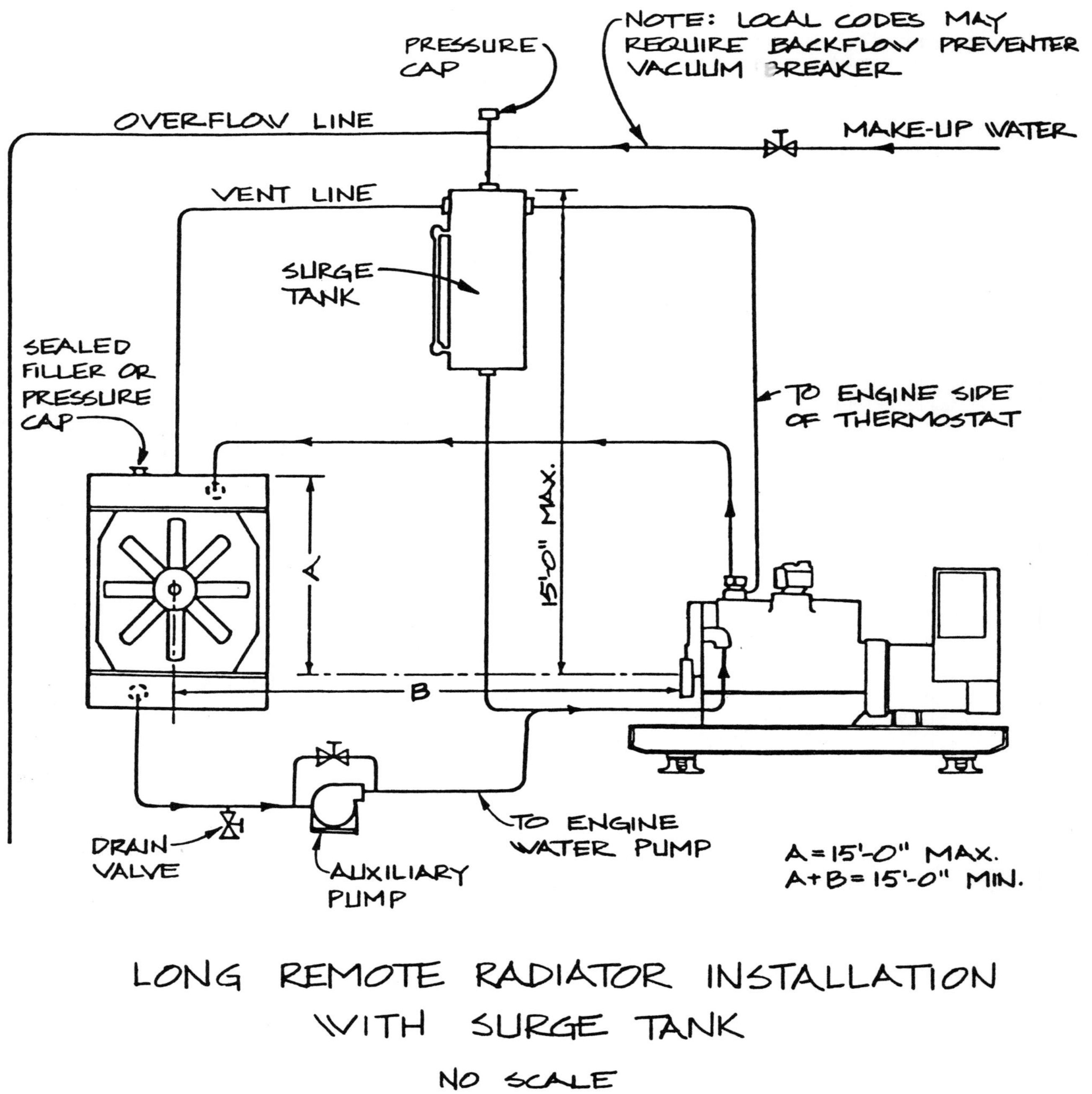

FIGURE 9-5

exceeds 15 ft *but the vertical distance alone does not exceed 15 feet*" is the criterion for this system. If, as shown in Figure 9-5, the values of those two lengths *do not* add up to 15 ft, you can delete both the auxiliary pump and the surge tank. In that case you merely have lines running from engine outlet to radiator and back from radiator to engine inlet. The drain valve on the return line is required for either situation.

The surge tank, in a somewhat different setup, now appears under the hood of many late-model cars. It is usually plastic with a removable cap. If you read your owner's manual you will know this is supposed to be about half full. The expansion and contraction of water being heated and cooled can take place in the surge tank. Also the surge tank provides a fill location as well as a supply of water to more readily balance the flow of a double-pumped water-circulating system for an emergency generator.

The other item is the remote pump which, as long as there is a positive head on the pump, can be located anywhere. Good practice would be to locate the pump near the remote radiator. Notice that there is a manual valved bypass that can be used when required. Not shown are two additional gate valves inside the supply and on the pump inlet and discharge openings. The system requires careful sizing of pipes, surge tanks, and pumps. The remote pump points up the fact that the built-in engine pump simply is not capable of pumping the water volume and rate required at that distance. Manufacturers do not design coolant-circulating pumps to handle the unusual or excessive piping or heat losses that this type of installation creates.

Air-Cooled Emergency Generator Ductwork Frequently the emergency generator selected is air-cooled and not water-cooled. Figure 9-6 shows the source of air supply and exhaust for this type of installation.

There are several points to be made in discussing Figure 9-6 which are all too frequently overlooked. The first point, not shown on this detail, is that it applies in part to every emergency generator installation. The generator, even if it is water-cooled, is an engine, just like the one in your car. And, if you carefully vented the engine exhaust out of your car by using a hose to the outside of a closed garage, you would still have a very hot garage if you kept the engine running any appreciable length of time. In the piping details that follow you will note that there is always a ventilation system of some sort for the generator room.

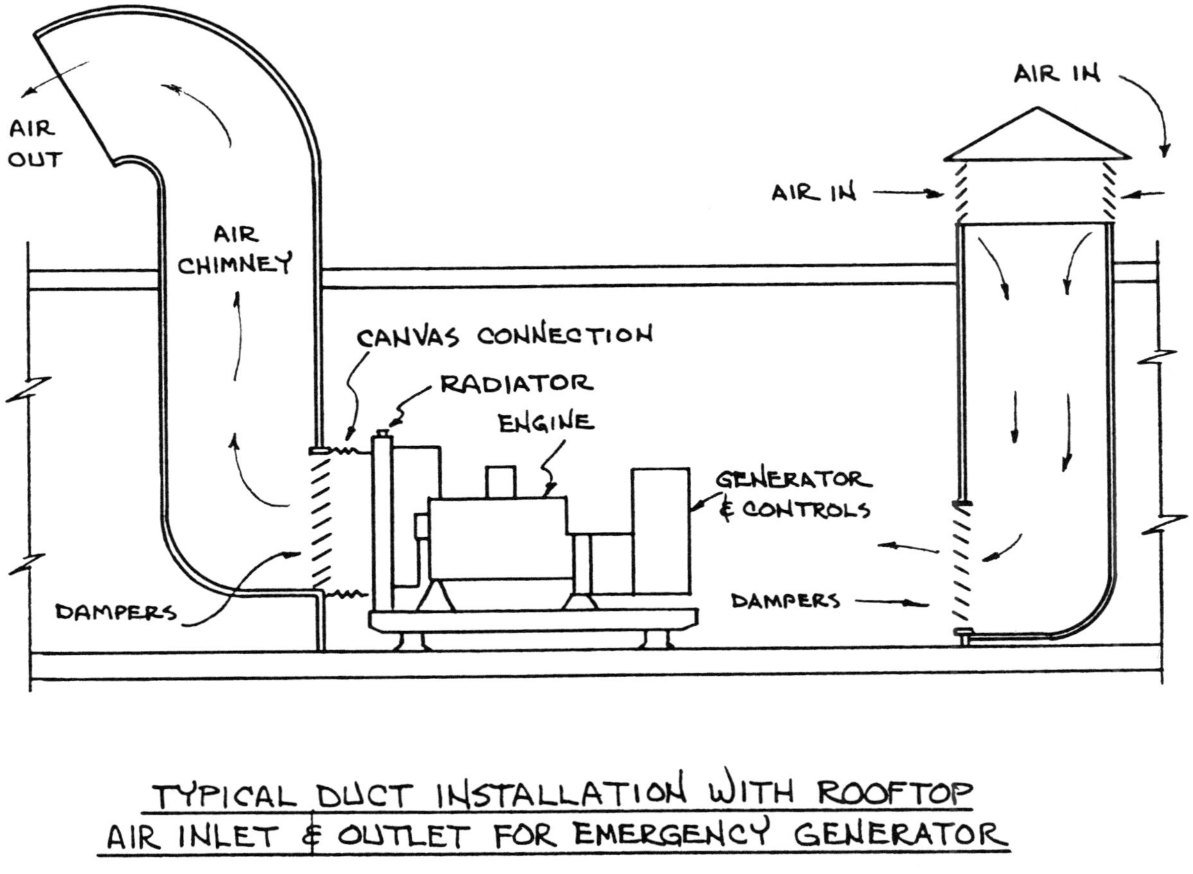

FIGURE 9-6

To determine the basic rate of air needed there is the standard formula, $V = (Q/1.08) \times$ temperature difference, where V is cubic feet per minute, Q is the heat to be removed in Btu/h., 1.08 is constant, and temperature difference is the permissible rise in room temperature. For example, if the heat rejected equaled 120,000 Btu/h. and the temperature was allowed to be only 20°F hotter than normal, $V = (120{,}000/1.08) \times 20$, or 5.555 ft^3/min. Since square feet of duct is based on a velocity of some 400 to 600 ft/min in the usual case, we would have both intake and exhaust ducts (using 500 ft/min average) equal to 5555/1111 or 5 ft^2. This would be approximately 2 ft 6 in $\times$ 2 ft 0 in.

In the detail shown the heat rejected is the use of radiator-dissipated heat and engine and generator radiated heat. The duct work as shown could in some situations easily be 6 ft 0 in $\times$ 8 ft 0 in or larger where large air supplies are required.

The material specification for the duct could be taken from your standard mechanical specifications. We have shown the intake and exhaust location to be on the roof. They could have been on the walls with vertical locations above ground in conformance with the building codes in your area. Dampers are required in both the intake and exhaust ducts and are normally of the self-closing type called *butterfly dampers*. The push or pull of air opens the damper. The dampers return to a closed position when the engine is shut down. To avoid transmission of vibration a heatproof flexible canvas connection is placed between the duct and the air-cooled radiator. The duct that connects to the radiator will be at least the full size of the radiator grillage with no reduction in size thereafter. While the air could have been brought in high on the wall, superior results are obtained by ducting as close to the floor and as close to the engine as is reasonably possible.

Pressure-Cooled Emergency Generator Figure 9-7 is the first of four details that appear very similar at first glance. To help you to appreciate the subtle differences, we have separated the details by some text material.

Figure 9-7 not only is air-cooled but also has an integrally mounted fan to assure engine airflow over the radiator. In this case it is not necessary (though it would desirable) to duct the outdoor air supply to the floor and near the unit. The fan assures that the generator will get the air it needs. And to make certain the engine air is properly removed, there is a solidly connected duct to discharge it directly to the outside. We have not shown whether this goes through a roof or sidewall, and we leave that to your judgment. We have also shown a relief louver to allow other generated heat to be dissipated outside.

Electrically we have shown a couple of conduits to the transfer switch. Their size, location, and arrangement must be a part of your design. A new element, the engine exhaust pipe, is shown. This is usually a 4-in schedule-40-welded iron pipe. You will need a muffler located where we show it, or nearer, for the generator exhaust. We have vaguely indicated a flexible connection on this line, which may be required.

For the fuel supply we have shown a buried tank with supply and return piping common to diesel fuel systems. In this type of fuel system the pump is usually designed to push more fuel than the engine requires. Note that the supply pipe extends into the tank (usually to within 6 in of the bottom) but that the return terminates at the top of the tank. The tank has both fill and vent connections which must be sized and located as per the codes in your area. Finally, all generators vibrate, and, to avoid problems, we have shown the generator mounted on vibration isolators.

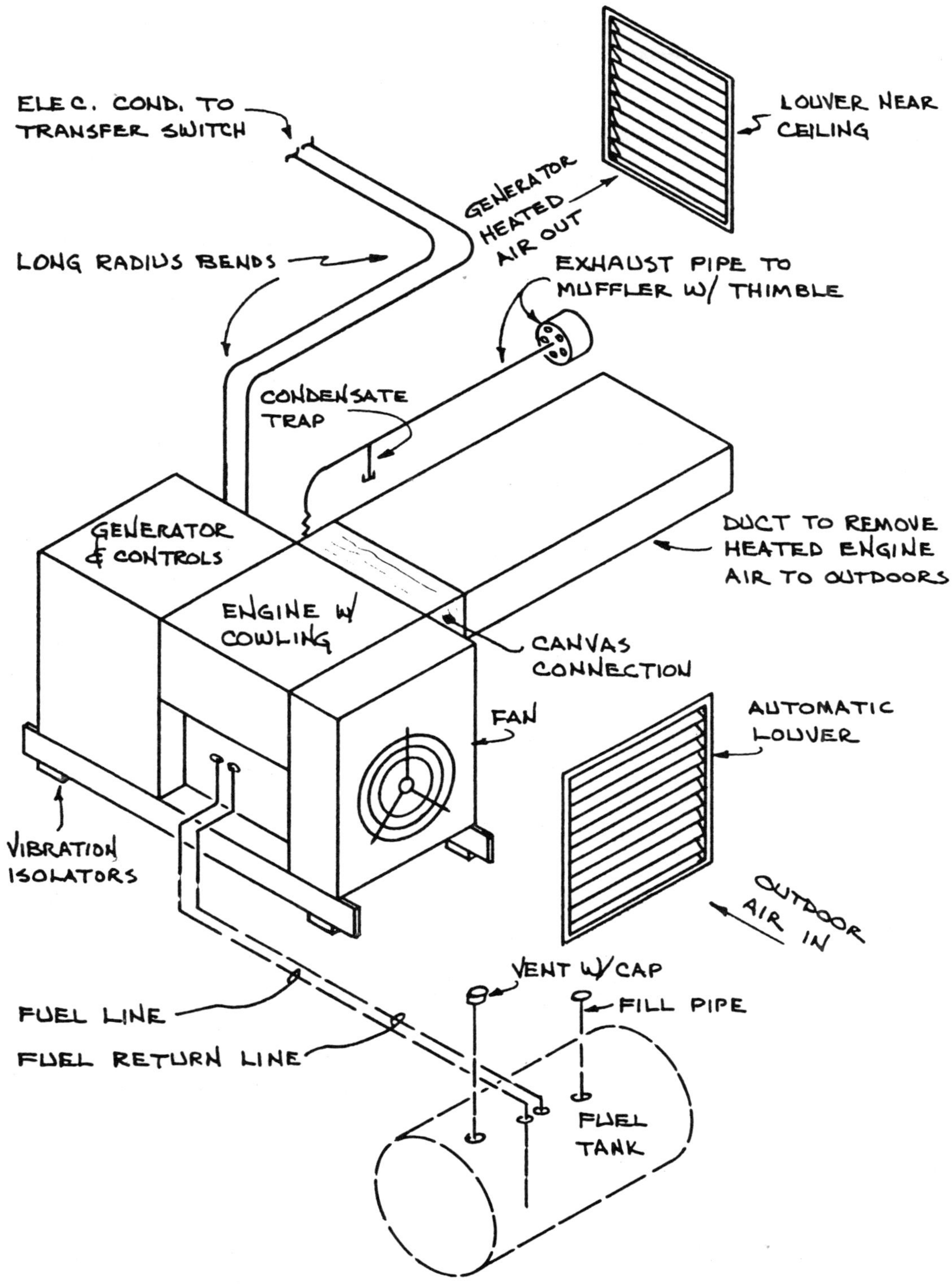

TYPICAL AIR COOLED (PRESSURE COOLED) ENGINE DRIVEN EMERGENCY GENERATOR

FIGURE 9-7

In general this detail applies to a large number of installations. The generator you select may not have the fan we have noted, but all other parts of this detail are quite common.

Radiator-Cooled Gas or Gasoline Generator Generally the catalogs of most generator manufacturers describe the installation shown in Figure 9-8 as a radiator-cooled unit, which is why we have used the same title. As can be readily seen, the radiator is cooled by air. The title could mislead one into thinking the radiator is water-cooled. The detail is a composite type and depicts the engine being fed by two sources. In the case of your design you should select *one* of the two fueling systems and *omit* the other.

While the cooling air enters the room in the same manner as in previous details, the air off of the radiator is ducted directly outside even though there is no special fan on the generator itself. While the normal radiator fan does not have the same capabilities as a special fan to propel air, the ducted air scheme shown is strongly recommended wherever possible.

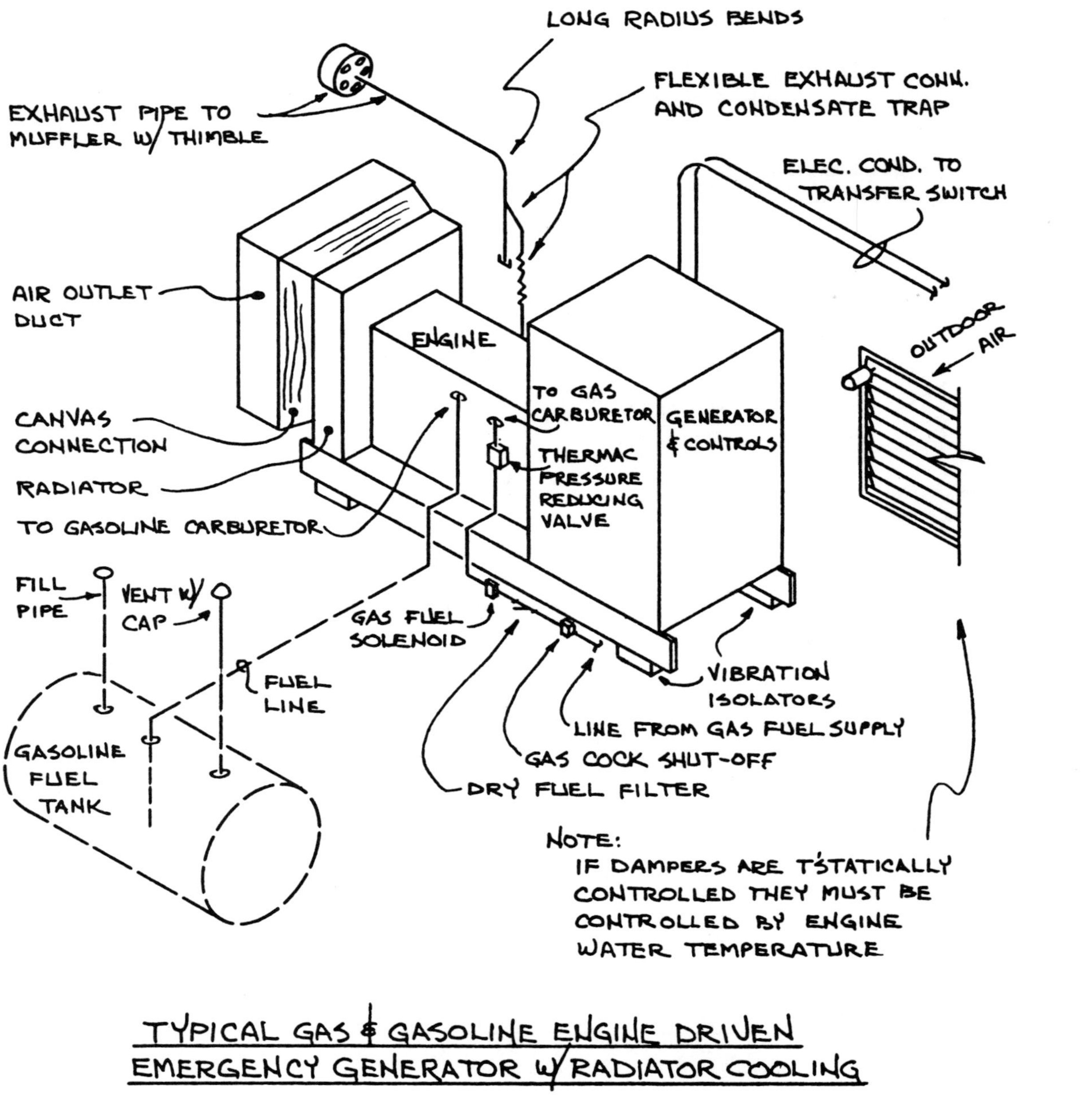

FIGURE 9-8

If you use the gasoline option you will need a buried tank of sufficient volume to enable the generator to run a considerable length of time without refill. Exact values and conditions should be checked against codes applicable in your area and with the requirements of your client. The same line of reasoning applies if your source of gas is bottled gas (propane) and not utility-supplied natural gas. Note also that in the gasoline option you need fill and vent connections that must comply with the codes applicable in your area. And further, the gasoline line from the tank to the generator goes into the tank and terminates about 2 in from the bottom of the tank. The gas-piping train may require additional special features, which are not shown in Figure 9-8.

In the engine exhaust line we have shown a flexible connection at the engine and a special wye connection to the line. The drip line shown may or may not have a drain valve to collect both dirt and moisture. A removable cap can serve the same purpose. The muffler is a definite requirement and is usually part of the generator specification. A generator without a muffler sounds like a car without a muffler—only louder.

Again we have shown the electrical conduits between the generator and the transfer switch, and finally the vibration isolators. Note the canvas connection between the radiator and the exhaust duct work. Note also that the outdoor-air damper indicates a possible motorized connection.

Water-Cooled Gasoline Generator Figure 9-9 shows an uncommon type of emergency generator installation in this era of energy and water conservation. The installation has a radiator cooled by city water. The water could be from any available source such as a nearby lake, pond, or stream.

The cooling system utilizes the standpipe installation which we presented in Figure 9-4. The water to the engine goes in and out of the standpipe and the city supply forces water into the standpipe and out the drain. This arrangement creates a balanced supply of water at power temperature for the engine. A solenoid valve stops water flow when the engine is off through its connection to the engine's starting system.

The details of the gasoline piping system are the same as they were for Figure 9-8, and if you use natural or bottled gas, you can substitute the gas-piping train shown in Figure 9-8 for the gasoline system depicted in this detail.

The engine exhaust piping, its flexible connection, drip leg, and muffler are the same as shown in Figure 9-8. The outdoor-air damper indicates a possible motorized connection. This is similar to Figure 9-8 with an added feature that we cover in the following paragraph.

There is always a discussion of air intake and exhaust openings in an emergency generator installation—especially when the system is water-cooled. Except for very cold climates, the chances of freezing the radiator are not very great, but they can and do occur—even in relatively *safe* locations. Our detail shows a propeller-driven fan behind a wall louver that is set to run when the generator runs. Normally it would have a self-closing damper on its discharge to close when the fan is off. To be certain that a problem does not arise you could further install a motor-operated damper on the intake air louver. When the fan starts it would also open the intake damper and let air in. When the fan stops it would close the intake damper to keep air from entering the room. The ultimate would be to also change the self-closing damper to a motor-operated damper. Now both are assured of only being open when the engine and fan are running. It is a question of just how safe you wish to be.

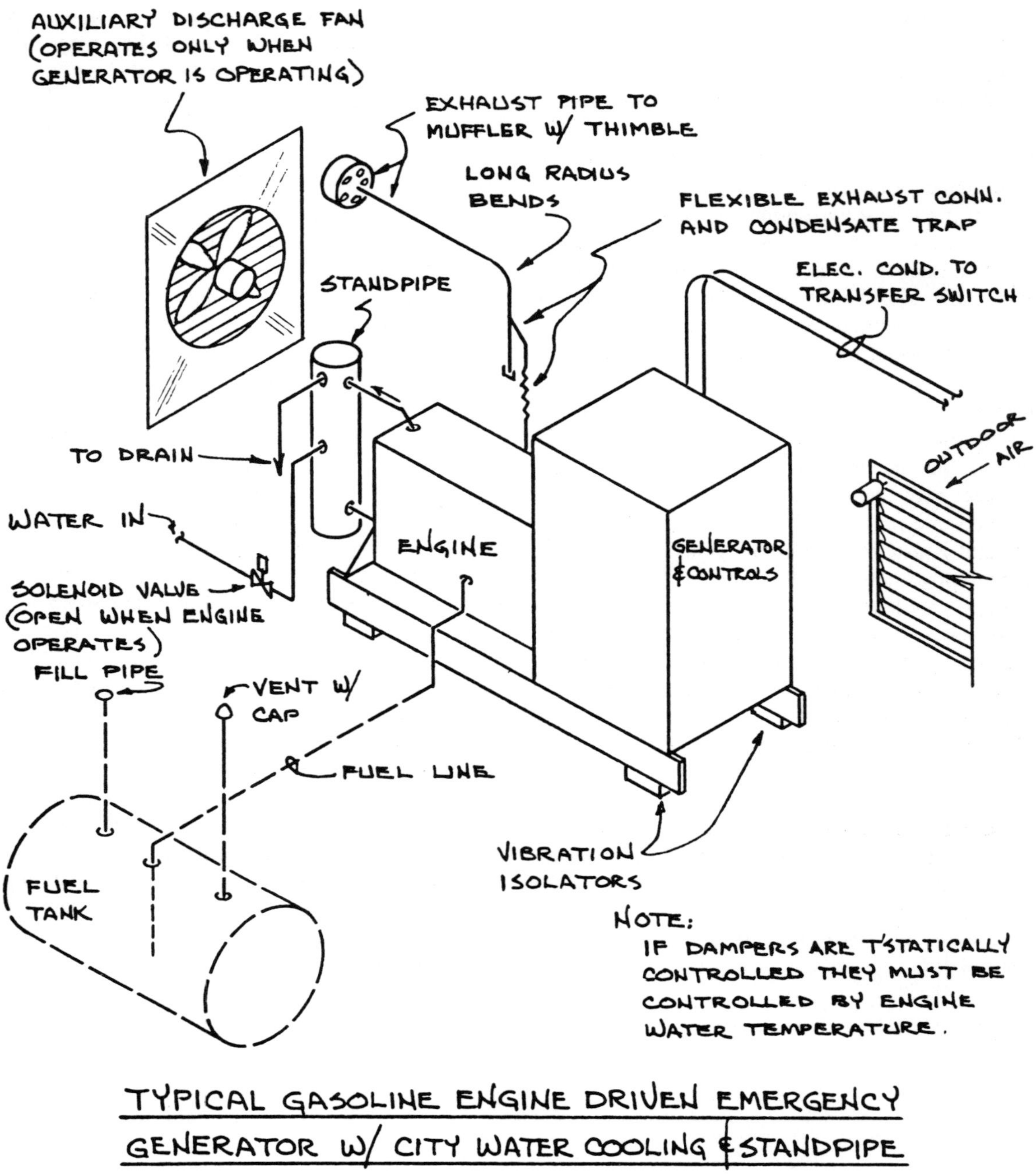

FIGURE 9-9

Finally, we show the electrical conduits and the vibration isolators as before.

Radiator-Cooled Diesel Engine Generator Figure 9-10 is another detail that has the words "radiator cooling" in its title. And again, this means air-cooled and not water-cooled. This detail is, at first glance, almost identical to the one shown for a gas- or gasoline-driven system. The details for the ducted radiator air exhaust, the air exhaust, the engine exhaust, the electrical conduits, and the vibration isolators are identical. But there are some special differences.

The air system indicates that the intake and exhaust louvers may have motorized dampers. And, to answer the control question, the dampers, as noted on the detail, must be controlled by the engine water temperature. This not only is a positive means of controlling intake and exhaust air but it also protects against freezing and is an energy

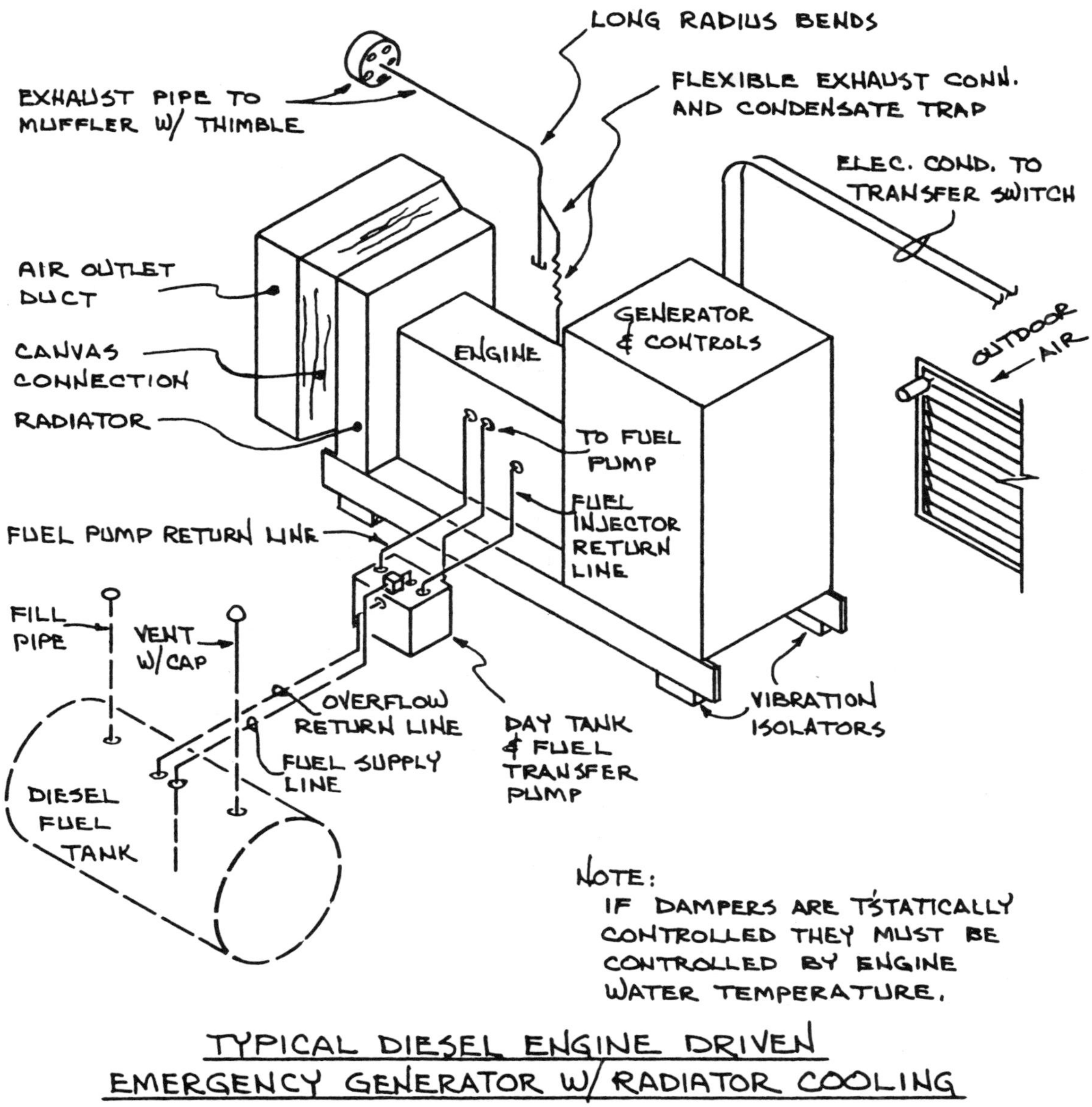

FIGURE 9-10

optimization measure. As we noted in a previous discussion of a similar system, the use of dampers is a design and safety choice of your firm.

The diesel tank has an added feature of a day tank and a transfer pump. There are slight differences among manufacturers as to how the day tank and transfer pump will be connected. Check with the manufacturer of your tank to be certain of the proper relationship of the connections shown. We did not show them in our previous general detail for that reason.

The day tank is used to solve a problem where the fuel pump does not have the capacity to draw fuel from the supply tank when it is considerably below the pump intake or when the supply tank is much higher than the engine intake and thus presents a problem of high pressure on the fuel return line. The transfer pump may or may not be needed. If it is needed, it is installed in the supply line *between* the main tank and the day tank. There are instances in overhead installations where there would be two tanks. One tank (the day tank) would be on the supply line to the engine and a return tank would take fuel from the engine return and then, via a return pump, send the fuel back up and into the overhead tank. One of our following *separate* details depicts this particular situation.

Using the four details just presented or parts of them, you can construct a detail that will fit most any situation.

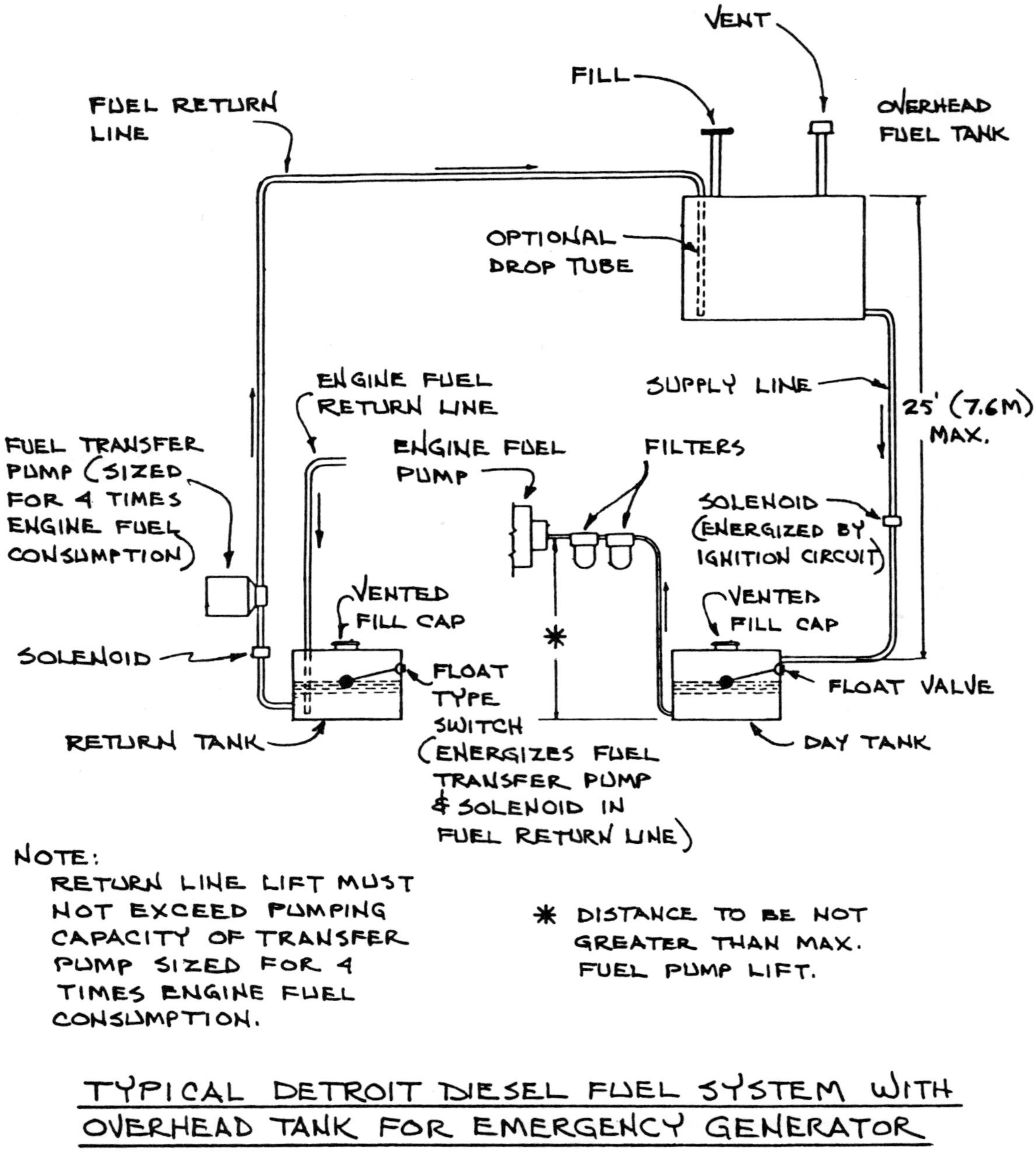

FIGURE 9-11

Overhead Tank Installation In a previous discussion of Figure 9-10 we alluded to a situation where not one but two tanks would be required. Figure 9-11 illustrates that situation.

The right-hand portion of the detail illustrates the typical diesel system with an overhead fuel tank. The day tank is installed relatively near to the generator's engine and within the engine's fuel pump lift capability and also below the fuel-injection system. The solenoid valve in the line stops fuel flow when the fuel level of the day tank reaches the preset optimum level. The solenoid controls the ignition circuit.

In Figure 9-11 the fuel which would normally be returned to the day tank cannot because it has been used for cooling engine components. Were it to be returned it would create a rise in the fuel temperature in the day tank and a loss in engine power. Thus a separate solution for the return line must be created.

The solution is to direct the returned fuel to a separate tank which is similar in design to that of the day tank. A

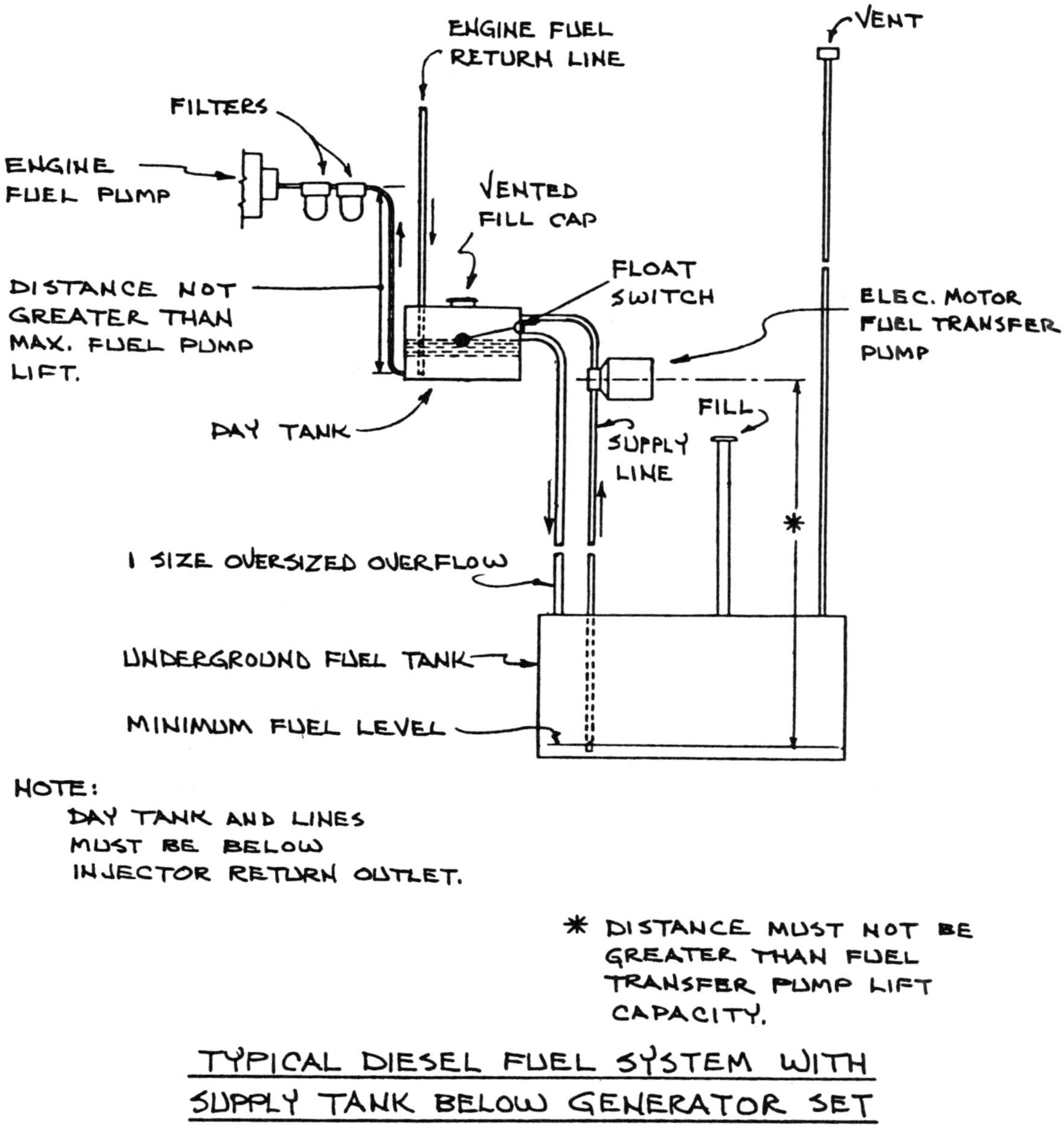

FIGURE 9-12

float switch in this tank energizes the shut-off solenoid and the fuel pump, which transfers the fuel back to the overhead tank.

It is important to note that the return fuel line enters the return tank and extends down below the minimum level in the return tank. This is to preclude a drain back from the engine fuel pump and filters when the generator is idle. The pumped return enters the overhead tank, and the extension to the bottom, as noted, is optional, not mandatory.

In closing we would like to emphasize the frequently mentioned subject of hydraulic lock. Because of the serious problem of fuel trapped on top of pistons (hydraulic lock) which can cause serious engine damage, you must not use a gravity-feed system tied directly to an engine fuel pump. Fuel *returns* from an injector system must gravity feed to a day tank located *below* the injectors.

To be certain of the piping arrangement of your particular generator, it is *most* advisable to consult with the engineering representatives of the manufacturer before finalizing your detail.

Typical Diesel Fuel System For many applications the diesel fuel system with its supply tank below the generator is a common. This is depicted in Figure 9-12. The underground tank has a supply of fuel sufficient for 36 hours of operation. The fill and vent lines are sized in accordance with local codes. The fuel-supply line has a transfer pump with a solenoid valve on the suction side of the pump (valve not shown) to assure the liquid is solid to the pump when the pump starts. The asterisk in the detail calls attention to an important point, that is, the suction lift capacity of the pump. Some manufacturers advise that this distance is a maximum of 15 ft. If a distance is noted on your detail, be certain it is within the limits of the pump you have selected.

The day tank is located near the generator's engine, below the fuel-injection system and within the limit of the fuel pump's lift capability. The supply pump is, in actuality, close to the day tank, and the float switch in the day tank controls its operation.

The supply tank top must be below the top of the day tank to preclude siphoning from the fuel supply tank to the day tank. The return line must be provided from the engine injection system to the day tank and extended to near the bottom of the day tank (below the minimum level of the fuel in the tank) to preclude drain back from the engine fuel pump and filters.

If your generator is made by Detroit Diesel, it is an exception to the above system. As in our previous detail, the Detroit Diesel, for reasons of heat transfer, must have its return go directly to the main tank. With the main tank below the generator this is pure gravity flow.

To keep the day tank from overflowing, a line one size larger than the supply line is installed as depicted. Since this line is already a part of the piping, the usual procedure is to connect the Detroit Diesel return line to the overflow line. There is no need to put in two lines when one will do. This creates a tee connection in the overflow line near the day tank. To make sure fuel does not take a wrong path, a check valve is placed in the overflow line as it leaves the day tank.

Primers and Antisiphon Devices If you use gasoline as a fuel for your emergency generator, your day tank system is very similar to the diesel system we have previously presented. The day tank is located *below* the carburetor and within the lift capability of the pump. Gasoline systems do not use a fuel return and you should carefully check your local code requirements. Do not vent the day tank's fill cap.

In Figures 9-13 and 9-14 we depict solutions to a couple of problems with gasoline systems. Figure 9-13 is a primer system. Since it is a light volatile fuel, gasoline can evaporate out of the carburetor bowl during long shutdown periods. The primer system depicted gravity feeds fuel to the carburetor. The solenoid valve is arranged to open during generator operation. It also prevents gasoline from draining into the engine.

The primer tank is pressurized by a bushing in the return line. The return line to the supply tank, acting as a vent, will not gravity feed if the line is in any way trapped, and thus it holds fuel which blocks air movement through the line.

Figure 9-14 is an antisiphon system which is frequently used when the gasoline tank is above the generator. The special piping described in the detail solves the problem in the following manner. Fuel from the supply tank fills a void created by the engine pump in a vacuum pipe. The vacuum pipe, as noted, is approximately four times as

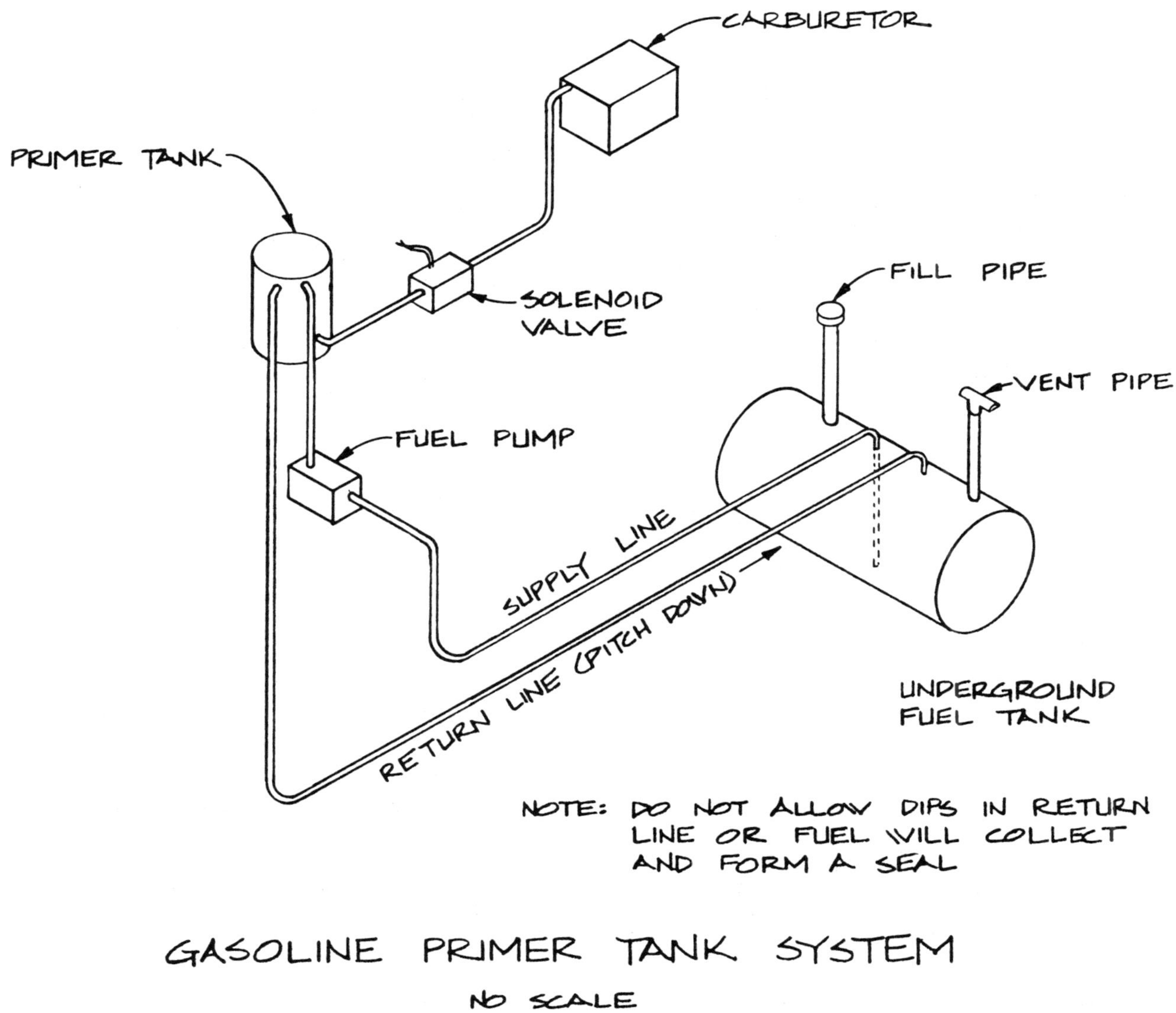
CARBURETOR
PRIMER TANK
SOLENOID
VALVE
FILL PIPE
VENT PIPE
FUEL PUMP
SUPPLY LINE
RETURN LINE (PITCH DOWN)
UNDERGROUND
FUEL TANK
NOTE: DO NOT ALLOW DIPS IN RETURN
LINE OR FUEL WILL COLLECT
AND FORM A SEAL
GASOLINE PRIMER TANK SYSTEM
NO SCALE

FIGURE 9-13

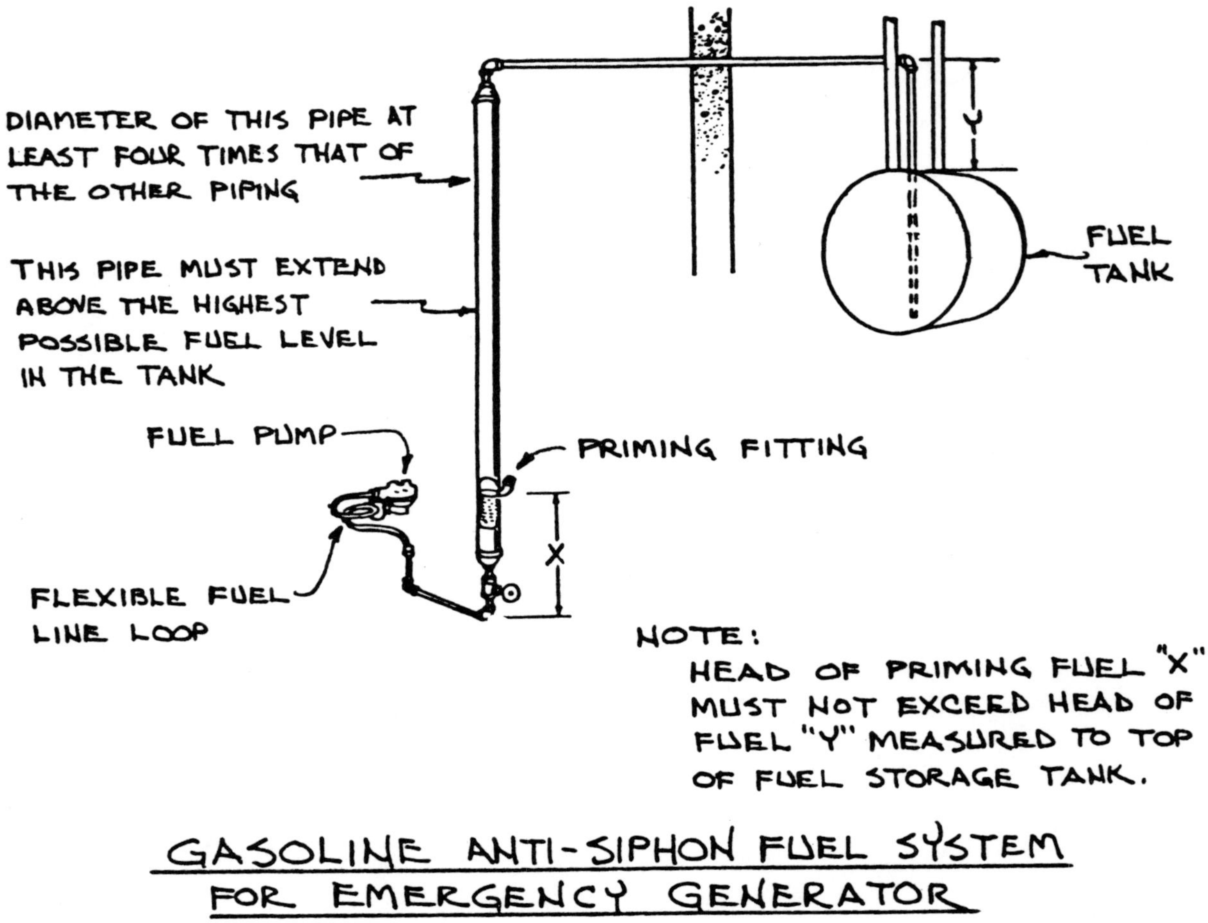
DIAMETER OF THIS PIPE AT LEAST FOUR TIMES THAT OF THE OTHER PIPING
THIS PIPE MUST EXTEND ABOVE THE HIGHEST POSSIBLE FUEL LEVEL IN THE TANK
FUEL PUMP
FLEXIBLE FUEL LINE LOOP
PRIMING FITTING
X
Y
FUEL TANK
NOTE:
HEAD OF PRIMING FUEL "X" MUST NOT EXCEED HEAD OF FUEL "Y" MEASURED TO TOP OF FUEL STORAGE TANK.
GASOLINE ANTI-SIPHON FUEL SYSTEM FOR EMERGENCY GENERATOR

FIGURE 9-14

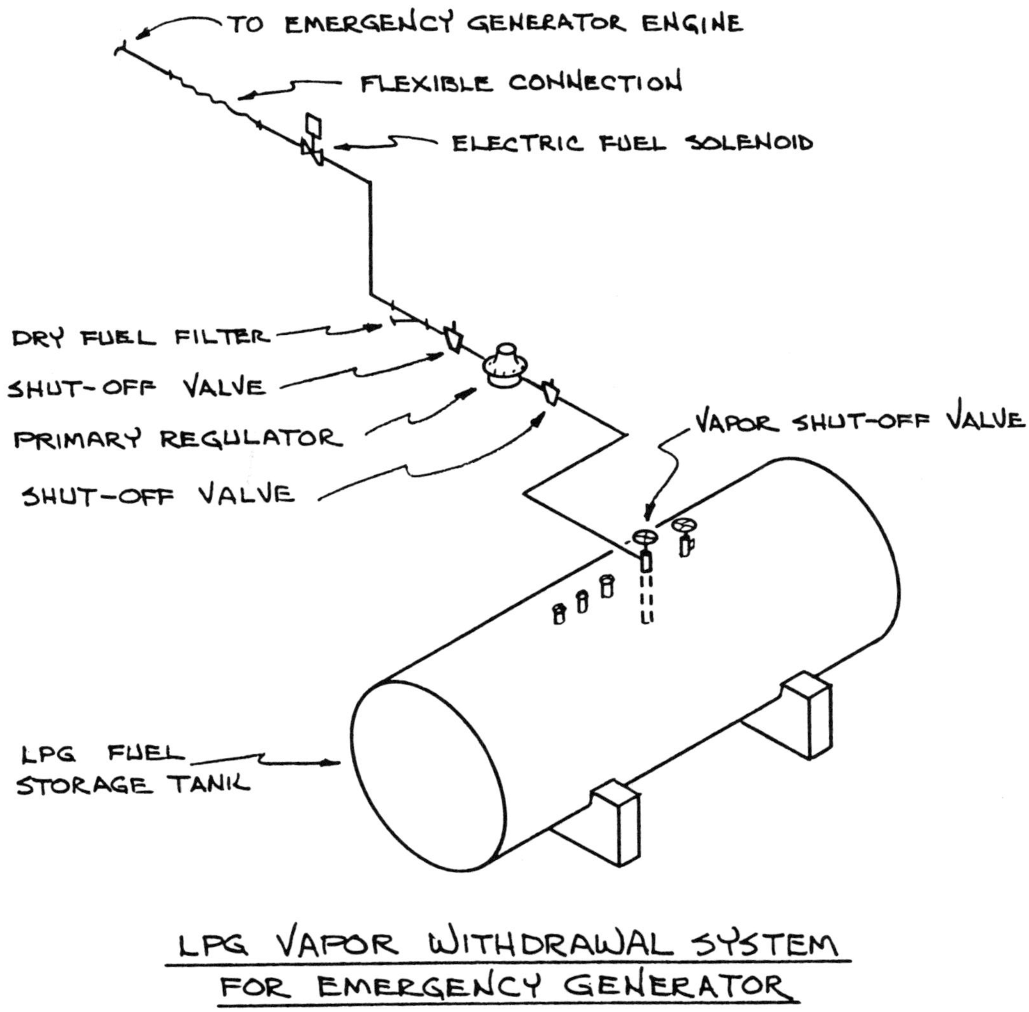

FIGURE 9-15

large in diameter as the line from the fuel tank to this pipe. It sounds simple and it is. Because we are dealing with pressure-sensitive situations, the head (height) of the priming fitting (X) must not exceed the head of the fuel line where it arises vertically out of the tank (Y).

General Details In bringing our discussion of emergency generator installation to a close, we are including five details covering certain specific points of the overall installation that may apply to your particular situation.

In some areas the choice of fuel may not be the usual diesel or gasoline. There are situations where bottled gas, sometimes called LPG (liquefied petroleum gas), or propane is the fuel. Figures 9-15 and 9-16 illustrate some pertinent points on the LPG installation. Figure 9-15 illustrates the general piping arrangement without respect to location of the devices. Coming from the tank shut off we have a gas-flow regulator isolated by valves on either side. The solenoid is the controlled, positive fuel stop wired into the engine ignition circuit. Figure 9-16 indicates other items that may or may not be a part of your system. We could call this a reminder detail. Check with the manufacturer of your generator and the supplier of your LPG fuel. Do you need the potential vaporizer (to make certain you get gas vapor) to your system? Is the delivery pressure such that you need a pressure-reducing valve? If you do, this is how you detail these devices. If you do not, you can simply forget them.

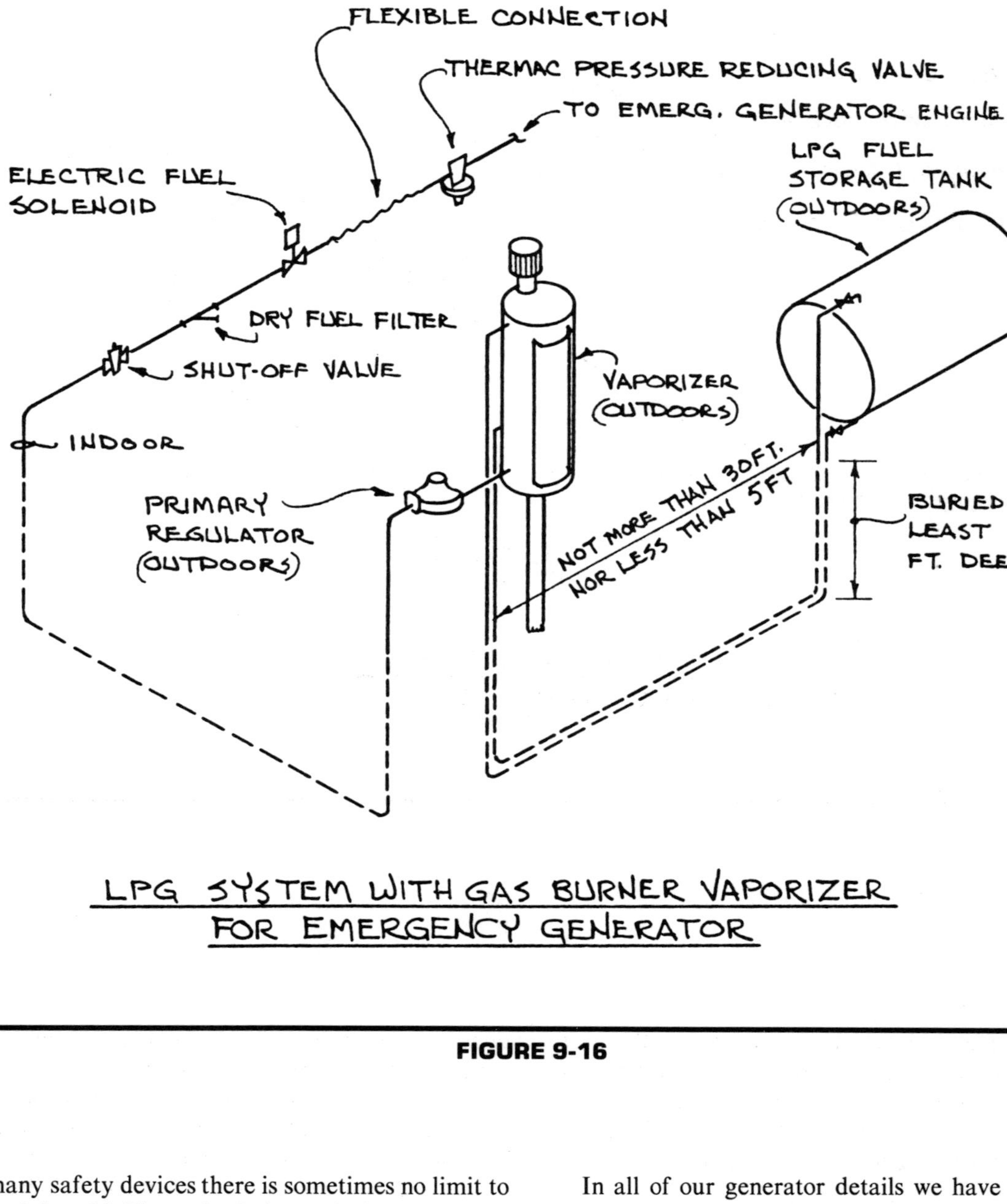

FIGURE 9-16

As in many safety devices there is sometimes no limit to what you might add. Figure 9-17 shows the installation of a low-level alarm switch in a fuel tank. This single-pole double-throw switch senses fuel level, and when the level is too low it sounds an alarm. (The alarm bell is by your specification or is a client-supplied sound unit.) The pipe is usually set within 1 in below the minimum fuel level. The note on the detail gives a generalized rule for this riser pipe.

In all of our generator details we have referred to the requirement for vibration isolators. Usually this requirement is noted in the specifications as "to be furnished as specified by the manufacturer." The two commonest types shown in Figure 9-18 are the spring type and the pad type. We could also describe this detail as a protection detail. You should show one or the other, depending on your requirements, to be certain that you get at least some kind of vibration isolation.

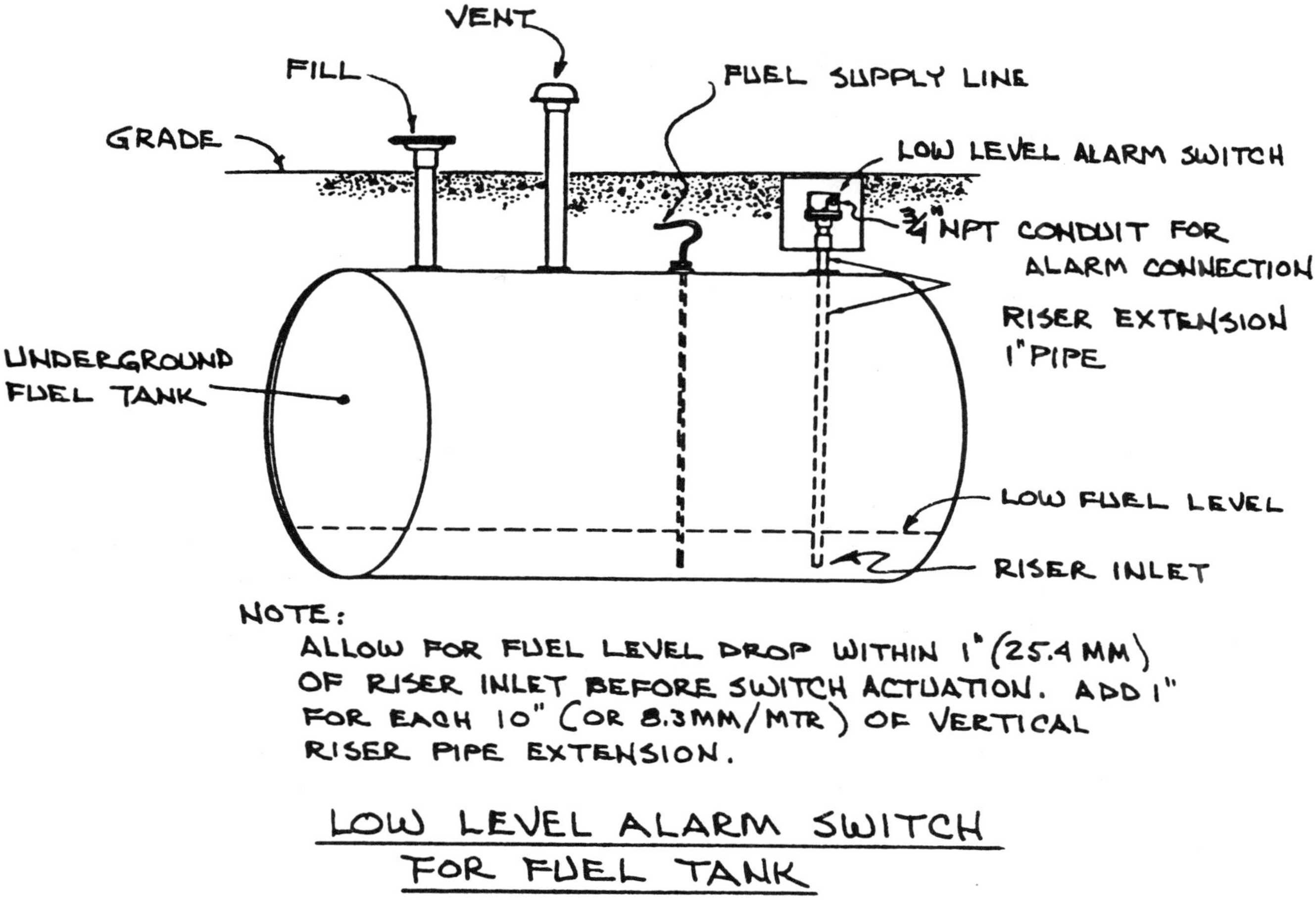

FIGURE 9-17

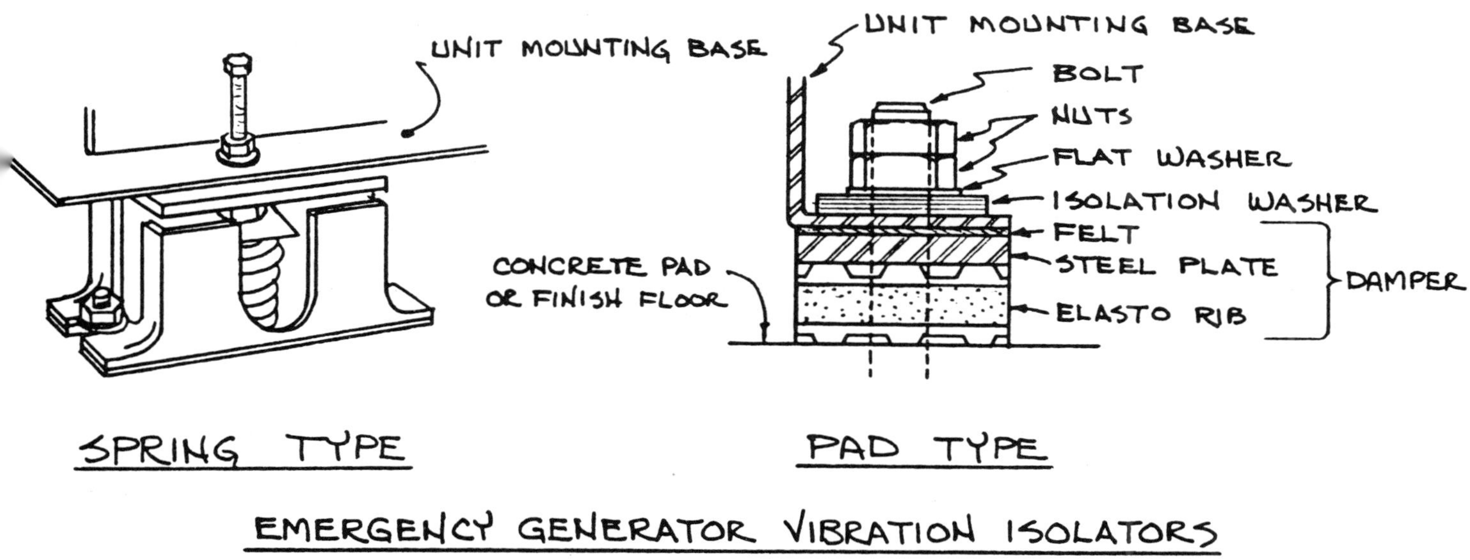

FIGURE 9-18

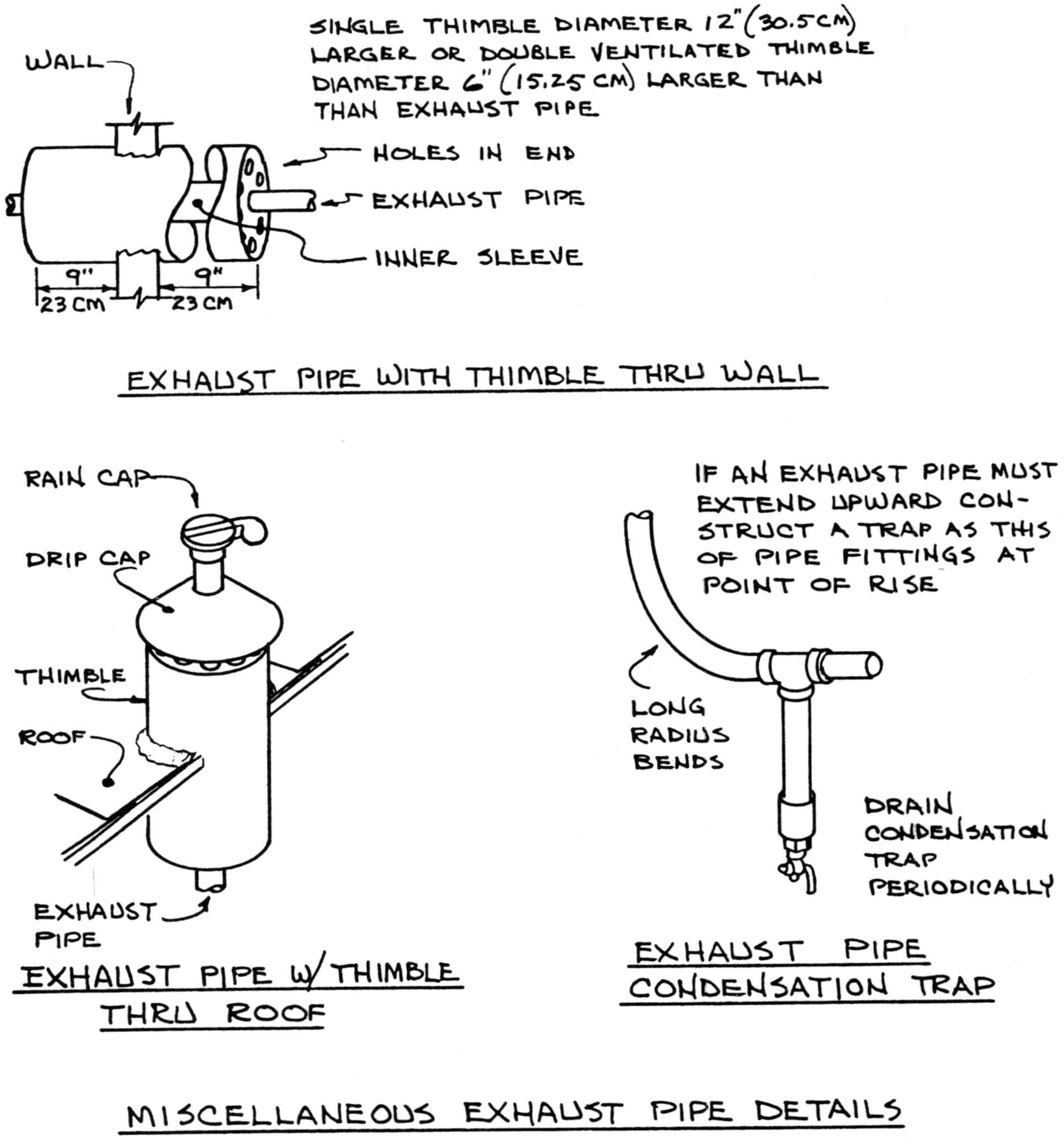

FIGURE 9-19

Finally, in Figure 9-19, described as miscellaneous details, we show portions of the exhaust-pipe installation. Real trouble can occur from a poorly planned exhaust, because exhaust gases are deadly. Where the pipe passes near flammable materials it should be covered with fireproof insulation and a metal retainer. Be sure to pitch exhaust pipes downward and away from the generator in a horizontal run and install a condensation trap where the vertical rise begins. Also be sure there is a flexible connection at the engine. An approved type of thimble must be used where pipes pass through walls or partitions. Our thimble does not show too clearly the required holes in the end of the inner sleeve. The thimble must meet local codes and should have a rain cap which automatically closes (like a flapper or butterfly valve) when the engine stops to protect the exhaust system from rain and snow.

TEN

Hospital Call Systems

Central Hospital Systems One of the highest priorities of a hospital is an efficient communication system between patients, nurses, doctors, and other staff members. The detail in Figure 10-1 is a generalized schematic of a central system.

Basically the system is a local call network interconnected with a public phone company. The local system includes a public address and sound system, frequently with background music piped to preselected areas, voice paging using terms such as Dr. Pyro (fire) or Dr. Rushpace or Dr. Quick (heart failure problem).

The master control panel also frequently includes the master clock control panel. It can also interrelate the fire alarm system. If it does, the fire alarm system will, at least, include a control panel, pull stations, alarm signals, and detector units and have remote monitoring to the control switchboard and the nurses' station.

Because this system is very complex and directly related to life and safety, it is extremely important to understand that this is not a complete, detailed presentation, but a composite detail to give a system overview. In Chapter 13 we cover fire alarm systems in greater detail. If you couple our clock system details with our fire system details, our word of caution should become even more clear.

Finally, the total detail that you can piece together from the pertinent details in this book must be subservient to a thoroughly detailed description in your specification of all the parts of all the systems involved.

Call Systems Call systems like the one shown in Figure 10-2 are sometimes called public address or paging systems. They are essentially a localized version of a telephone system. In their simplest form they would include a central microphone and amplifier with wired connections, which in effect describe an ordinary speaker system in a hall. If, in lieu of a hall, we had speakers in a series of areas or rooms, we would have an announcement system. We could have a console with an all-call button, as well as a series of individual push buttons which would enable us to talk to one room, a number of rooms, or all rooms. Going one further step we could arrange the individual speakers so that some or all areas could talk back.

The above situation presents the "big brother is listening" concept. In the past, a lot of schools had this sort of system. Whether the speaker-microphone setup is clearly visible or cleverly concealed, it is, or at least has the potential to be, a surveillance system. So to solve that problem, at least in friendly environments, a toggle switch was installed in reach of the teacher or other persons. Now the speaker could be used for any call or announcement. Both talk-back and listen-in portions of the device were under the control of the person or persons in the room.

But the above-described system still had two drawbacks: everyone in the room could hear both ends of the conversation, and a call to the central area from the room could not be made. So a further refinement was introduced, a wall phone with a speaker and push button. If a personal message were to be sent, the room occupant was advised to pick up the phone; this disconnected the speakers and connected the phone so that only the person on the phone heard what the speaker said. If you pushed a button on the phone after picking it up, it would set up a visual and/or audible signal at the central control desk. The person at the central control desk picked up the phone, pressed your area button, and you had a phone call in operation. In essence this is the generalized description of any modern call system.

Hospitals have probably more need than any other type of building for a workable call system. Figure 10-2 illustrates, in general form, the nurses' call system. This is the prime need of the hospital, and it involves a sort of three-way requirement. There is a need for communication be-

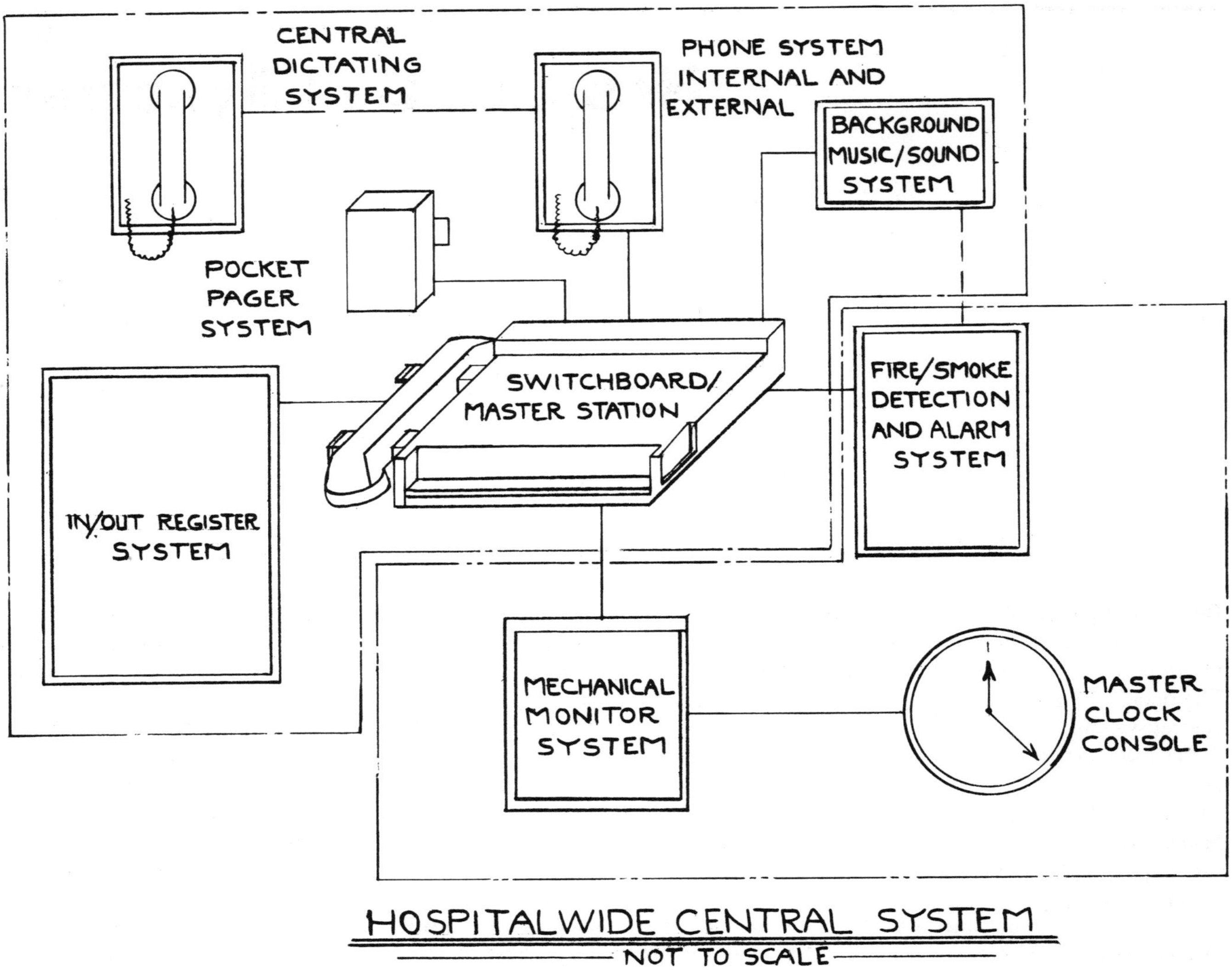

FIGURE 10-1

tween the patient and the nursing staff as well as a need to communicate between members of the staff. And the means of these communications involve a combination of voice, visual, and audio outputs, speak, see, and hear. Finally, there has to be an indication of whether this is a routine or emergency transmission.

The master station is the heart of the system. It gives both audio and visual signals when a call is placed or a connection is made to a closed-loop page system. It controls communications between patient and staff and from staff to staff. Built-in indicators differentiate between routine and emergency calls.

Familiar to anyone who has been a patient in a modern hospital is the bedside station. This unit has a calling-cord jack, a button to clear the call when answered by the nurse, an indicator light that comes on when a call is placed by the patient, and a monitor light that is energized when the patient is being monitored at a control station or when a call is being answered. The deluxe version includes a pillow speaker with radio and/or TV control and a private indicator when the line is open. Cord sets can also be of the explosionproof cord-action type for people in oxygen tents.

In bathrooms there are emerency stations with call push

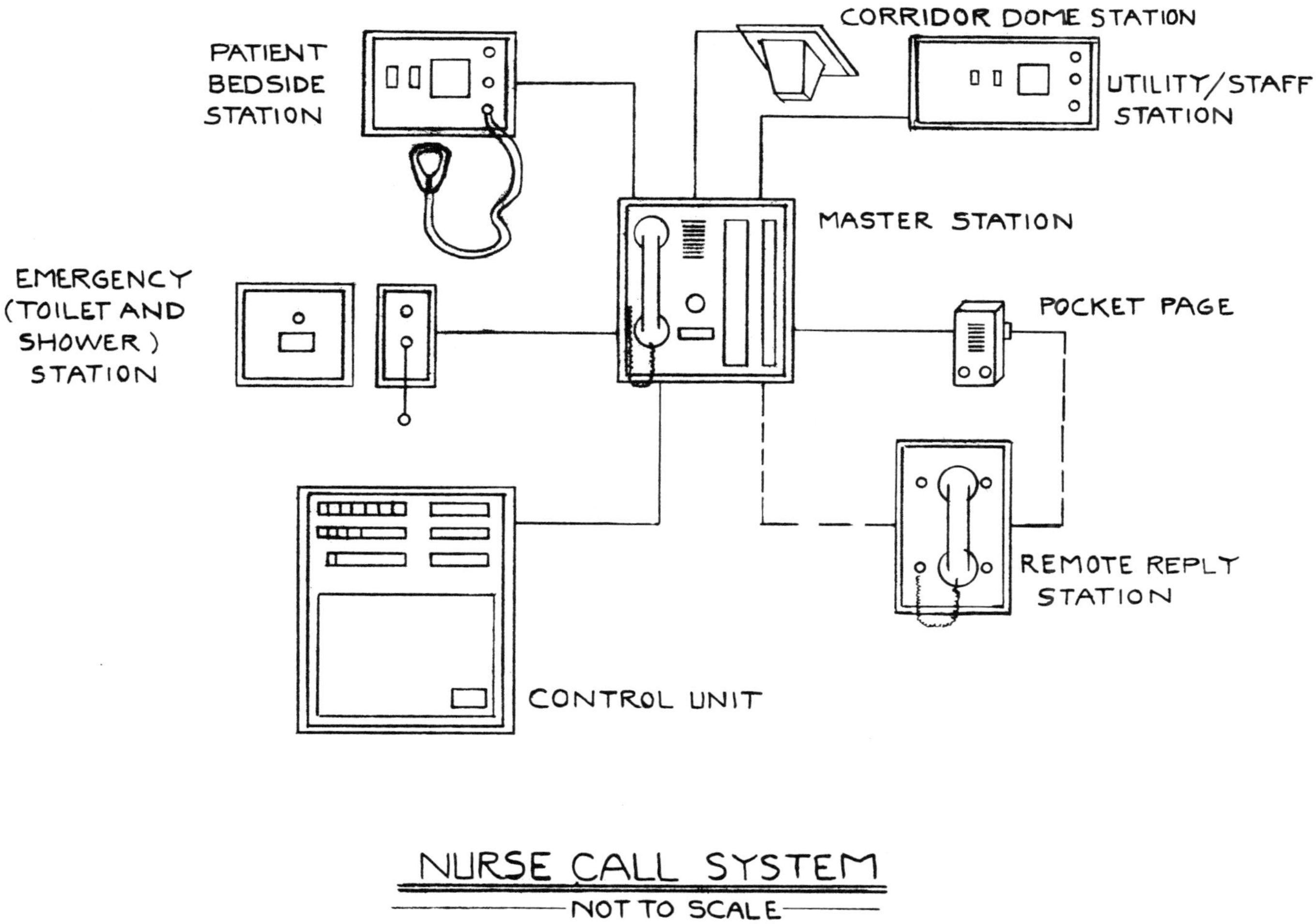

FIGURE 10-2

buttons and reset buttons. In the bath or shower the toggle switch is controlled by a cord with a ball on the end. Energizing these units usually creates an audio signal and a flashing light at the nurses' station as well as causing the corridor dome light to flash intermittently.

In duty or staff stations audio and visual signals are sent to indicate that a patient needs help and to provide a source of two-way conversation. Items include speaker-microphone, privacy lamp cancel button, patient call lamp, and nurse call button. The difference between a staff and a duty station is that patient calls do not go to the staff station.

The familiar dome light, usually on the ceiling outside the hospital room door, comes on when you make a call and lets everyone who sees it know that a call is in progress. The control unit shown is the system checker and houses the speech amplifier, a motor flasher for buzzers and lights, the power supply, and the wiring terminal-board connectors.

The above description is not a system specification and the detail does not begin to take the place of a complete and detailed description of all items. It is a tool to show the various devices on the drawings and their general relationship to each other.

ELEVEN

Security Systems

Security Systems Four simple details of security systems are shown in Figure 11-1. The primary reason behind the simplistic details is that the true operation of many security systems is not only highly proprietary but usually carefully guarded information. The engineer receives enough only to write a performance-type specification of scope, intent, and operation, but not an explanation of how the system works.

Because we are precluded from really detailing these systems, we will seek to clarify the subject by describing functions. A security system controls access; provides silent, visual, or audible intrusion alarms; identifies people, friendly versus intruder; and does a fire watch. As professionals will tell you, the best system consists of trained, sufficient, and efficient security personnel.

Your first design concern is what type and degree of security you need. Two extreme examples are grocery stores and maximum-security prisons. An in-depth analysis with your client is required. Among the many factors is one of public image. No business wants to display proudly the fact that it thinks its customers, employees, vendors, and the like are crooks or thieves.

Building design plays a large role in security, not only in minimizing doors and windows but also in relationship to other structures, parking lots, high crime areas, and the like. Ease of access is both good for the public and good for the criminal.

All crime response and apprehension is affected by time. How fast can security or police forces be notified? How fast is tbe response time? How fast can the criminal get from the scene of the crime to a reliable getaway vehicle? Can this action be slowed? In a large system the police department must be consulted prior to design. Certain systems work better with certain police departments.

There are a large number of possible elements in a security system. Common to many is a card reader which electronically identifies a precoded card. Another common device is a coded combination lock. To open, press in the right numbers in the correct sequence. Another similar idea is the time-sequenced security guard's tour key system. Within a given time period the guard must visit specific areas, insert the key, and turn the switch. If this is not done, a remote or built-in timer will automatically sound an alarm. Another simple device is the set door switch. The door has two locks, one of which is an alarm. Once it is set, if you unlock the door without using a second key to first shut off the alarm, the alarm will sound. Similar devices, with or without taping, can protect windows.

Photoelectric-cell automatic door openers can also sound alarms. An infrared detection device works in the same manner except that infrared light is not visible. Sound can be used to do the same work. The intruder can unbalance an ultrasonic beam or, for longer corridors, the beam could be a microwave beam. A somewhat similar device is a vibration detector attached to a safe. Any vibration, or noise, will sound the alarm. This is the implied device in television commercials showing the burgular who does everything right until he forgets himself and sneezes. This sneeze sets off a contact microphone system attached to the safe.

Video cameras can monitor customers, employees, intruders, as well as ordinary, innocent visitors. Video cameras are usually connected to a remote television screen on a security officer's desk. Finally there are not only sound systems that can detect approaching intruders but also buried lines common to many fenced areas that pick up the pressure of a footstep. The systems can light up the area and sound an alarm before the intruder even touches the fence, which may have its own alarm system.

A truly integrated monitoring system is usually very elaborate and controlled from a central console. This is a built-up unit containing all types of building control and

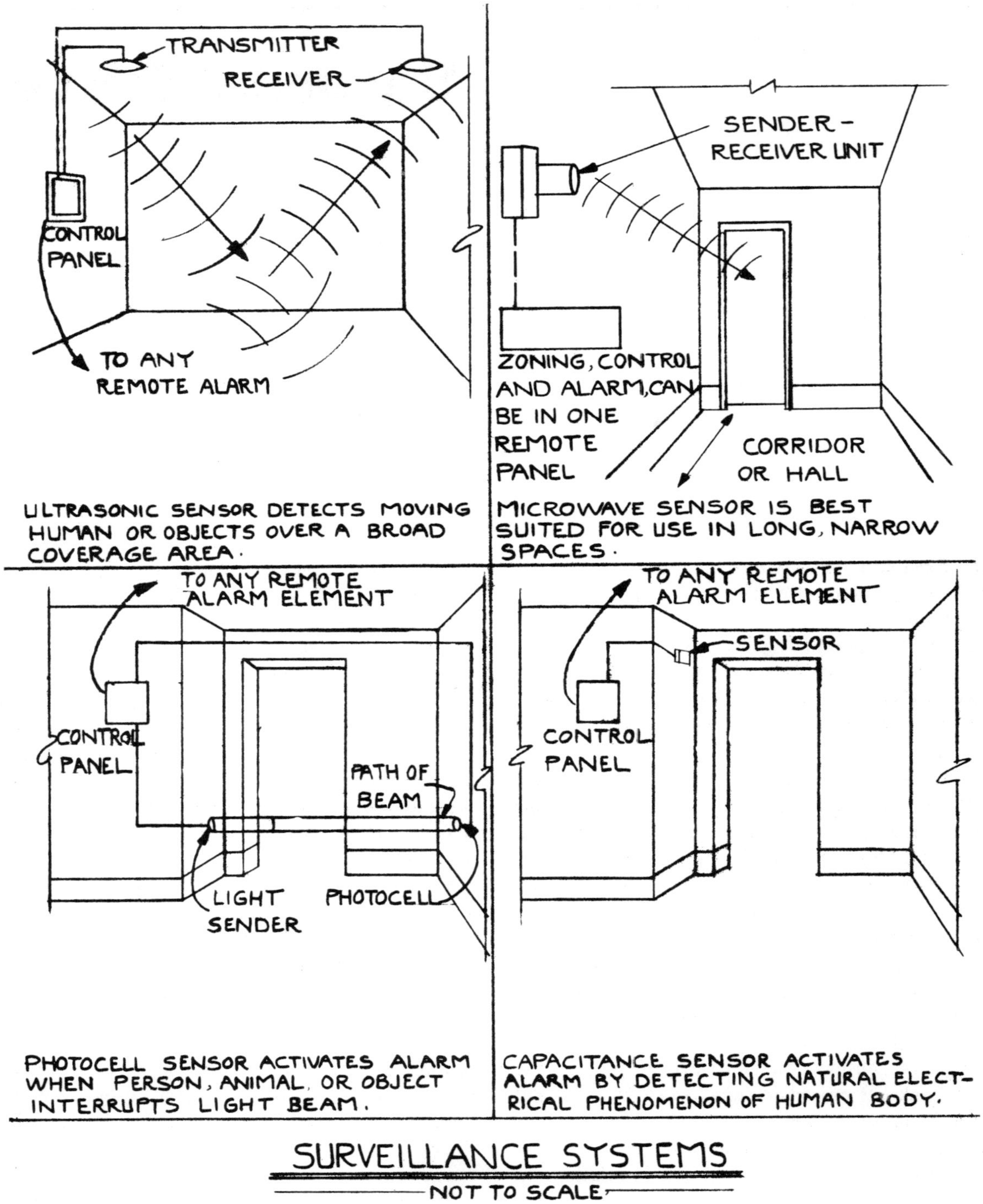

FIGURE 11-1

monitoring systems. It normally includes closed-circuit television monitors; all alarms and controls; public and private telephone and sound two-way communication; communication and paging systems; fire alarm monitors; heating, ventilating, and air-conditioning monitors, signals, and controls; even piped-in music. This is custom design on a very large scale and requires considerable input from each of various system manufacturers. However, many details shown in various areas of this book illustrate most of the parts of these large systems. What it takes in both design and detail is the careful integration of these parts into a composite system.

TWELVE

Clock Systems

The term *clock system* can mean almost anything your specifications intend it to mean. If, for example, you have three ordinary plug-in 120-V ac clocks in your house, you have a clock system. We have gone to this ridiculous extreme not to be facetious but to emphasize a point. Far too frequently the detail on the plan does not agree with the system described in the specifications. In the rush to complete the project, the schematic diagram depicting the clock system frequently is one of the last details put on the drawings. And, because they appear very similar in schematic form, an electronic clock schematic may be detailed while a wired system is specified.

Disregarding our somewhat silly example, the normal understanding of a clock system is a group of clocks controlled from a central source. To clarify this further, the proper title of the specified system should be "master clock system" rather than "clock system." Further, while there is no need to draw full clock faces and elaborate pictures of the various system components, it does help if your schematic details at least resemble the item depicted and that a clock, whether shown round or square, has the familiar hands of a standard clock symbol. In our details we do not indicate clocks that may have shatterproof glass faces or wire guards. If only a few of your clocks need this protection it might be well to note this fact on the detail. Generally, a good specification will more than adequately cover this requirement, but a further note on the detail would not be out of order.

The object of the master clock is to maintain the time very accurately and to send its time signal to all the clocks connected to it, so that every clock in the building will depict not only the same time but also the correct time. In many situations the master clock can also be preset to ring bells and sound horns at stated intervals, as well as to input the correct time to such devices as factory payroll time-card-stamping machines and postage machines. If your system can perform all of those functions, it should be called a "master clock and program system" instead of merely a master clock system to further call attention to its functions.

The master clock is the heart of the system and the way it distributes its time signals defines its type. There are two normally used ways of distributing the time signal. One is electronic and the other is synchronous-wired. There also can be systems that combine both methods.

The electronic system has no special wiring between the individual clocks and the master clock. The individual clocks are connected to the nearest electrical outlet. The electronic master clock has a built-in transmitter which emits high-frequency signals. These coded signals are picked up by receivers in the clocks which decode them and adjust the hands on the clock accordingly. The beauty of this system is that you can locate and relocate any clock and its receiver anywhere within the signal reception range. You do not even have to correct the time on the clock unless you need an instant answer, since most master clock systems have both hourly and twelve-hourly correction signal devices that set the time on all clocks at those points. The problem comes with stray electronic signals that interfere with the clock signals and create false information. There are devices to suppress stray signals, but they may not work in all cases.

To avoid the signal interference problem, the second type of master clock system does not depend on electronic signals. Each clock and program device is wired in separate conduit back to the master clock. The signal is carried over these three-wire circuits. The clocks themselves are driven (controlled) by synchronous motors.

In both cases the system can grow to a point where booster panels are required to further expand the system.

A variation of the normal synchronous system is the minute-impulse system which also uses a master clock and remote clocks. Instead of a synchronous motor the clock is driven by a dc signal transmitted at 1-minute intervals from the master clock. Normally there is also an hourly correction signal. In this case the three-wire system becomes a two-wire system.

A further variation of the above schemes is a dual-motor system. It has no master clock. Each clock has two motors, a synchronous motor for normal operation and a high-speed motor for correction. The basic premise is that the 60 cycles of electric current is always present and the high-speed drive moves the clock hands to correct for power loss. This setting can be manually or automatically controlled from a central point. Being separately wired, it is easy to set the system for such things as daylight savings time changes.

The master clocks themselves contain various methods of maintaining time when power is lost. These can be spring-operated sources that will, on power loss, be activated and run for at least 12 hours, somewhat similar to a windup clock. Frequently, there are electric motors designed to rewind the spring when the power comes on. There are systems where the ac supply is converted to dc through a battery system, and the time signal comes from quartz crystal oscillator. In power failure the always-charged batteries can keep the system going for 24 hours. To be certain the time on the master clock is truly accurate, there are units with antennas and radio receivers that get their time signal from the National Bureau of Standards (NBS) time-signal generator.

The above description is an abbreviated outline of master clock and master clock and program systems. Before your detail is put on the drawings, it is necessary to be thoroughly aware of what is being specified.

Electronic Systems Figure 12-1 shows a master electronic system diagram. In order to get this detail on a single page and still illustrate the points we want to make, we had to compress it. Normally the secondary transformer for a building would have some secondary switchgear associated with the local distribution panel and thence to other panels. The various devices shown might well be distributed from lines of several panels.

The connection to our electronic master clock and electronic transmitter is made via a fused disconnect because we want our system to be undisturbed by other installations in the building that may cause voltage and current variations. This line goes to the transmitter. The system also requires a constant supply of 120-V power for clocks and relays. This is obtained from the local distribution system via a separate fused 120-V disconnect.

In using this detail only show the devices that are in your system. Figure 12-1 may or may not show all of them. As we described in the preceding general discussion, the system has no wires between devices and master clock. The note on wiring from the transmitter to the secondary bus should always be a part of your detail.

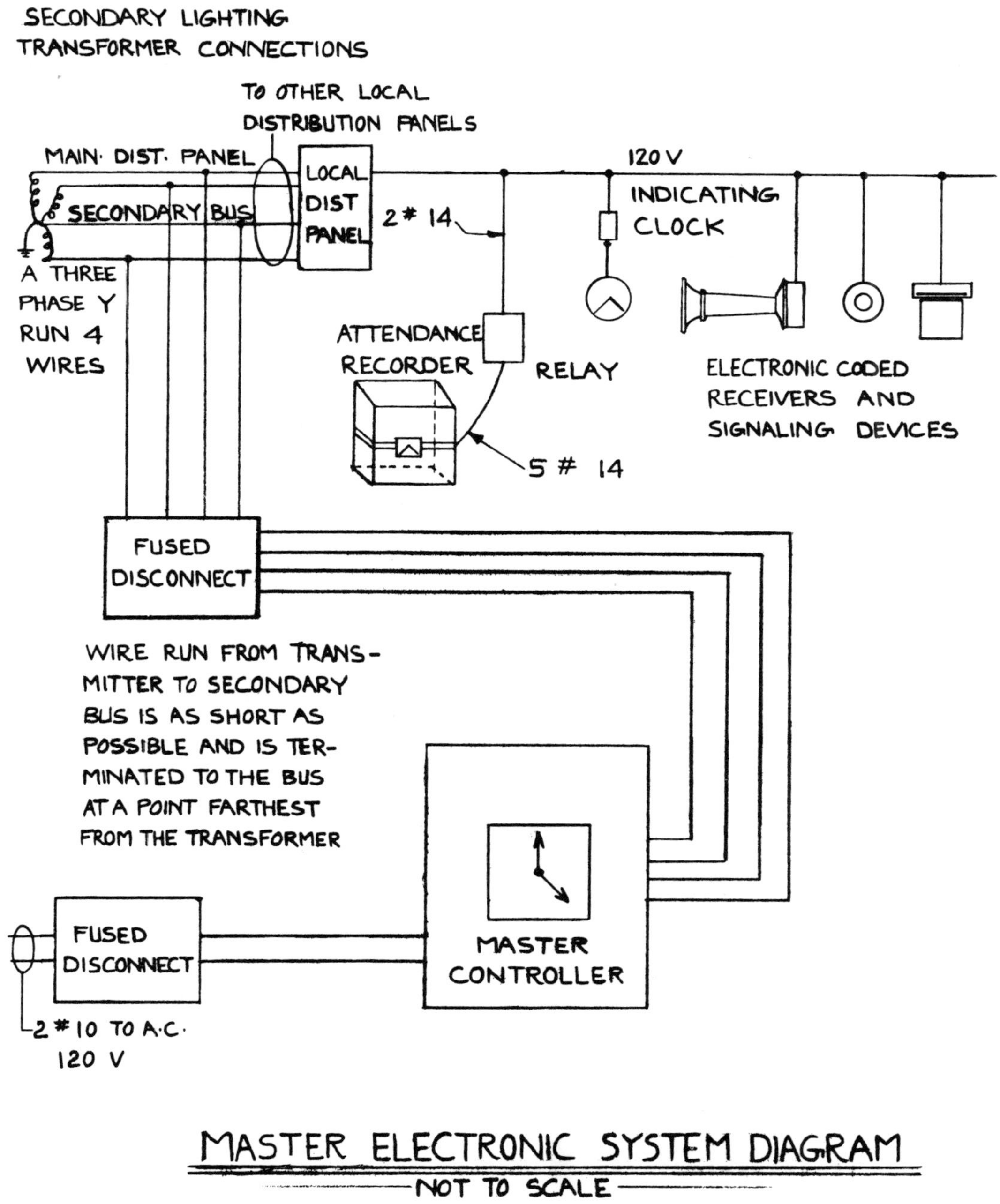
SECONDARY LIGHTING
TRANSFORMER CONNECTIONS
TO OTHER LOCAL
DISTRIBUTION PANELS
MAIN DIST. PANEL
LOCAL
DIST
PANEL
120 V
SECONDARY BUS
2 # 14
INDICATING
CLOCK
A THREE
PHASE Y
RUN 4
WIRES
ATTENDANCE
RECORDER
RELAY
ELECTRONIC CODED
RECEIVERS AND
SIGNALING DEVICES
5 # 14
FUSED
DISCONNECT
WIRE RUN FROM TRANS-
MITTER TO SECONDARY
BUS IS AS SHORT AS
POSSIBLE AND IS TER-
MINATED TO THE BUS
AT A POINT FARTHEST
FROM THE TRANSFORMER
FUSED
DISCONNECT
MASTER
CONTROLLER
2 # 10 TO A.C.
120 V
MASTER ELECTRONIC SYSTEM DIAGRAM
NOT TO SCALE

FIGURE 12-1

Wired System As can be seen in Figure 12-2, master synchronous wired system diagram, the wired system means that all devices are directly connected back to the master clock and program device. Again, the master clock and program system requires a source of 120-V power. But in this case the 120-V source that serves the master clock also serves all the other clocks and relays.

In terms of system depiction this is by far the simpler system to detail. All of the devices have three wires, and this can be indicated as we have done, or you can merely note on a single line that it represents three #14 wires.

Again we have shown a number of devices, all of which may not be a part of your project. Check which apply and delete or add as required.

Electronic and Wired System Figure 12-3 is a composite detail that tries to indicate all of the features of a combined electronic and wired system, including impulse control of clocks and an antenna to receive the NBS signal. The detail presumes a very large commercial or industrial installation.

As you may recall from our opening discussion, the

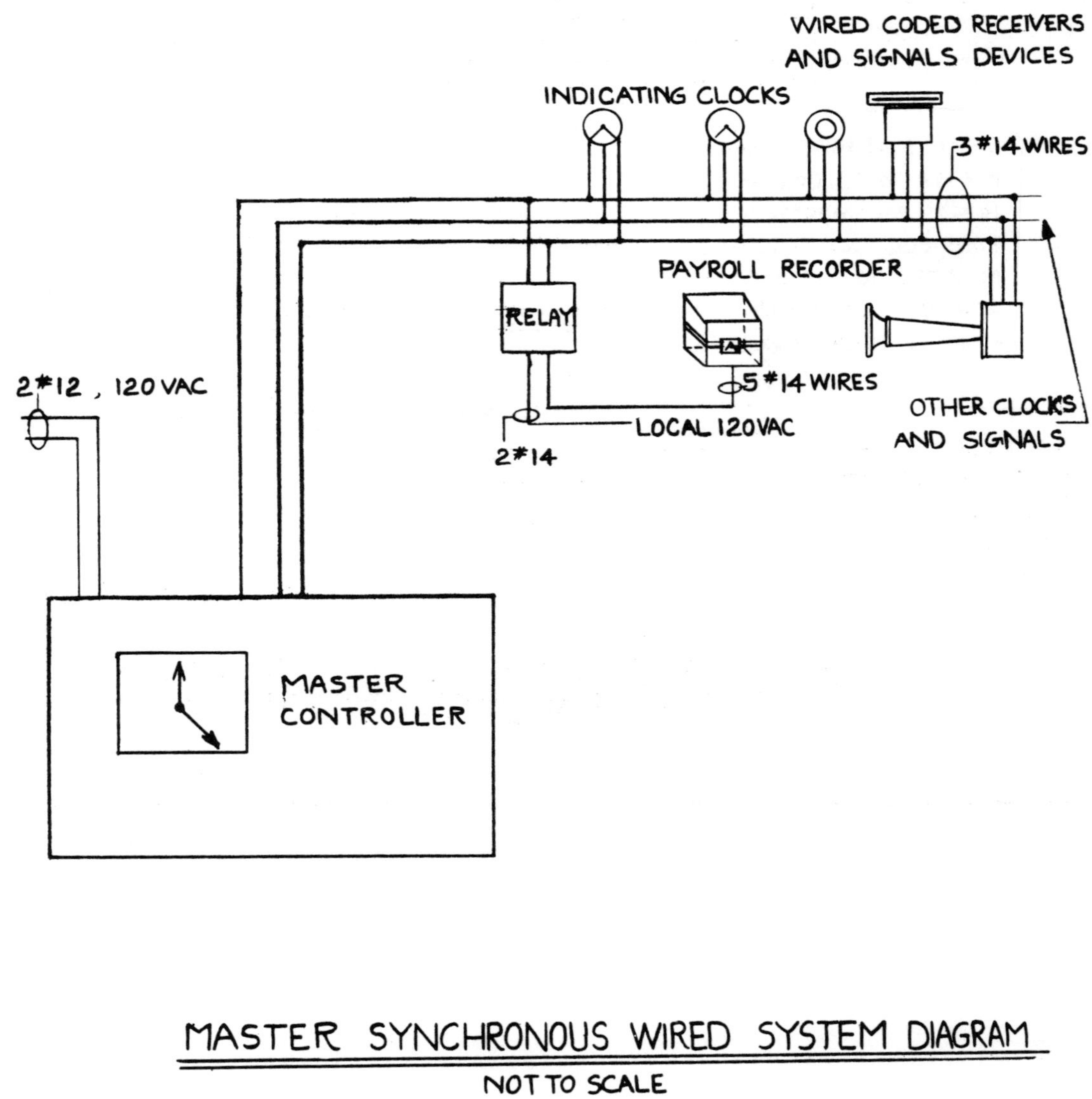

FIGURE 12-2

clocks are driven by a dc signal transmitted at 1-minute intervals from the master clock. These clocks are shown in the upper half of the detail. The system has capacitors to couple high-frequency pulses into the main switchboard and are connected there to our electronic transmitter from the generator control unit. If the lines in the building or plant are at voltages above 120 V, we will need a step-down transformer for our clock wiring.

In the main program panel we interconnect the wired and electronic systems through to the master clock. For purposes of illustration, our detail also shows recorders and clocks on a three-wire system. Our master clock, as we previously noted, always needs a separate 120-V source at least for its clock, and this too is shown. And, since our elaborate system has a NBS signal antenna this, too, is shown.

This scheme can be modified to suit your particular needs. However, each item is shown in its proper relationship to other items and should be used in the arrangement depicted.

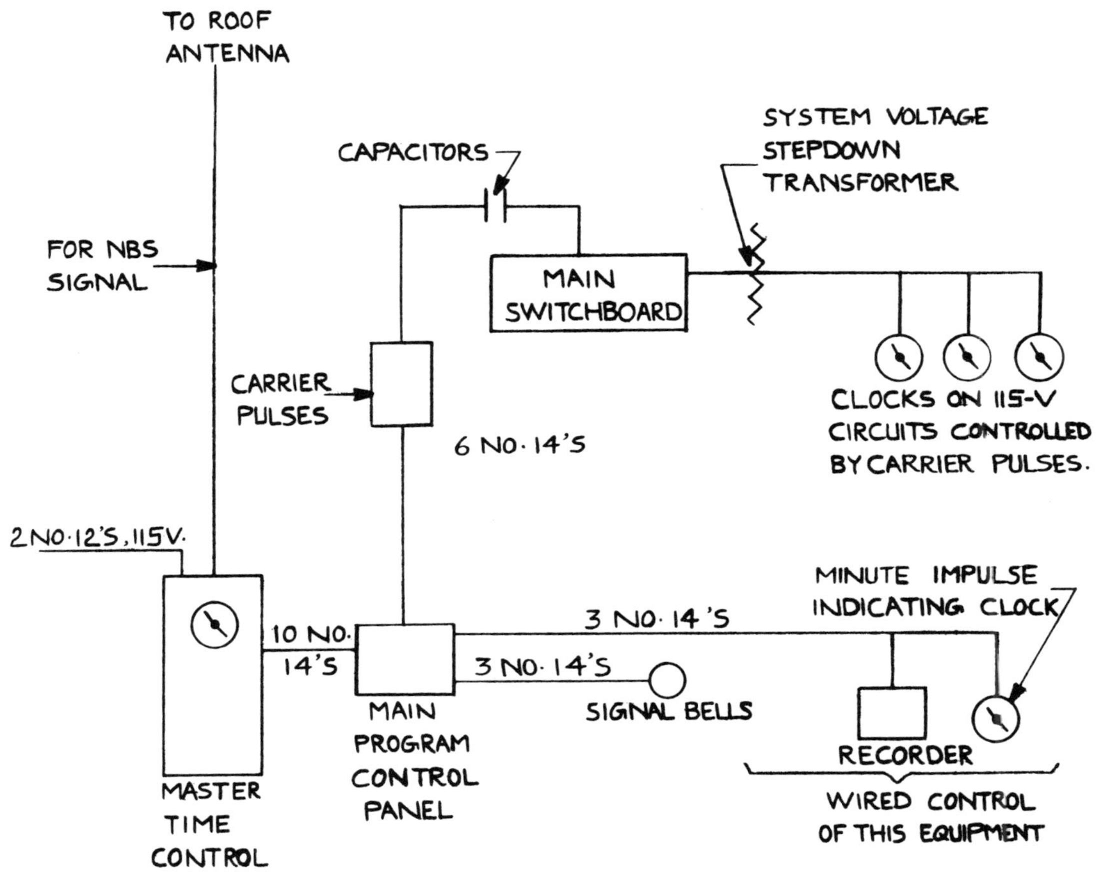

SCHEMATIC WIRING OF ELECTRONIC AND WIRED CONTROL SYSTEM FOR CLOCKS

NOT TO SCALE

FIGURE 12-3

THIRTEEN

Fire Alarm Systems

Fire alarm systems can be categorized in a number of ways, but before categorization, we should understand what all of the systems have as basic components and what the functions of those components are. One can set off an alarm when a source of fire is detected. But, waiting until flames are visible is very late indeed to sound an alarm. Thus, heat- and smoke-detecting devices which can automatically sound an alarm are basic required components of any modern fire alarm system.

The reporting devices of any fire alarm system consist of pull boxes or pull stations where, if you see a fire (or smoke), you break the glass and pull, push, or press the alarm signal device. In addition there are detecting devices located in selected areas that sense smoke and/or heat and automatically send a signal which activates an alarm-sounding device. How these signals are processed and dispersed by the master box (master controller) defines the type of alarm system.

The most common type of system is the noncode continuous sound 24- or 120-V ac alarm system. In this system any signal from any source will cause all horns and bells to sound until the master panel control is reset. The next step in design is a box code. Now any device sending a signal causes all bells and horns to sound, for example four times, then stop, but a trouble signal will sound at the panel until it is reset. In a selected-code version of this system, a remote submaster panel will get the trouble signal if the problem is in its area. There are also further refinements, such as sounding only a horn or bell in the fire-, smoke-, or heat-affected area. If, upon investigation, a general alarm is required, the fire investigator inserts a key, turns a device, and sounds a general alarm.

At the top of this succession of complexities is the system where the signal activates a punch register creating a time and date recording and a coded signal that sounds four times. Horns and bells will sound in a gong in a related or selected area or all over. Priority areas are usually programmed into this device so if two separate areas send alarms the top-priority area is handled first. The system also has its own troubleshooting system built into it. If a system fault occurs, a trouble signal will sound and the punch register will strike the time, date, and four rounds of the trouble code without sounding evacuation signals. A test arrangement can also be built into the system so that each device can be activated without setting off an evacuation alarm.

Finally, your alarm system can have a fire-zone location annunciator panel clearly visible to incoming firefighters and building fire safety personnel. In large buildings the areas are defined on this panel, and when a signal is sent an indicator light adjacent to the noted area is activated so anyone can quickly note the general location of the signal. Depending on both the system and your local fire department rules, your fire alarm system may be directly wired to the city or community fire alarm system. When this is not permitted, a city fire alarm device is located just outside the main control area on the exterior building wall or lawn which must be separately pulled. Connection in many city systems is now made through dedicated telephone lines in place of box alarms.

Alarm-Call Systems At first glance the presentation of a fire alarm system for anything other than a very simple situation appears very complex and difficult to understand or follow. While in a sense that feeling is normal, the fact remains that the complex system is merely a collection of rather simple straightforward components.

Figure 13-1 illustrates a simple panic alarm system which could be a component of a fire alarm or protection system.

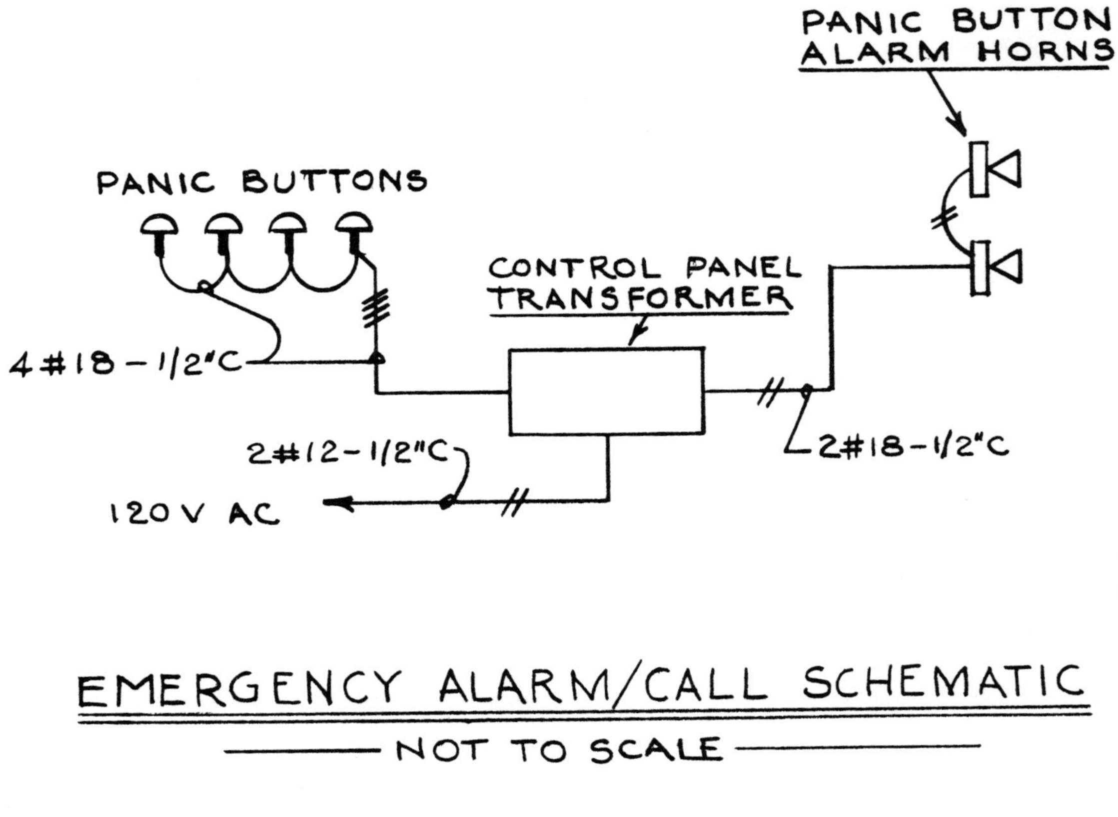

FIGURE 13-1

What is depicted is a collection of four mushroom-type push buttons located in strategic, *dry* locations around an indoor swimming pool. The chances of fire in the pool area are obviously remote, but, if this were a YMCA, the chances of an older member coming in for his or her noontime swim and having an unexpected heart attack, or a sudden attack of indigestion that results in near drowning, or any other of a number of life-threatening situations are higher than average.

At the time of such a situation, expert help may not be readily available. This is a real-life and not a joke situation where you really do "hit the panic button." The buttons send a low-voltage signal through a transformer which energizes a couple of alarm horns. One of these horns usually sounds off in the pool director's office and the other is usually located at the main desk of the facility. Normally they would have reset shut-off devices in both locations. This simple, usable pool panic alarm is very similar to a simple noncoded fire alarm system which, through its detector or pull station, also energizes a signal, in this case a fire bell or horn.

Fire Alarm Riser Diagram As we stated in our description of an alarm-call system, the usual fire alarm schematic diagram seems a lot more complicated than it actually is. The detail in Figure 13-2 is a fairly typical city-connected fire alarm system. The city system is a Gamewell system and thus all of the devices within the building must be compatible with it.

Not all the devices are described on the detail, although the symbols used are fairly standard. The symbol S in a six-sided enclosure is a smoke detector. The F in a square is a fire station. Where the F also has a triangle attached, it is a fire station and fire horn. Heat detectors are shown by an *open* circle. Future stations are dotted squares. The numbers adjacent to certain devices relate to a special area coding peculiar to this installation. This is an ac-dc system with all wiring terminated at a central control cabinet. Also connected to this cabinet is a dc battery and charger for system operation in case of power failure, which is also the way the city system is arranged. Interconnected also is a supervisory panel and a fire alarm master box.

The connection from the master box flows both to the building system and to the city system. An alarm within the building will show up at the main control unit. It will *not* be transmitted directly to the city fire alarm system. If this result is desired, all the building personnel need do is to pull the master box alarm. If and when the city decides to integrate this building into its overall system, an arrangement will be made to let the signal flow directly through the master box to the city system. As presently arranged the master box has one other function. Pulling it will set off both the building's general alarm system and the city alarm system.

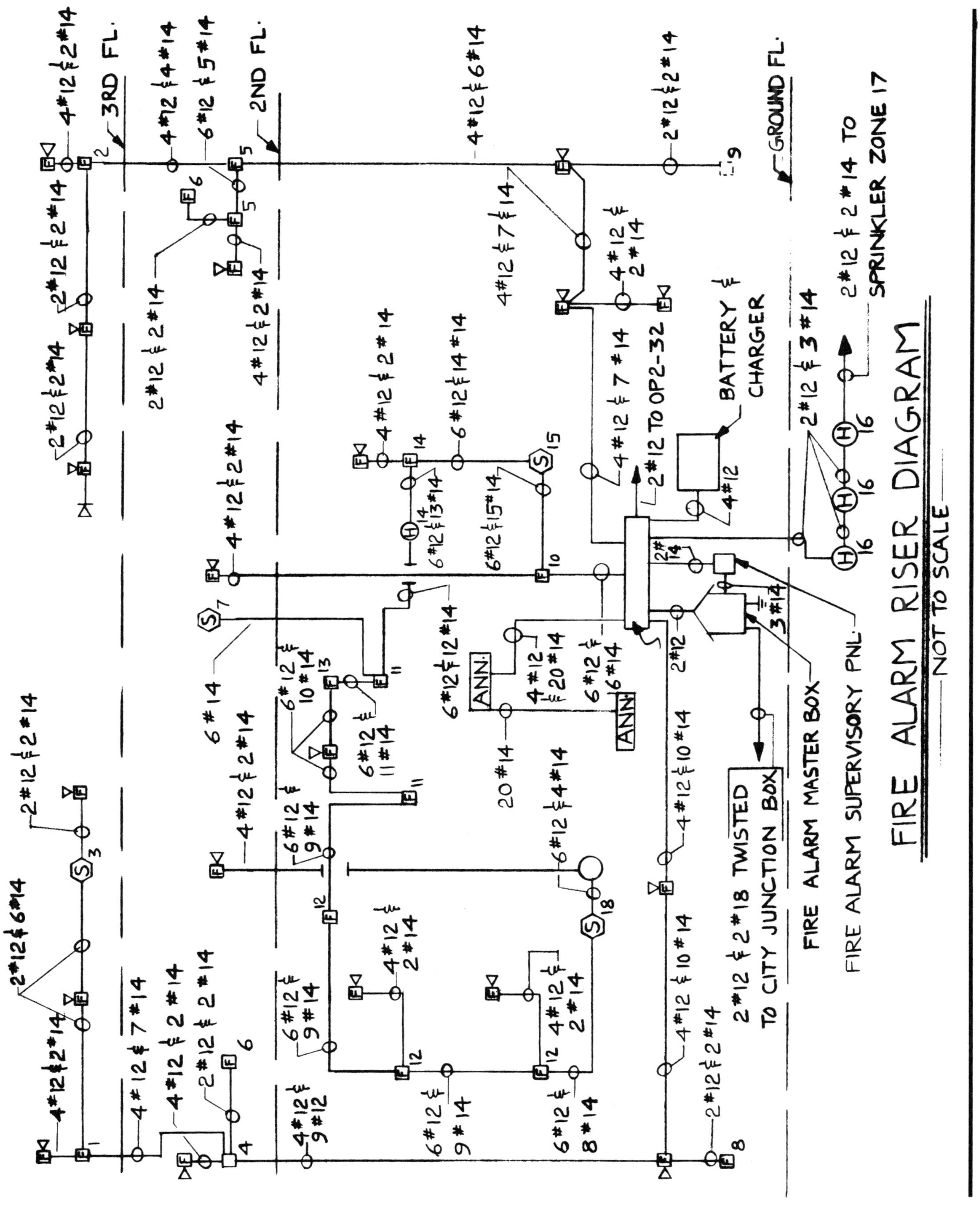

FIRE ALARM RISER DIAGRAM
NOT TO SCALE
3RD FL.
2ND FL.
GROUND FL.
BATTERY & CHARGER
FIRE ALARM MASTER BOX
FIRE ALARM SUPERVISORY PNL.
TO CITY JUNCTION BOX
2#12 & 2#18 TWISTED
2#12 & 2#14 TO SPRINKLER ZONE 17
2#12 TO OP2-32
ANN.

FIGURE 13-2

This detail serves to illustrate the basic system outline. Zoning, coding, and the like are features of the system. Primarily because certain well-defined areas were selected to be coded, the specifications accompanying this detail describe the systems and requirements involved individually.

Some engineers prefer to prepare a series of smaller diagrams illustrating each system and its functions and coding, and an overall master block flow diagram to indicate the relationship of major components to the central source. There is nothing wrong with this approach, but it still requires the same descriptive data in the specification.

In theory, a well-written specification plus locating the items on the plans covers the detailing of a fire alarm system. In practice this places a large burden on the contractor and creates an opportunity for numerous misunderstandings. The objective of the overall schematic is to show how the various parts connect to each other and the order in which they return to central control. A detail such as this should show all the major components and ideally should, for example, show four stations if there are four and not one followed by the word "typical." This is also an excellent way to double-check items connected to each coded gong, which are also noted on the plans.

Complex details such as the one depicted can be time-consuming to prepare. Ours was crowded on one 8½- × 11-in sheet for book presentation. It was more than twice this size in its original form and was accompanied by a series of notes and a separate fire alarm riser symbol list.

Fire Alarm Control Following our fairly complex detail in Figure 13-2, we are presenting in Figure 13-3 a typical detail to illustrate the usual elements of a fire alarm system. Note that we have used the words "usual elements" since there are other items in other systems which will be separately portrayed.

In Figure 13-3 the heart of the system is a control unit which can provide from 1 to 12 zones for various types of alarm devices. The control unit is complete, with a local energy trip circuit, a battery-charging unit, a trouble buzzer, a 24-V dc circuit for smoke detectors, accessory alarm contacts from each zone, and double-throw contacts for both common alarm and trouble conditions.

Activating any alarm will sound all devices until silenced by a control unit "alarm off" switch or by restoring the tripped device to normal. The tripped alarm will also light up the associated zone lamps. The system can further operate a local energy master box and automatically transmit the fire call to the local fire department. There is also a trouble alarm system supervisory circuit that monitors breaks or ground faults in the system wiring.

The system illustrated is basic, fairly flexible in application and cost, and can be used in many types of institutional, industrial, and commercial projects. It is a 120-V single-phase system which easily fits most available wiring systems. It can also be operated at 208-V single-phase current.

We have, as noted, not illustrated all the devices possible including bells, horns, etc. In addition we have not shown all the circuits possible. Finally, there could be one or more circuits for auxiliary output as our detail suggests. These circuits are normally open accessory alarm contacts from each zone and they can be combined in the way that we have noted.

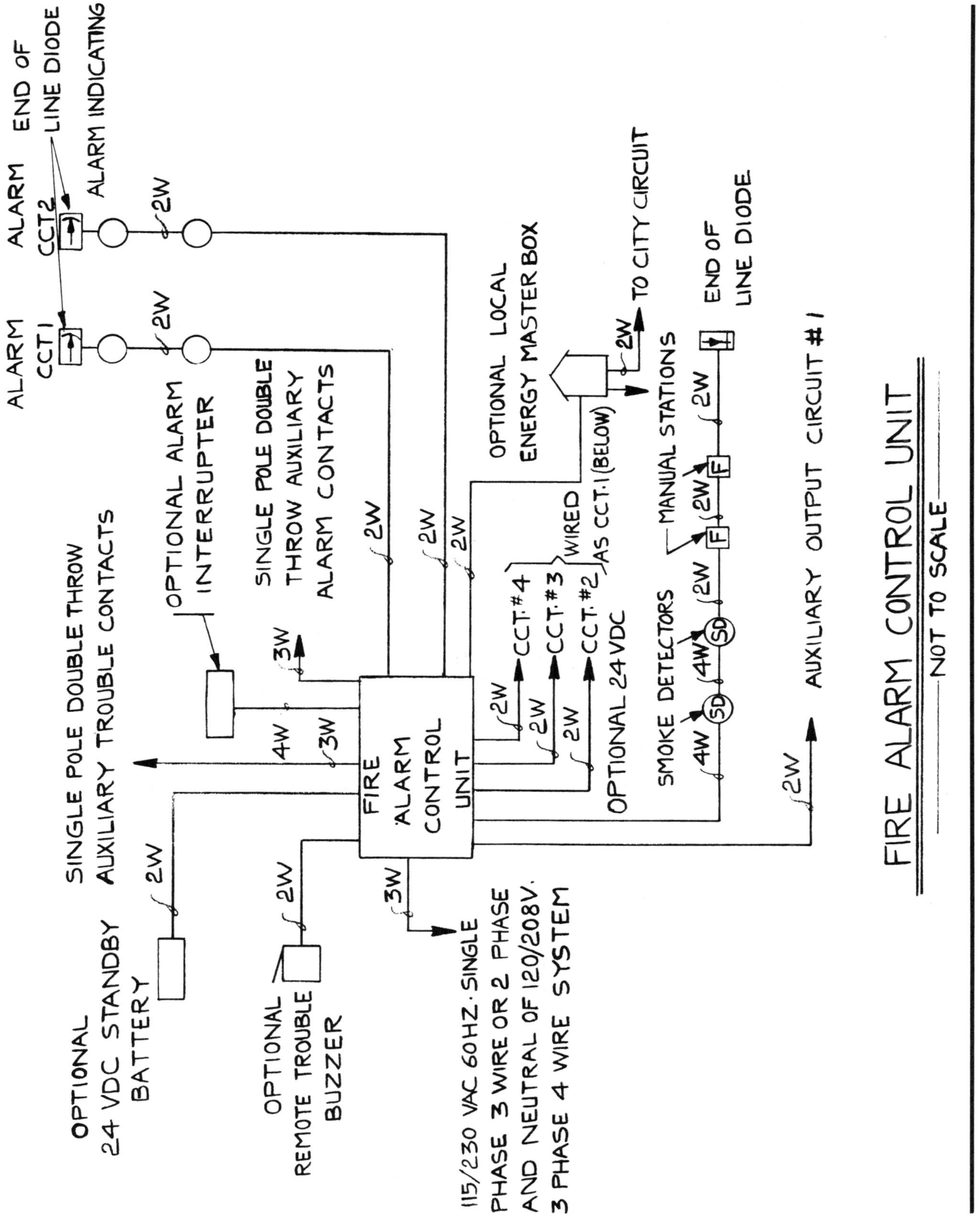
ALARM CCT 1
ALARM CCT 2
END OF LINE DIODE
ALARM INDICATING
2W
OPTIONAL 24 VDC STANDBY BATTERY
SINGLE POLE DOUBLE THROW AUXILIARY TROUBLE CONTACTS
OPTIONAL ALARM INTERRUPTER
SINGLE POLE DOUBLE THROW AUXILIARY ALARM CONTACTS
OPTIONAL REMOTE TROUBLE BUZZER
FIRE ALARM CONTROL UNIT
3W
4W
115/230 VAC 60 HZ. SINGLE PHASE 3 WIRE OR 2 PHASE AND NEUTRAL OF 120/208V. 3 PHASE 4 WIRE SYSTEM
CCT. #4
CCT. #3
CCT. #2
WIRED AS CCT. 1 (BELOW)
OPTIONAL 24 VDC
OPTIONAL LOCAL ENERGY MASTER BOX
TO CITY CIRCUIT
SMOKE DETECTORS
SD
MANUAL STATIONS
F
END OF LINE DIODE
AUXILIARY OUTPUT CIRCUIT # 1
FIRE ALARM CONTROL UNIT
NOT TO SCALE

FIGURE 13-3

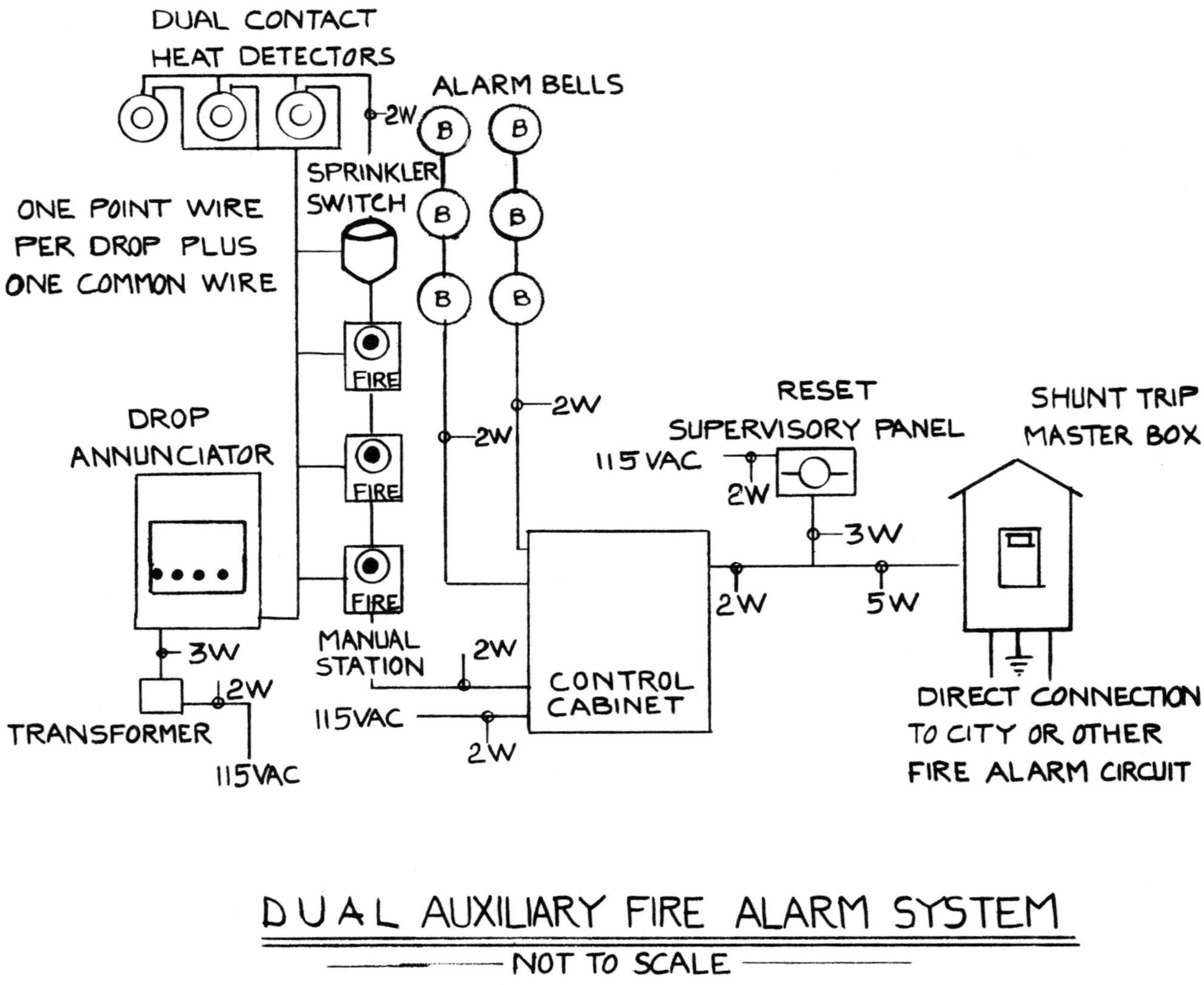

FIGURE 13-4

Dual Fire Alarm System As we noted in the previous discussion, the fire alarm system can have a number of additions and/or improvements made in its overall application. Figure 13-4 illustrates an application describing two very important (in some cases) functions. When any box is pulled, or any device senses and originates an alarm, two things happen simultaneously. First, the fire horns and bells sound throughout the system. Second, a coded signal is transmitted directly to the city fire headquarters through the shunt trip master box. The signal can also be a noncoded signal.

The firefighter answering the call will see from the lighted annunciator panel in the building entry just where the signal originated. Since this is a sophisticated system used in military bases, state and municipal institutions, large public buildings, large schools, hospitals, high-rise buildings, and the like there usually exists a fire program with personnel trained to react immediately to the coded audible signal. Each manual station or transmitter has a distinctive code number which will be sounded on all stations four times.

The master box circuit runs on both ac and dc and thus will work even in a building power failure. The system also is arranged to superimpose a lower-voltage ac on the dc shunt loop so that should the municipal or headquarters' circuit be broken the master box will be tripped and the local alarm will still sound. Dual circuits are provided to bells and horns so that if one fails the other will still provide current to sound the alarm.

Again, the detail does not indicate every circuit and every device that may be involved in your project, but additional circuits and devices would be merely additions to an enlarged version of our detail.

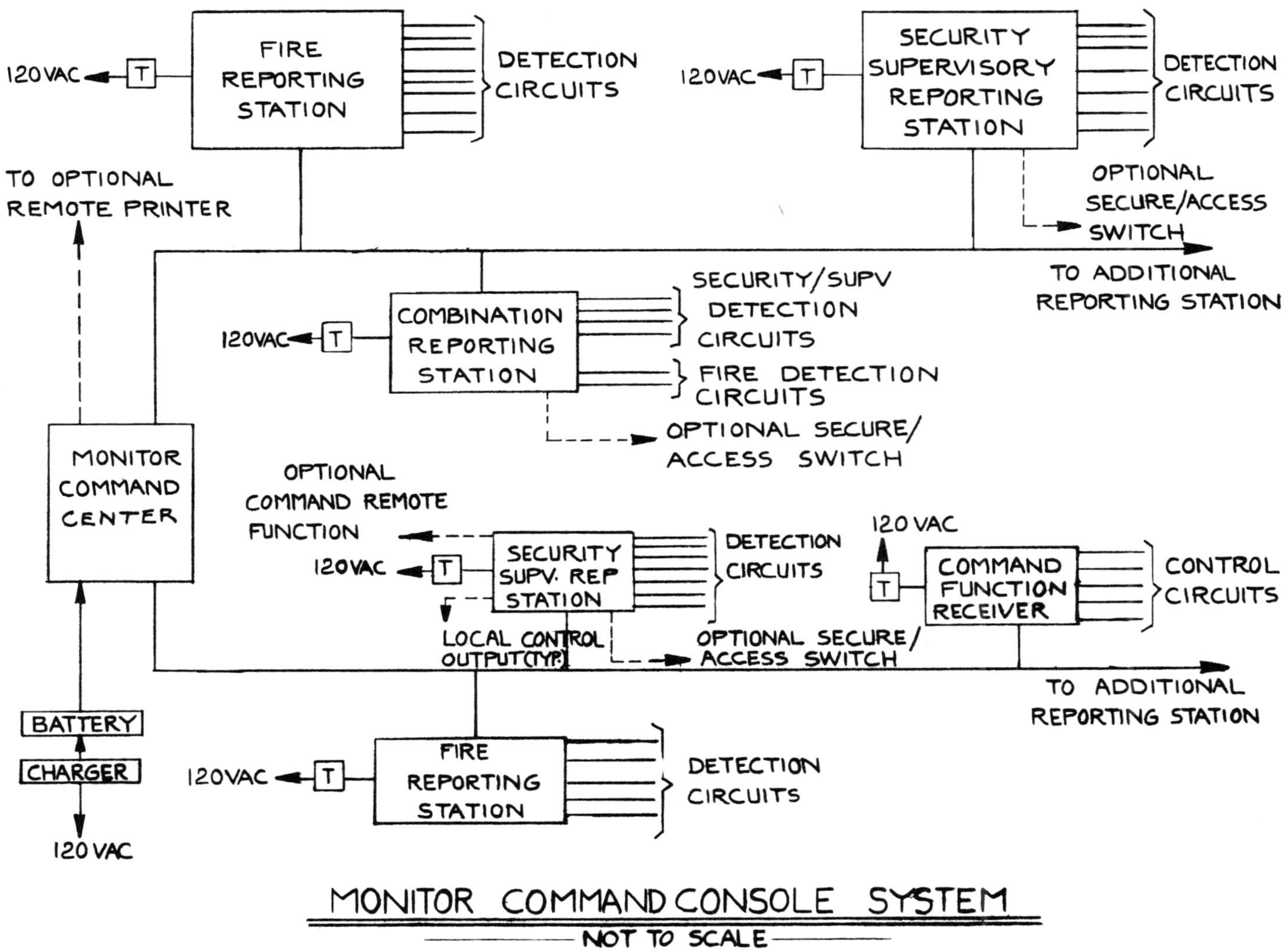

FIGURE 13-5

Monitor Command System As with every other type of control and signal system, the fire alarm system can grow and grow, increasing in complexity. Figure 13-5 is not a detail in the true sense. More properly it would be described as a flow diagram. In effect it is the master diagram delineating how information from various subsystems would flow back to a central collection point.

The objective of this system is to control and monitor not only the fire alarm system but also security and surveillance systems. Any change in status on any system is noted, recorded, and reported, including trouble, tampering, and power failure. Implied in the arrangement is a remote printout and ac-dc operation.

Note that for each subsystem there is a power source noted as 120-V ac plus a transformer noted as a square with a T inside. The system flow diagram also implies an alternative, separate or dual, monitoring, printing, and/or control of the security supervision function.

Control and monitoring in the age of computers has become a very complex subject. Without doubt one large computerized master command console could control every possible function in any structure. The real problem of the engineer is not to design the system (although this is certainly a problem unto itself) but rather to arrange the output(s) of the system to suit the needs of the client's operating and control personnel.

Generally, the building(s) (if large and expensive) have an in-house or separately hired security force. Fires, whether accidental or intentional, as well as smoke- and heat-related problems can and do occur when the building is unoccupied or staffed with only minimal personnel. The first question is how well can these security personnel handle fire and incipient fire problems when they are re-

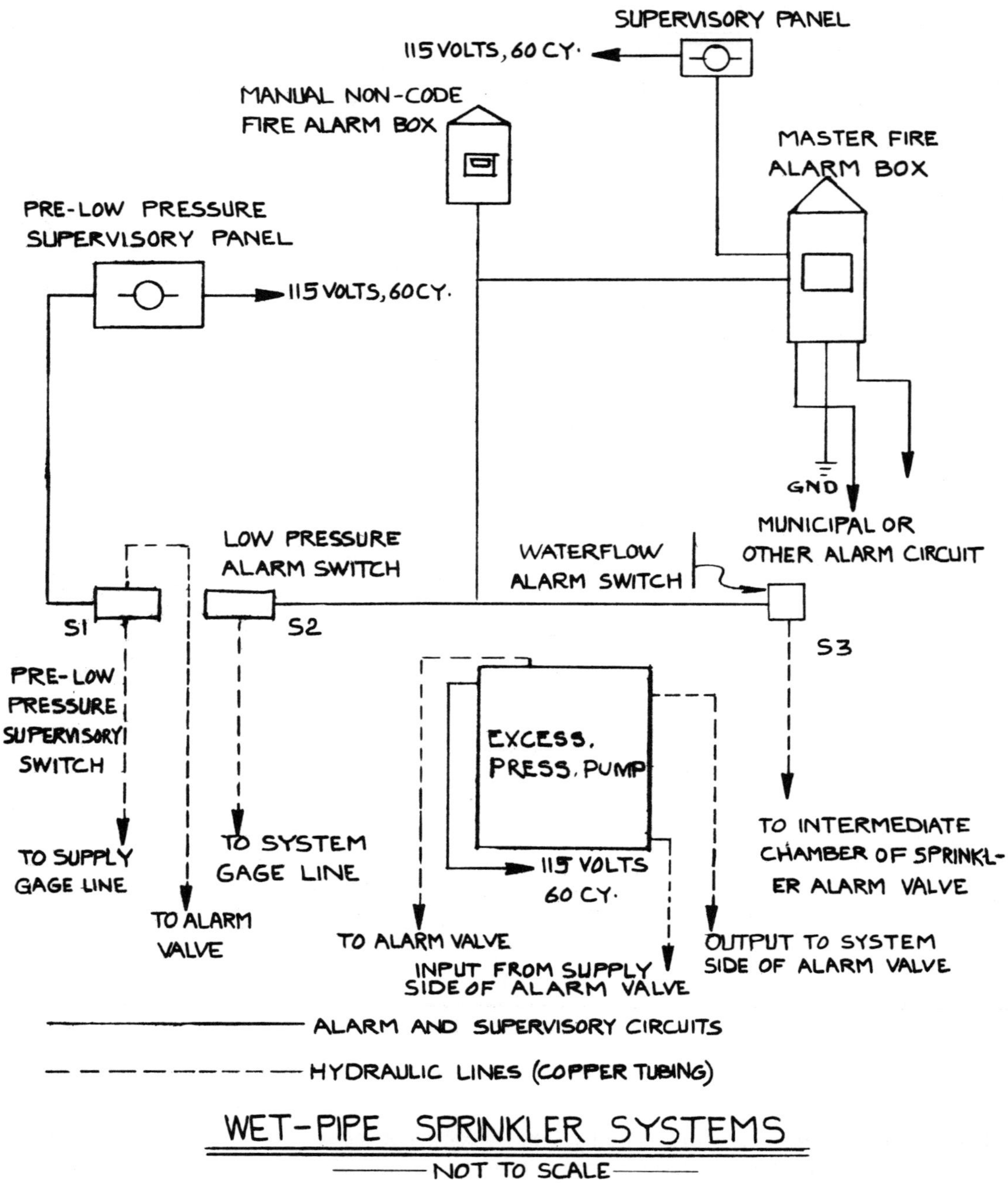

FIGURE 13-6

ported? There is little logic in alerting untrained personnel in a fire or incipient fire situation. It is also not necessarily good practice to call the city fire department every time smoke or heat is detected. The flow system depicted must presume that either the security personnel are properly trained in fire prevention and safety or there are personnel instantly available who are so trained.

Sprinkler Fire Alarm Systems One of the items than can be overlooked when designing sprinkler and fire alarm systems is the integration of the two. As is well known, when a sprinkler head is activated there are loud and continuous alarms being sounded on an exterior wall of the building. But this is a sprinkler alarm and not a fire alarm. Figures 13-6 and 13-7 illustrate how, with either a wet-pipe or a dry-pipe system, the sprinkler system can also be a fire alarm system. It can thus serve the purposes of containing and/or extinguishing the fire and sounding the local fire alarm system and calling the city fire department. If the sprinkler water supply is shut off, accidentally or on purpose, the opening of a sprinkler head will still sound the alarm.

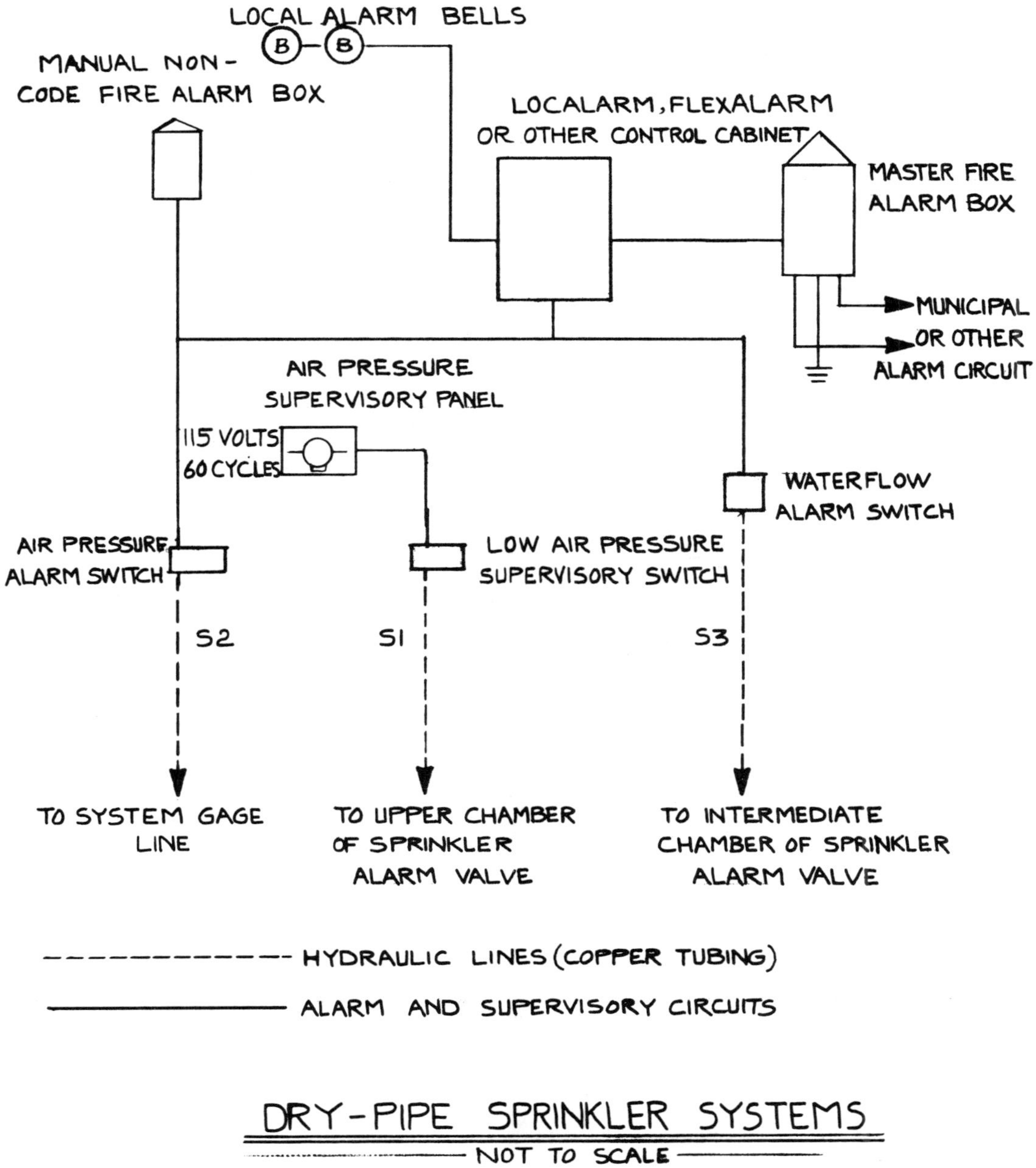

FIGURE 13-7

As can be readily comprehended, the system interconnect requires a device in the sprinkler system that will activate an electrical signal. Many such devices have been tried. The three most common are vaned-flow, diaphragm-pressure, and excess-pressure pumps. The vaned-flow and diaphragm-pressure solutions read flow movement in the sprinkler system. The excess pressure system senses a pressure via its pump which is considerably above maximum sprinkler system pressure. The pump usually is very low volume, such as 1 gal/min. The release of a head in the sprinkler system will activate the pump and the related fire alarm system.

The dry-pipe system, as can be seen in Figure 13-7, is the same except there is no excess-pressure pump. There will be an alarm initiated when a sprinkler head opens causing the dry-pipe alarm clapper to rise.

Sprinkler–Fire Alarm Interconnections Most of the work shown in Figures 13-8 to 13-10 is not electrical work at all. As you can see, the small double-line work is ⅜- and ½-in copper tubing transmitting fluid pressures from one device to another. The purpose of these

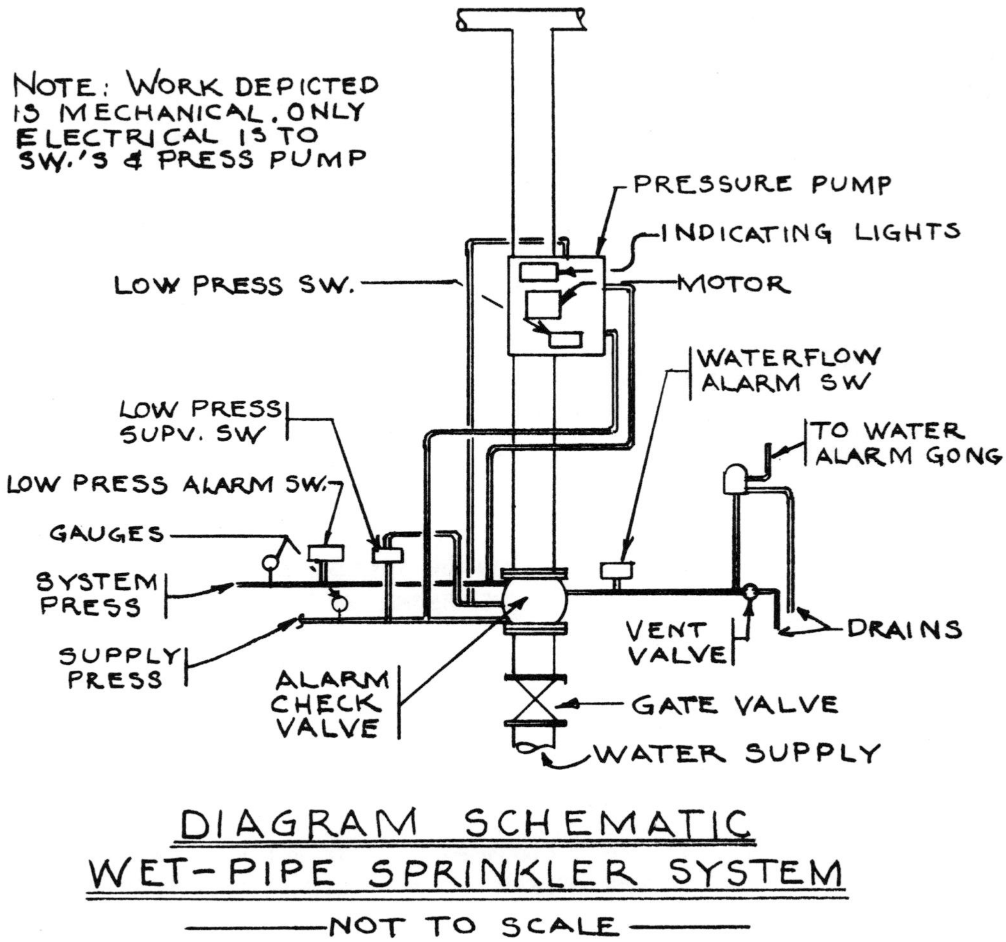

FIGURE 13-8

details is to enlarge the area of interconnect so that the relationship of electrical and mechanical work can be depicted.

Figure 13-8 shows that only a wet-pipe system has a pressure pump. The pressure pump has a series of lights to indicate normal and low pressure, a motor to drive the small pressurizing pump, and a low-pressure switch. All of these devices must be wired and interconnected to the fire alarm system.

Common to all of the systems are three switches: one to provide a low-pressure alarm contact, one to provide a low-pressure supervisory contact, and one to activate the alarm

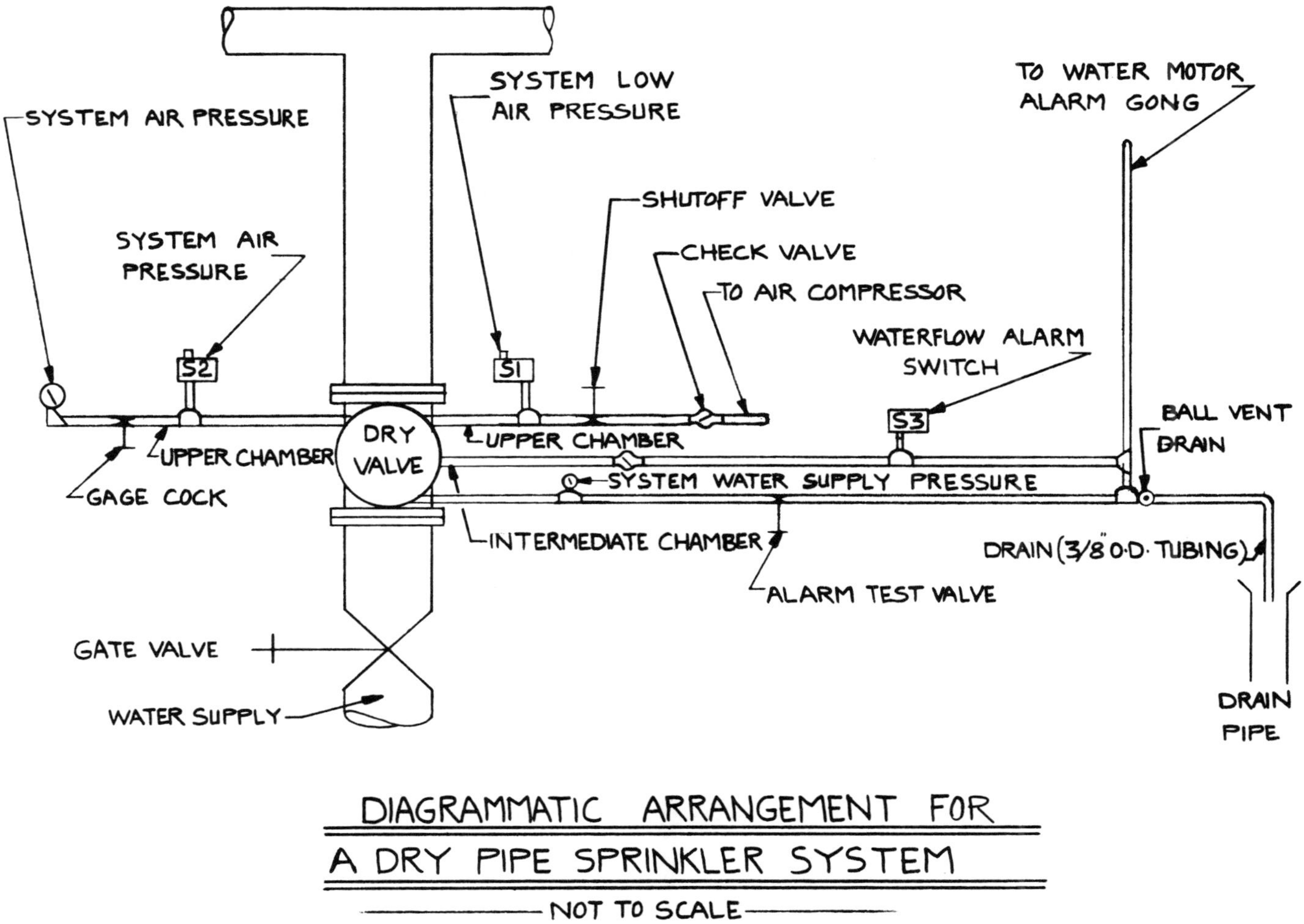

FIGURE 13-9

system when a sprinkler head opens.

Normally, all of this equipment would be installed by the sprinkler contractor. Depending on local requirements as well as the mandate of your specifications, the sprinkler contractor may or may not provide the interconnect wiring between this system and the fire alarm system.

Your firm may or may not deem it necessary to elaborate detailing to the extent shown, since most of the work is repetitive if one presumes the sprinkler system is fully detailed. The point made here is that the details should appear on some detail sheet and be made a part of the overall contract drawings prepared by your firm.

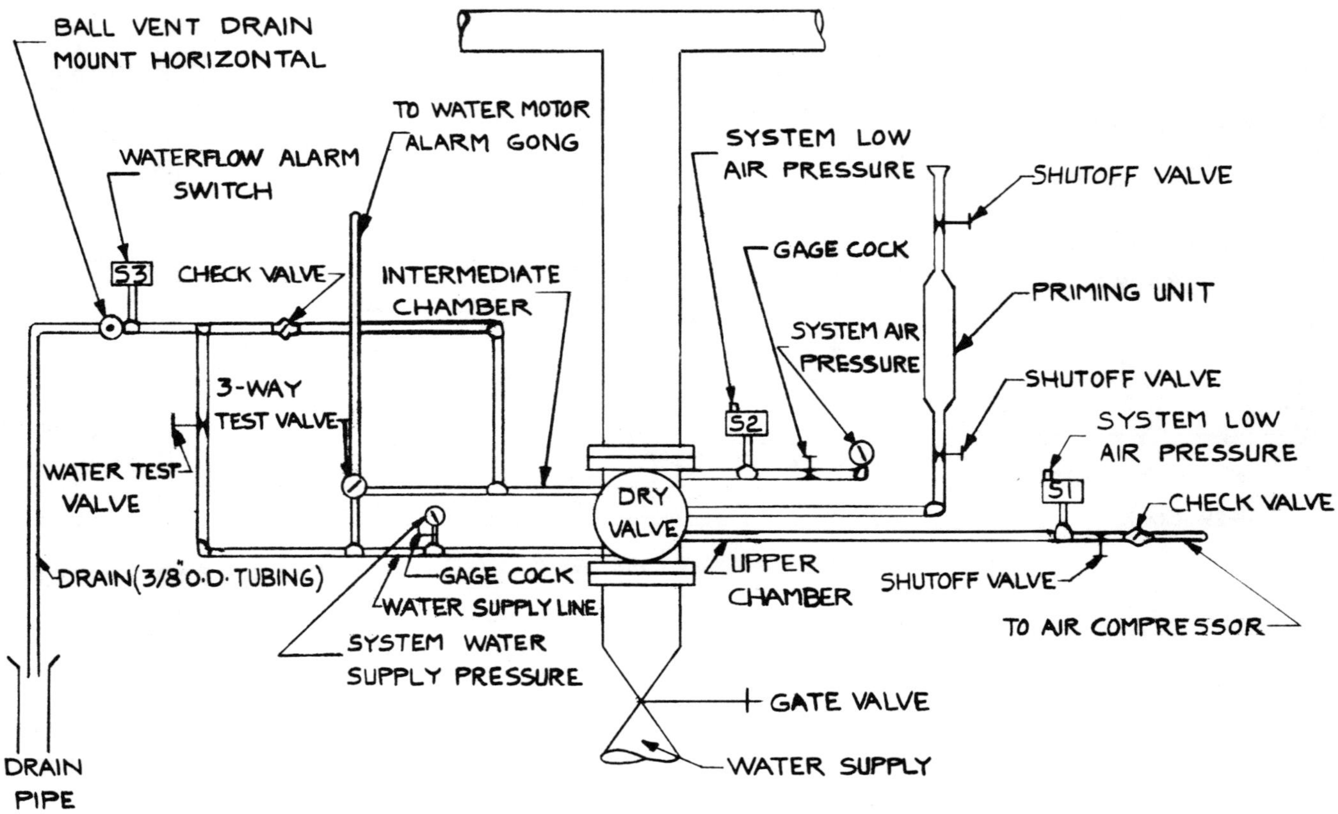

DIAGRAMMATIC ARRANGEMENT FOR A DRY PIPE SPRINKLER SYSTEM USING A 3-WAY TEST VALVE

NOT TO SCALE

FIGURE 13-10

FOURTEEN

Television Systems

The details presented in the illustrations that follow were reasonably up to date in early 1982. This is probably the only place in this book where we mention a date related to the time of manuscript preparation. This is not an excuse, but rather a sober admission that by the time of publication the details may be at least partially out of date.

The materials and methods used by the installers of on-site television distribution systems are in a period of very rapid change. These systems are a part of the explosively growing cable television business. Not many years ago there were literally no standards for anything. In the rush to get community and small-site (such as spread out housing for the elderly) systems intalled, almost anything available was the standard. You put up some antennas on some kind of pole or tower, ran coaxial cable or cables in some fashion to a television outlet or junction box, and then you had a system.

Like calculators and computers, things are continuing to change and to improve very rapidly. For example, two-year-old books on computers and programming refer to items no longer used, and current ways of connecting and installing television distribution systems are also making even two-year-old systems obsolete. The details that follow include text material that is vague or incomplete because we felt it was better to not describe a feature in detail that may be obsolete by the time this book is printed.

On the other hand, there are details and connections that, by their nature, will remain fairly standard. We hope we can illustrate these data plus the evolutionary items together into meaningful, usable collection of information.

Antenna Tower In order not to squeeze our antenna tower detail and thus destroy some essential features, we cut the detail into two parts. Figure 14-1 shows the lower half and Figure 14-2 shows the upper half.

The heart of any television signal receiving and disbursing system is the tower and its antennas. Figures 14-1 and 14-2 delineate the installation of a 50-ft-high tower with receiving antennas commonly used on one- or two-story garden apartments and on single-family housing and schools. The first word of caution concerns the concrete base and supporting guy wires. If the soil in the area of the tower is good undisturbed earth which permitted the associated structures to be erected with no unusual foundation problems, it is more than likely that the tower base will suit your application—*if you require a 50-ft tower.*

If your tower is higher than 50 ft you will require a larger base. Do not use the one shown in our detail. Conversely, if your tower is installed in good undisturbed earth and is only 40-ft high, the base shown is ample. For those designing systems in areas where hurricane-force winds or earthquakes are a problem, further structural analysis of the base and supporting guy wires is required, because our detail may not be satisfactory.

The actual steel bracing and framework in the tower is diagrammatic. As we note in our details, the tower framework is a standard fabricated item which can be produced by more than the one manufacturer. This is also a departure from other details in this book where a given manufacturer was clearly implied, since this sort of situation is frequently difficult to avoid. We make two points in this clear-cut departure from our previous presentation.

The first is that the tower, even one completely structurally detailed, will most likely be a standard manufactured item. Second, you could create a lot of problems for yourself if your structural detail was not really complete. Thus, the simpler and safer approach is to name a manufacturer in your area who furnishes, installs, and erects prefabricated television antenna towers.

As you can see from the two details, it was difficult to

ANTENNA PLACEMENT SCHEDULE		
MAST	UPPER ANTENNA	LOWER ANTENNA
A	CH. 3	CH. 8
B	CH. 24	CH. 40
C	CH. 30	CH. 57
D	–	CH. 22

FINAL ARRANGEMENT & SELECTION OF ANTENNAS TO BE DETERMINED BY SIGNAL SURVEY

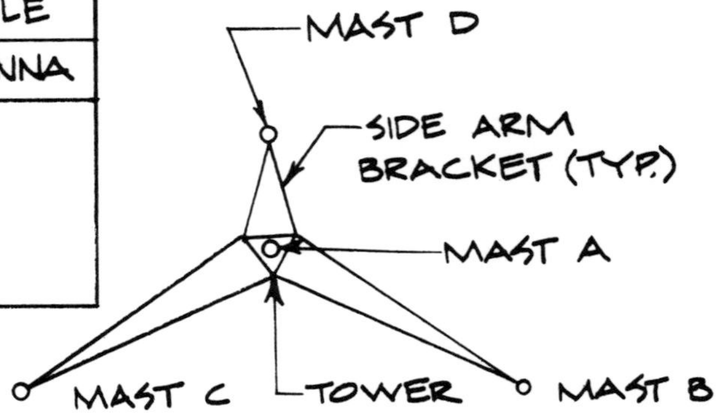

ANTENNA MASTING KEY PLAN

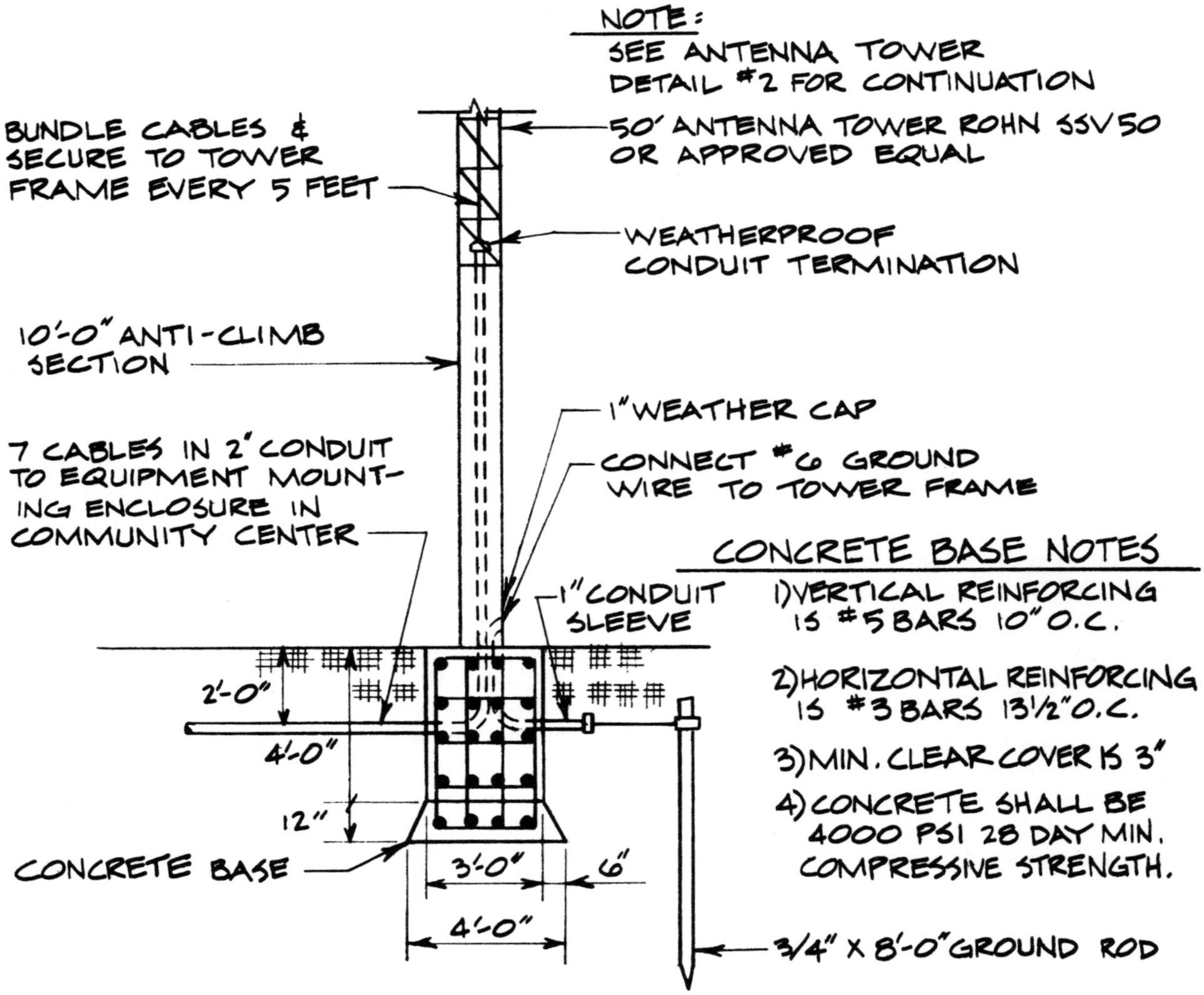

ANTENNA TOWER DETAIL #1

NO SCALE

FIGURE 14-1

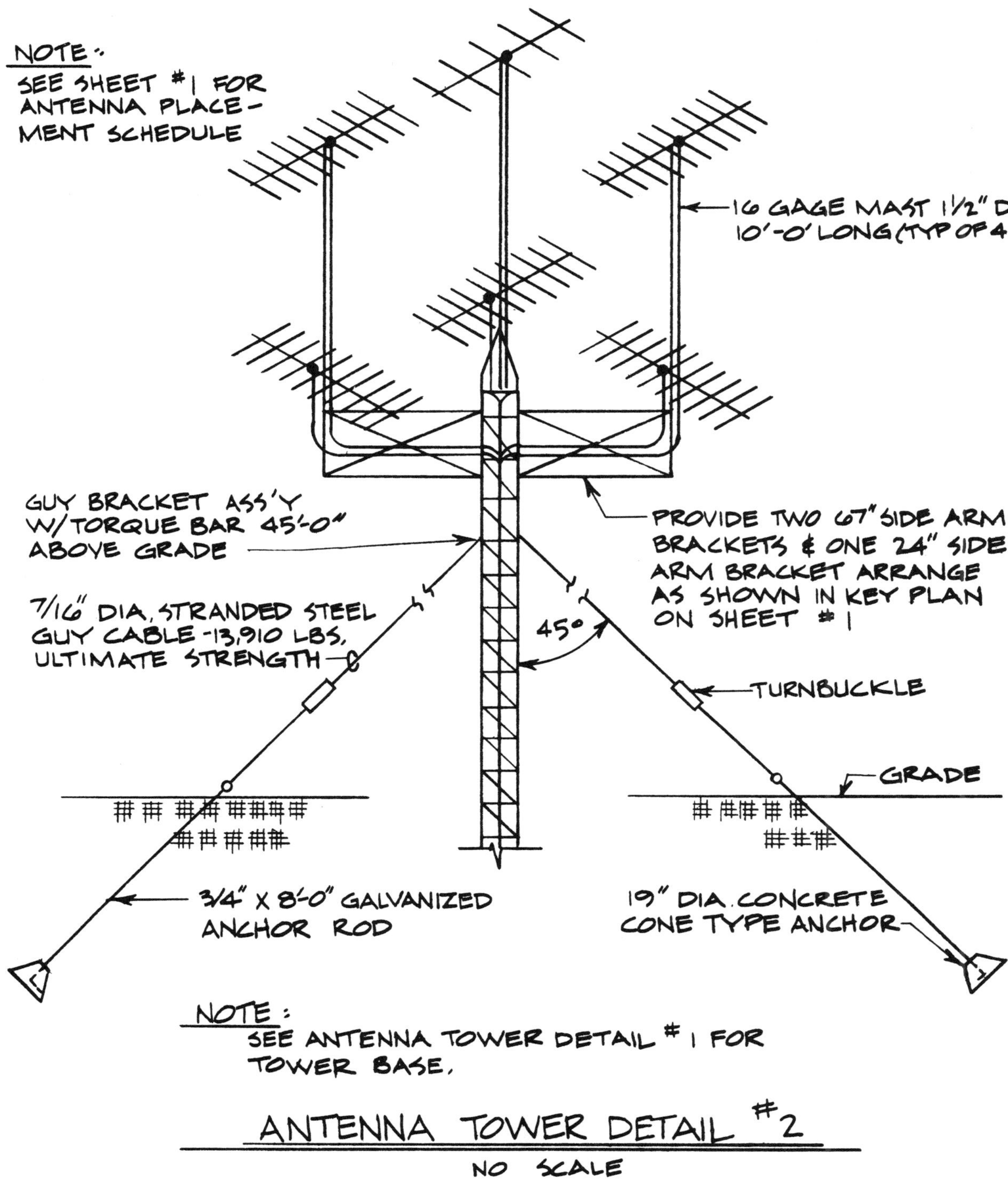

FIGURE 14-2

locate all the related notes properly. In some instances a note on one detail refers to a situation illustrated on the other detail. The most glaring example is the antenna schedule, which is shown on Figure 14-1 and referred to on Figure 14-2. This schedule is a very important item. You should carefully review the signal reception in your area. In all probability the maximum distance from a television station sending tower to your tower is going to be 50 mi or less. Since television reception is basically line of sight, your tower and its antennas are not going to pick up signals blocked by mountains or buildings.

Before we leave Figure 14-1, we call your attention to the lower 10-ft section described as an "anticlimb" section. Usually this section of the tower is covered by a smooth enclosure that makes it difficult to climb. A determined person could overcome this obstacle easily, but a climbing animal such as a squirrel would never make it up.

Finally, any tower and antenna is a target for a stray bolt of lightning. We show one driven ground rod and connections to it. Usually, though not always, this will suffice. We also indicate that the television signal cables are traveling in an underground conduit to the building that houses all the signal-reception equipment. You do not have to go underground with this line, but it is more secure if you can.

Figure 14-2 shows the upper half of the tower and antenna installation. Again, we have some problems in presenting data. The masting key plan is back in Figure 14-1, while the depiction of the masts is in this detail.

In order to show the guying detail completely, we invented some earth which properly should have been shown on Figure 14-1. We hope you will accept this sort of odd arrangement, because we felt it was important to cover the guying all in one piece. Unfortunately, the consultant frequently forgets just how much area is taken up by guys. We have shown the standard 45° angle guy wires. These must be at least three guys spaced at 120° included angles. The better installations have four spaced at 90° included angles. Since the guy wires are at a 45° angle to the tower and are connected 45 ft above ground, they obviously terminate 45 ft out from the tower. If your tower is closer than 45 ft to some building, walk, or road, you could easily have a very troublesome guying problem. Since the best way to anchor the guys is to connect them to an anchor rod 8 ft 0 in long terminating in a concrete conduit, we are sure you can visualize the problem. With a 45° angle the terminal concrete cone is approximately 5 to 8 ft below grade.

The brackets that support the actual antennas are arranged so that signal interference is minimized. Six antennas are shown, while seven appear in the schedule. There was a problem making the pictorial representation clear, and the specification contained a note of explanation tied to the note on the antenna placement schedule. In the actual installation the signal test indicated that all seven stations could not be satisfactorily received and only six were installed.

The bracket assemblies, side-arm brackets, and other devices are fairly standard and should be covered, whenever possible, by reference to standard manufacturers' data. Little items that seem almost irrelevant, such as turnbuckles used as tension-adjustment devices, should *not* be eliminated from your detail. Even though their need is obvious, they can be readily forgotten or omitted by installers.

Figures 14-1 and 14-2 are details for a specific installation which, when coupled with a proper specification, can be readily modified to suit your particular situation. Each of the points mentioned in this text may need further elaboration.

We have not said too much about the antenna and perhaps you should not also. What your specification should require is an antenna specifically designed for maximum signal reception of the particular television station you are trying to receive. Let the installing contractor be responsible for the type and arrangement of this antenna. Finally, a signal survey should be specified as a preinstallation requirement. You cannot always get a decent signal in the antenna location you have selected, even though less than a mile away the signal reception is loud and clear.

Television Signal Distribution The distribution of the television signal received at the antennas requires some modification in most cases. The methods used to amplify, modify, and clarify the signal will vary from location to location. Figure 14-3 illustrates some of the devices and methods that can be used. The arrangement shown is diagrammatic and is not intended to be used directly. The various devices we portray *may be more or less* than your particular situation requires.

The strength of the signal received generally needs to be reinforced or increased. This is the purpose of the amplifier. The amplifier is usually a specially tuned or arranged device designed to reinforce or increase the strength of a television signal. Sometimes this amplification can create signal interference or amplify reception problems. To help resolve this problem some channels may need further help. This is the primary task of the preamplifier. There frequently are problems of signal distortion which must be reduced or "trapped out."

Common to many signal systems is the attenuator. This device can be used to dampen or attenuate out signal noise. To provide the proper voltage to all of these devices requires the use of specially designed voltage and current regulators. These devices are described in Figure 14-3 as power supply.

The triangular-shaped objects shown, some of which contain numbers, are specially designed signal-distribution splitters that are selected to ensure equal signal strength to all locations. From these devices the cable lines extend to specifically selected outlets in each chosen area.

In explaining the flow of electrical current, a comparison to the flow of water in pipes is usually made. Even though we are talking about very low current and voltage values, the analogy still holds. If all systems were connected and operating and everyone except the person on the end of the line shut their systems off, there would be a surge in force similar to everyone shutting off all their faucets except the fellow on the end of the water line. In water systems the device to take the surge shock is called a *water hammer arrester*. In the television cable system the device is called a *terminator*.

In keeping with our plan to avoid being an electrical engineering text and to make the details more readily understood to the detailer, we have taken some liberties with electronic equipment descriptions. If we seem to have oversimplified this discussion, we hope it will be understood and accepted by those more technically trained. There are some excellent texts published by McGraw-Hill

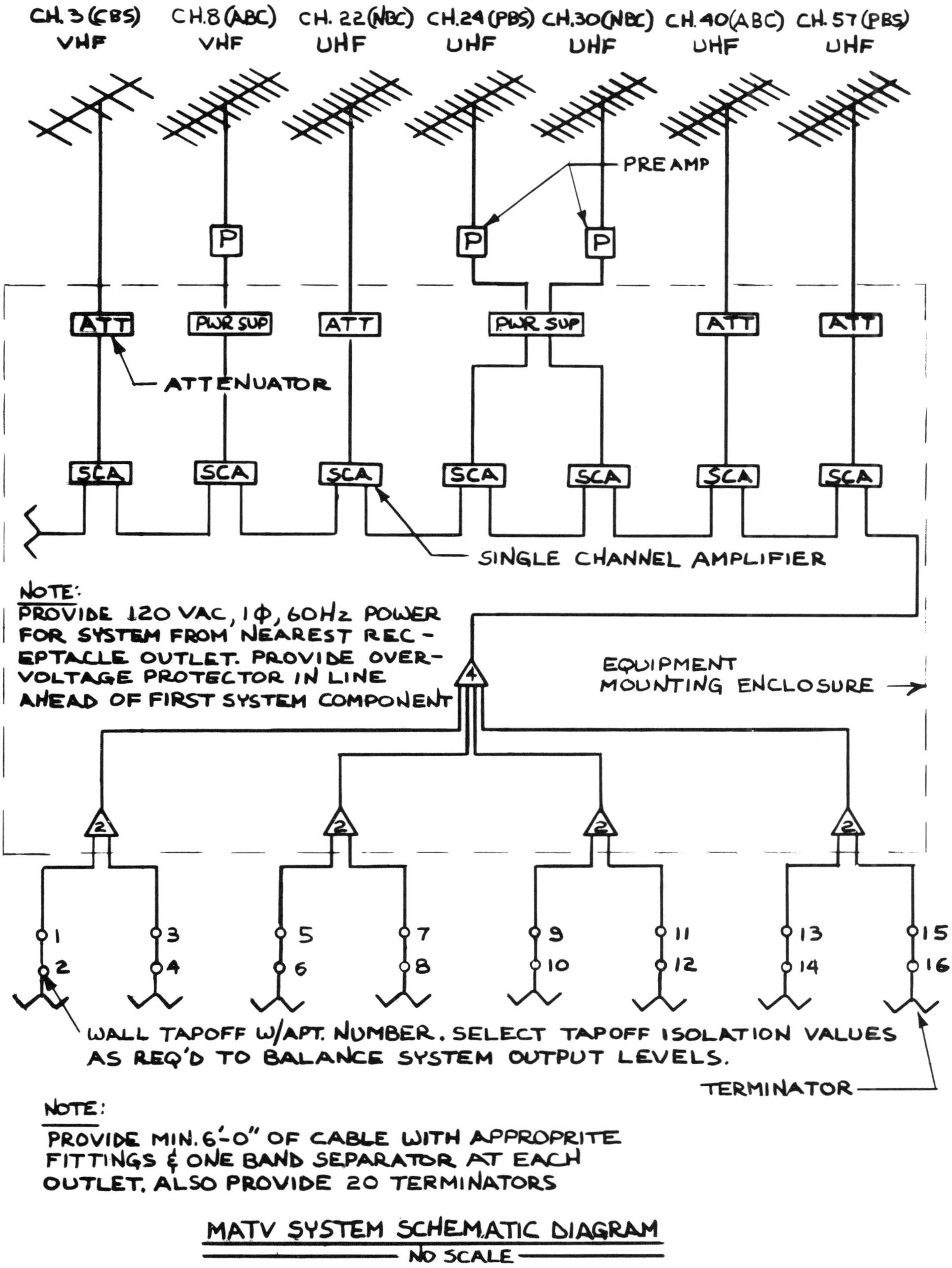
CH. 3 (CBS) VHF
CH. 8 (ABC) VHF
CH. 22 (NBC) UHF
CH. 24 (PBS) UHF
CH. 30 (NBC) UHF
CH. 40 (ABC) UHF
CH. 57 (PBS) UHF
PREAMP
P
ATT
PWR SUP
ATTENUATOR
SCA
SINGLE CHANNEL AMPLIFIER
NOTE:
PROVIDE 120 VAC, 1 Φ, 60 Hz POWER FOR SYSTEM FROM NEAREST RECEPTACLE OUTLET. PROVIDE OVER-VOLTAGE PROTECTOR IN LINE AHEAD OF FIRST SYSTEM COMPONENT
EQUIPMENT MOUNTING ENCLOSURE
4
2
1 2 3 4 5 6 7 8 9 10 11 12 13 14 15 16
WALL TAPOFF W/APT. NUMBER. SELECT TAPOFF ISOLATION VALUES AS REQ'D TO BALANCE SYSTEM OUTPUT LEVELS.
TERMINATOR
NOTE:
PROVIDE MIN. 6'-0" OF CABLE WITH APPROPRITE FITTINGS & ONE BAND SEPARATOR AT EACH OUTLET. ALSO PROVIDE 20 TERMINATORS
MATV SYSTEM SCHEMATIC DIAGRAM
NO SCALE

FIGURE 14-3

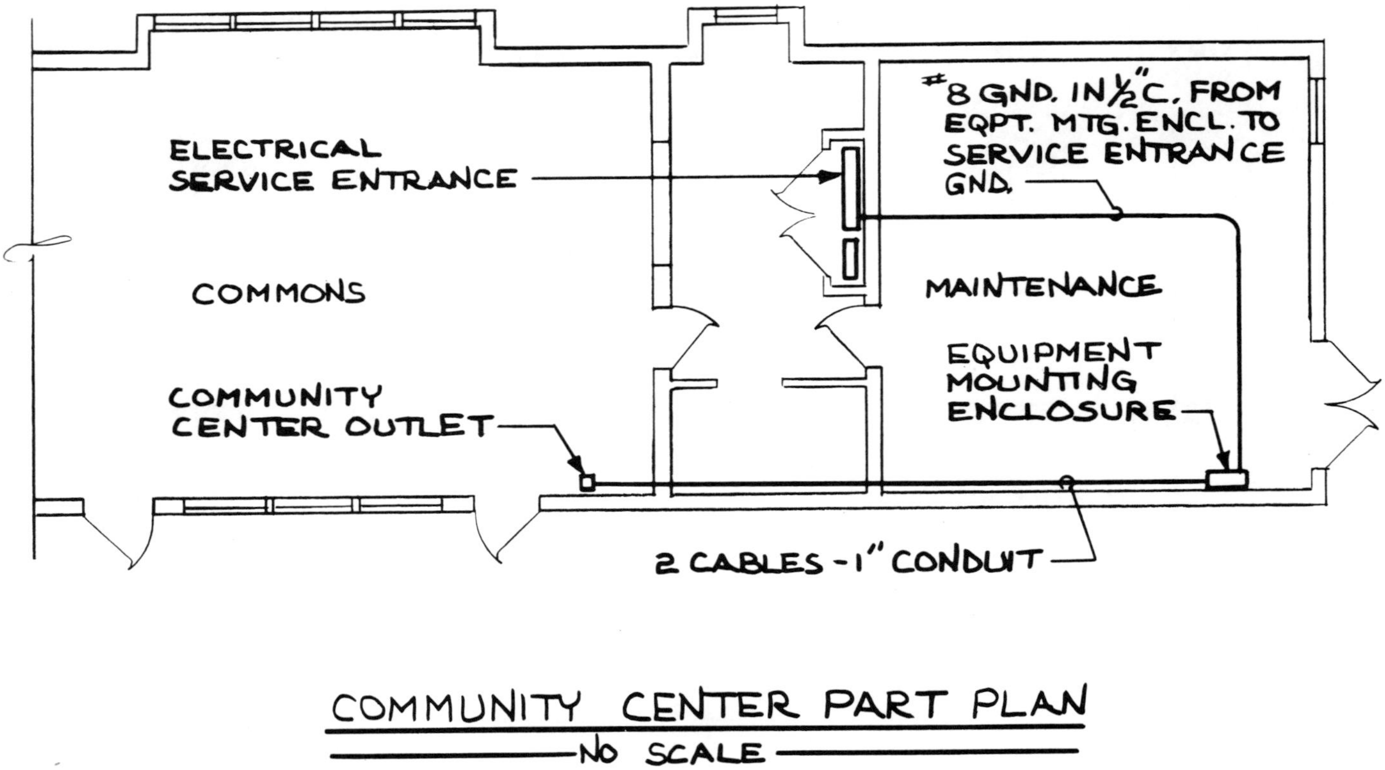

FIGURE 14-4

and others on the subject of electronic engineering, and we refer you to those texts for further amplification of our explanation.

Signal System Equipment One of the problems of all schematic diagrams is that at a quick glance one can easily be misled as to the actual location of the equipment. Figure 14-3 is an excellent detail if its use is understood, which is to show clearly all the parts and their relationships to each other.

Presumably the cable splitters, individual outlet boxes, and terminators can be readily seen to be in each building or apartment. But where, physically, are the preamplifiers, attenuators, amplifier power sources, and the like? Figure 14-4 is not a detail at all. It is a partial floor plan designed to clarify a point that would also be repeated and further explained in the system specification text. All of our devices are located in a locked, specifically designed enclosure in the maintenance office of the community center, which is centrally located in the apartment complex. It may appear that we are making a lot out of not very much, but we feel it is important that the detailer make the installer clearly understand where to locate all the items. The combination of Figures 14-3, 14-4, and 14-5 should clarify all these installation location problems.

Cable Distribution Inside as well as outside the building television cable could be run without any enclosure in a similar manner to what is commonly done for any single-family residence. In a house the installation of the antenna is secured to a pole. The pole is in turn fastened to a pipe, chimney, or other high point on the building, and the cable dangles down the side of the house into the room where the television set is located and is fastened directly to the back of the set. Commonly there are two cables—one for VHF and one for UHF.

Figure 14-5 is very simple and presumes that a cable system is being added to a solidly constructed flat-roof masonry building. Your installation might not be as simple as the one depicted, but, simple or complex, it would be very poor practice to either say in your specification to put all cable in conduit and let the contractor figure it out or, worse still, say nothing and hope for the best.

Where the path of the cable is subject to abuse and/or where it must conform to the building's aesthetic appearance, some plan or plans illustrating where and how you want the cable installed should be made. Our typical case is a simple situation and we took the simple way out. A partial isometric of a typical garden apartment was prepared and the conduit run delineated. It may seem that we have gone a bit far with the depiction of weatherproof

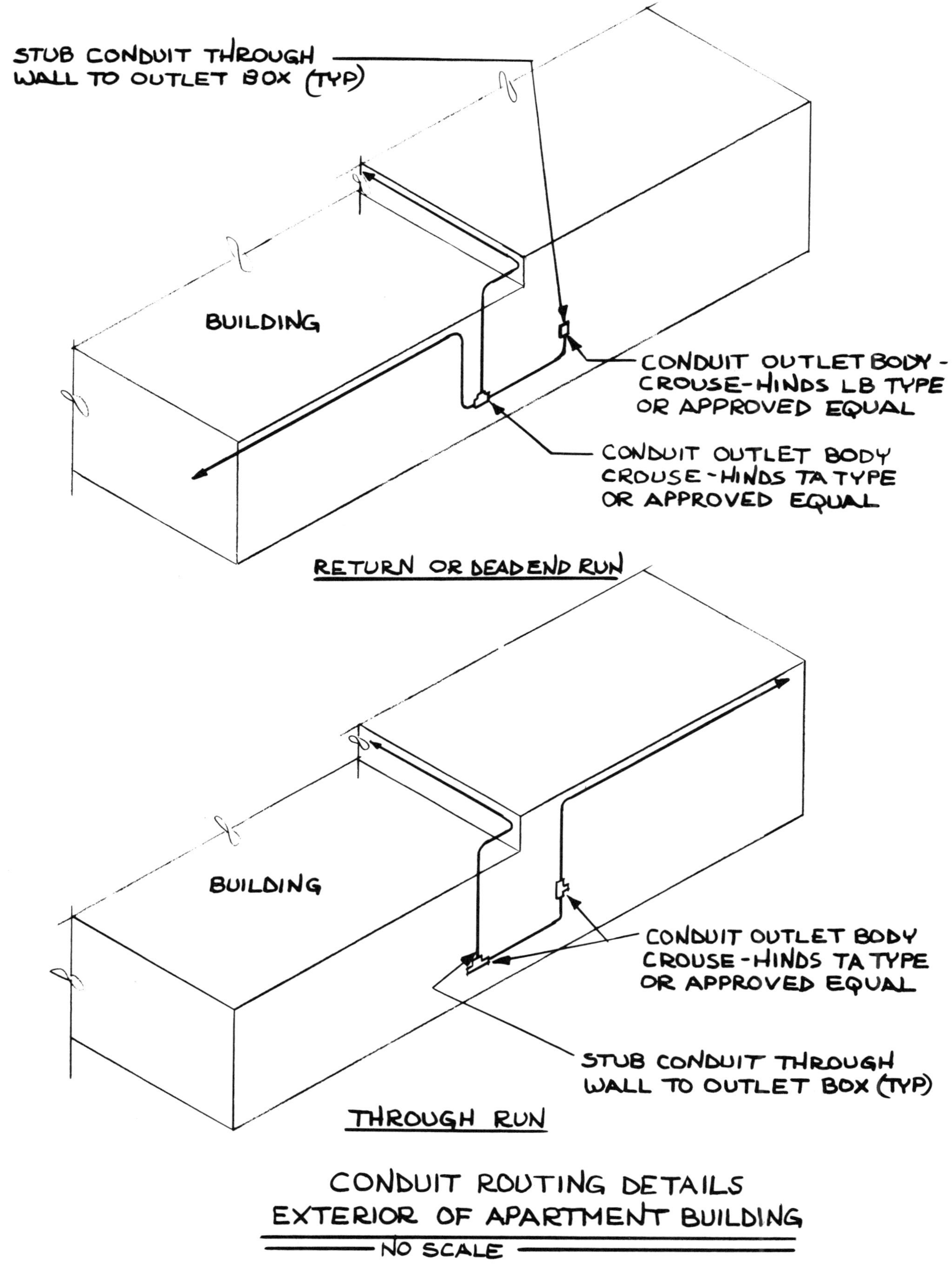
STUB CONDUIT THROUGH WALL TO OUTLET BOX (TYP)
BUILDING
CONDUIT OUTLET BODY - CROUSE-HINDS LB TYPE OR APPROVED EQUAL
CONDUIT OUTLET BODY CROUSE-HINDS TA TYPE OR APPROVED EQUAL
RETURN OR DEADEND RUN
BUILDING
CONDUIT OUTLET BODY CROUSE-HINDS TA TYPE OR APPROVED EQUAL
STUB CONDUIT THROUGH WALL TO OUTLET BOX (TYP)
THROUGH RUN
CONDUIT ROUTING DETAILS
EXTERIOR OF APARTMENT BUILDING
NO SCALE

FIGURE 14-5

outlet boxes, but this is a sure way of getting what we want.

This is not the only way to do this part of the design. It could be done by using plans and elevations and carefully noting where the lines will be run. Whatever way you choose, it is important that you pick some method and show the conduit runs—especially those on the outside of the structure.

Pole-Type Tower As is well known, a flag pole has no guy wires; a tall, sometimes very tall, highway lighting standard has no guy wires; and, certainly the luminaires and their supports on a 100-ft or higher highway lighting standard are much heavier and create much greater overturning forces than a lightweight television antenna. So why cannot a television antenna be installed without guy wires? Figure 14-6 illustrates that a pole tower can be used. Like every good consultant says "For enough money we can design most anything." And that is exactly why you rarely see television sending or receiving antennas on poles. There is usually no need and it is much more expensive.

In Figure 14-6 the pole itself is the major departure from our previous detail. We would like to caution you that this pole and its base represent a structural problem that should be carefully approached. We have deliberately left off any mention of the diameter of the pole, although it was taken from an actual installation. On the other hand, we included the dimensions of the concrete base. Our reasoning was that showing the relationship of the base to pole would give you some reasonable guidance as to what the answer to your design problem will probably be.

Although the individual antenna mast-arm brackets are not dimensioned, they are obviously fairly good sized U-bolt clamps. The mast arm's horizontal section attaches to its vertical section, and ultimately the antenna becomes attached as we have depicted in our antenna attachment sketch in Figure 14-6. These can usually be standard specified items.

Because the pole is hollow, we have a nice weather-protected path for the coaxial cables. In this case the UHF-VHF mixers were located at the antennas. And finally, the cables were to be enclosed in a conduit since the pole was close to the building, but they were supported on a messenger cable to the building. Once inside the building the cables were enclosed in conduit raceways. As can be easily understood, there is some movement in the pole because of wind. A solid conduit between the pole and the building would have been a problem. This is a somewhat older installation and is rarely used today.

Interior Television Distribution Television systems not only include signal reception but also include closed-circuit systems and sometimes interrelated sound and public address systems. The wiring of these systems can get exceedingly complex. Not infrequently the engineering consultant, attempting to do the job as thoroughly as possible, gets in trouble over the number and type of wires going from one device to another.

The perfect solution is, obviously, the totally correct wiring schematic showing each and every wire, properly sized, numbered, and counted, to connect each and every device. But this is frequently not practical or possible.

Figure 14-7 is one method of trying to come to a reasonable resolution of the problem. It is, rather obviously, an overview type of diagram such as might appear in part of a system covering a series of separated buildings. The exterior distribution is underground between buildings in communications conduits. The detail implies that signals can be carried in the proper number of conductors to an exterior manhole which is a junction point for some exterior system that could go anywhere. Signals could come from this location to the area covered by the schematic.

Judging by the notes appearing on the plan, you would guess that the diagram covers a physical education building in a college. This is indeed the case. Events can be picked up by public address systems and closed-circuit television cameras and displayed on television and sound monitors in this building or in other distant buildings connected by *the cable* system. As a side comment, note the end of the line resistor (terminator) which was a noted feature in our previous discussion of television distribution for Figure 14-3.

Certainly a lot of wiring is required between the various devices illustrated. The quantity would undoubtedly vary, depending on who were the selected suppliers, vendors, and contractors. The decision was made that this schematic detail, as well as others not depicted, would show all the required conduit runs sufficient in size to allow any of the probable suppliers, vendors, and contractors to pull their wires from one device to another. The specifications would call for the equipment suppliers to furnish and install the interconnect wiring between devices, since they alone, not the consultant or the contractor, would know exactly how much of what type of wiring was required to interconnect their equipment.

This may seem like something of a cop-out, but we do not agree. We might add, no one complained before, during, or after the bidding and subsequent installation. In theory, if you show 11 wires between two devices you have selected and specified and the actual number of devices selected by the system contractor requires 16, that is not your problem. In practice, the electrical contractor may bid on the wiring using your figure of 11 wires and have some arguments when the equipment you approve for installation requires 16. Whether right or not, we feel this system of showing conduits only is the better procedure.

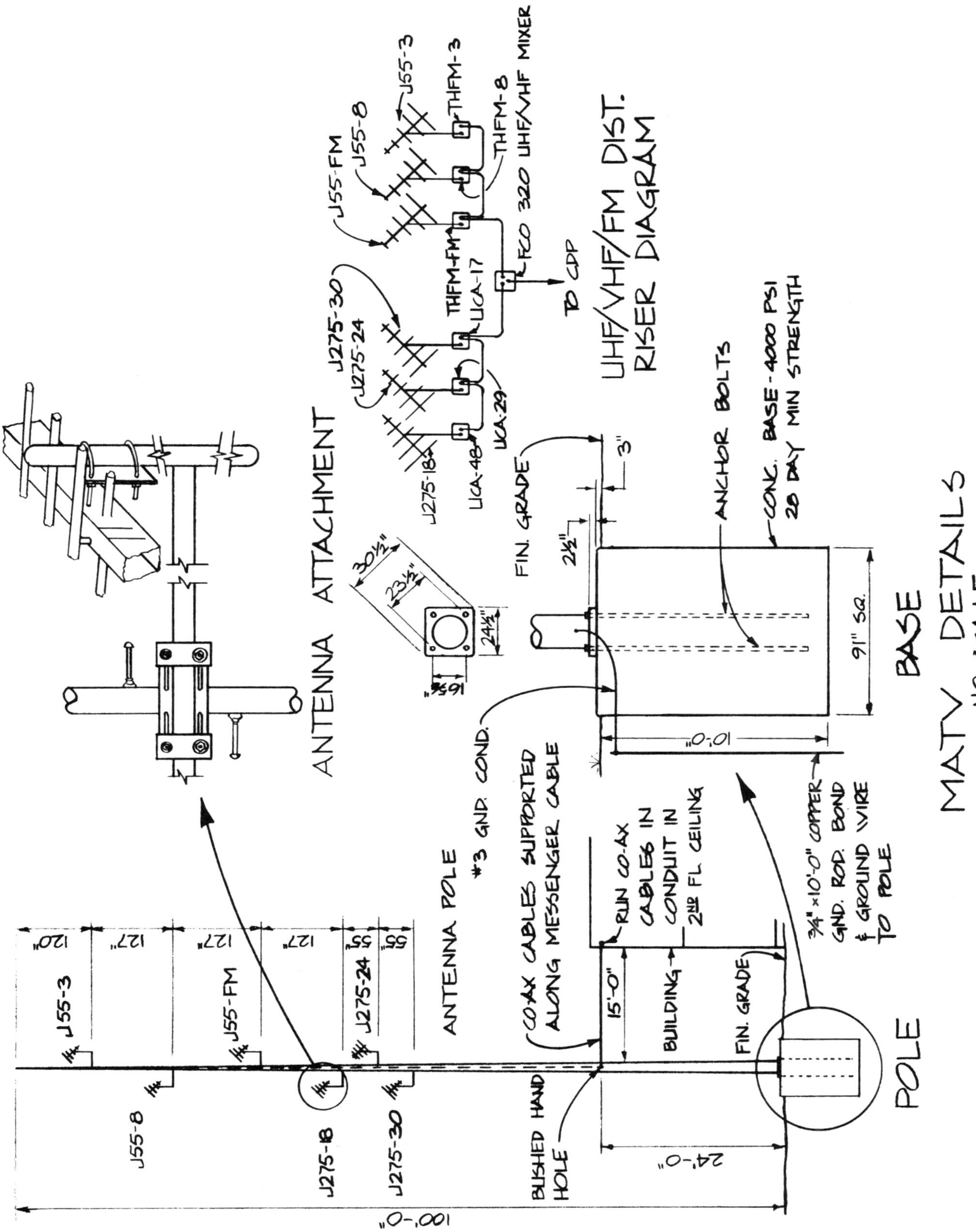
MATV DETAILS
NO SCALE
POLE
ANTENNA ATTACHMENT
BASE
UHF/VHF/FM DIST. RISER DIAGRAM
ANTENNA POLE
#3 GND. COND.
CO-AX CABLES SUPPORTED ALONG MESSENGER CABLE
RUN CO-AX CABLES IN CONDUIT IN 2ND FL CEILING
BUILDING
FIN. GRADE
BUSHED HAND HOLE
3/4" x 10'-0" COPPER GND. ROD. BOND & GROUND WIRE TO POLE
ANCHOR BOLTS
CONC. BASE - 4000 PSI 28 DAY MIN STRENGTH
91" SQ.
10'-0"
100'-0"
24'-0"
15'-0"
J55-3
J55-8
J55-FM
J275-18
J275-24
J275-30
THFM-3
THFM-8
THFM-FM
UCA-17
UCA-29
UCA-48
FCO 320 UHF/VHF MIXER
TO CDP

FIGURE 14-6

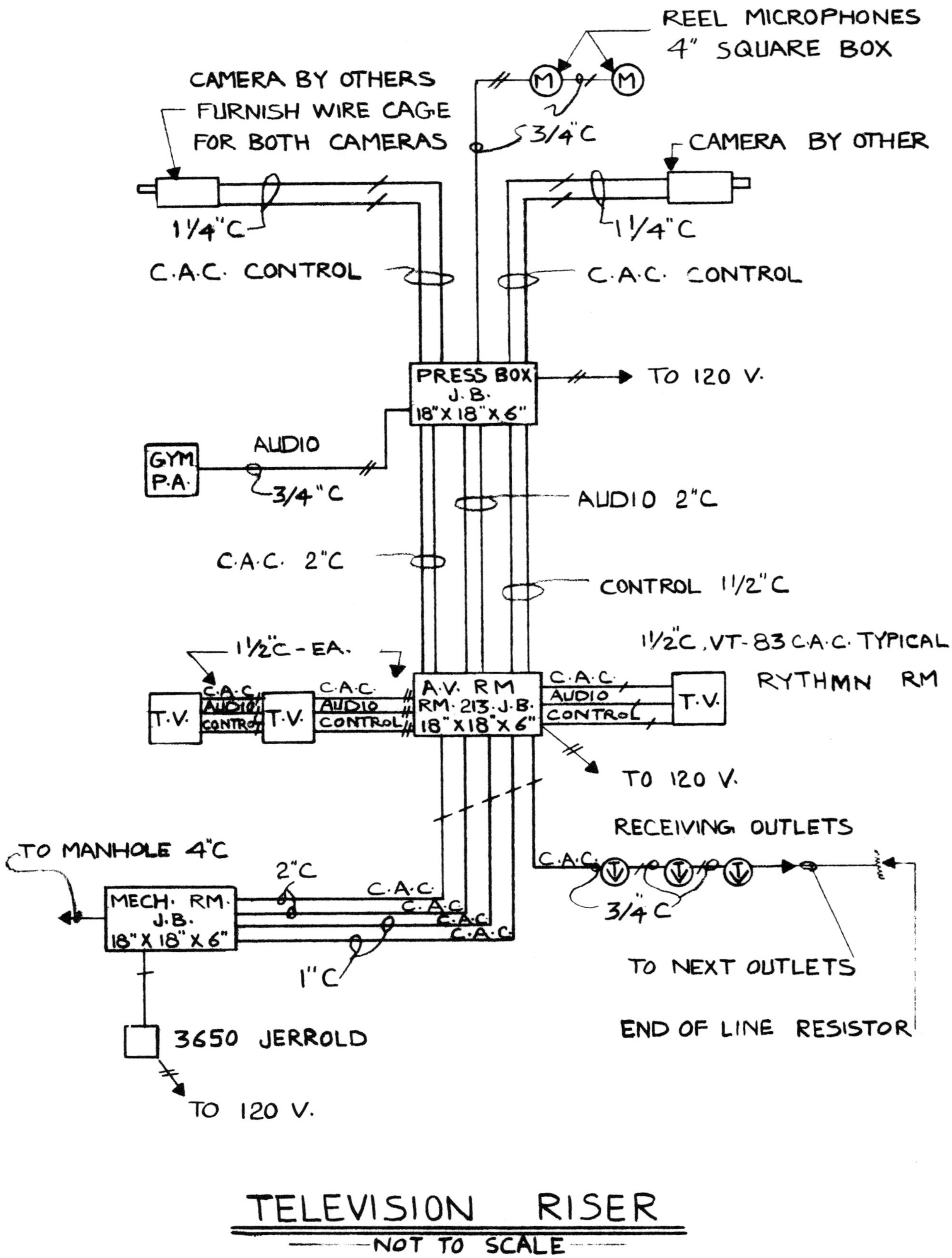
REEL MICROPHONES
4" SQUARE BOX
CAMERA BY OTHERS
FURNISH WIRE CAGE
FOR BOTH CAMERAS
3/4"C
CAMERA BY OTHER
1 1/4"C
1 1/4"C
C.A.C. CONTROL
C.A.C. CONTROL
PRESS BOX
J.B.
18"X18"X6"
TO 120 V.
AUDIO
GYM
P.A.
3/4"C
AUDIO 2"C
C.A.C. 2"C
CONTROL 1 1/2"C
1 1/2"C - EA.
1 1/2"C, VT-83 C.A.C. TYPICAL
RYTHMN RM
C.A.C.
AUDIO
CONTROL
T.V.
A.V. RM
RM. 213. J.B.
18"X18"X6"
TO 120 V.
RECEIVING OUTLETS
TO MANHOLE 4"C
2"C
MECH. RM.
J.B.
18"X18"X6"
1"C
3/4"C
TO NEXT OUTLETS
END OF LINE RESISTOR
3650 JERROLD
TO 120 V.
TELEVISION RISER
NOT TO SCALE

FIGURE 14-7

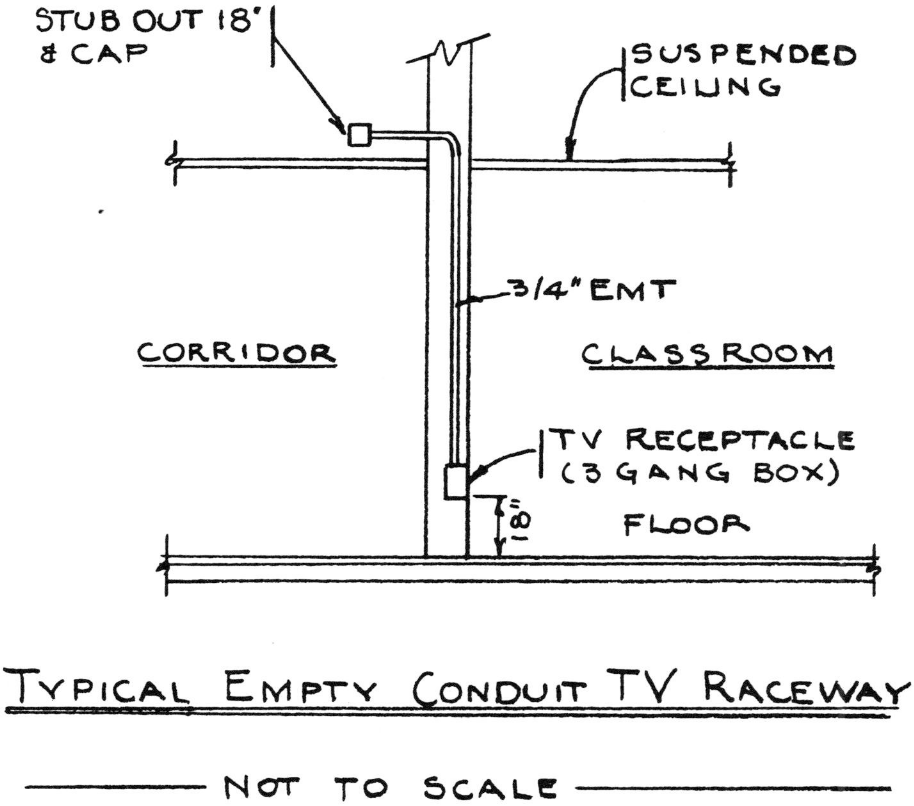

FIGURE 14-8

Figure 14-8 is very small and very simple. One might ask why we even included it. Being so small and simple it is the sort of thing frequently covered by a brief note in the specification or an even briefer note on the plans. A fast-paced construction project, a school, for example, could be half-complete before someone noticed there was no raceway for the television system. The accompanying second detail makes it loud and clear that permission is going to be given for open wiring above the dropped ceiling. But to get from that ceiling space down a block, brick, or other type of wall, the contractor is to provide a raceway to a television outlet. And to make sure the outlet box is large enough, the recessed outlet box is specified as a three-gang box. Sounds simple and it is. If you forget to show Figure 14-8, the complaints will make it seem as if this installation is one of the world's most difficult tasks.

FIFTEEN

Stage and Television Lighting Systems

The vast majority of stage lighting projects involving the practicing consultant are the stages of high schools, junior highs, and elementary schools. Usually, if it is an elementary school, there will be a single pipe row holding probably three standard sections of border interspersed with two or three downlights in between the border lights and on each end. At a not-too-distant point in front of the proscenium opening, on the ceiling or sidewalls, are some additional spotlights, usually in clusters of three. Normally each type of lighting contains repeating sets of the three basic colors—red, blue, and green. These three colors combine to make a white light and in selected combinations can produce other colors.

The entire lighting system could be controlled by single-pole switches or by a small packaged dimmer module. Technically and legitimately, this system is stage lighting. It is by no means "lighting the stage" as the term is generally used. Depending on the exact type of fixture selected, the beam spread of the overhead lighting results generally in a coverage of some 20 ft of stage depth (front to back as seen by the audience). As everyone knows, transition from light to dark is not a sharp line. So if you stand farther than 10 ft 0 in in front of or behind the overhead border lights, you will not be in the dark. You will, however, be in an area of rapidly decreasing light intensity, sort of in the shadows.

Lighting the stage not only concerns lighting the area used by the performers but also the curtains or scene (back drop) behind them, the front curtain that closes when the performance is over, and the hall or auditorium where the audience is seated. It also includes the front lobby or entrance. Lighting of all these areas is normally controlled from a master control board. Part or all of the entry and hall or auditorium lighting may also be separately controlled by switches or by a combination of master and submaster contact switches that are located for ease of control by hall operating and maintenance personnel.

A well-illuminated hall or auditorium begins with an attractive, inviting lighting design of the lobby. This, and all the other lighting we will describe, is incandescent. Rarely is a fluorescent, mercury, or sodium-vapor fixture used in any hall or auditorium.

Inside the hall itself, regardless of whether it is used for meetings, theater, motion picture, or television production, the primary problem is first to have good general lighting so that the audience can safely enter and leave. If a requirement is that the audience is expected to read printed documents easily (not merely a large-type printed playbill) the footcandle level of general illumination has to be enough to accomplish this task. A further improvement consists of small luminaires built into the arms of the aisle seats. When the house lights are turned off these luminaires provide a lighted path for those who leave while the show is in progress. Usually, these aisle lights are also turned down or selectively off during a performance. Theater operators usually leave a few dimmed but not off in strategically selected locations.

The true stage-lighting design is a much more elaborate

version of the grade-school stage-lighting system described in our opening paragraph. When the curtain is opened what the audience sees are the ends of the curtains on each side of the stage and a top, short curtain called a valance which creates a top border closure. Behind this valance are rows of border and Fresnel lights that light the stage. The first row is really two rows—one of continuous border lights and the other holding a series of Fresnel spotlights. On approximately 7-ft 0-in spacing are additional rows of border lights and downlights. At the rear of the stage, mounted on the floor or from a pipe overhead, are cyclo rama (rear scene or curtain) lights.

On the sides of the stage, hidden from the audience, are the special-effect lights or spotlights. Finally, out in the front of the stage beyond the front curtain and near the front edge are the footlights, which are either of the recessed or disappearing type. The footlights are frequently used to light up the closed curtain which opens when the show begins.

In the ceiling of the hall or auditorium at the proper angle to the stage, or on the sidewalls, are located six or more elliposidal spotlights. These lights are on approximately 6-ft 0-in centers. As you have probably seen, the orchestra usually sits in a depressed area (pit) in front of the stage. The orchestra has no special lighting and relies on a series of outlets around the pit to plug in cords to light their music stands.

Somewhere in the hall, usually high in the rear, is a special cubicle containing an operator with a high-powered adjustable (both to direction and color) follower spotlight. He or she is the person who highlights the individual performer, or scene, familiar to all who have attended almost any major stage show. Sometimes, in large auditoriums, this location may be built into the first balcony in a cove.

Finally, all of this lighting, as we noted at the start of this discussion, must be controlled from a master control panel which can become a very large and complex device. The three locations for this panel, in order of preference, are behind the curtains and proscenium wall to one side of the stage, in the orchestra pit, or over the auditorium ceiling in a special constructed and supported housing with a clear view of the stage.

A television production creates a different set of problems. There are so many camera connections and people all over the area surrounding the set that footlights are simply not usable. Nor does border lighting play much of a role, if any. The set is usually lighted from a series of carefully located and directed spot- and floodlights. A large modern stage can readily be divided up into several small sets where the performers can readily walk from one scene to the next.

Stage Lighting Composite To try to illustrate our opening text as well as to present a composite detail of a stage lighting system, we illustrate in Figures 15-1 to 15-3 most of the elements of a stage lighting system. To avoid clutter we have not included the auditorium overhead or aisle lighting, or the vestibule lighting, since these were assumed to be more obvious forms of lighting.

Most of the lighting on the stage should be presumed to be overhead floodlight-type illumination. Spotlighting from overhead can be occasionally used for special effects. Sidewall-mounted spotlights serve as a close-up form of highlighting performers and scenes. The ceiling ellipsodial spotlights can also be used (and are commonly used) for the same purposes. Keep in mind that these lights are not set up to follow a moving performer. Instead the performance, arrangement, and area are all based on lighting a certain preselected position on the stage. Only the balcony or projection booth follower spotlight, which can be continuously aimed and directed, can keep a performer in the spotlight at all times.

All of the lighting fixtures have cords (pigtails) with a quick-disconnect feature. The overhead lighting system begins with steel pipes held up on steel cables. Attached to these pipes are the electrical lines and outlets. The overhead fixtures (individual lights and border lights) are hung by wires and chains from the same pipe. (The pipe may or may not be able to be raised and lowered). The fixtures are plugged into electrical outlets on the pipe. The wiring is attached to the pipe and carried back (usually as open wiring) to the master controller.

Fixtures on walls and ceilings are bracketed off of supporting pipes or brackets and connected in the same fashion. Keep in mind that all stage lighting must be readily movable, adjustable, and removable. You will require floor outlets (covered floor pockets) so that portable floor-lamp-type spotlights or floodlights can be properly plugged in and controlled.

All of the wiring from all of the outlets for all types of lighting returns as individual circuits to a master control (dimmer) board. This is the point where the lighting director controls, mixes, and arranges the intensity (output) of all lighting for each area of each scene in the show. In a complex system a number of circuits may be "preset," that is, their intensity, color, and quality are set and then locked into a master control lever. This may be one setting for the entire show or a series of settings. One lever sets all lighting and only the spots are separately controlled.

The Dimmer System As we mentioned in our opening discussion, what really makes stage lighting separate and different from all other types of lighting are basically two items: first, the strategic location and color selection of the lighting fixtures, and second, and even more important, the control of light output by the dimmers. This is where engineering ends and lighting as an art begins. Poor lighting and control may not cause a show that is otherwise well written, produced, directed, and acted to fold, but it certainly will negate these areas to a very large degree. Invariably hit shows also score very well in lighting awards, and in some reviews the only positive comment on a new

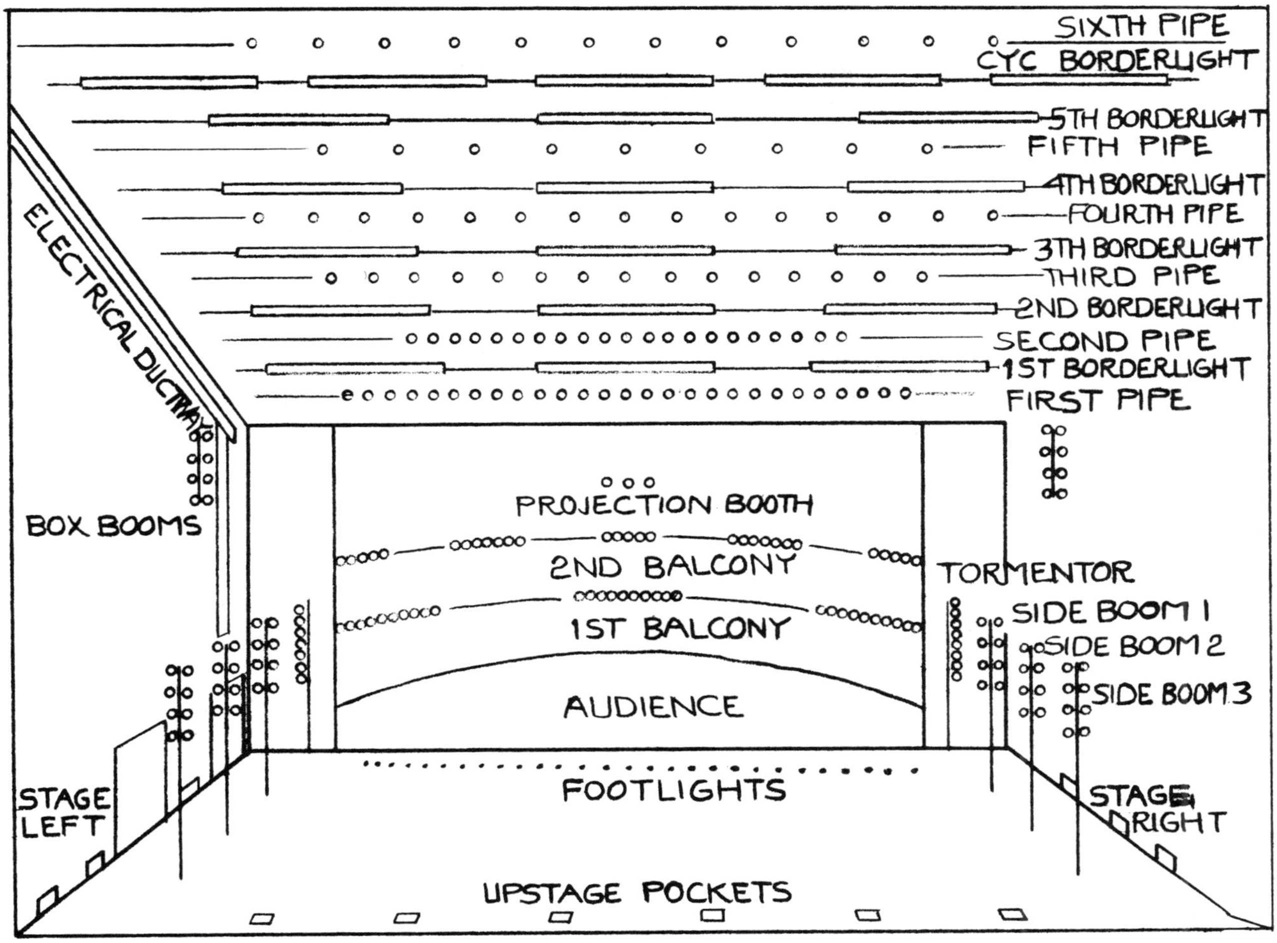

EXPANDED STAGE VIEW
LOOKING TOWARDS AUDIENCE
NO SCALE

FIGURE 15-1

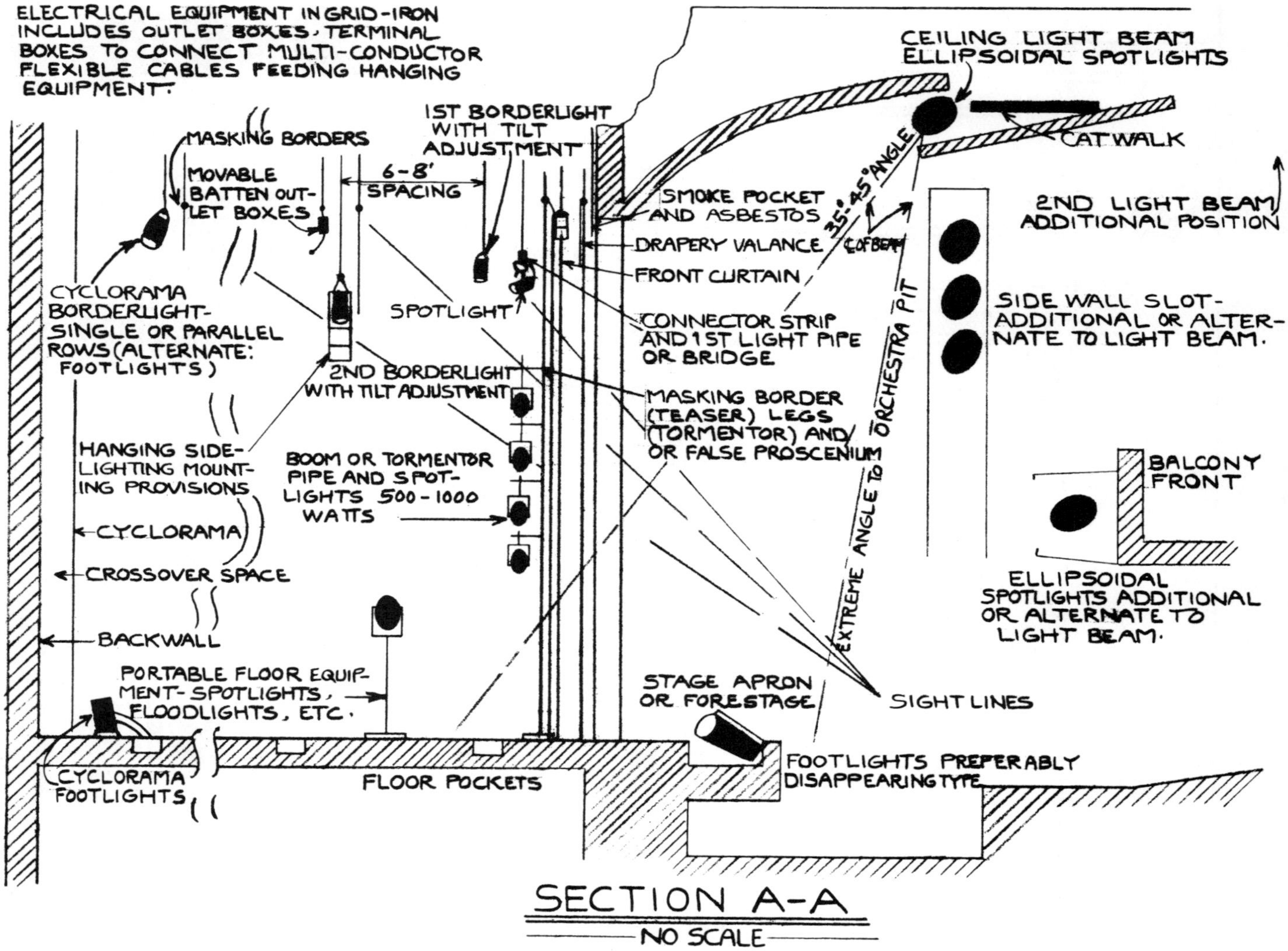
ELECTRICAL EQUIPMENT IN GRID-IRON INCLUDES OUTLET BOXES, TERMINAL BOXES TO CONNECT MULTI-CONDUCTOR FLEXIBLE CABLES FEEDING HANGING EQUIPMENT.
CEILING LIGHT BEAM ELLIPSOIDAL SPOTLIGHTS
1ST BORDERLIGHT WITH TILT ADJUSTMENT
MASKING BORDERS
CATWALK
MOVABLE BATTEN OUTLET BOXES
6-8' SPACING
SMOKE POCKET AND ASBESTOS
35°-45° ANGLE
2ND LIGHT BEAM ADDITIONAL POSITION
DRAPERY VALANCE
℄ OF BEAM
FRONT CURTAIN
CYCLORAMA BORDERLIGHT-SINGLE OR PARALLEL ROWS (ALTERNATE: FOOTLIGHTS)
SPOTLIGHT
CONNECTOR STRIP AND 1ST LIGHT PIPE OR BRIDGE
SIDE WALL SLOT-ADDITIONAL OR ALTERNATE TO LIGHT BEAM.
2ND BORDERLIGHT WITH TILT ADJUSTMENT
MASKING BORDER (TEASER) LEGS (TORMENTOR) AND/OR FALSE PROSCENIUM
HANGING SIDE-LIGHTING MOUNTING PROVISIONS
BOOM OR TORMENTOR PIPE AND SPOTLIGHTS 500-1000 WATTS
EXTREME ANGLE TO ORCHESTRA PIT
BALCONY FRONT
CYCLORAMA
CROSSOVER SPACE
ELLIPSOIDAL SPOTLIGHTS ADDITIONAL OR ALTERNATE TO LIGHT BEAM.
BACKWALL
PORTABLE FLOOR EQUIPMENT-SPOTLIGHTS, FLOODLIGHTS, ETC.
STAGE APRON OR FORESTAGE
SIGHT LINES
CYCLORAMA FOOTLIGHTS
FLOOR POCKETS
FOOTLIGHTS PREFERABLY DISAPPEARING TYPE
SECTION A-A
NO SCALE

FIGURE 15-2

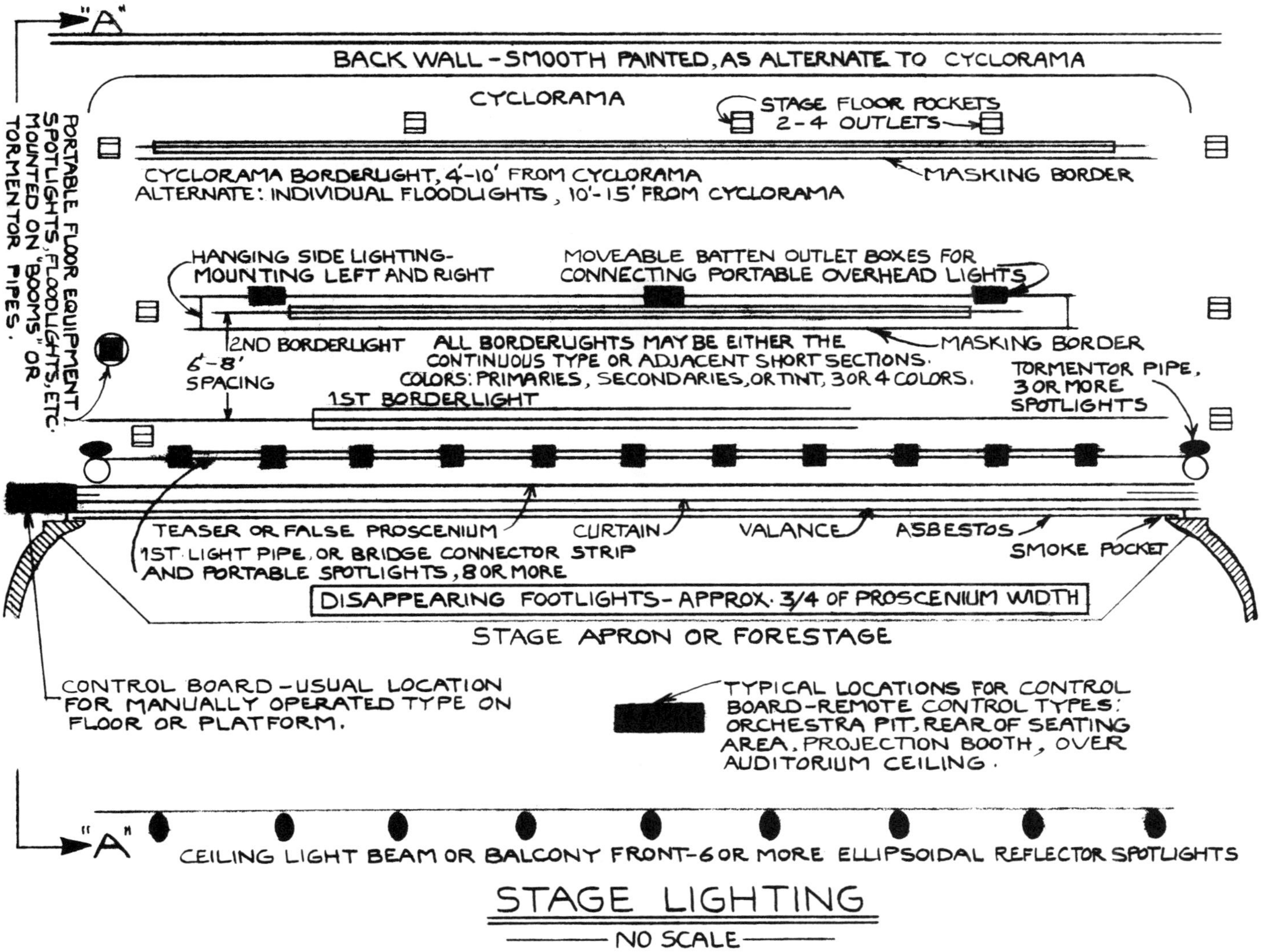
"A"
BACK WALL - SMOOTH PAINTED, AS ALTERNATE TO CYCLORAMA
CYCLORAMA
STAGE FLOOR POCKETS 2-4 OUTLETS
CYCLORAMA BORDERLIGHT, 4'-10' FROM CYCLORAMA
ALTERNATE: INDIVIDUAL FLOODLIGHTS, 10'-15' FROM CYCLORAMA
MASKING BORDER
PORTABLE FLOOR EQUIPMENT, SPOTLIGHTS, FLOODLIGHTS, ETC. MOUNTED ON "BOOMS" OR TORMENTOR PIPES.
HANGING SIDE LIGHTING-MOUNTING LEFT AND RIGHT
MOVEABLE BATTEN OUTLET BOXES FOR CONNECTING PORTABLE OVERHEAD LIGHTS
2ND BORDERLIGHT
ALL BORDERLIGHTS MAY BE EITHER THE CONTINUOUS TYPE OR ADJACENT SHORT SECTIONS.
COLORS: PRIMARIES, SECONDARIES, OR TINT, 3 OR 4 COLORS.
MASKING BORDER
6'-8' SPACING
TORMENTOR PIPE, 3 OR MORE SPOTLIGHTS
1ST BORDERLIGHT
TEASER OR FALSE PROSCENIUM
CURTAIN
VALANCE
ASBESTOS
1ST LIGHT PIPE, OR BRIDGE CONNECTOR STRIP AND PORTABLE SPOTLIGHTS, 8 OR MORE
SMOKE POCKET
DISAPPEARING FOOTLIGHTS - APPROX. 3/4 OF PROSCENIUM WIDTH
STAGE APRON OR FORESTAGE
CONTROL BOARD - USUAL LOCATION FOR MANUALLY OPERATED TYPE ON FLOOR OR PLATFORM.
TYPICAL LOCATIONS FOR CONTROL BOARD - REMOTE CONTROL TYPES: ORCHESTRA PIT, REAR OF SEATING AREA, PROJECTION BOOTH, OVER AUDITORIUM CEILING.
"A"
CEILING LIGHT BEAM OR BALCONY FRONT - 6 OR MORE ELLIPSOIDAL REFLECTOR SPOTLIGHTS
STAGE LIGHTING
NO SCALE

FIGURE 15-3

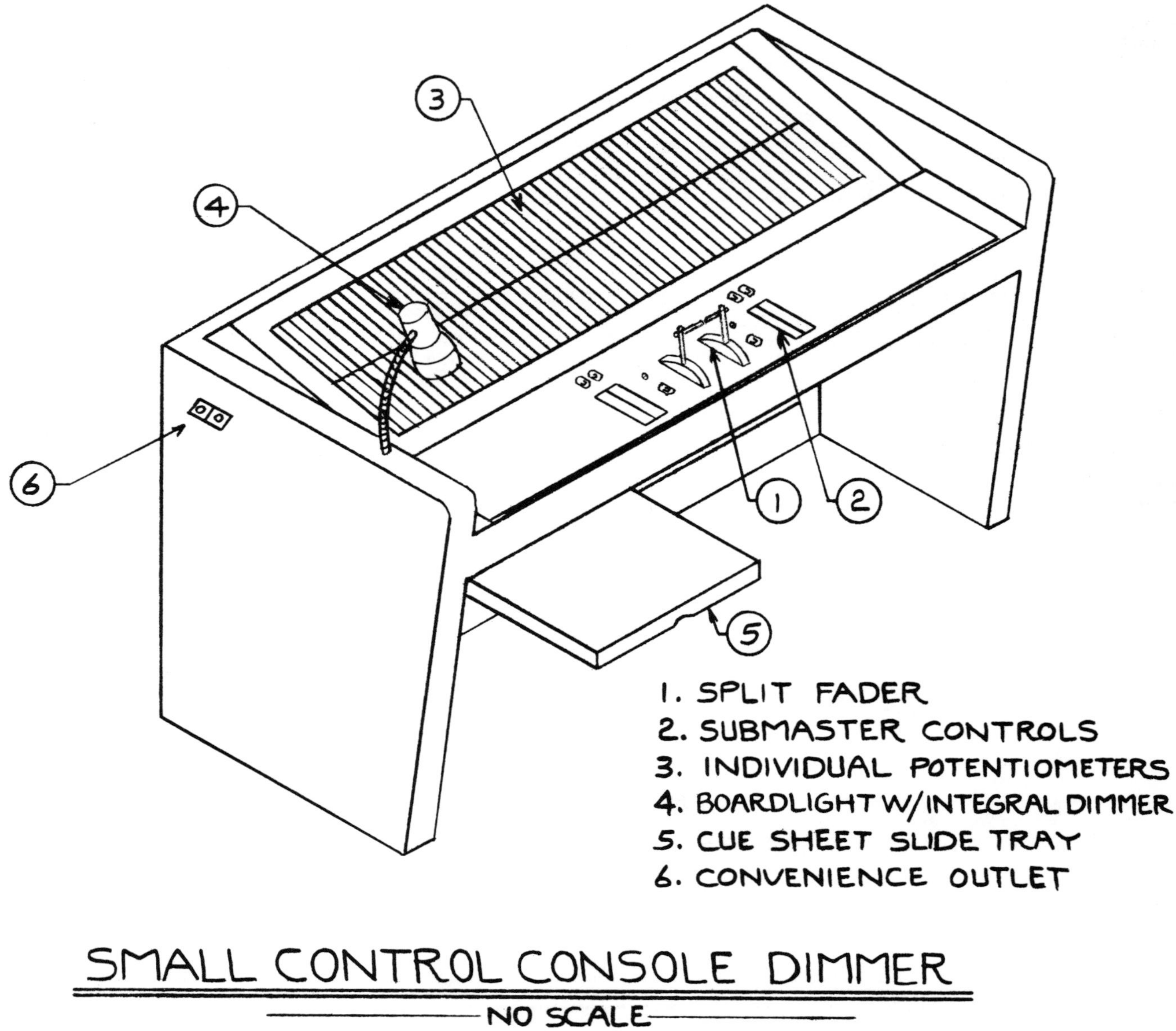

FIGURE 15-4

show may be for its lighting and general effects.

Figure 15-4 is a smaller standardized dimmer unit to which the lighting circuits can be attached and the lighting output raised or lowered. The type of system using this simplified dimmer control is known as manual preset lighting control system. The entire stage lighting is tied to this system and preset to desired levels of intensity. The horizontal rows of controllers are termed *presets*. These presets are tied to submaster and master faders which provide for dipless, smooth cross fades between presets. One row, for example, is set for one effect and the other row for another. Moving the split fader levers transforms the lighting effect from one type to another in a smooth, eye-pleasing transition. Obviously, this system is limited as to type and selection, but for many small systems it is more than adequate.

Figures 15-5 and 15-6 illustrate the well-known "patch" panel. If you have seen an old-fashioned telephone switchboard, you know that your incoming call is on the plug the operator picks up. When you give your calling number request the operator plugs the cord into the hole and makes your connection.

The patch panel operates in the same manner. All the lighting units terminate in cords at the plug counter.

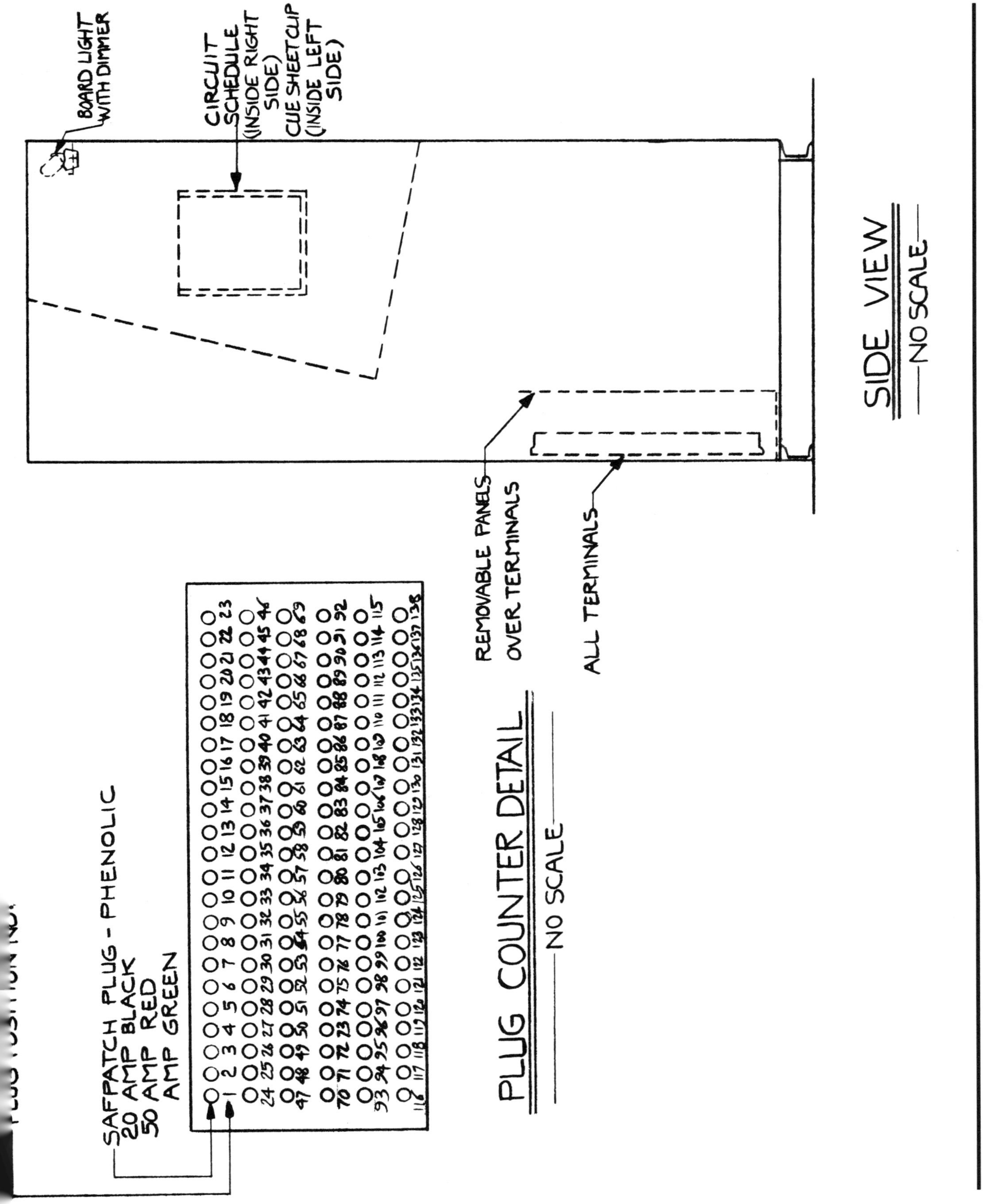
BOARD LIGHT WITH DIMMER
CIRCUIT SCHEDULE (INSIDE RIGHT SIDE)
CUE SHEETCLIP (INSIDE LEFT SIDE)
REMOVABLE PANELS OVER TERMINALS
ALL TERMINALS
SIDE VIEW
NO SCALE
SAFPATCH PLUG - PHENOLIC
20 AMP BLACK
50 AMP RED
AMP GREEN
PLUG COUNTER DETAIL
NO SCALE

FIGURE 15-5

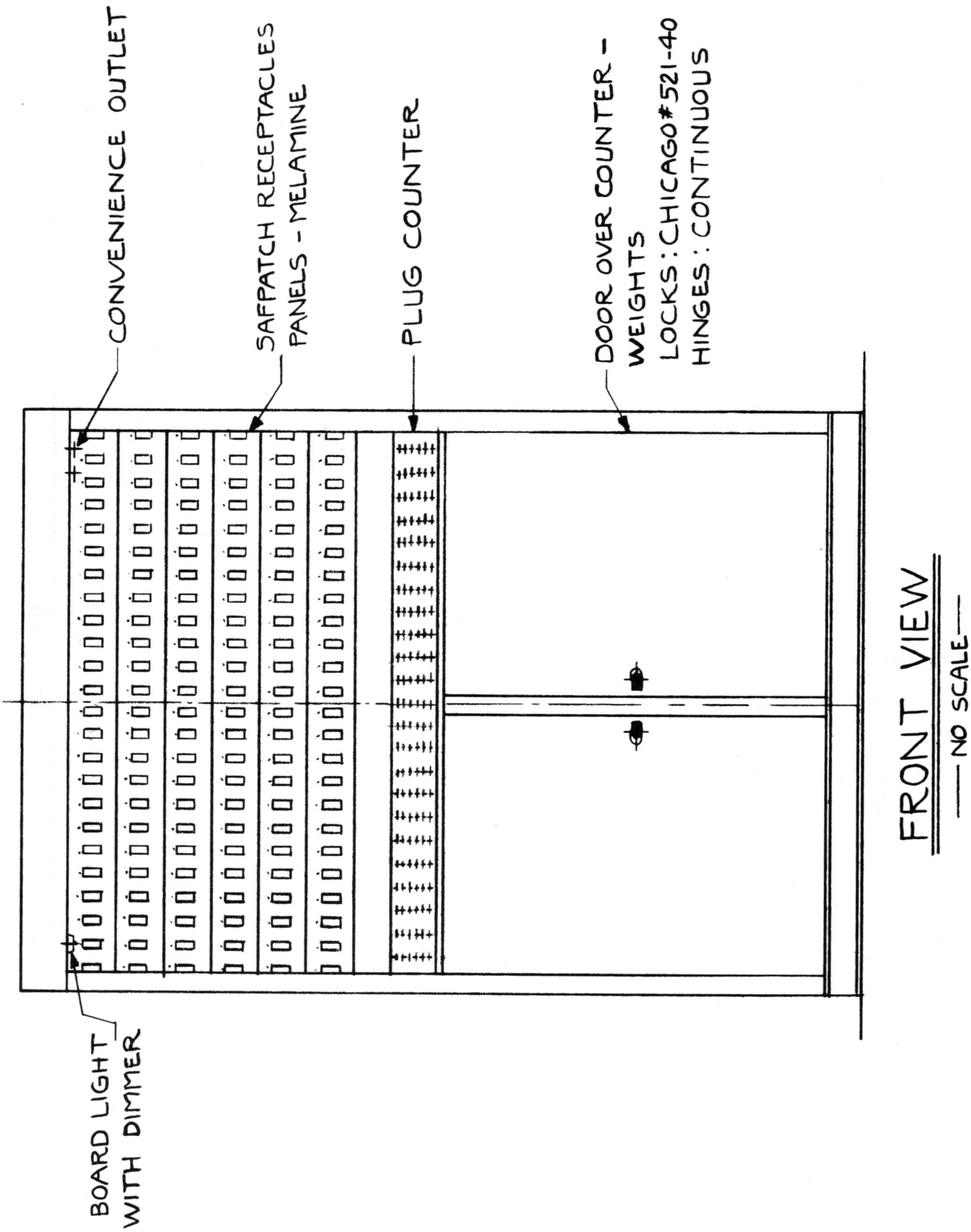
CONVENIENCE OUTLET
SAFPATCH RECEPTACLES
PANELS - MELAMINE
PLUG COUNTER
DOOR OVER COUNTER -
WEIGHTS
LOCKS: CHICAGO #521-40
HINGES: CONTINUOUS
BOARD LIGHT
WITH DIMMER
FRONT VIEW
NO SCALE

FIGURE 15-6

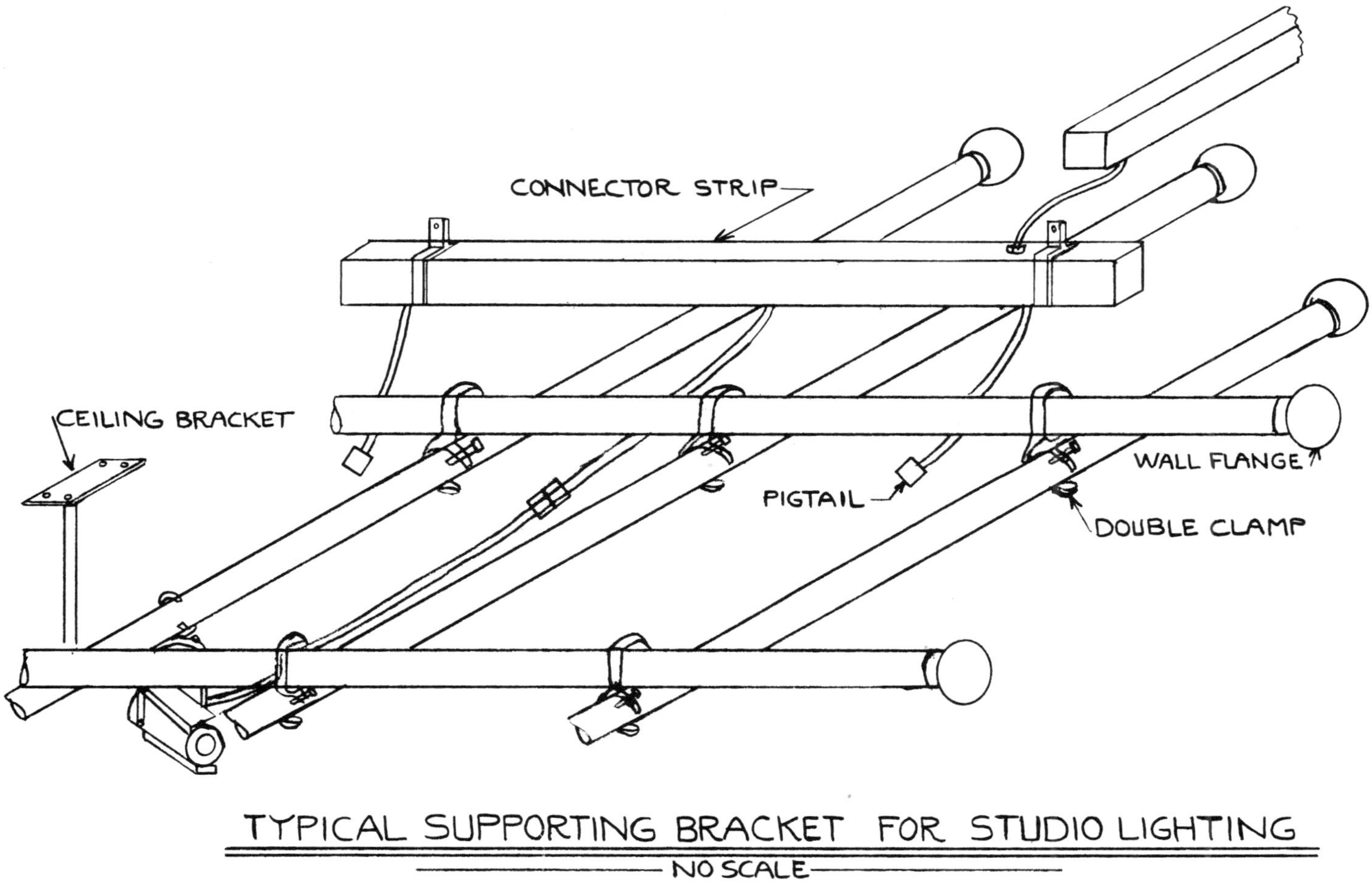

FIGURE 15-7

Plugging the cords into the patch receptacles puts the circuit under dimmer control. The dimmer bank (not shown) controls the flow of current to specific receptacles. The variety of results is endless, because the dimmer bank can contain dimmers and groups of dimmers arranged in a variety of sequences. Lights not involved in a given sequence are simply not plugged into the patch panel. The dimmers still operate but those lights not plugged in are off.

Given this freedom of choice the action on the set can be anywhere and the proper patching of the stage lights enables the dimmer bank operator to control the light output separately or jointly.

The dimmer bank was not repeated on the drawings as it still is basically the same as the smaller bank except for many more dimmers controlled individually, by submaster, and by overall master switches. The whole system can be quickly rearranged to suit act after act and scene after scene.

The only lighting not controlled in this fashion is spotlighting. This is a separate manual operation located in the rear of the theater and designed to highlight a performer or specific object. In the case of the star performer the spotlight is manually maintained on the star and follows his or her movement. Commonly, through special cue cards, the spotlight changes color to intensify the mood of the scene.

The dimmer bank and its operating controls are usually in the general location of the patch panel so that coordinated changes can readily be made by the system operator(s).

Television Studio Lighting Lighting a television set is totally different from stage lighting. Stripped to its essentials, the lighting is first of all uniform. The lighting source generally used is an incandescent source with a fixed color temperature of 2900 to 3350 K. The prime criteria is that the lamps should remain within a plus or minus 150 K of this design temperature for accurate color rendering by the video camera. The television monitor mixer provides for the fade-out scene-change feature.

Figures 15-7 and 15-8 illustrate the subject-lighting technique and the general arrangement of overhead luminaire support. It is not within the realm of this book to delve very deeply into lighting design, so we will advise

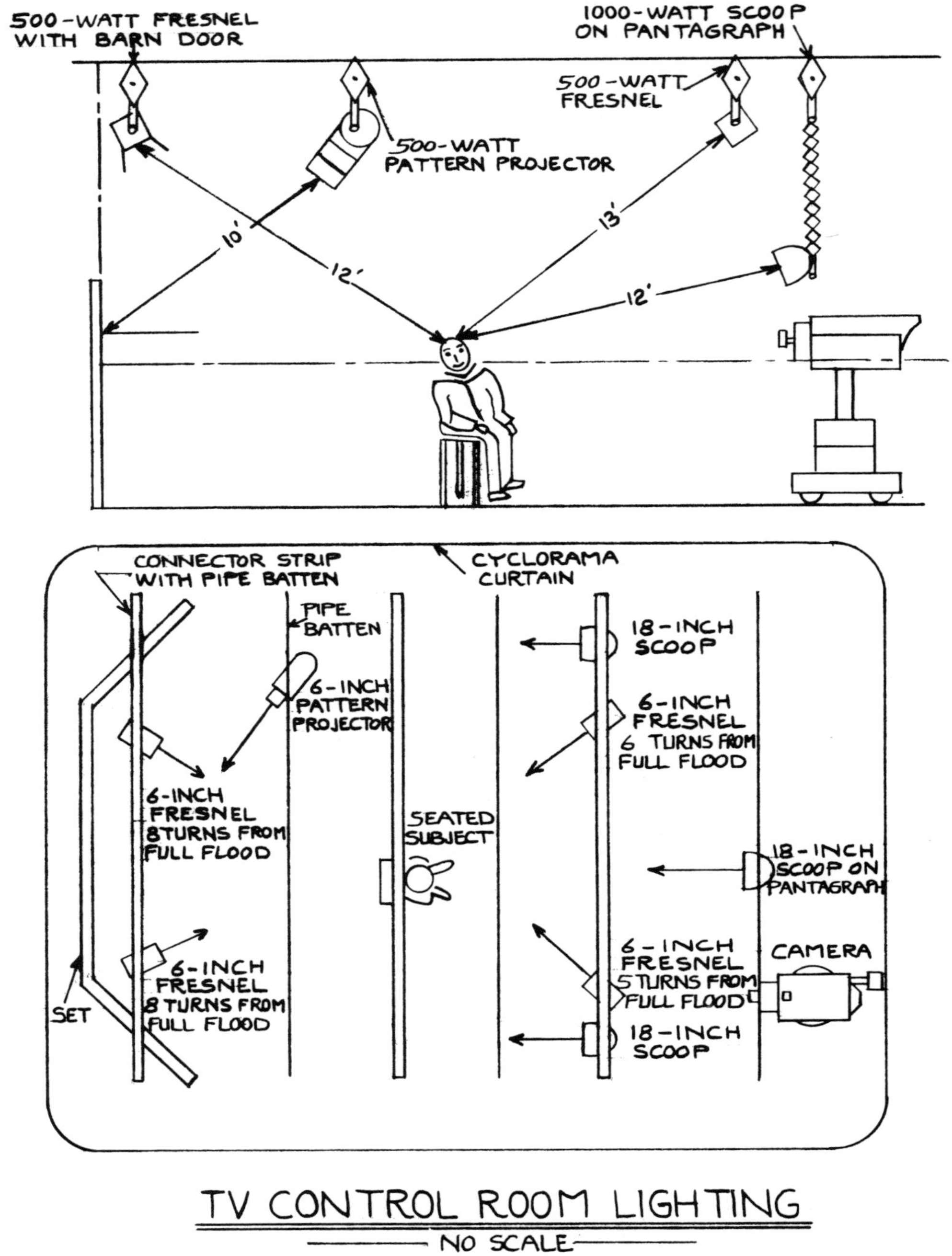

FIGURE 15-8

that, in general, the basic illumination level of the area is 50 fc and the performer, scene, or subject illumination level is about 250 fc. Our details indicate that not only is the subject illuminated but also a background can (and usually is) highlighted.

Generally, the lighting design of the television studio requires a variety of preset lighting arrangements. As such, Figure 15-8 illustrates a subject or product illumination. Although the television set is not like a stage, it does have a multiplicity of lighting requirements. Figure 15-7 illustrates a grid system of supports to provide for ease of relocation of fixtures. Here a series of pipes attached to walls and ceilings creates a grid. Running over and through the grid are standard connector strips with standard pigtail connectors. Within reason the luminaries can be fastened to any pipe and plugged into the nearest pigtail. Large grid systems in high-ceiling areas are usually not firmly attached but are suspended from chain-operated raising and lowering systems both for relamping and to create special effects.

SIXTEEN

Electric Heating Systems

The use of electrical energy for heating encompasses a large area, which, in today's energy conservation atmosphere, is being considered to a far greater degree than before. The most rapidly growing area of application is the heat pump.

As has been repeatedly stated in this book, our presentation is focused on engineering details for the electrical detailer. However, the heat pump is a subject that should be understood by everyone. In essence, any air-conditioning system absorbs heat from the area to be cooled and dissipates this heat through an air- or water-cooled evaporative condenser or cooling tower. The energy driving this system is electricity. Not only must the heat removed from the space be dissipated but also the energy (in the form of heat) used in the system compressors must be dissipated.

Since heat is a relative term (anything above absolute zero, $-459.69°F$, is heat) it follows that theoretically heat can be absorbed at any temperature above absolute zero. In the spring and fall and on milder winter days, if we reverse the air-conditioning system by letting it try to further cool the already cool air and deliver that to the space we can partially supply heat to the space. Heat can, in this manner, be recovered from any body of water or air. Temperature is the sole criterion.

The heat pump is a mechanical, not electrical, design function and system. For further explanation there are the books published by the American Society of Heating, Refrigeration and Air Conditioning Engineers (ASHRAE); the texts published by McGraw-Hill and others; and the many manufacturers' catalogs on equipment and design. Generally, the electrical design department's function is merely to provide the service and disconnect device for this system.

Many of the other systems we will cover are themselves excellent examples of one of the gray areas of engineering practice. Frequently, they are designed, specified, or otherwise alluded to in the mechanical design section, and how they are to be installed is covered nowhere.

The simplest of all the electrical heating systems that requires little or no detailing (except for proper size and location of service wire and disconnect) is the electric unit heater, force-flow heater, unit ventilator, air-handler radiation, or duct heater. These are standard devices appearing in many product engineering catalogs and merely require proper conduit, wire, and disconnect switch. But they are not quite as simple as that. There are two code requirements. The first is that contained in the current NEC and the second is contained in Space Heating Standard No. 1096 of the UL. Although both of these codes are uniform throughout the country, local codes may deviate from the NEC.

Both UL and NEC require two types of overtemperature protection. The primary protection consists of an automatic reset thermal cutout wired to the magnetic contactors in the temperature-control circuit. For secondary protection many manufacturers use heat limiters or fusible links which require field replacement when an excessively high temperature condition occurs.

UL and NEC require that a heater drawing more than 48 A be subdivided into circuits that draw no more than 48 A. And each circuit must be protected by fuses or circuit breakers. Generally, all the heaters are regarded as a continuous load. Overcurrent devices must be rated for at least 125 percent of the circuit load. Thus, while circuits are limited to 48 A, fuses are limited to 60 A. While the fuses must be accessible, they need not be readily accessible. Most local authorities will accept fuses that can be reached by a stepladder. A check with your local authority is required on this point.

Again, both NEC and UL require that each heater be

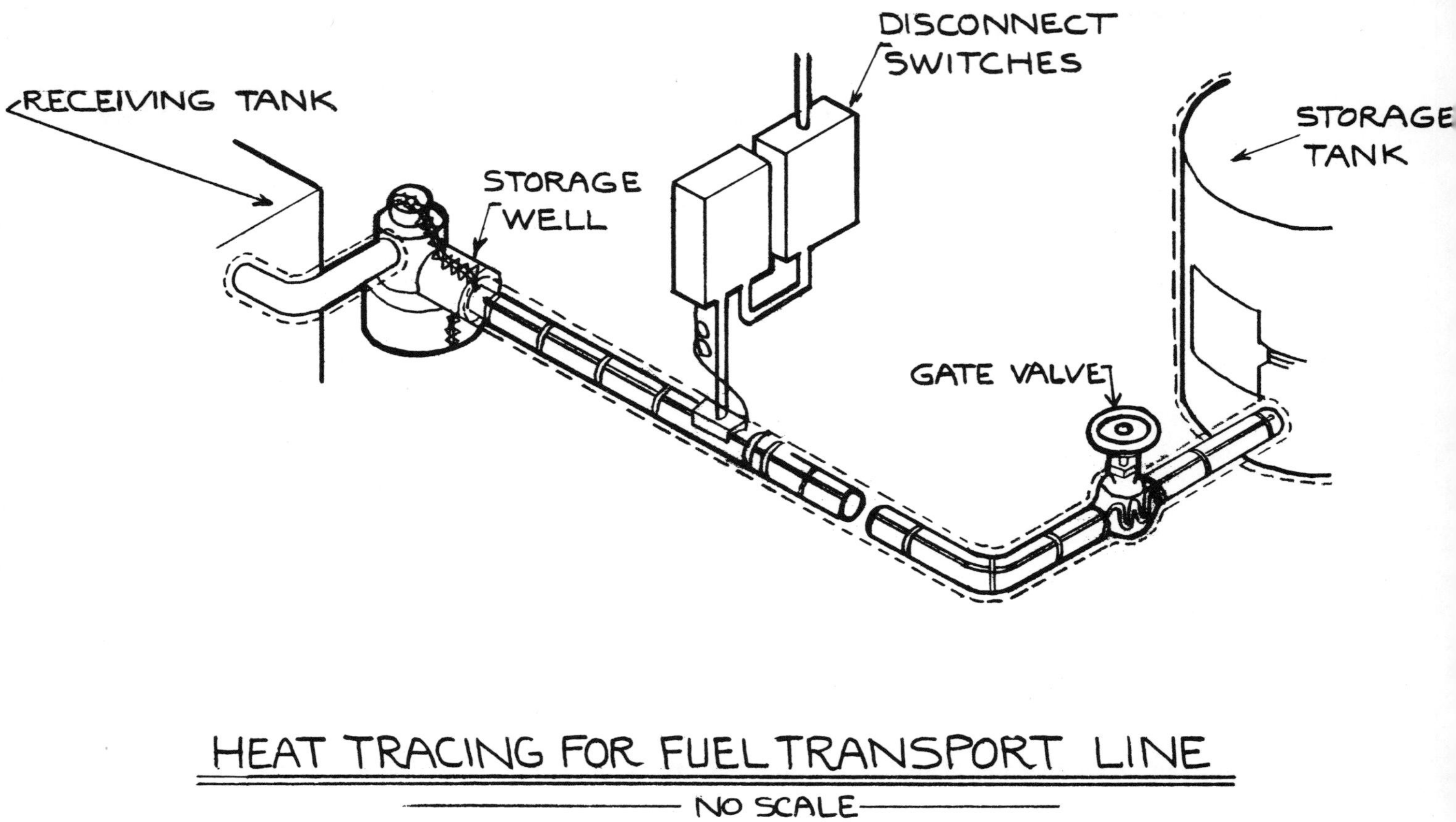

FIGURE 16-1

interlocked with the associated device that must function for the system to be operative. This device is usually a built-in differential-pressure switch or a built-in fan relay.

Where heaters are furnished with fuses and contactors, the requirement of the NEC for a means of disconnecting the heater and overcurrent protection devices can be met with one disconnect switch. This switch can be a field-mounted disconnect within sight of the heater or a built-in unfused disconnect switch. The better solution would include a switch locked in the open or closed position and interlocked with the terminal box door so that the door cannot be opened unless the switch is off.

Areas that are, or should be, designed by the electrical department of your firm are primarily specialized resistance heating systems such as snow-melting systems, heat tracing of piping systems, and specialized fluid-heating systems from domestic hot water to specialized industrial and laboratory applications.

Many of these systems require mechanical, electrical, as well as some structural design input. The mistake commonly made, especially where these systems are a minor part of a major design effort, is to treat their design in an off-hand, casual, last-minute manner. Granted their designs are generally not very involved. At the same time it is very easy to make a simple, expensive mistake if care is not taken.

Heat Tracing Heat tracing generally refers to a system carrying a liquid (hot or cold) that either requires that its temperature be maintained at a given value or that its temperature be kept at least above freezing. Figure 16-1 illustrates a common simple problem, its detail, and its solution.

We will not cover total design but will highlight and outline some major items in the design. First of all, the line is presumed to be out of doors and subject to all the vagaries of the weather. One of the commonest misunderstandings is that because the line is insulated the problem is solved. Nothing could be further from the truth. Certainly enough insulation would solve the problem, but the thickness would be unbelievable. A normally insulated pipe in temperatures below 32°F will eventually freeze its contents as has been painfully proved many times.

In your design analysis do not fail to calculate the temperature value so dear to the hearts of all television weather forecasters—the wind-chill factor. In a 15 mph wind and 10°F your poor pipe and its insulation suffer, just like you, a temperature whose actual value is −44°F.

You are bitterly cold and so is the pipe. Thus if you are using, say, 56°F as an ideal fluid transport temperature, your heat loss is based on (56 + 44) or a 100°F spread and *not* (56 + 10) or a 66°F spread. That error alone is more than a 50 percent error in load sizing, and it is a very common mistake.

If your pipe has valves they are counted separately and added to your load requirement. A rough (very rough) rule of thumb is that each valve adds 1 percent to your load calculation. Using our stated conditions of 56°F design with −10°F outside air with a 15-mph wind, you would get an electrical load of some 15 W/ft and some 2 ft of cable per foot of pipe.

The above solution creates two very simple possibilities for your detail. To get 2 ft of cable per foot of pipe, you can wrap the cable in a spiral fashion around the pipe, or you can do as we have illustrated and run two lines of cable on opposite sides of the pipe.

Since we have illustrated a solution that will work for many situations in many parts of the country, the temptation to use it as is and forget about design will be very considerable. Since we are not really covering design of any system we strongly urge that you make a thorough analysis of the problem.

Our analysis is even more usable. For example, to transport oil at 160°F would double the temperature spread and four lines of heat trace would bring the temperature to 156°F.: close to the desired preheat temperature to transport #6 fuel. But before you leap to that conclusion, we would like to point out that there is an entire patented system which should be investigated by your mechanical design department that does the same job by making the pipe part of the system.

In Figure 16-2 we show a system that has limited use in a few parts of the country, called roof or gutter deicing. Both mineral-insulated (MI) cable and insulated wire are used to prevent heavy snow and ice accumulations in gutters and downspouts.

Wire, or MI cable, used in this situation generally has a rating of 6 to 16 W per linear foot. Generally, 1 ft of cable per foot of gutter or downspout is all you will need. If the edge and/or gutters are to be heated, the downspout should also be heated. Lead wires should be spliced or plugged into a waterproof junction box and a ground wire installed from the downspout or gutter to a driven ground rod. While the system can be thermostatically controlled, a simple on-off switch is commonly used.

Snow-Melting Systems Basic systems using electricity for snow melting usually include MI cable, a resistance-wire assembly embedded in the paving materials, or a high-intensity infrared source. Figure 16-3 illustrates the MI-cable system that can be used in driveway or walkway material.

Again, we are not going into engineering calculations to any great degree, as there are many texts on this particular phase of the design, including the previously mentioned ASHRAE guide, as well as books by McGraw-Hill and other publishers. We will point out that system output depends on resistance and voltage of the equipment installed, and it cannot be altered or overfired.

Where the system is installed in concrete, the slabs containing electric heating cable or wire must be designed and constructed with a subbase, expansion joints, and reinforcement and drainage to preclude slab cracking. Cables must not run through expansion joints. Slabs containing heating cables are commonly poured in two layers with the top layer containing the heating cables. The top layer does not contribute to slab strength, it is designed for maximum "weatherability" and is usually rated at 4000 to 5000 psi compressive strength. Thus, the lower layer must carry the total load. When the snow is melted the slab system must be arranged so that the water does not run off and immediately freeze in the adjacent area or create frost heave problems for the slab. Drains should either be heated or below the frost lines.

The basic MI-cable layout is used to illustrate the system. The actual cable layout and its spacing require proper calculations. The cable is a standard product fabricated with a non-heat-generating cold lead cable attached to the heating cable which must be of sufficient length to reach a dry location. The junction box installation must be arranged to remain dry. Preferred locations are indoors; on the side of a building, utility pole, or wall; or inside a manhole on its wall. The boxes should have a hole in the bottom so any condensation can drain out.

If only one pour is used (not the best choice) the cable must be fastened on top of the reinforcing mesh before concrete is poured. In the preferred method, the cable is placed on top of the first structural slab and then the top slab is poured. Generally, the civil engineers will advise that the slabs be poured within less than 24 hours for bonding purposes and that a bonding grout be applied.

As can be seen in the detail the cable forms a serpentine pattern and prepunched copper strips, nailed to the lower slab, control spacing. Silicone grease, a glass tape, or a nylon jacket (one of these) is usually applied to the cable to protect it against slab movement. Note in the detail the method used to pass an expansion joint, as well as the protective cable cover which is filled with a sealing compound. If the top layer were asphalt instead of concrete, the installation is similar except a layer of mesh is placed over the cable and anchored to the concrete to keep the cable from floating in the asphalt topping.

Embedded-wire systems, mentioned in our opening paragraph, are commonly prefabricated, factory-assembled mats available for embedding in specified paving materials. These systems can be designed to deliver a variety of watt densities to match desired snow-melting capacities. For details on the installation of these proprietary products you should consult the manufacturer's literature. There are certain criteria that should guide your design. First, select wires of the longest length that will fit between control or expansion joints. In asphalt applications, as

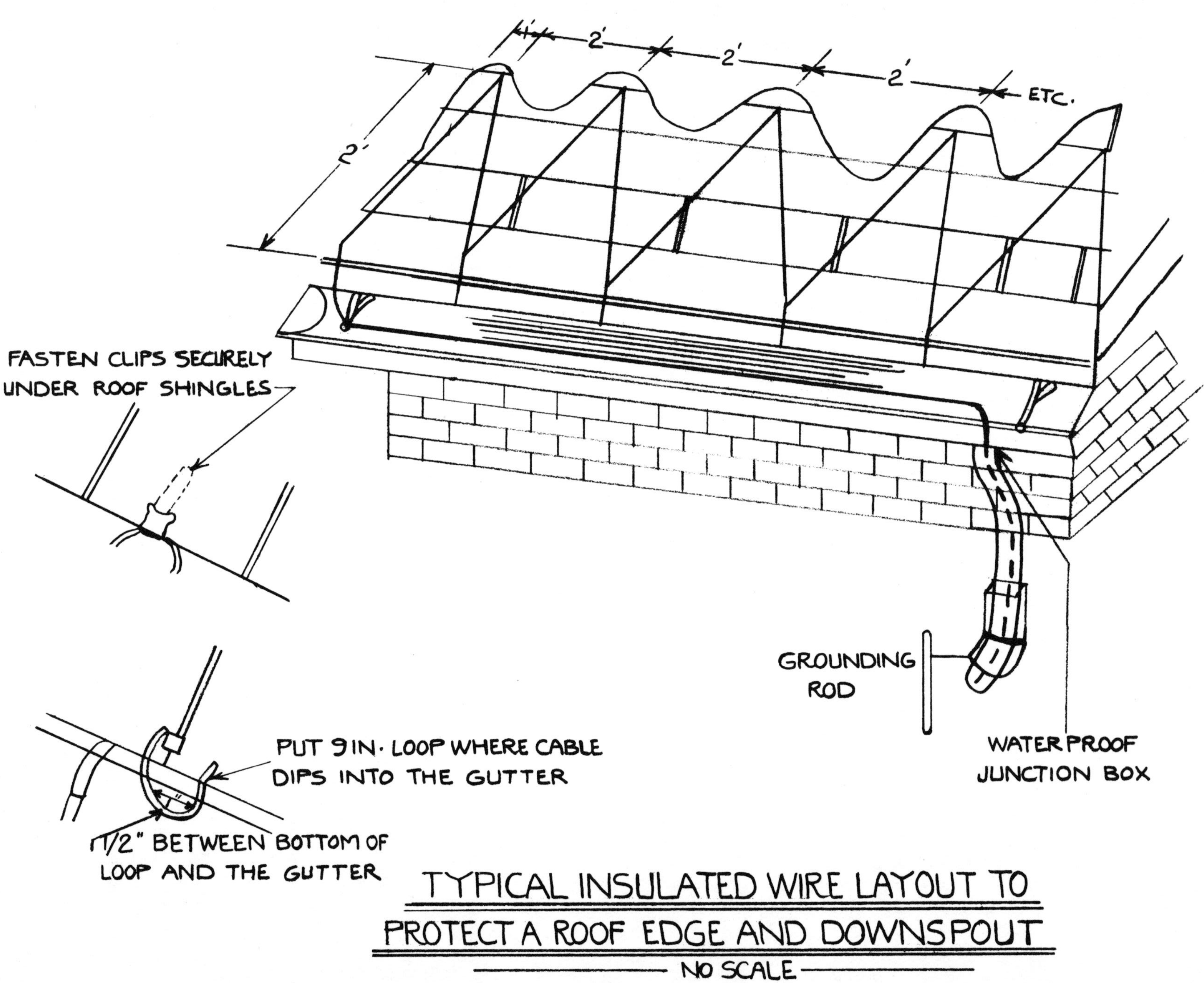
1'
2'
2'
2'
2'
ETC.
FASTEN CLIPS SECURELY
UNDER ROOF SHINGLES
GROUNDING
ROD
WATER PROOF
JUNCTION BOX
PUT 9 IN. LOOP WHERE CABLE
DIPS INTO THE GUTTER
1 1/2" BETWEEN BOTTOM OF
LOOP AND THE GUTTER
TYPICAL INSULATED WIRE LAYOUT TO
PROTECT A ROOF EDGE AND DOWNSPOUT
NO SCALE

FIGURE 16-2

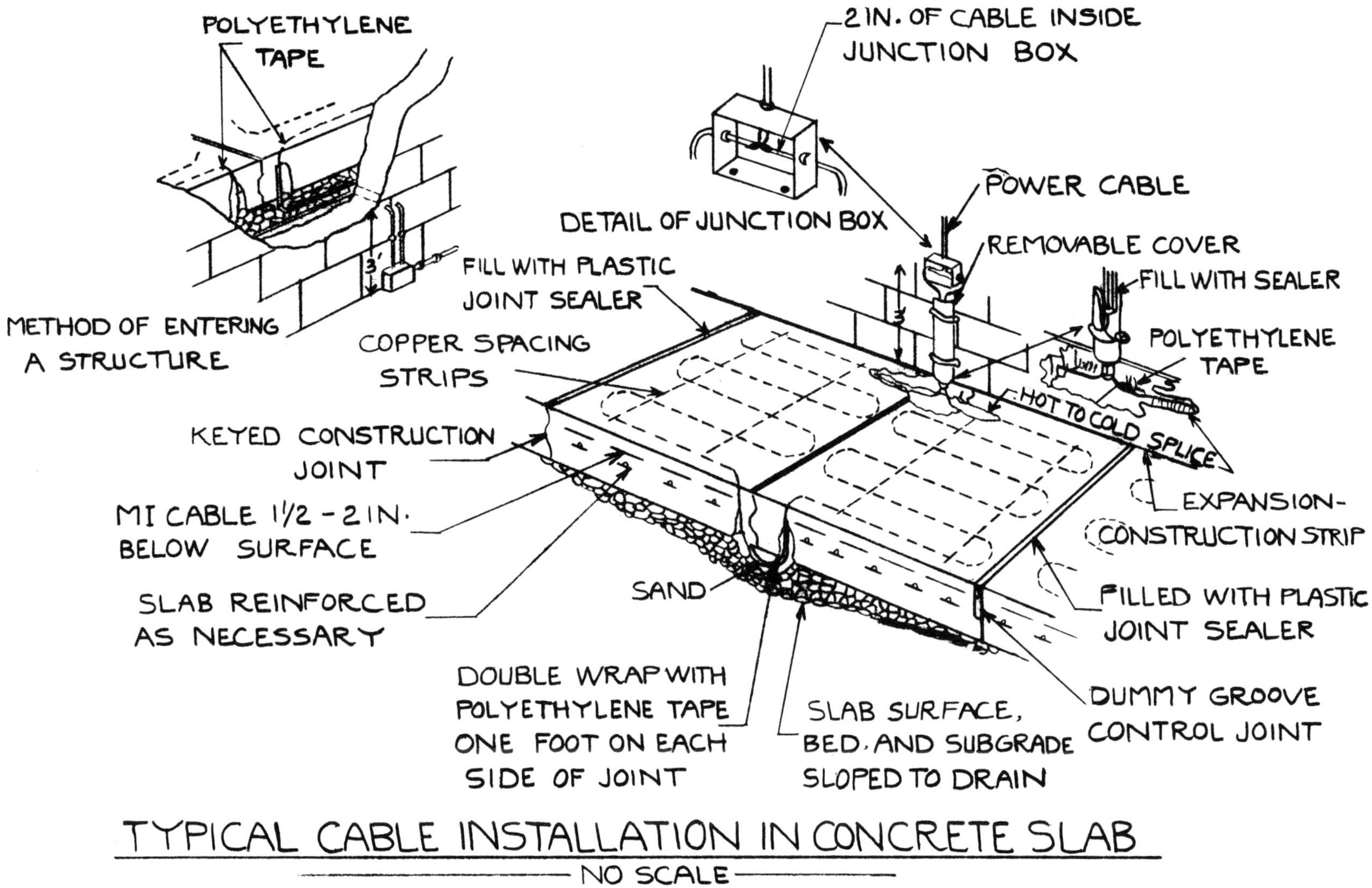
POLYETHYLENE TAPE
METHOD OF ENTERING A STRUCTURE
3'
2 IN. OF CABLE INSIDE JUNCTION BOX
DETAIL OF JUNCTION BOX
POWER CABLE
REMOVABLE COVER
FILL WITH SEALER
POLYETHYLENE TAPE
FILL WITH PLASTIC JOINT SEALER
COPPER SPACING STRIPS
KEYED CONSTRUCTION JOINT
MI CABLE 1½ - 2 IN. BELOW SURFACE
SLAB REINFORCED AS NECESSARY
SAND
HOT TO COLD SPLICE
EXPANSION-CONSTRUCTION STRIP
FILLED WITH PLASTIC JOINT SEALER
DOUBLE WRAP WITH POLYETHYLENE TAPE ONE FOOT ON EACH SIDE OF JOINT
SLAB SURFACE, BED, AND SUBGRADE SLOPED TO DRAIN
DUMMY GROOVE CONTROL JOINT
TYPICAL CABLE INSTALLATION IN CONCRETE SLAB
NO SCALE

FIGURE 16-3

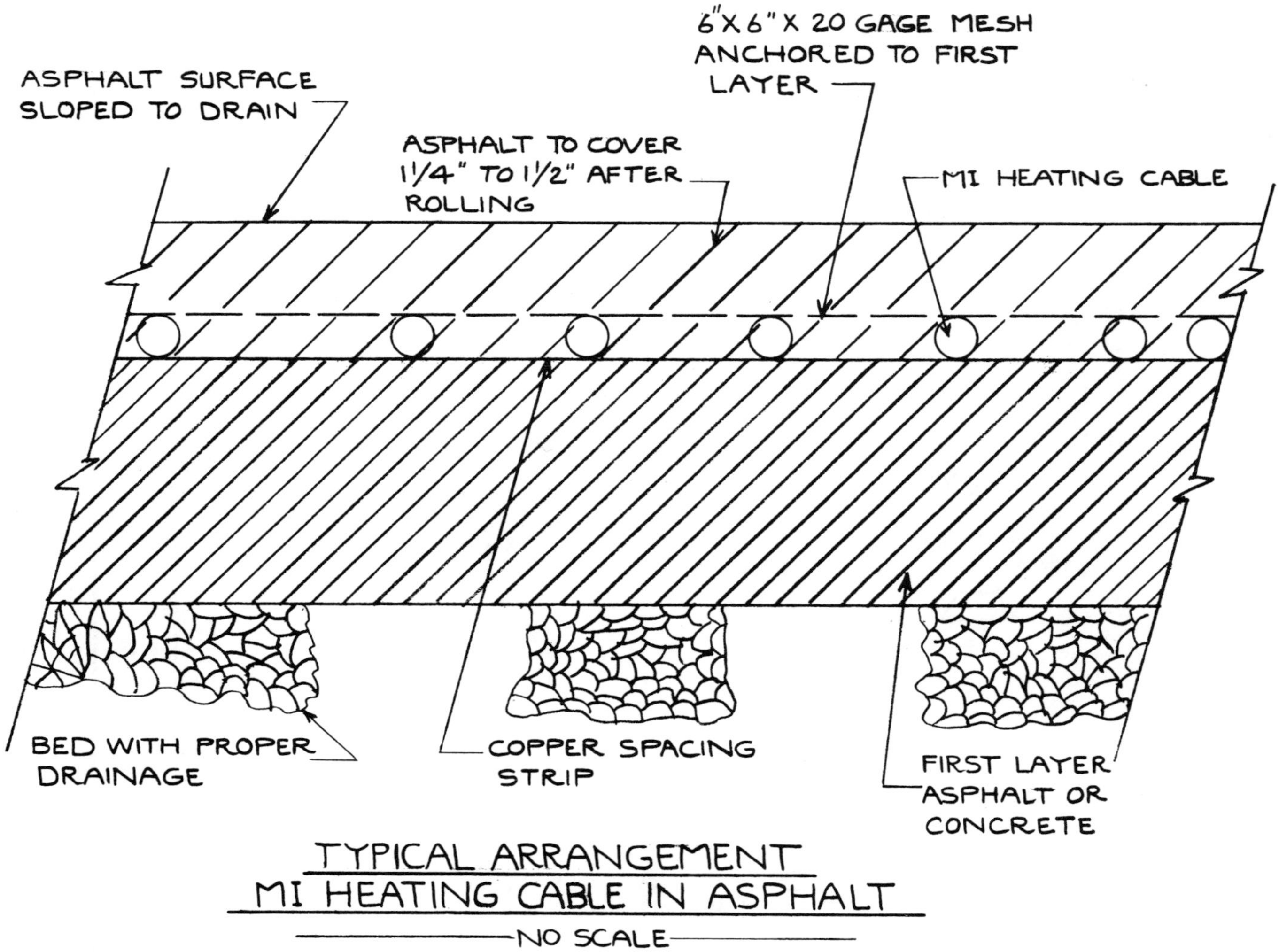

FIGURE 16-4

illustrated in Figure 16-4, select the longest wire which can be used on straight runs. Keep mats at least 1 ft in from the edge of the pavement. Do not overlap mats.

There is another choice, infrared heating, not detailed in our book. These systems are usually selected not only because they can melt snow but also because they can keep people warm who work at the hospital entrances or plant loading docks. If you add to the heat needed for creature comfort and direct it more positively to the horizontal surface, you have a snow-melting capability.

Generally, the infrared fixtures are mounted in a row parallel to the longest dimension. If the area is more than 8 ft 0 in wide two rows are used. It should be noted that theoretical wattage densities in infrared systems are for radiation incident to the parallel surface. Simply multiplying the recommended wattage density by the area does not produce a satisfactory result. The point method of breaking the area into 1-ft squares and cumulatively adding the watts per square foot contributed by each fixture is the proper method. The necessary data can be obtained from the fixture manufacturer.

One highly desirable feature of the infrared system is its fast warm-up capability. An area can be preheated to snow-melting temperatures in 20 to 30 minutes. This

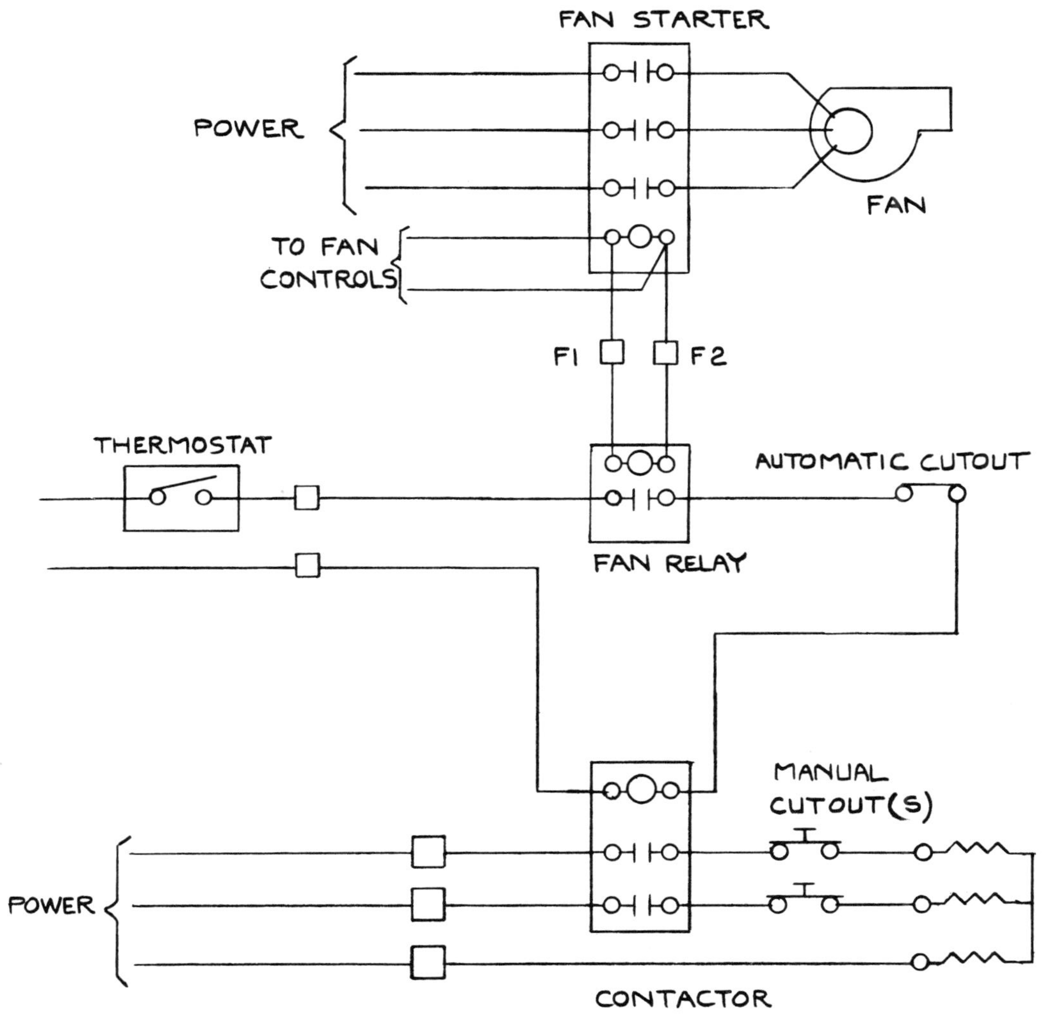

TYPICAL SCHEMATIC DIAGRAM OF HEATER WITH FAN RELAY

NO SCALE

FIGURE 16-5

makes it unnecessary to turn on the system until snow begins to fall. Usually the system is kept on 4 to 6 hours after the snow has stopped to allow time to dry the pavement completely.

Heating Control Diagrams In the opening discussion of electrical heating we described control requirements for these installations. Frequently, the actual control diagrams are implied or somewhat described in the specifications, but they are not depicted on the contract drawings. It probably is not essential that a detail should appear on the drawings. However, for a number of reasons, the installing contractors can, and frequently do, raise questions about what is actually wanted.

Those familiar with control diagrams know there is no such thing as a standard diagram. There are fairly standard methods of handling fairly standard situations. To illustrate these fairly standard methods we are including five details. We feel it is important that you realize these are *not standard* details. They are close to standard, but they are not unchangeable.

Figure 16-5 is a typical balanced three-phase fan controlled in typical fashion by a magnetic contactor and a fan relay in accordance with NEC and UL standards. It is a common arrangement.

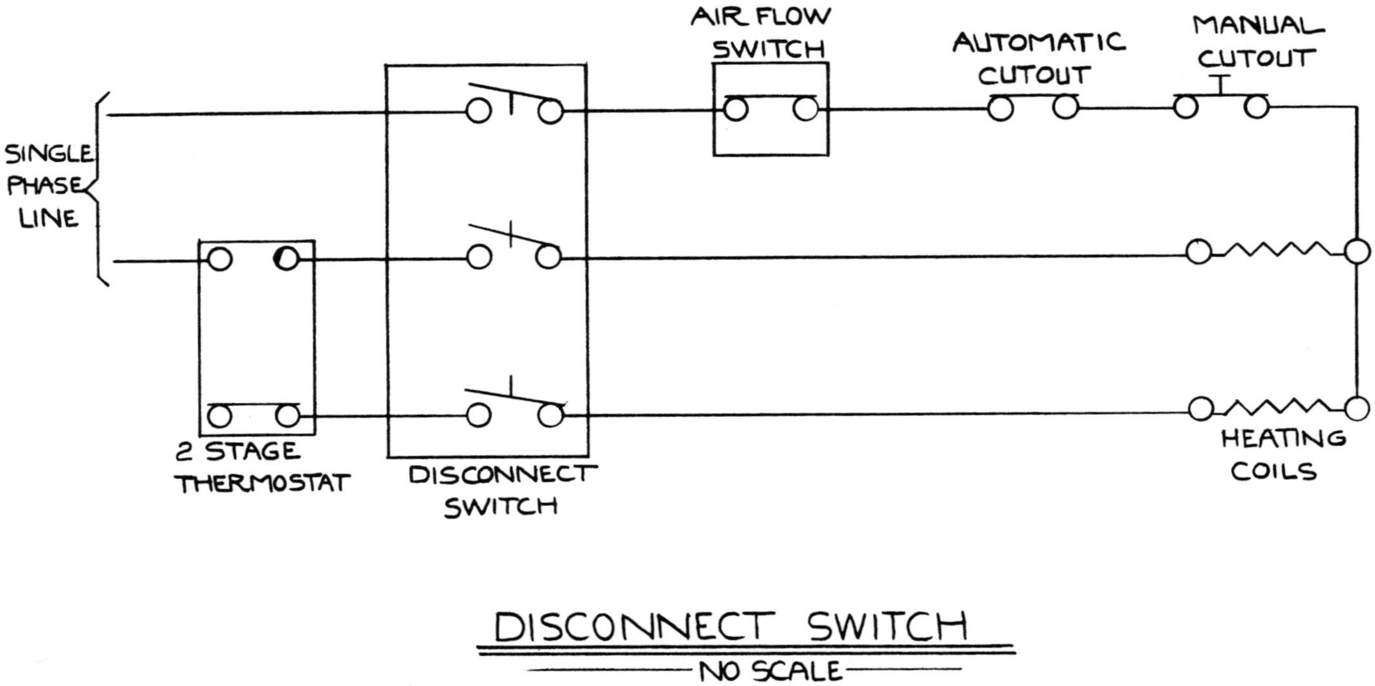

FIGURE 16-6

Figures 16-6 to 16-9 are typical wiring diagrams for individual electric duct heaters. In Figure 16-6 all safety and control devices carry the load directly. This eliminates the need for a separate control circuit. The thermostat is a line voltage thermostat.

The system in Figure 16-7 controls both coils and fans. It illustrates how two drawings can be interrelated. Notice that there is an airflow switch at the top of the drawing. *In lieu of* the airflow switch you could show the fan relay and fan control as we show in our box. This function is identical to the upper part of Figure 16-5. Now you can begin to see why we say there is no standard detail.

Figure 16-8 features another rather typical arrangement. It has one special feature. In addition to the usual devices, this drawing illustrates contactors breaking all lines. Figure 16-9 covers a larger (in terms of kilowatts produced) heater where the airflow switch and the automatic cutout operate safety contactors. If the voltage is *less* than 480 V the control voltage comes directly from line voltage. On a 480-V system a separate control circuit transformer is required.

These five details can and should be modified where necessary to suit your particular requirements. They are presented both as a guide and as an illustration of how the diagram appears and should be depicted and not traced directly.

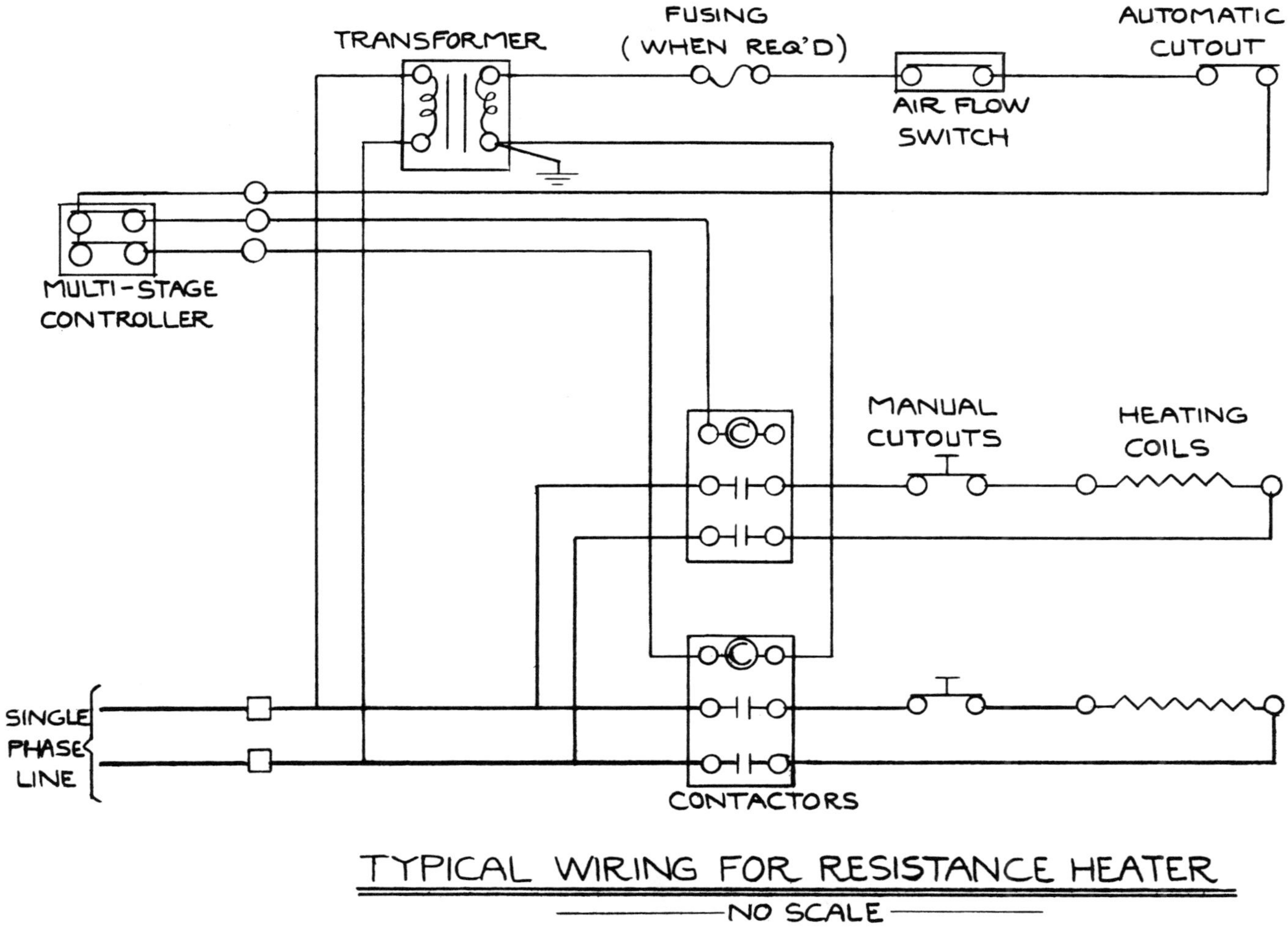
FUSING
(WHEN REQ'D)
TRANSFORMER
AUTOMATIC
CUTOUT
AIR FLOW
SWITCH
MULTI-STAGE
CONTROLLER
MANUAL
CUTOUTS
HEATING
COILS
SINGLE
PHASE
LINE
CONTACTORS
TYPICAL WIRING FOR RESISTANCE HEATER
NO SCALE

FIGURE 16-7

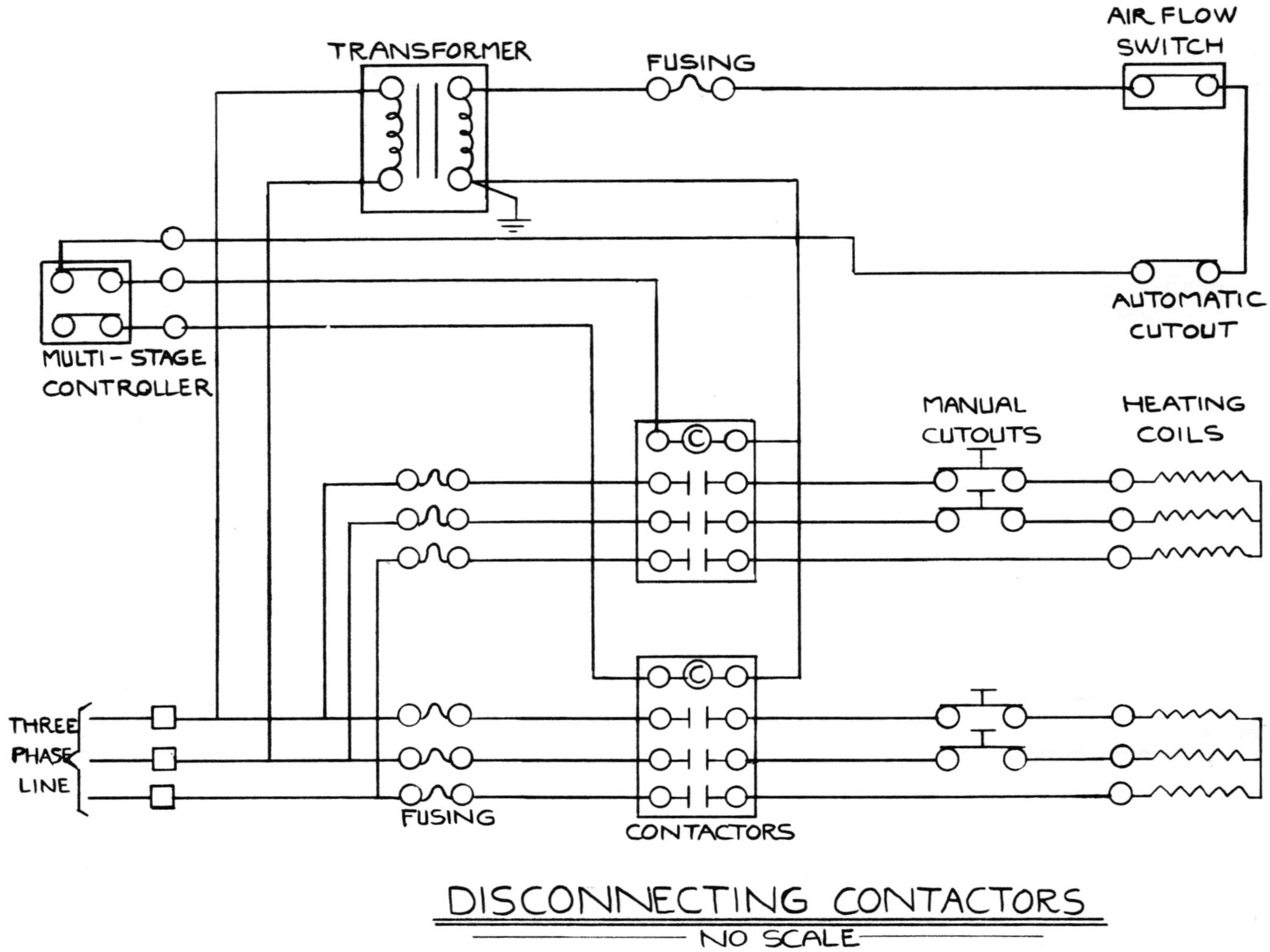
TRANSFORMER
FUSING
AIR FLOW SWITCH
MULTI-STAGE CONTROLLER
AUTOMATIC CUTOUT
MANUAL CUTOUTS
HEATING COILS
THREE PHASE LINE
FUSING
CONTACTORS
DISCONNECTING CONTACTORS
NO SCALE

FIGURE 16-8

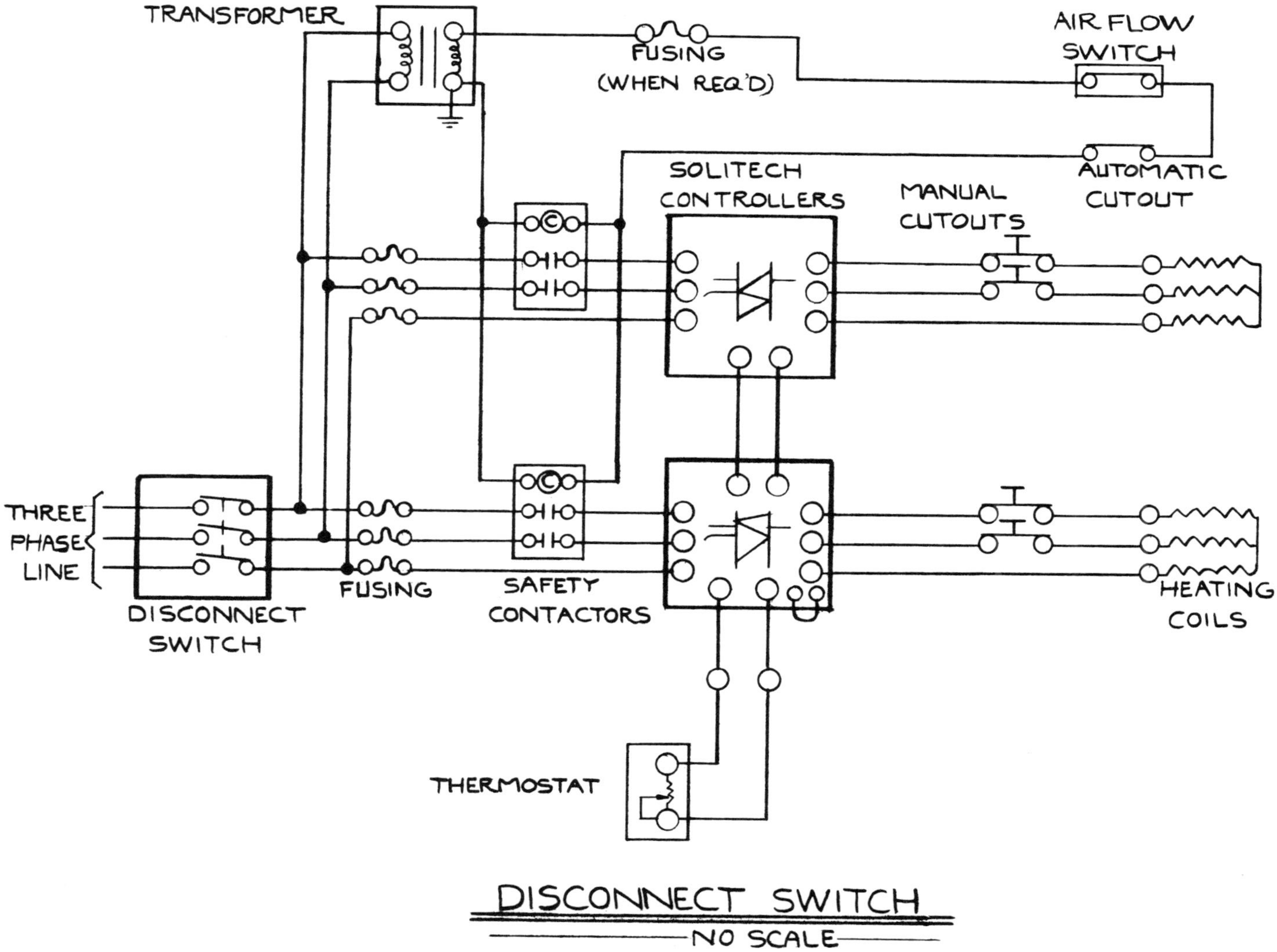

FIGURE 16-9

Interior Space Heating Commonly, interior space heating systems using electrically powered heating elements are thought of as systems using duct heaters, baseboard radiation, converters, unit heaters, forced-flow heaters, or unit ventilators. These are all standard devices that frequently come with their own built-in controls and thermostats. Where a remote thermostat is required, this device can also be supplied as a standard manufactured item. The only work required is to run a line or low-voltage wiring from the remote thermostat to the readily recognizable point of termination on the unit to be controlled.

Radiant heating systems, however, can also be electrically powered (in lieu of hydronically powered). Because of the total flexibility of these electrical panel systems there is no limitation as to where they can be installed (walls, floors, ceilings, etc.). For many situations a factory-assembled and -fabricated unit can be selected. The accompanying factory design and installation data remove a lot of the design effort from your work. Cables for these panels should have coverings resistant to medium temperature, water absorption, aging effects, and chemical reaction with plaster cement used on ceiling lath. Normally, the insulation is polyvinyl chloride (PVC) with a nylon jacket.

Some prefabricated panels can be used in wall installations. However, because of the damage that may result from picture nails and building alterations, the majority of

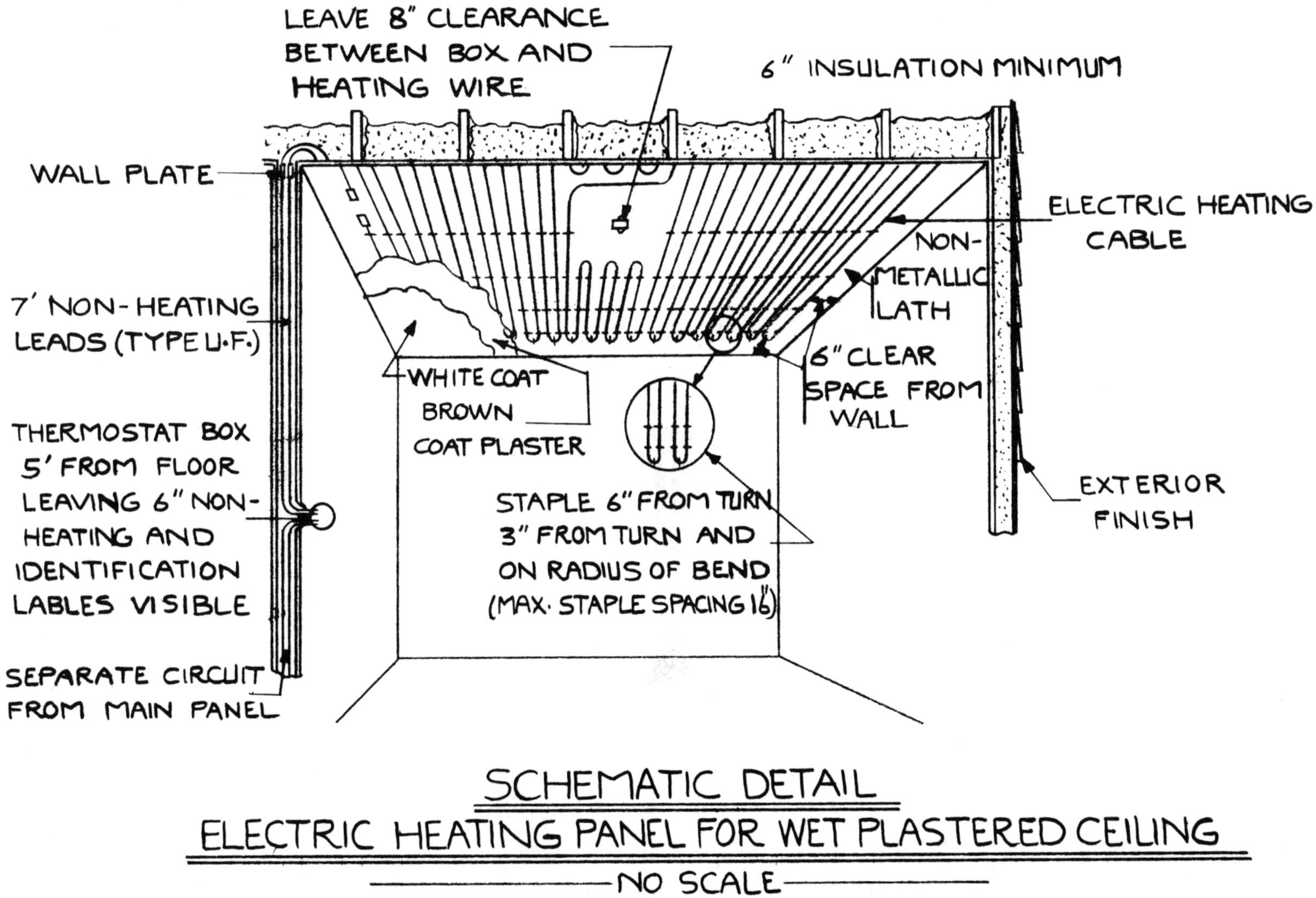

LEAVE 8" CLEARANCE
BETWEEN BOX AND
HEATING WIRE
6" INSULATION MINIMUM
WALL PLATE
ELECTRIC HEATING
CABLE
NON-
METALLIC
LATH
7' NON-HEATING
LEADS (TYPE U.F.)
6" CLEAR
SPACE FROM
WALL
WHITE COAT
BROWN
COAT PLASTER
THERMOSTAT BOX
5' FROM FLOOR
LEAVING 6" NON-
HEATING AND
IDENTIFICATION
LABLES VISIBLE
EXTERIOR
FINISH
STAPLE 6" FROM TURN
3" FROM TURN AND
ON RADIUS OF BEND
(MAX. STAPLE SPACING 16")
SEPARATE CIRCUIT
FROM MAIN PANEL
SCHEMATIC DETAIL
ELECTRIC HEATING PANEL FOR WET PLASTERED CEILING
NO SCALE

FIGURE 16-10

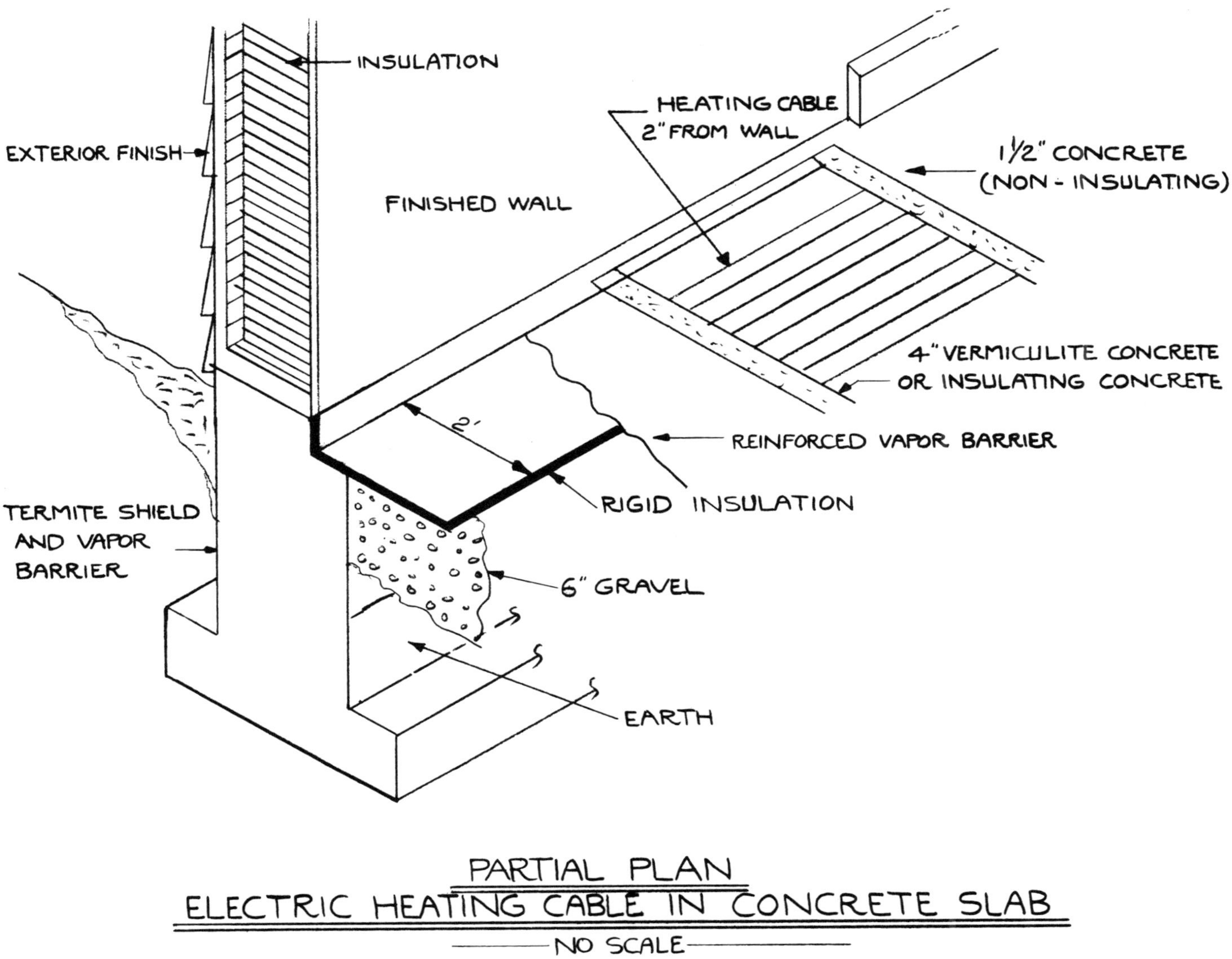

FIGURE 16-11

American building codes prohibit wall installations.

Although prefabricated panel systems can be used for floors and ceilings, the customized cable installation is more common. Figures 16-10 and 16-11 illustrate features of the custom installation. Before discussing the details, but not going into engineering design, we have listed 12 points that should be thoroughly considered in the design.

1. The space hourly heat loss.
2. The available panel area.
3. The required panel output.
4. The size, location, and connection of units.
5. The required panel temperature.
6. The insulation for the reverse side and edge of the panel.
7. Panel loss and panel input.
8. Other temperatures required.
9. Use of standard design parameters.
10. Avoidance of panels near cold areas of unusual heat loss.
11. Limitation of ceiling surface temperatures especially in low-ceiling areas.
12. Avoidance of floor surface temperatures above 85°F (29°C).

Ceiling panels in plastered ceilings can be stapled to gypsum board, plaster lath, or other *fire-resistant* materials using rust-resistant staples. If metal lath is used, a first coat (brown coat) of plaster must completely cover the metal lath or conducting surface before the cable is installed. Before covering (plastering over) the cable, test the insulation to ensure at least a 100-MΩ resistance to ground. Finish with noninsulating sand plaster of ½ to ¾-

in thickness. Do not energize the system until the plaster is thoroughly dried.

If the ceiling is drywall, the cable is placed between two layers of drywall and stapled directly to the first layer. In this case the cable will keep the two layers separated and the voided space must be filled with insulating plaster. Not only does plaster hold the cables in position but also ensures proper heat transfer and prevents the cable from overheating and failing in void areas.

In the wet-plastered as well as in the drywall ceiling, these points are reasonably depicted. Figure 16-10 also illustrates several other points to be noted. Cables normally end 6 in from the walls and you must clear outlet boxes by at least 8 in. Note also the insulation requirements, the leads to the thermostat, and the separate home runs. Cables must be installed in areas *not* covered by partitions, cabinets, and similar coverings. You are permitted to have a single embedded run over a partition. All general power and light wiring must run above thermal insulation, or at least 2 in above the heated ceiling surface or be derated. In drywall construction cable runs should be parallel to ceiling joists.

For floor slab panels much of what has been said about snow-melting systems also applies here. Figure 16-11 is a part plan of a typical installation. The slab is installed in two pourings. The first pour is to be at least 3 in thick. Cables are usually MI or PVC jacket cables. Where PVC is used, the cable is attached to predetermined, fixed wood nailing strips with spacing maintained by cement, plaster of paris, or tape. In light-weight, or uncured concrete installations the cable can frequently be stapled directly to the slab with hand-operated or electric-powered stapling machines. There are special anchor devices available which hold the cable in position while the second pour is made. The second pour should be made within 24 hours of the first pour, should be 1½ to 2 in thick, and not insulating concrete. Cables should be kept 2 in away from walls and partitions.

MI cable can be installed in single- or double-poured slabs. For single pours the cable is installed on top of the concrete reinforcement before the pour is begun. Preformed mats can be embedded in the concrete in a continuous pour after proper positioning between expansion joints and then be electrically connected to a junction box. Usually the slab is poured 1½ to 2 in below finished floor level, the mats placed on the screeded surface, and a final cap immediately applied. In Figure 16-11 we also call your attention to the requirements for an insulation vapor barrier and termite shield.

While these systems can provide excellent heating results and we have covered many points, it still is very important to check your local codes as well as the latest edition of the NEC, especially in the areas of panel watt density and other special criteria and installation requirements. We have shown edge insulation where a major loss of heat in the slab frequently occurs. There may be further special requirements in your area on this item as well.

SEVENTEEN
Lightning Protection

The design of the lightning protection system is, as we stated for other designs, not a part of our purpose. Because so many sets of contract plans only vaguely cover lightning protection, we hope the details here will create a desire for more thorough coverage in your plans.

As in other areas of your electrical design, lightning protection plans, specifications, and requirements are covered by specific codes. The NEC, under Article 280, covers the lightning arresters. There are further brief references in Articles 501 and 800. The two major treatments of lightning protection are contained in the UL bulletin UL-96A entitled "Master-Labeled Lightning Protection Systems," and in volume 7 of the National Fire Protection Association (NFPA) books under NFPA78-1980, *Lightning Protection Code.*

As stated in the above documents and in the literature of various manufacturers, the purpose of this equipment is to provide a path whereby lightning can enter or *leave* the earth without damage or loss to life or property. Note that we said "enter or leave" the earth. This brings up the relatively unknown fact that lightning can begin on the earth as well as in the sky.

Disregarding lightning bolts between or within clouds, the bottom of a lightning cloud can have a negative field, positive field, or both. Commonly negative charges build up in the lower part of the cloud with positive charges on the earth and in the upper cloud. The thunderstorm creates an upward flow from the earth and keeps the earth's charge negative. Like a giant oval racetrack, current flows up from the earth, through the cloud, and back down to the earth.

Nobody has precisely measured lightning. The best know approximation is that the bolt is ½ to ¾ in thick surrounded by a 4-in-thick channel of superheated air. The average bolt is 4000 ft long and may contain a force equal to 100 MV and 200,000 A. No area in the United States is free of lightning strikes. The idea that lightning never strikes the same place twice is totally false. The Empire State Building in New York City has been hit as many as 48 times in 1 year. And there is a record of the Washington Monument having been hit five times in a single storm.

Basically, there are two categories of protection. Class 1 is for ordinary buildings not exceeding 75 ft in height. Class 2 is for heavy-duty stacks and buildings over 75 ft in height or for a building of any height whose structural steel framing may be substituted for the lightning system down conductors. Tall stacks for smoke or vent purposes are sometimes referred to as Class 2 modified (a third class). The NFPA describes their protection in Class 2 modified in Chapter 5 in their NFPA78-1980 bulletin.

The above-mentioned codes describe in considerable detail the proper quantity and location of lightning protection rods and the interconnection of the air terminals by cables of the proper size. Careful reading of this material is required since there are frequently items such as roof drains to be connected as part of the system. Where the building is tall, it probably has a structural steel frame. With proper investigation and arrangement the steel frame can play an important role in the overall system design.

Important in the overall scheme are the details of how the various parts are fastened and/or connected, as well as how the grounding system is installed and connected. Our details are basically concentrated in these two areas.

Class I System Details A Class I building is generally described as a building of normal, conventional design and construction and used for an everyday purpose such as a commercial, farm, industrial, institutional, or residential application. It is less than 75 ft high. The basic protection system consists of air terminals, down conductors, interconnection of metallic parts, ground connections, and grounding electrodes. Figures 17-1 to 17-5 illustrate some of the points to be detailed.

In general the basic requirement for air terminals is that they must not project less than 10 in nor more than 36 in above the object they are protecting. Following our custom of not acting as a book on design, we have not shown the quantity of air terminals required in Figure 17-1. In general, air terminals are placed on 20-ft 0-in centers around the perimeter of the roof. Terminals are required where an interior point is more than 50 ft 0 in from an exterior perimeter terminal. This is not the total design requirement but rather a generalized summary. Since terminals are required on the ridge of pitched roofs, and many roof designs have relatively gentle pitches to avoid some of the flat-roof drainage problems, a gently pitched roof is classified as a flat roof and its design treated accordingly.

Figures 17-2 and 17-3 illustrate the usual collection of objects found on the roofs of many buildings. All of them should be connected to the lightning conductor system as

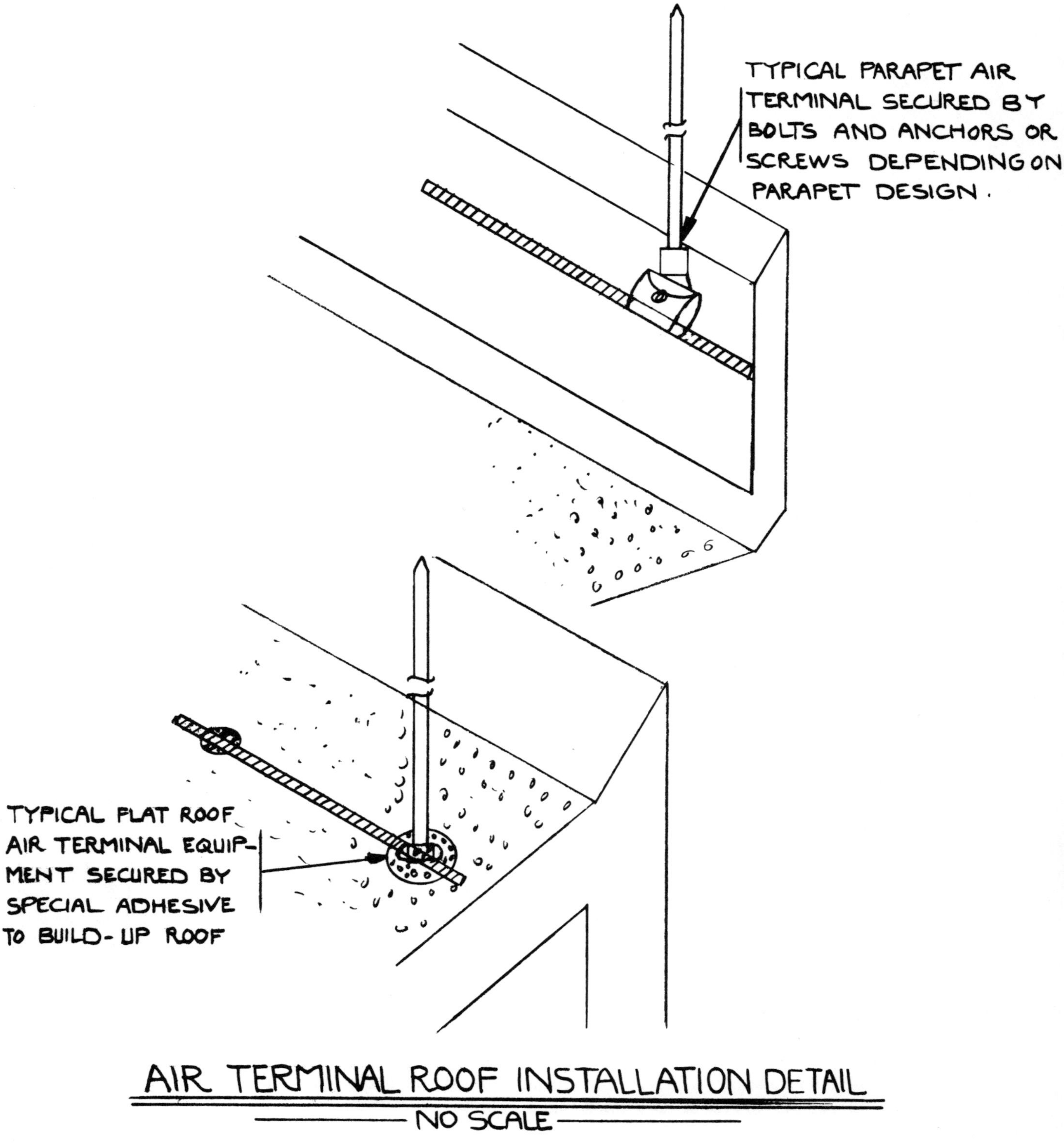

FIGURE 17-1

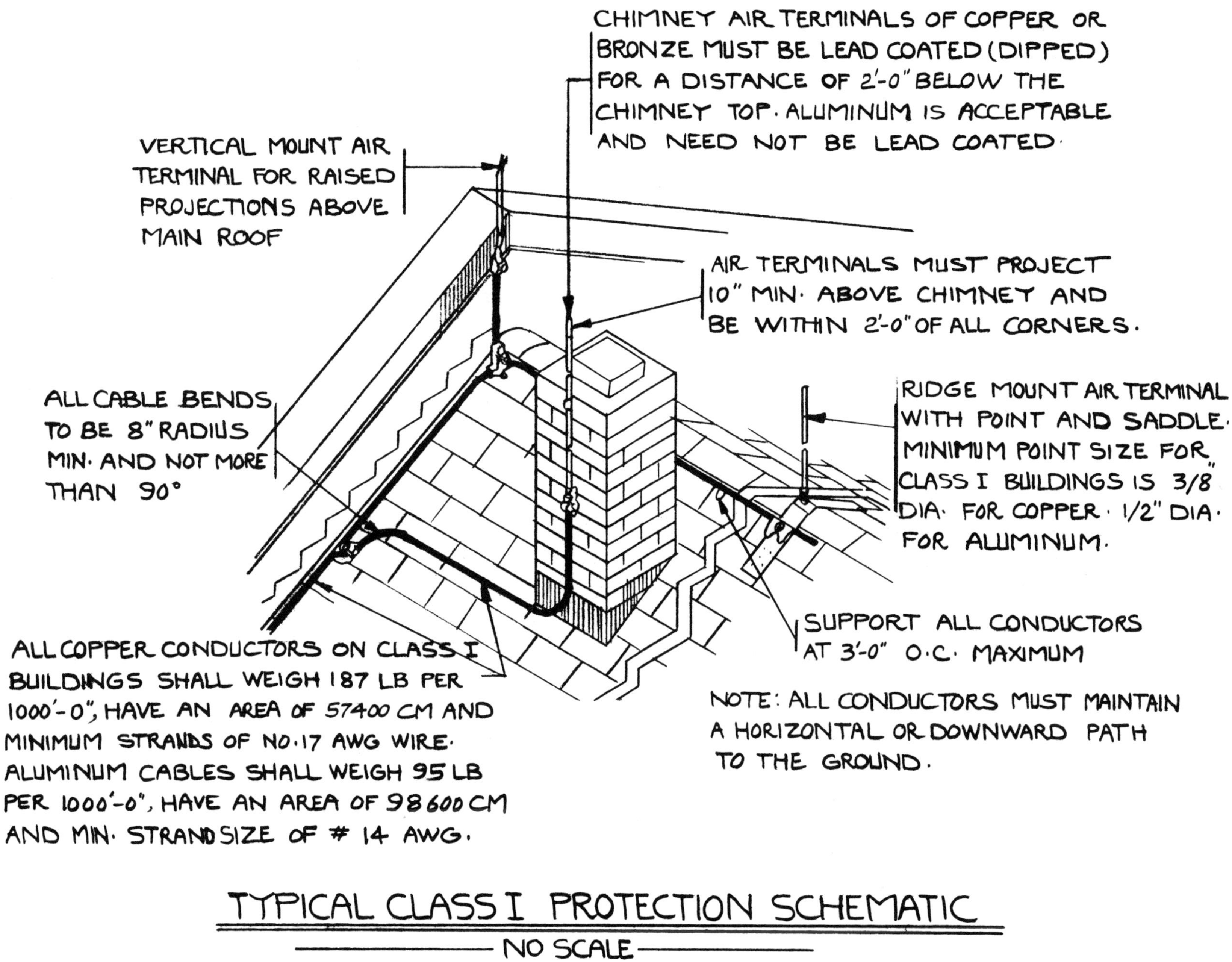
CHIMNEY AIR TERMINALS OF COPPER OR BRONZE MUST BE LEAD COATED (DIPPED) FOR A DISTANCE OF 2'-0" BELOW THE CHIMNEY TOP. ALUMINUM IS ACCEPTABLE AND NEED NOT BE LEAD COATED.
VERTICAL MOUNT AIR TERMINAL FOR RAISED PROJECTIONS ABOVE MAIN ROOF
AIR TERMINALS MUST PROJECT 10" MIN. ABOVE CHIMNEY AND BE WITHIN 2'-0" OF ALL CORNERS.
ALL CABLE BENDS TO BE 8" RADIUS MIN. AND NOT MORE THAN 90°
RIDGE MOUNT AIR TERMINAL WITH POINT AND SADDLE. MINIMUM POINT SIZE FOR CLASS I BUILDINGS IS 3/8" DIA. FOR COPPER. 1/2" DIA. FOR ALUMINUM.
SUPPORT ALL CONDUCTORS AT 3'-0" O.C. MAXIMUM
ALL COPPER CONDUCTORS ON CLASS I BUILDINGS SHALL WEIGH 187 LB PER 1000'-0", HAVE AN AREA OF 57400 CM AND MINIMUM STRANDS OF NO.17 AWG WIRE. ALUMINUM CABLES SHALL WEIGH 95 LB PER 1000'-0", HAVE AN AREA OF 98600 CM AND MIN. STRAND SIZE OF # 14 AWG.
NOTE: ALL CONDUCTORS MUST MAINTAIN A HORIZONTAL OR DOWNWARD PATH TO THE GROUND.
TYPICAL CLASS I PROTECTION SCHEMATIC
NO SCALE

FIGURE 17-2

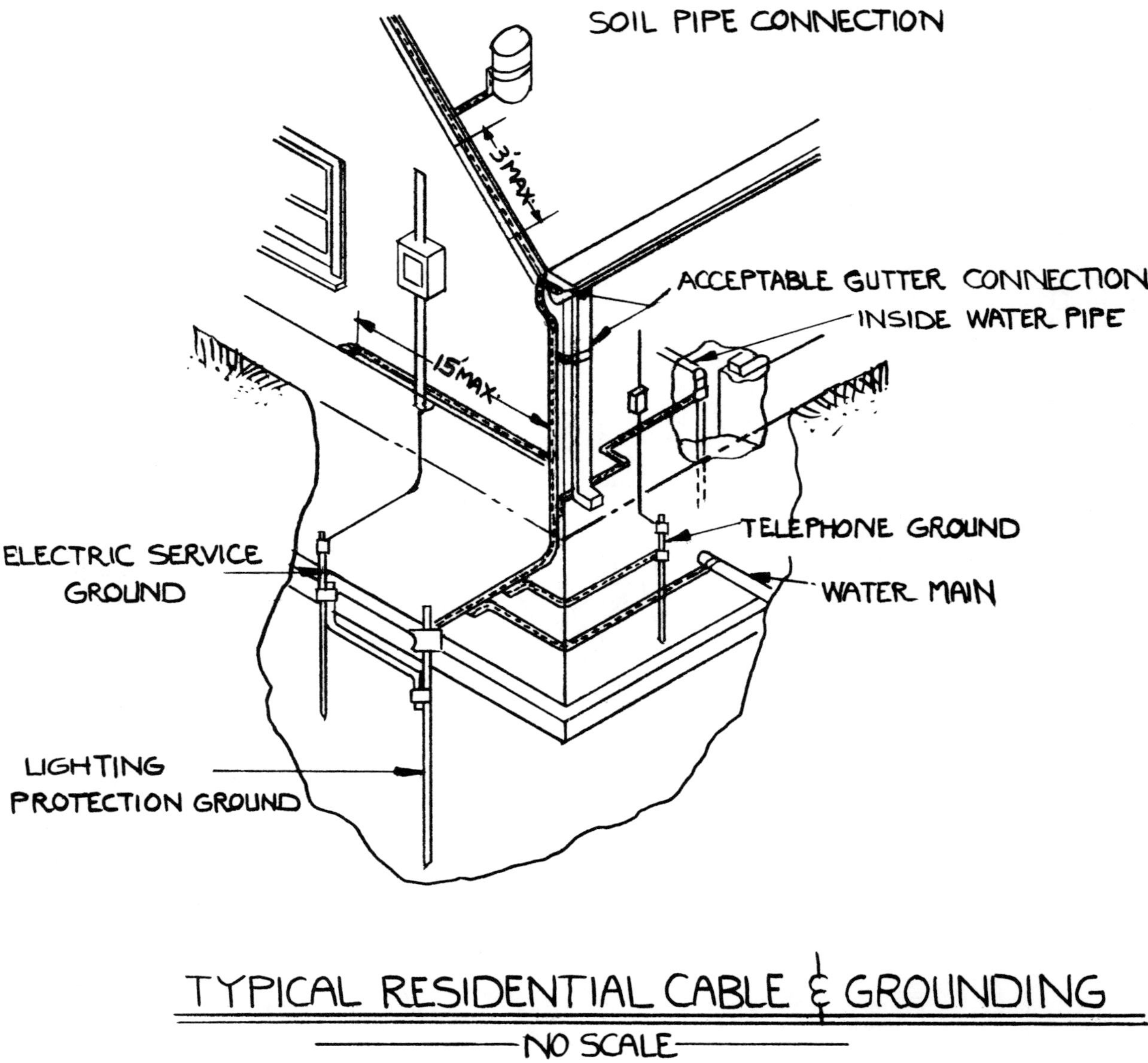

FIGURE 17-3

we have illustrated. As the conductors run down the building, note that we are picking up other items along the way to the ground rods. Note also that among those items is the building electrical service.

Frequently, the down conductors are located in places where they are subject to physical abuse. To protect them, run the conductor in iron pipe or steel tubing. Where this method is employed, the conductor must be bonded to the top and bottom of the pipe or tubing.

There is really no firm rule on how many ground rods to employ. Figures 17-4 and 17-5 illustrate two choices. In a soil of normal resistivity (up to 50 Ω), two rods will serve a small building. Another generalized rule would place ground rods about 100 ft apart on all four sides of the building. Underground metal water pipes are probably the best grounds and at least one down conductor should be connected to the pipe immediately outside the foundation wall. If more than one pipe serves the building, all such systems must be interconnected. In a master-labeled system common grounding, which is the most effective, must be employed. Here all grounding mediums are bonded together. Piping systems involved include water service, gas piping, underground conduits, underground LPG piping, and so forth.

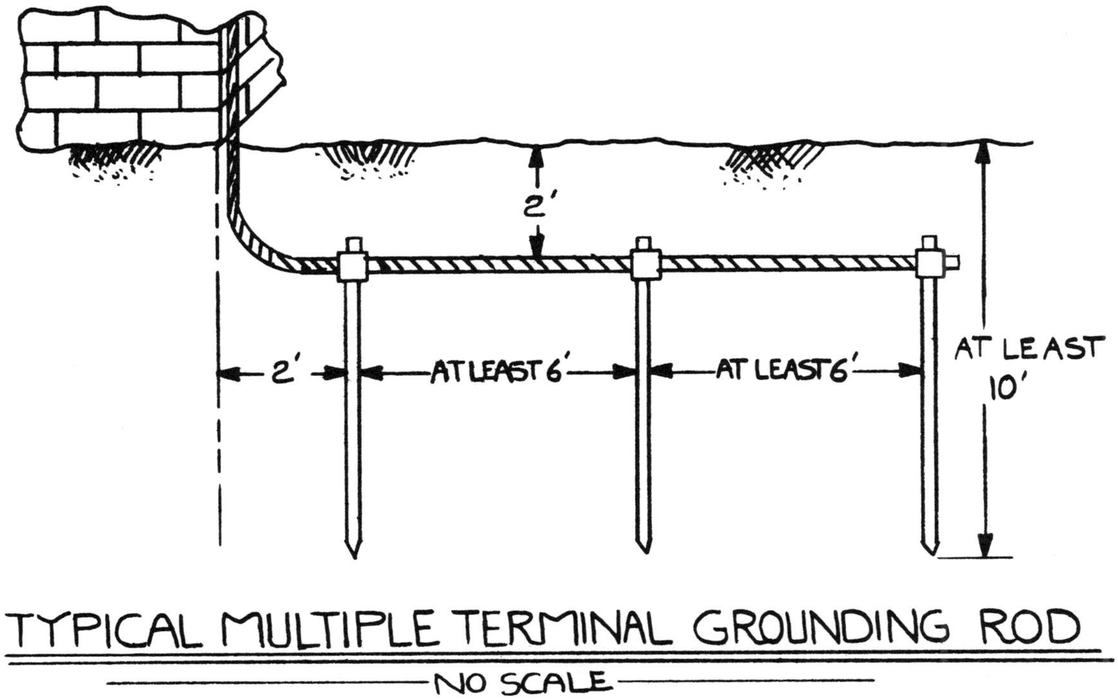

FIGURE 17-4

Class II System Details Recalling our description of the difference between Class 1 and Class 2 systems, the primary criterion is that Class 1 is for buildings under 75 ft in height and Class 2 is for buildings over 75 ft in height. Generally, buildings over 75 ft in height have flat, or relatively flat, roofs. There are, of course, special-purpose structures such as cathedrals, capitol buildings, and the like that have pitched roofs. Commonly, these taller buildings have a structural steel frame although a reinforced concrete structure may sometimes be used up to 120 ft in height. The applicable codes generally state that Class II structures must follow all the requirements of Class I plus some added ones. Figures 17-6, 17-7, and 17-8 break no really new ground, since the basics of lightning protection remain the same.

For your detailing there are, however, some real differences. In Figure 17-6 we show a typical partial roof plan where the air terminals are not on the roof but on the parapet wall. The chimney has two terminals because of the rule that requires a terminal to be within 2 ft 0 in of all four corners, and the chimney shown in the detail is presumed to be too large to permit coverage by a single terminal. The connection details as well as the sizes of the various portions of the overall systems are also more strin-

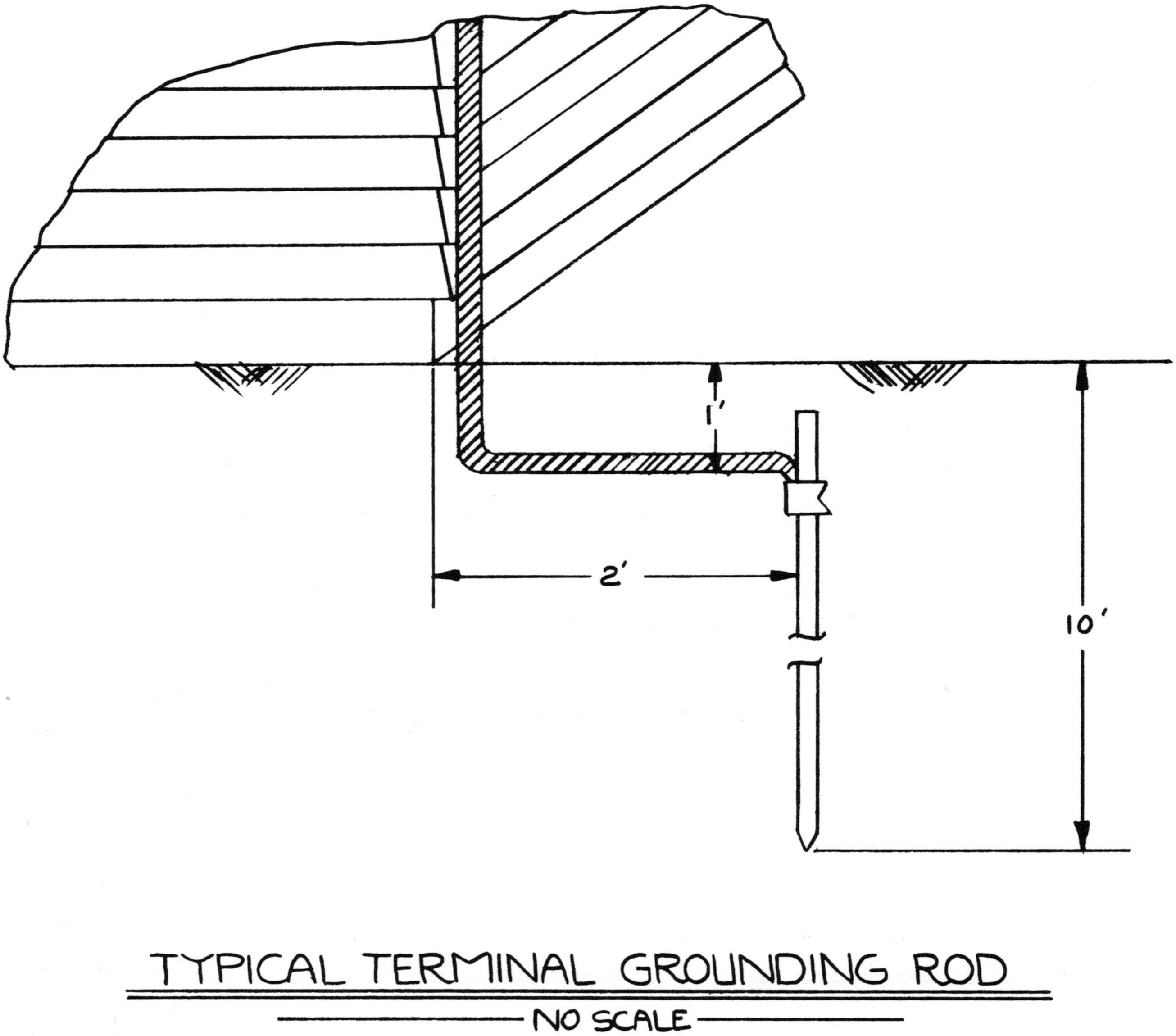

FIGURE 17-5

gent in Class 2. The various parts in Class 2 are larger and stronger. Only welded or bolted metal splicers and connectors are acceptable for Class 2 structures.

There are commonly special appearance considerations in Class 2 installations. This tends to create a situation where the use of the structural frame as a conductor is common. Figures 17-7 and 17-8 illustrate where this is allowable if the frame is electrically continuous or modified by bonding to be made continuous. Terminals must be bonded directly to the framework or connected together via a conductor that is bonded to the framework. Every other column around the perimeter of the building must be grounded. In no case should ground rods be more than 60 ft 0 in apart.

All of the rules previously stated in our other discussions of bonding roof-located metal objects still apply. Typical examples are plumbing stacks, exhaust fans, air-conditioning units, skylight frames, and roof hatches. However, where the framework is the conductor, and these items are electricity continuous with the steel framework, they need to be separately bonded.

Large Separate Stacks By large separate stacks we mean, as the codes state, a stack that is taller than 75 ft and has a flue area greater than 500 in^2. This does not mean that your two-story building with an architecturally appealing separate stack of 40 or 50 ft should not be protected. It definitely should be. This text and Figure 17-9

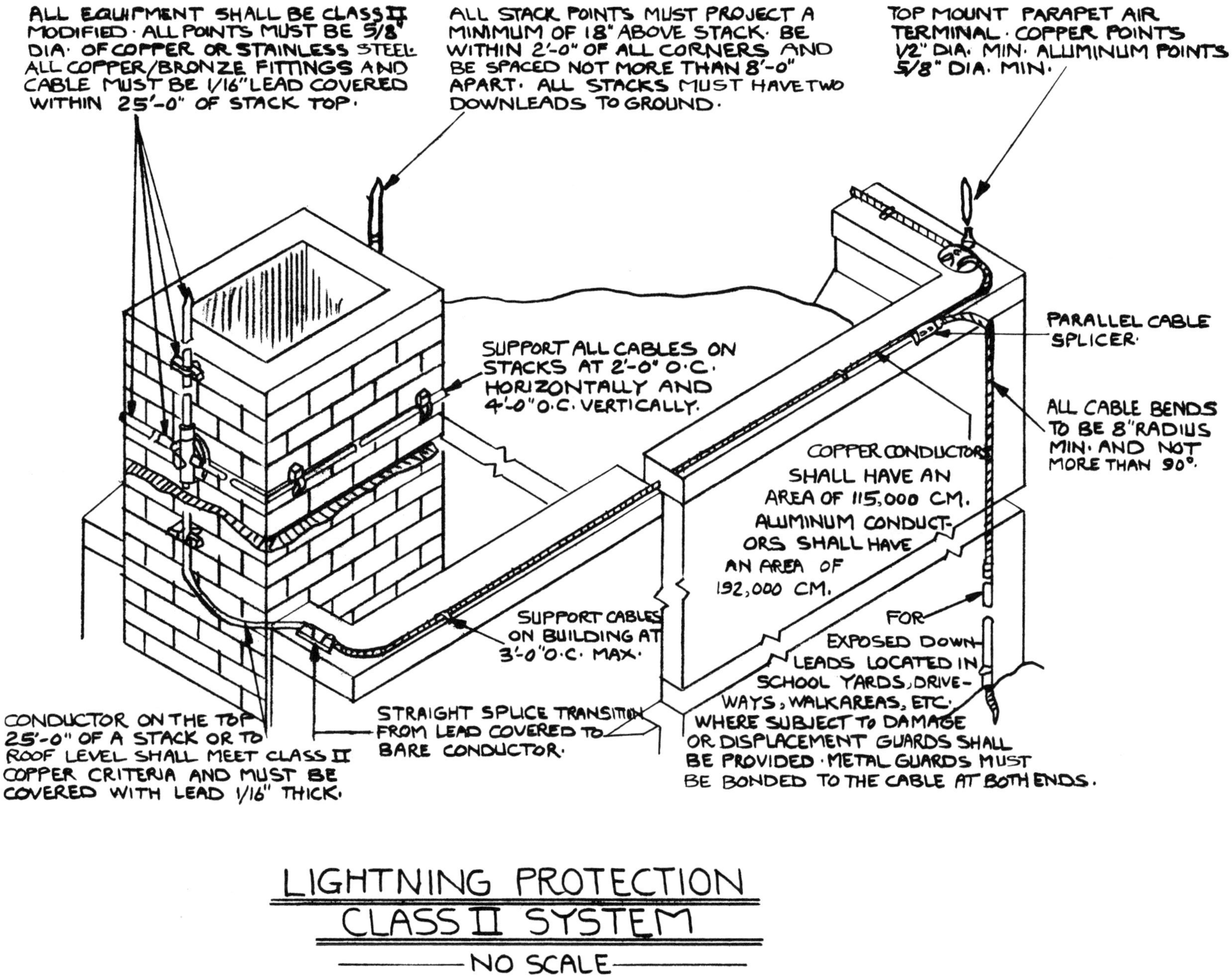
ALL EQUIPMENT SHALL BE CLASS II MODIFIED. ALL POINTS MUST BE 5/8" DIA. OF COPPER OR STAINLESS STEEL. ALL COPPER/BRONZE FITTINGS AND CABLE MUST BE 1/16" LEAD COVERED WITHIN 25'-0" OF STACK TOP.
ALL STACK POINTS MUST PROJECT A MINIMUM OF 18" ABOVE STACK. BE WITHIN 2'-0" OF ALL CORNERS AND BE SPACED NOT MORE THAN 8'-0" APART. ALL STACKS MUST HAVE TWO DOWNLEADS TO GROUND.
TOP MOUNT PARAPET AIR TERMINAL. COPPER POINTS 1/2" DIA. MIN. ALUMINUM POINTS 5/8" DIA. MIN.
PARALLEL CABLE SPLICER.
SUPPORT ALL CABLES ON STACKS AT 2'-0" O.C. HORIZONTALLY AND 4'-0" O.C. VERTICALLY.
ALL CABLE BENDS TO BE 8" RADIUS MIN. AND NOT MORE THAN 90°.
COPPER CONDUCTORS SHALL HAVE AN AREA OF 115,000 CM. ALUMINUM CONDUCTORS SHALL HAVE AN AREA OF 192,000 CM.
SUPPORT CABLES ON BUILDING AT 3'-0" O.C. MAX.
FOR EXPOSED DOWN-LEADS LOCATED IN SCHOOL YARDS, DRIVEWAYS, WALKAREAS, ETC. WHERE SUBJECT TO DAMAGE OR DISPLACEMENT GUARDS SHALL BE PROVIDED. METAL GUARDS MUST BE BONDED TO THE CABLE AT BOTH ENDS.
CONDUCTOR ON THE TOP 25'-0" OF A STACK OR TO ROOF LEVEL SHALL MEET CLASS II COPPER CRITERIA AND MUST BE COVERED WITH LEAD 1/16" THICK.
STRAIGHT SPLICE TRANSITION FROM LEAD COVERED TO BARE CONDUCTOR.
LIGHTNING PROTECTION
CLASS II SYSTEM
NO SCALE

FIGURE 17-6

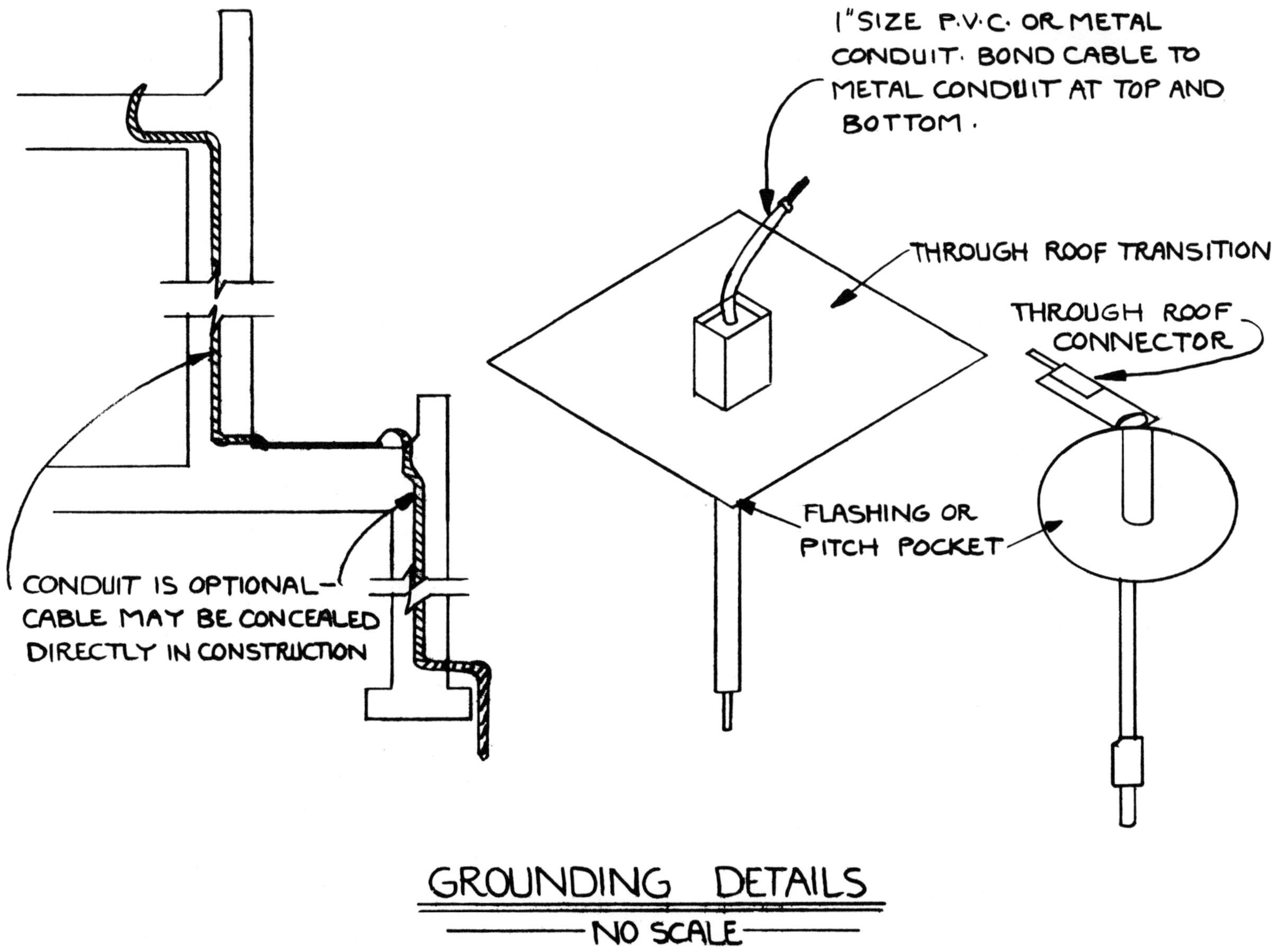
1" SIZE P.V.C. OR METAL CONDUIT. BOND CABLE TO METAL CONDUIT AT TOP AND BOTTOM.
THROUGH ROOF TRANSITION
THROUGH ROOF CONNECTOR
FLASHING OR PITCH POCKET
CONDUIT IS OPTIONAL—CABLE MAY BE CONCEALED DIRECTLY IN CONSTRUCTION
GROUNDING DETAILS
NO SCALE

FIGURE 17-7

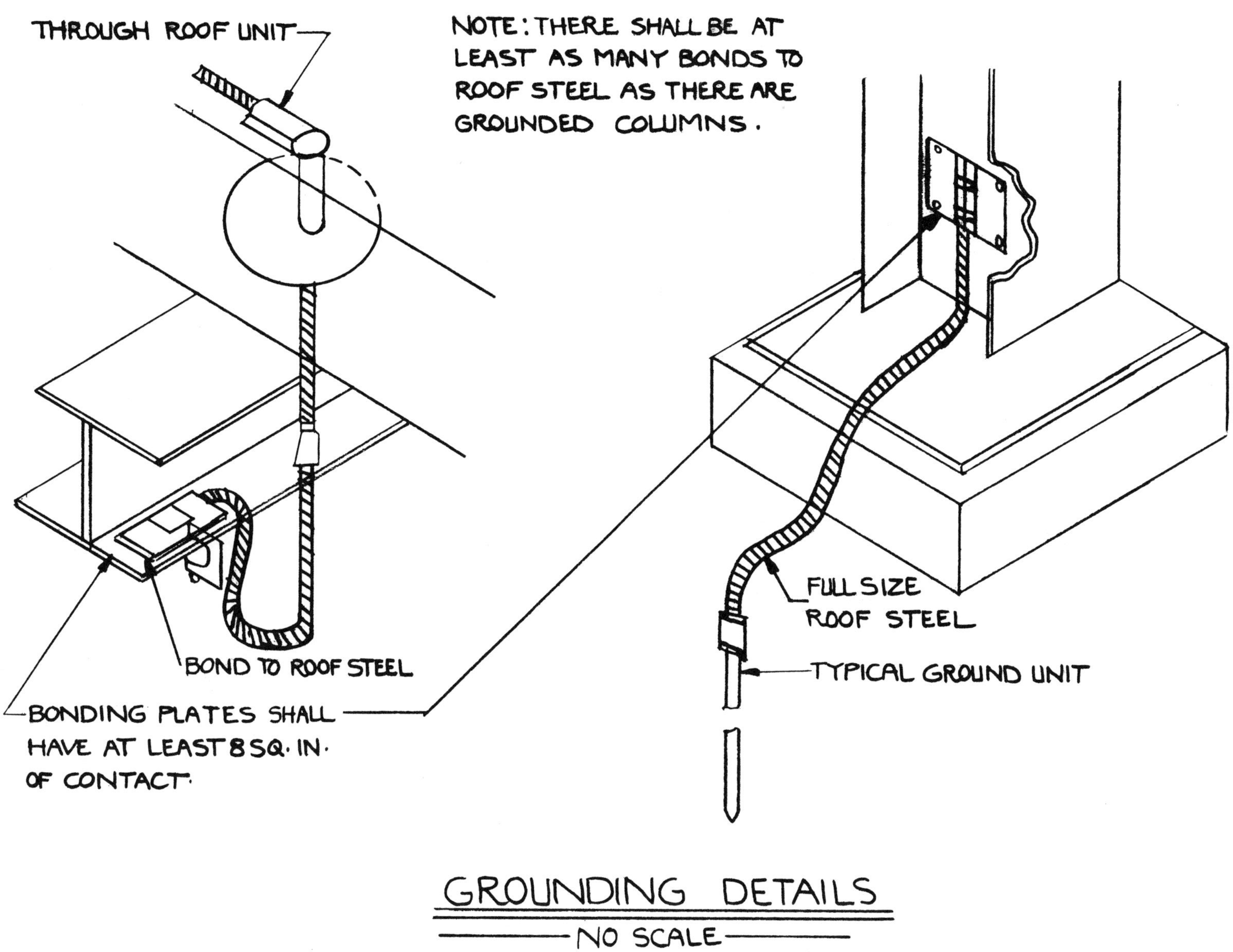
THROUGH ROOF UNIT
NOTE: THERE SHALL BE AT LEAST AS MANY BONDS TO ROOF STEEL AS THERE ARE GROUNDED COLUMNS.
BOND TO ROOF STEEL
BONDING PLATES SHALL HAVE AT LEAST 8 SQ. IN. OF CONTACT.
FULL SIZE ROOF STEEL
TYPICAL GROUND UNIT
GROUNDING DETAILS
NO SCALE

FIGURE 17-8

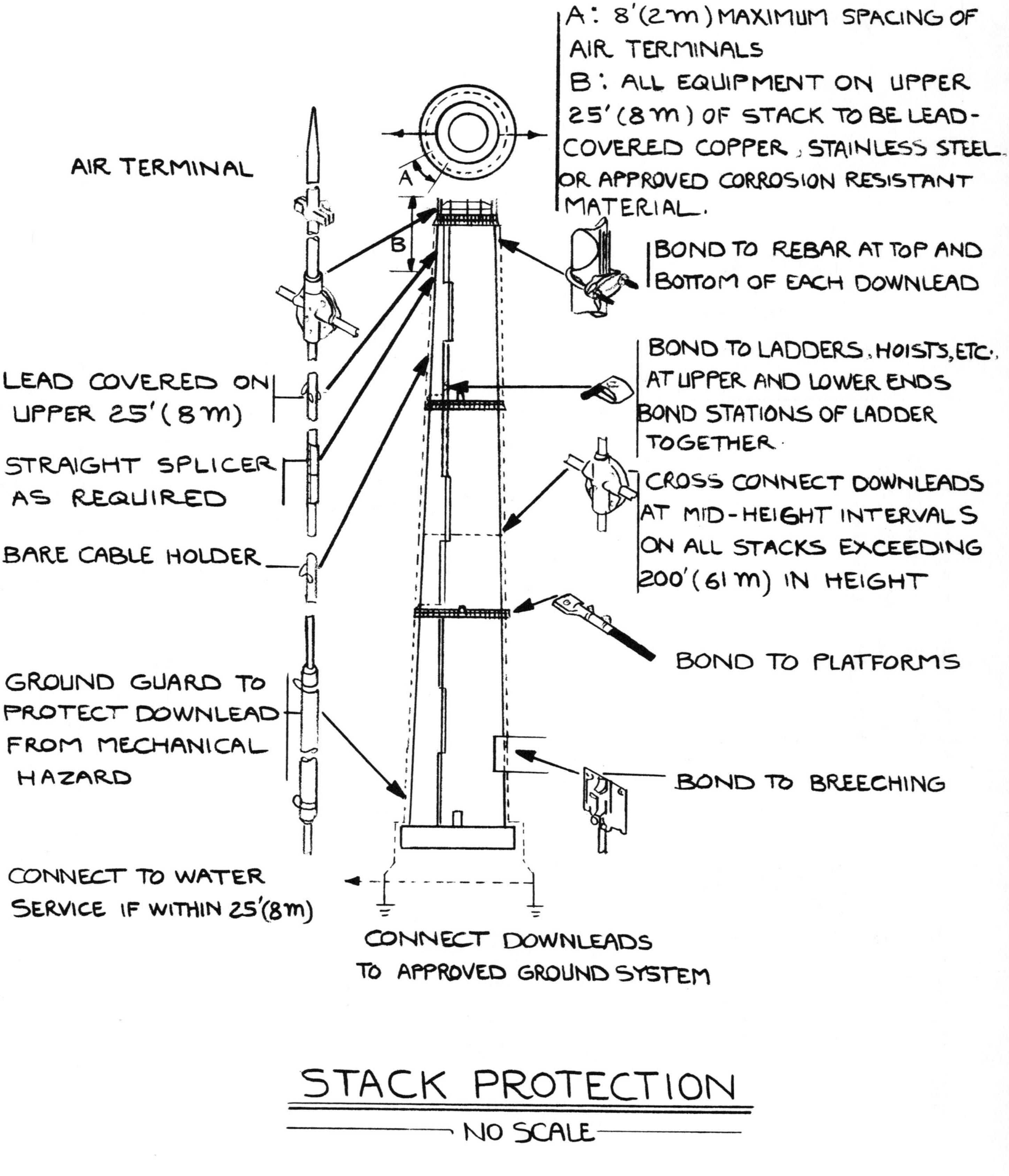
A: 8'(2m) MAXIMUM SPACING OF AIR TERMINALS
B: ALL EQUIPMENT ON UPPER 25'(8m) OF STACK TO BE LEAD-COVERED COPPER, STAINLESS STEEL OR APPROVED CORROSION RESISTANT MATERIAL.
AIR TERMINAL
A
B
BOND TO REBAR AT TOP AND BOTTOM OF EACH DOWNLEAD
BOND TO LADDERS, HOISTS, ETC. AT UPPER AND LOWER ENDS BOND STATIONS OF LADDER TOGETHER
LEAD COVERED ON UPPER 25'(8m)
STRAIGHT SPLICER AS REQUIRED
CROSS CONNECT DOWNLEADS AT MID-HEIGHT INTERVALS ON ALL STACKS EXCEEDING 200'(61m) IN HEIGHT
BARE CABLE HOLDER
BOND TO PLATFORMS
GROUND GUARD TO PROTECT DOWNLEAD FROM MECHANICAL HAZARD
BOND TO BREECHING
CONNECT TO WATER SERVICE IF WITHIN 25'(8m)
CONNECT DOWNLEADS TO APPROVED GROUND SYSTEM
STACK PROTECTION
NO SCALE

FIGURE 17-9

refer to the special rules that apply to the large power plant or industrial smokestacks.

We will not repeat all of the material requirements contained in the various codes that apply in this situation. However, it should be noted that no fewer than two ground conductors are required. The detail we have shown is fairly well covered and is fairly explicit. There are a few special code items that should be brought to your attention. Reinforcing steel in a concrete stack must be electrically bonded to the conductors at the top and bottom of the down leads. Metal stacks *do not* require air terminals or ground conductors. Ground the base of the stack on opposite sides. Metal guy wires or cables set in concrete and attached to buildings must be grounded at their lower ends. If a metal water pipe is within 25 ft 0 in of the stack it can be used as a ground.

Flagpoles, Masts, Spires In general these types of structures usually, where construction materials dictate, require a terminal, a down conductor, and a ground rod. This does not really require a detail and can easily be spelled out in the contract specifications. Our detail and the accompanying text refer to an electrically continuous metal structure that *does not* require an air terminal or down conductor but *does* require a ground terminal.

This frequently forgotten detail is shown as Figure 17-10. It is quite complete and we feel that little could be added by further discussion.

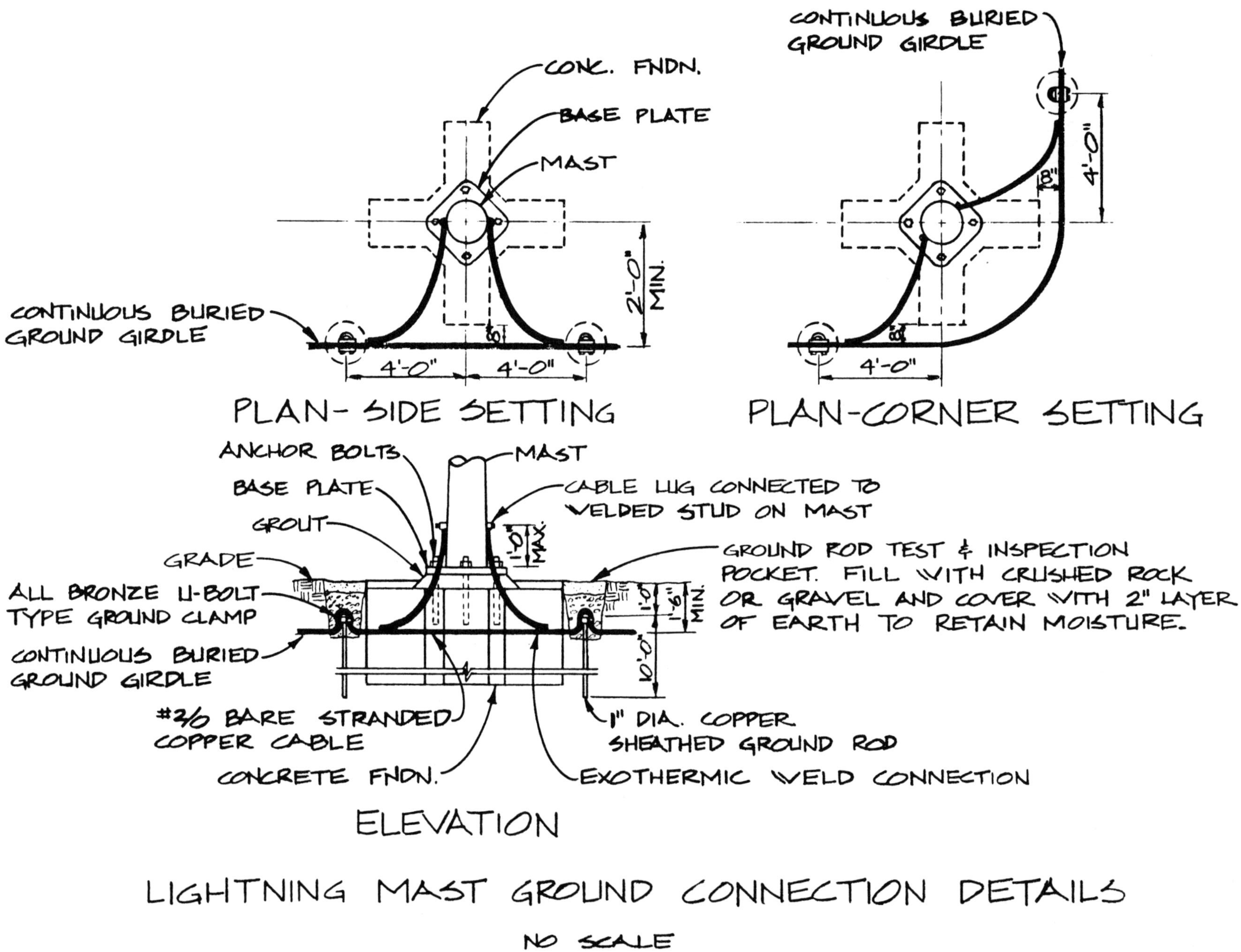
CONC. FNDN.
BASE PLATE
MAST
CONTINUOUS BURIED GROUND GIRDLE
2'-0" MIN.
8"
4'-0"
4'-0"
PLAN- SIDE SETTING
CONTINUOUS BURIED GROUND GIRDLE
4'-0"
8"
4'-0"
PLAN-CORNER SETTING
ANCHOR BOLTS
BASE PLATE
GROUT
MAST
CABLE LUG CONNECTED TO WELDED STUD ON MAST
1'-0" MAX.
GRADE
ALL BRONZE U-BOLT TYPE GROUND CLAMP
CONTINUOUS BURIED GROUND GIRDLE
GROUND ROD TEST & INSPECTION POCKET. FILL WITH CRUSHED ROCK OR GRAVEL AND COVER WITH 2" LAYER OF EARTH TO RETAIN MOISTURE.
1'-0"
1'-6" MIN.
10'-0"
#2/0 BARE STRANDED COPPER CABLE
CONCRETE FNDN.
1" DIA. COPPER SHEATHED GROUND ROD
EXOTHERMIC WELD CONNECTION
ELEVATION
LIGHTNING MAST GROUND CONNECTION DETAILS
NO SCALE

FIGURE 17-10

EIGHTEEN

Cathodic Protection

Cathodic protection is a part of the very large and complex subject of corrosion control. Although this subject is generally considered part of the chemical or mechanical engineering area, corrosion is frequently an electrochemical reaction which, by definition, involves a flow of electric current.

Certain physical processes that occur during the corrosion process relate to our details and are the reasons for including this complex subject in our book. When a metal is corroding, certain well-defined anodic and cathodic areas exist on the surface. At the anodic areas the metal goes into solution as positive ions. At the cathodic areas. a reducing reaction occurs which usually does not involve the metal. When, for example, an element loses ions, the process is called *oxidation* (nesting). *Reduction* is the process by which a metal gains ions.

In electrochemical theory corrosion is accompanied by both a chemical change and a production of electrical energy. Common to all types of corrosion are certain basic premises. Current flowing between dissimilar metals is a common example of galvanic corrosion. Differential environmental conditions such as a long pipeline passing through different soils can create major corrosion cells wherein several amperes of current may flow in the pipe. Corrosion will occur where current leaves the pipe surface. Steel piles and sheet piling in seawater are classic examples. Stray alternating current that accidentally rectifies into stray direct current, as well as stray direct current itself, are other sources. There are also bacteriological and chemical sources of corrosion.

We would be venturing far beyond our book's stated purpose if we continued discussing corrosion theory. Our point is that where an electrical current is the source or related cause of corrosion, then some form of counter electrical force is obviously part of the solution. The purpose of cathodic protection is to halt or reduce corrosion.

There are two basic types of cathodic protection: sacrificial and impressed current. Both have basic components which consist of a power source, a continuous electrolyte, and an external electrical connection between the subject and the source to conduct current back to the source.

In the sacrificial system the sacrificial anode is connected to the structure to be protected. The anode is more negative than the structure. This arrangement creates a galvanic cell. Current flows through the electrolyte from the anode to the cathode. The structure in this case becomes the cathode. Properly designed, this system creates sufficient current to suppress all local action current on the surface of the protected structure. The electrolyte is the earth between the anode and the structure.

In the impressed current installation, the anode is any suitable conducting material. Any continuous source of direct current may be used. Normally, this means a rectifier, but the source could be a motor-generator set, a gasoline-powered generator, or even a wind-driven generator. The direct current flows from the anodes through the electrolyte to the structure and returns through a metal circuit.

Cathodic protection can be used on all kinds of underground conduit, pipe, tanks, and pilings. It also can protect underwater transformers, sewage and water treatment plant equipment, pilings, and the like. Its aboveground uses are primarily for tanks, sewage and water treatment facilities, and air washers in corrosive atmospheres. Cathodic protection works on both new and existing structures.

Sacrificial systems primarily use magnesium anodes, although, under certain conditions, the anode can be zinc or aluminum. For impressed current systems the two most common anodic materials are graphite and high-silicon-content cast iron. The graphite anode is made by fusing

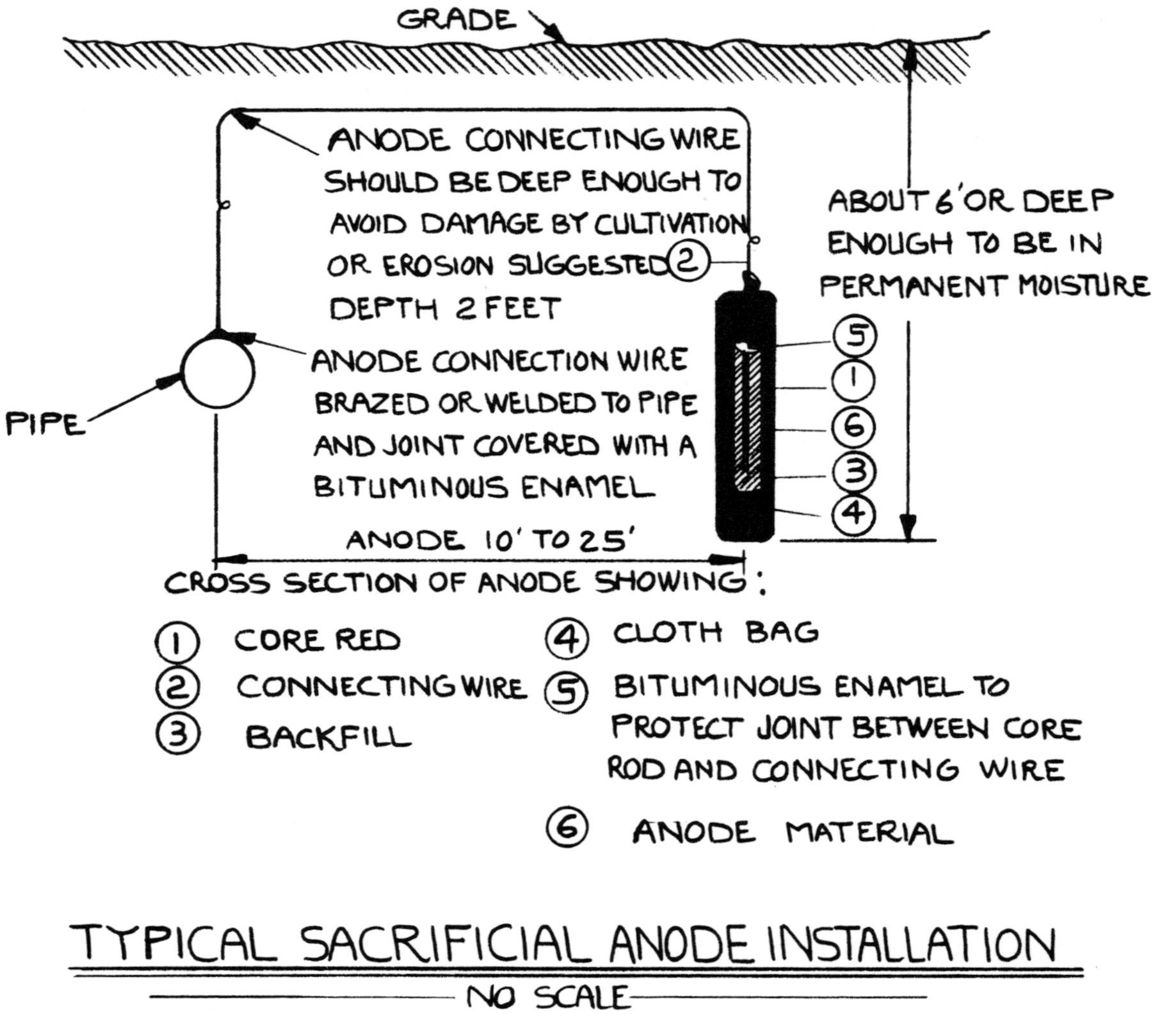

FIGURE 18-1

coke or carbon particles at a very high temperature. Rectifiers that are used in cathodic protection consist of a transformer, rectifier, stacks, meters, and switch all in one enclosure.

If ever there was a subject where mere detailing will not cover the situation, cathodic protection is it. Considerable knowledge and expertise is involved. Fortunately, or perhaps unfortunately, the details we will present make the installation and related design look rather simple. In fact, the details are fairly uncomplicated. The design is very complicated. Unless this special expertise exists in your firm, the best advice we can give is to secure the services of a corrosion expert for the design.

Among the data items which you or your hired expert will require are soil analysis, soil resistivity in the area involved, and structure-to-soil potentials. After these and other data have been collected and digested, a basic design of the type of system to be applied can be made.

Galvanic Protection The galvanic, or sacrificial, anode protection system provides protective current to metals by creating an electrical coupling. The metal being protected is referred to as being more anodic, or more negative. Because of this potential difference current flows from negative to positive. The creation of sufficient current through the electrolyte to suppress all local action currents on the metal creates the resultant cathodic protection. This system requires no electrical power supply and uses magnesium, zinc, or aluminum anodes. The electrochemical reaction causes the anode to slowly disintegrate or decay, hence the term "sacrificial."

Figures 18-1 to 18-3 show three typical installation details. Determining the correct number of anodes is a matter of considerable engineering investigation and calculation. Figure 18-2 shows a plan view of galvanic protection and is correct for the particular problem calculated. It is very important to understand that merely because this de-

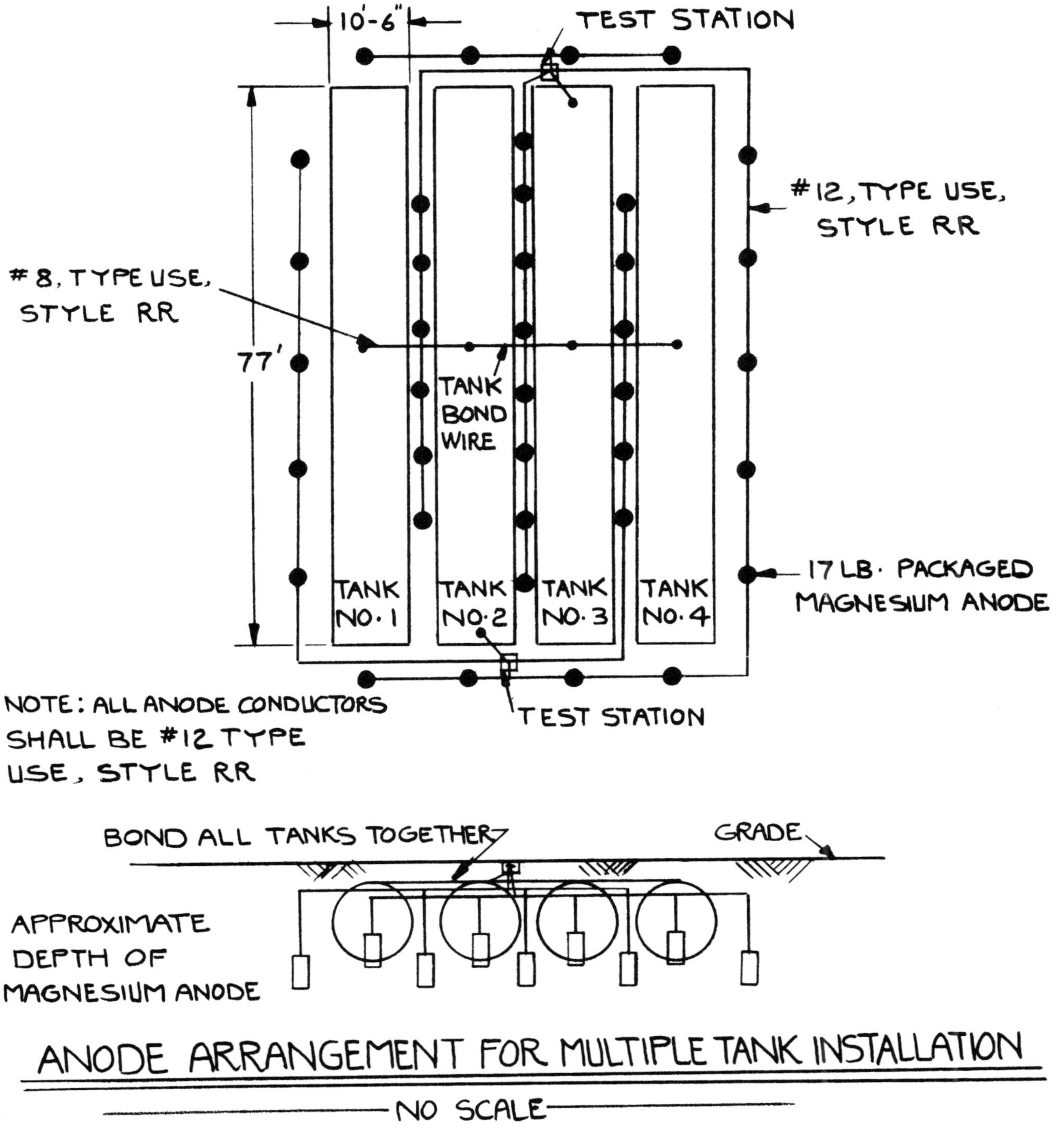

FIGURE 18-2

tail is correct in a certain situation, it very definitely cannot be used without modification as a solution to your particular design problem.

There are basic premises, advantages, disadvantages, and applications of this type of sacrificial cathodic protection system. First of all, a cathodic protection system is installed underground, or underwater, and is something that should not be changed, repaired, or replaced except after long periods of time. The life of a sacrificial anode is determined by its weight, current output, and ampere-hour rating per pound of metal. Since 10 to 15 years is a reasonable system life expectancy, the operating stability of the anode is crucial.

The advantages of the galvanic, or sacrificial, anode

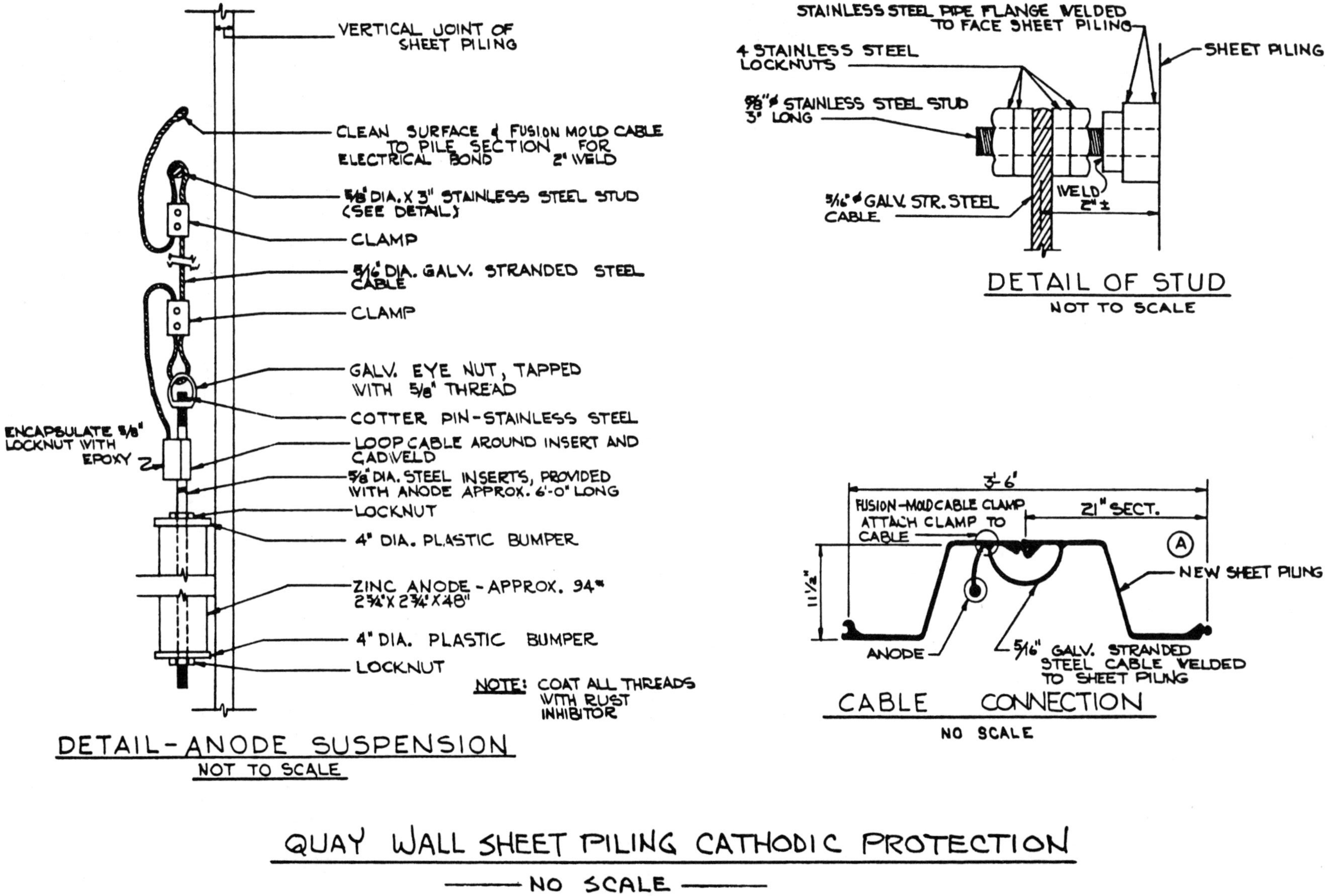
VERTICAL JOINT OF SHEET PILING
CLEAN SURFACE & FUSION MOLD CABLE TO PILE SECTION FOR ELECTRICAL BOND 2" WELD
5/8" DIA. X 3" STAINLESS STEEL STUD (SEE DETAIL)
CLAMP
5/16" DIA. GALV. STRANDED STEEL CABLE
CLAMP
GALV. EYE NUT, TAPPED WITH 5/8" THREAD
COTTER PIN-STAINLESS STEEL
LOOP CABLE AROUND INSERT AND CADWELD
5/8" DIA. STEEL INSERTS, PROVIDED WITH ANODE APPROX. 6'-0" LONG
LOCKNUT
4" DIA. PLASTIC BUMPER
ZINC ANODE - APPROX. 94# 2 3/4" X 2 3/4" X 48"
4" DIA. PLASTIC BUMPER
LOCKNUT
ENCAPSULATE 5/8" LOCKNUT WITH EPOXY
NOTE: COAT ALL THREADS WITH RUST INHIBITOR
DETAIL-ANODE SUSPENSION
NOT TO SCALE
STAINLESS STEEL PIPE FLANGE WELDED TO FACE SHEET PILING
4 STAINLESS STEEL LOCKNUTS
SHEET PILING
5/8" ø STAINLESS STEEL STUD 3" LONG
5/16" ø GALV. STR. STEEL CABLE
WELD 2" ±
DETAIL OF STUD
NOT TO SCALE
3'-6"
FUSION-MOLD CABLE CLAMP ATTACH CLAMP TO CABLE
21" SECT.
A
NEW SHEET PILING
11 1/2"
ANODE
5/16" GALV. STRANDED STEEL CABLE WELDED TO SHEET PILING
CABLE CONNECTION
NO SCALE
QUAY WALL SHEET PILING CATHODIC PROTECTION
NO SCALE

FIGURE 18-3

system include the absence of an external power source, simple installation, minimal cathodic interference and maintenance, and, by judicious arrangement, the ability to evenly distribute current over long structures. The disadvantages are inherent and self-evident. Each anode has a limited current output, and frequently it takes a lot of these anodes to solve the problem. Another disadvantage is that they will not work in highly resistant electrolytes (soil or liquid between the anode and the metal to be protected).

Sacrificial anodes are usually applicable in the protection of small to moderately sized situations such as small isolated tanks (interior and exterior), moderate-sized tank farms, local problem areas on pipelines, and overall protection on well-coated lines, or where electrical power is not available. They are also quite usable in explosionproof areas, in restricted areas, and in marine areas. There are many situations where the simplicity and usually lower cost of this system make it the best solution.

Because, as was pointed out, soil resistivity and anode life are very important, the failures that occur can frequently be traced to the lack of a proper site investigation.

Impressed Current Protection In impressed current protection, an ac source and a dc rectifier, or a dc generator, creates a flow of current to ground through a group of anodes. The current is then transmitted through the ground electrolyte to the structure to be protected. The anodes are normally made from graphite, carbon, iron, steel, or aluminum.

In general, an impressed current system can be used almost anywhere a sacrificial system can be used (provided there is a source of power), as well as in all the places where a sacrificial system cannot be used. There are a couple of disclaimers that should be added to the above statement. First, the underwater installation and its connections may present special waterproofing requirements that are not easily overcome, and second, a hazardous location requires explosionproof switches, electrical equipment, and low-resistance grounding. Figures 18-4 to 18-13 depict fairly common situations and illustrate a number of points in this type of protection. Figure 18-4 illustrates how the current flows and the use of the earth as the electrolyte. Since tanks, pilings, and underground distribution are primary areas of possible application, the details cover these somewhat classroom-type examples. Again, we caution you not merely to copy these details directly but rather to use them as guides to be modified to suit your particular requirements.

The advantages of the impressed current system include no limitation on current output, the ability to vary voltage and current output, and the flexibility to handle large or small situations. The disadvantages are that it is a much more complex design and could create cathodic interferences, it requires electrical maintenance and inspection and has a higher first cost, and it could cause hydrogen embrittlement in higher-strength metals or alloys. And, as we pointed out previously, the impressed current system commonly is not applicable in hazardous locations. Our details illustrate some major applications, but the list is literally unlimited except as noted above.

There are some general application notes that should be followed. These apply not only to the impressed current system but also to the sacrificial system. Where pipelines are protected, insulation from the rest of the system may be required. If, in gas piping, dresser couplings are used, provide electrical bonds across them. Insulate underground structures from pipelines or other structures before you apply cathodic protection. When protecting cable, use sectionalized anodes and regulate the amount of current to the section by central resistors. If you are protecting transformers, sectionalizing boxes, or steel enclosure equipment in a manhole or vault that may become flooded, carefully set your anodes to operate properly in either environment.

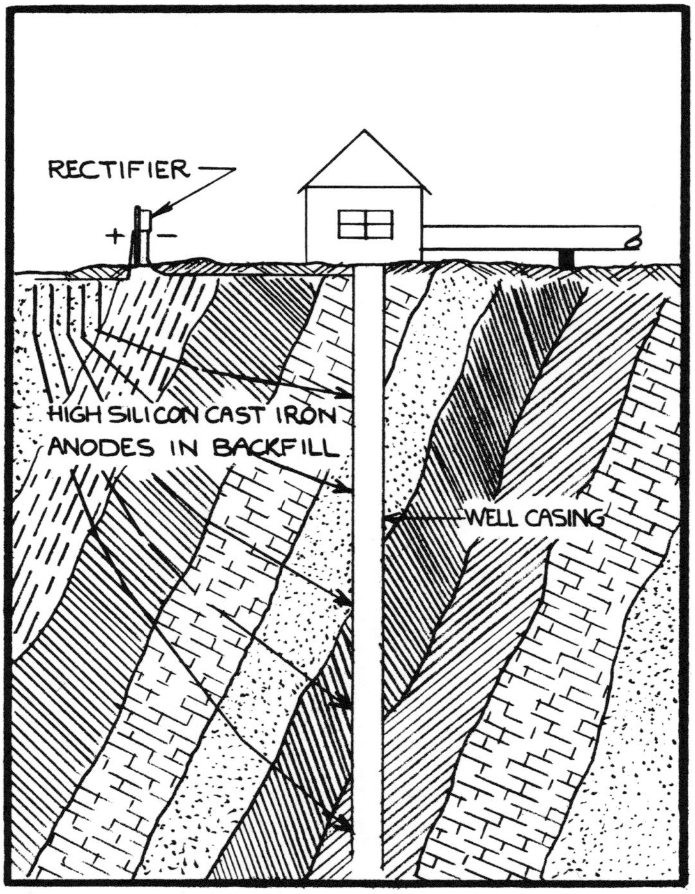

IMPRESSED CURRENT PROTECTION WITH
HIGH SILICON CAST IRON ANODES

NO SCALE

FIGURE 18-4

Figures 18-5 to 18-13 are self-explanatory in terms of the parts of the system and their locations. Depicted are fairly common applications of tanks and pipes above and below ground as well as the way the protective devices are applied. Your solutions will result in details that resemble ours but are *modified for your requirements.*

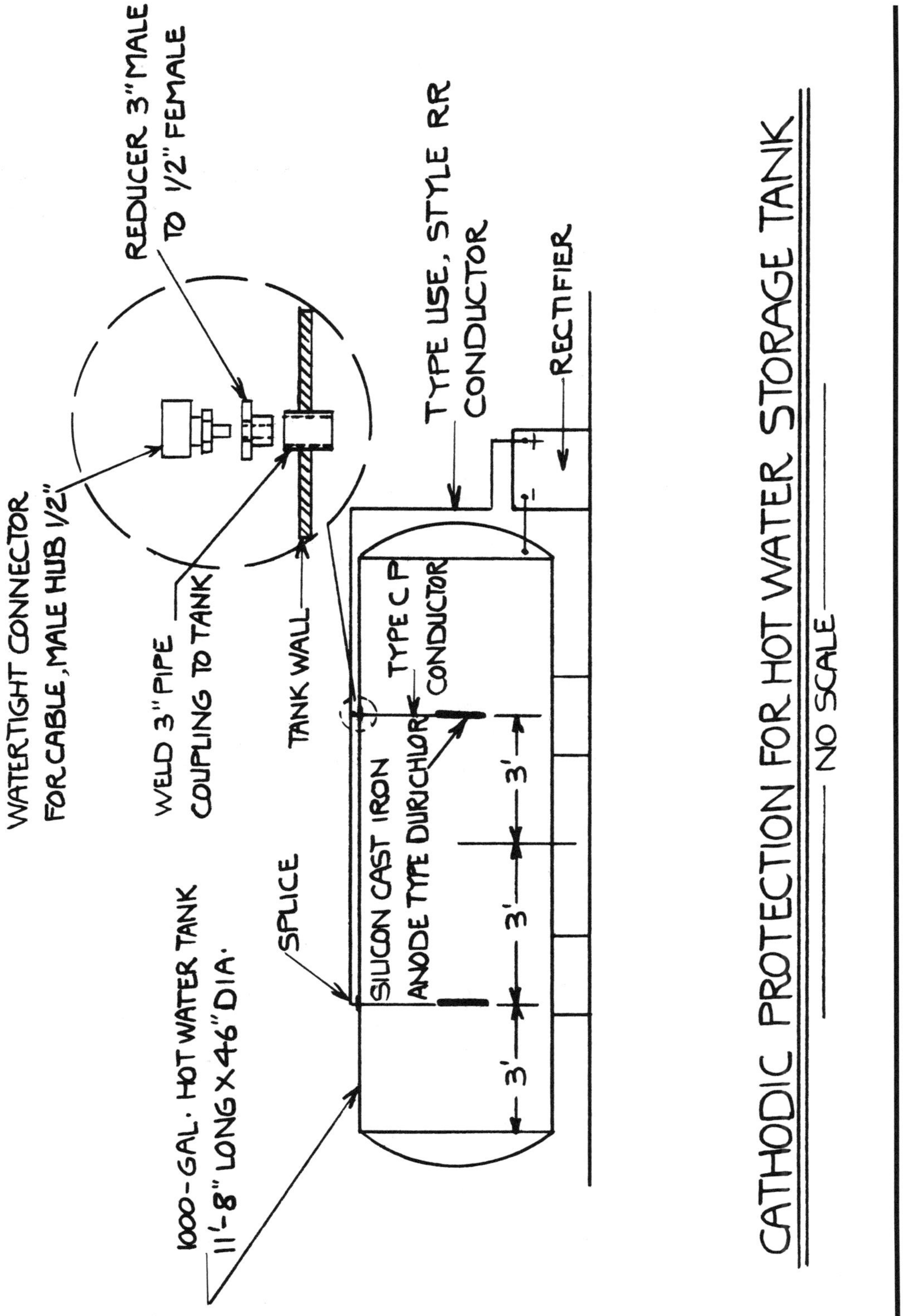

FIGURE 18-5

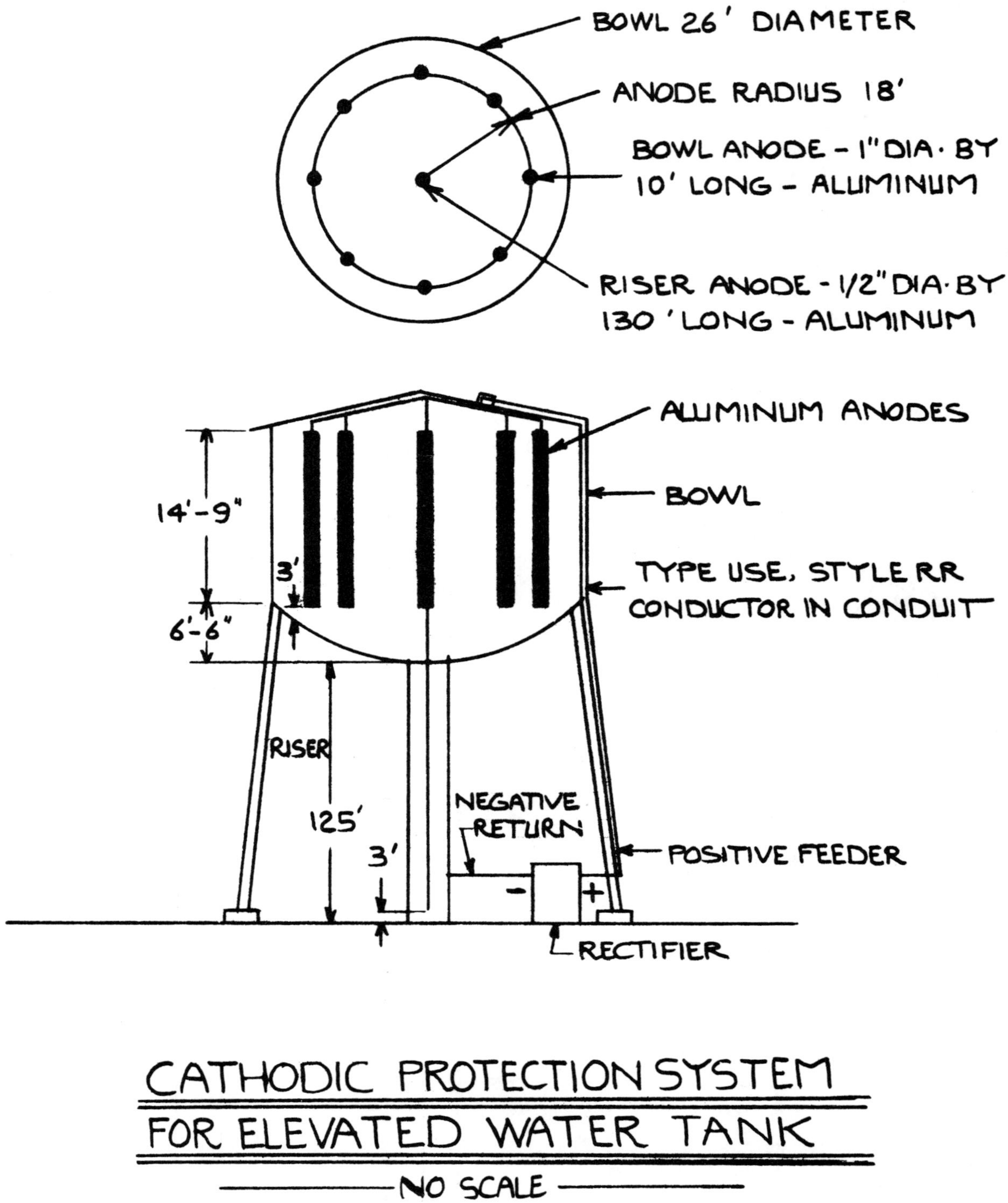

BOWL 26' DIAMETER
ANODE RADIUS 18'
BOWL ANODE - 1" DIA. BY 10' LONG - ALUMINUM
RISER ANODE - 1/2" DIA. BY 130' LONG - ALUMINUM
ALUMINUM ANODES
BOWL
14'-9"
3'
6'-6"
TYPE USE, STYLE RR CONDUCTOR IN CONDUIT
RISER
125'
3'
NEGATIVE RETURN
POSITIVE FEEDER
-
+
RECTIFIER
CATHODIC PROTECTION SYSTEM FOR ELEVATED WATER TANK
NO SCALE

FIGURE 18-6

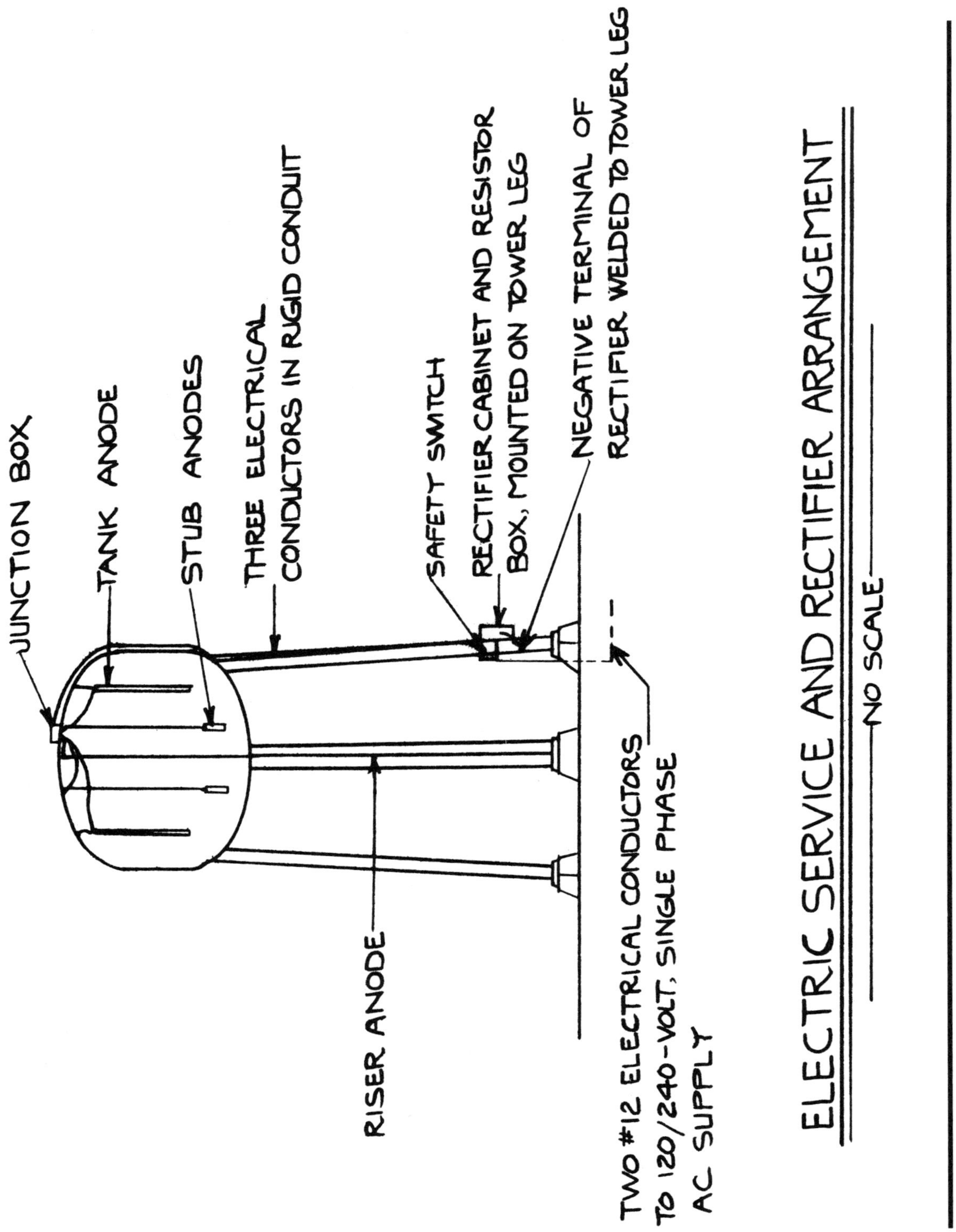
JUNCTION BOX
TANK ANODE
STUB ANODES
THREE ELECTRICAL CONDUCTORS IN RIGID CONDUIT
SAFETY SWITCH
RECTIFIER CABINET AND RESISTOR BOX, MOUNTED ON TOWER LEG
NEGATIVE TERMINAL OF RECTIFIER WELDED TO TOWER LEG
RISER ANODE
TWO #12 ELECTRICAL CONDUCTORS TO 120/240-VOLT, SINGLE PHASE AC SUPPLY
ELECTRIC SERVICE AND RECTIFIER ARRANGEMENT
NO SCALE

FIGURE 18-7

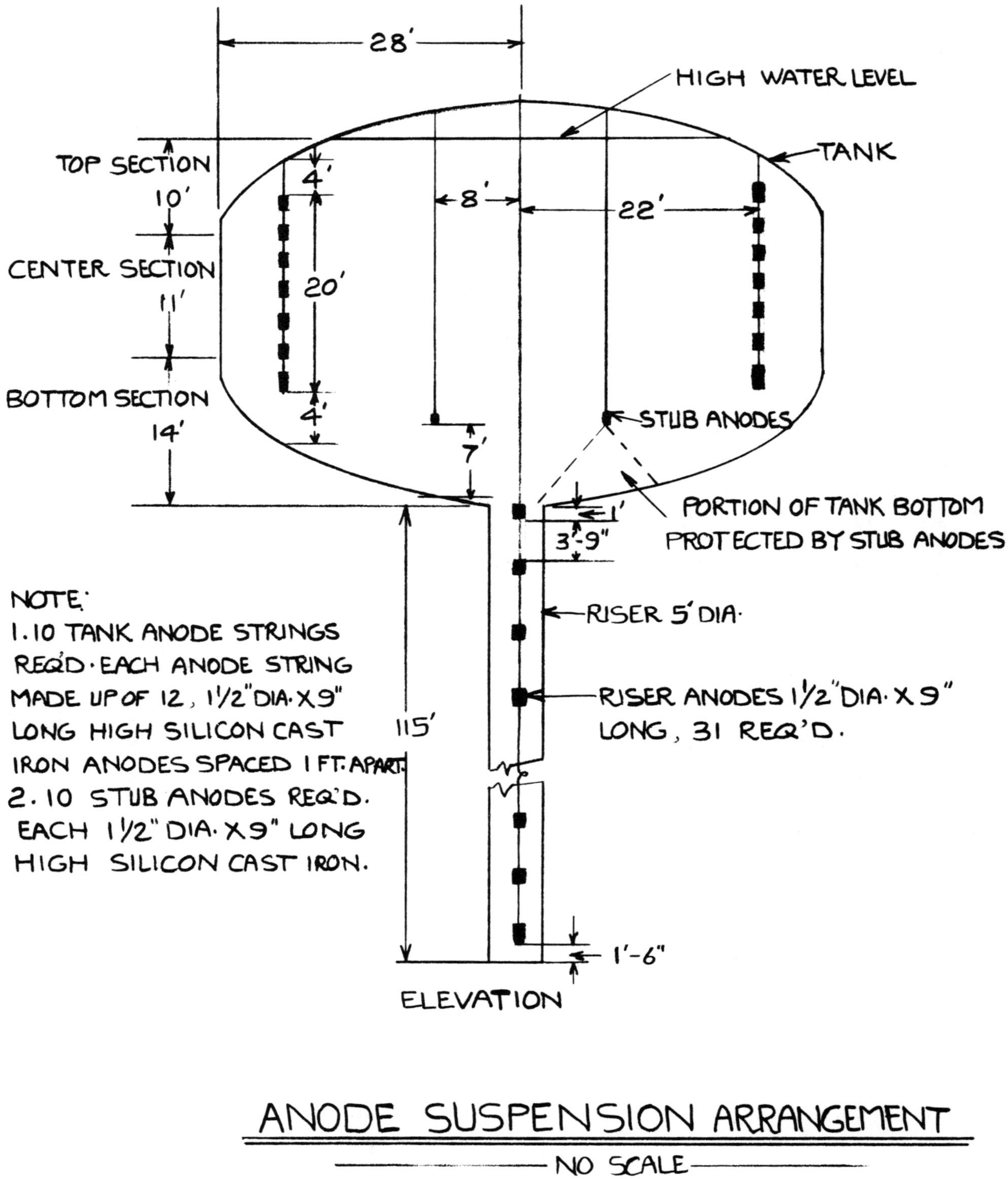

28'
HIGH WATER LEVEL
TANK
TOP SECTION
10'
4'
8'
22'
CENTER SECTION
11'
20'
BOTTOM SECTION
14'
4'
7'
STUB ANODES
PORTION OF TANK BOTTOM
PROTECTED BY STUB ANODES
1'
3'-9"
RISER 5' DIA.
NOTE:
1. 10 TANK ANODE STRINGS
REQ'D. EACH ANODE STRING
MADE UP OF 12, 1½" DIA. X 9"
LONG HIGH SILICON CAST
IRON ANODES SPACED 1 FT. APART.
2. 10 STUB ANODES REQ'D.
EACH 1½" DIA. X 9" LONG
HIGH SILICON CAST IRON.
115'
RISER ANODES 1½" DIA. X 9"
LONG, 31 REQ'D.
1'-6"
ELEVATION
ANODE SUSPENSION ARRANGEMENT
NO SCALE

FIGURE 18-8

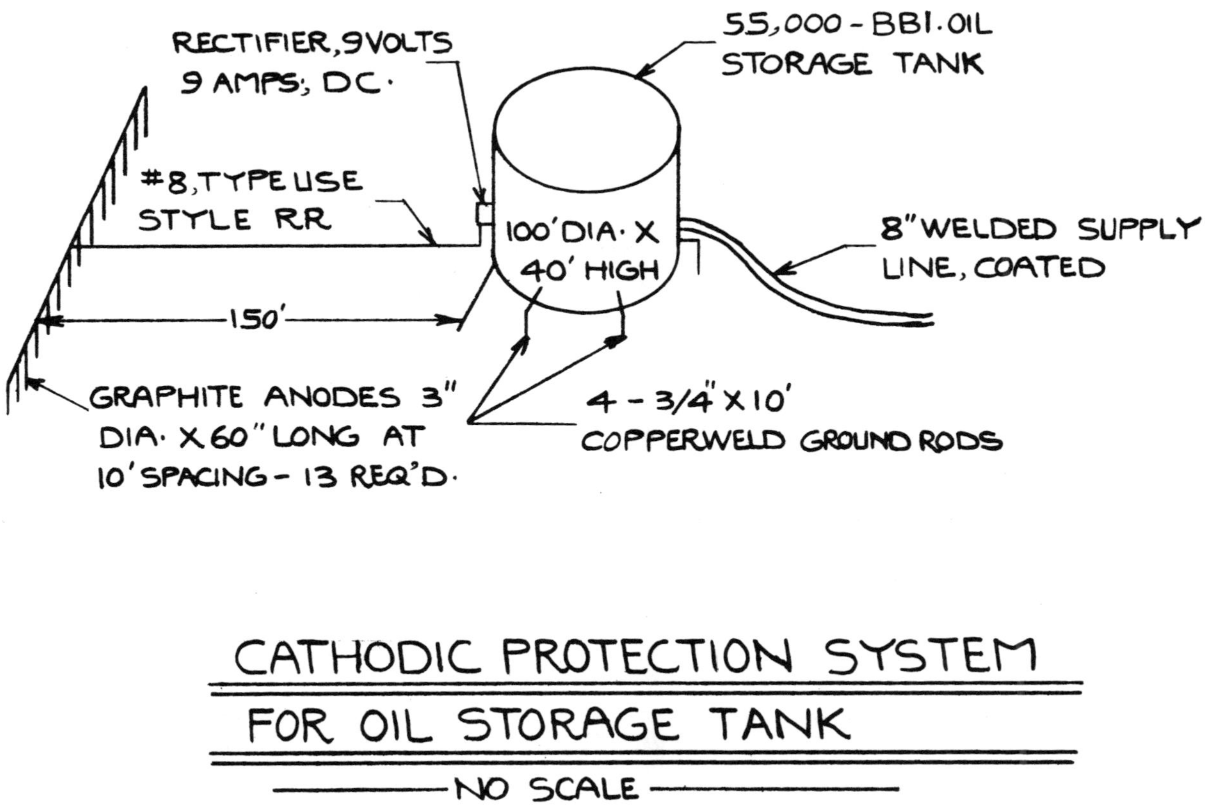
RECTIFIER, 9 VOLTS
9 AMPS; DC.
55,000 - BBl. OIL
STORAGE TANK
#8, TYPE USE
STYLE RR
100' DIA. X
40' HIGH
8" WELDED SUPPLY
LINE, COATED
150'
GRAPHITE ANODES 3"
DIA. X 60" LONG AT
10' SPACING - 13 REQ'D.
4 - 3/4" X 10'
COPPERWELD GROUND RODS
CATHODIC PROTECTION SYSTEM
FOR OIL STORAGE TANK
NO SCALE

FIGURE 18-9

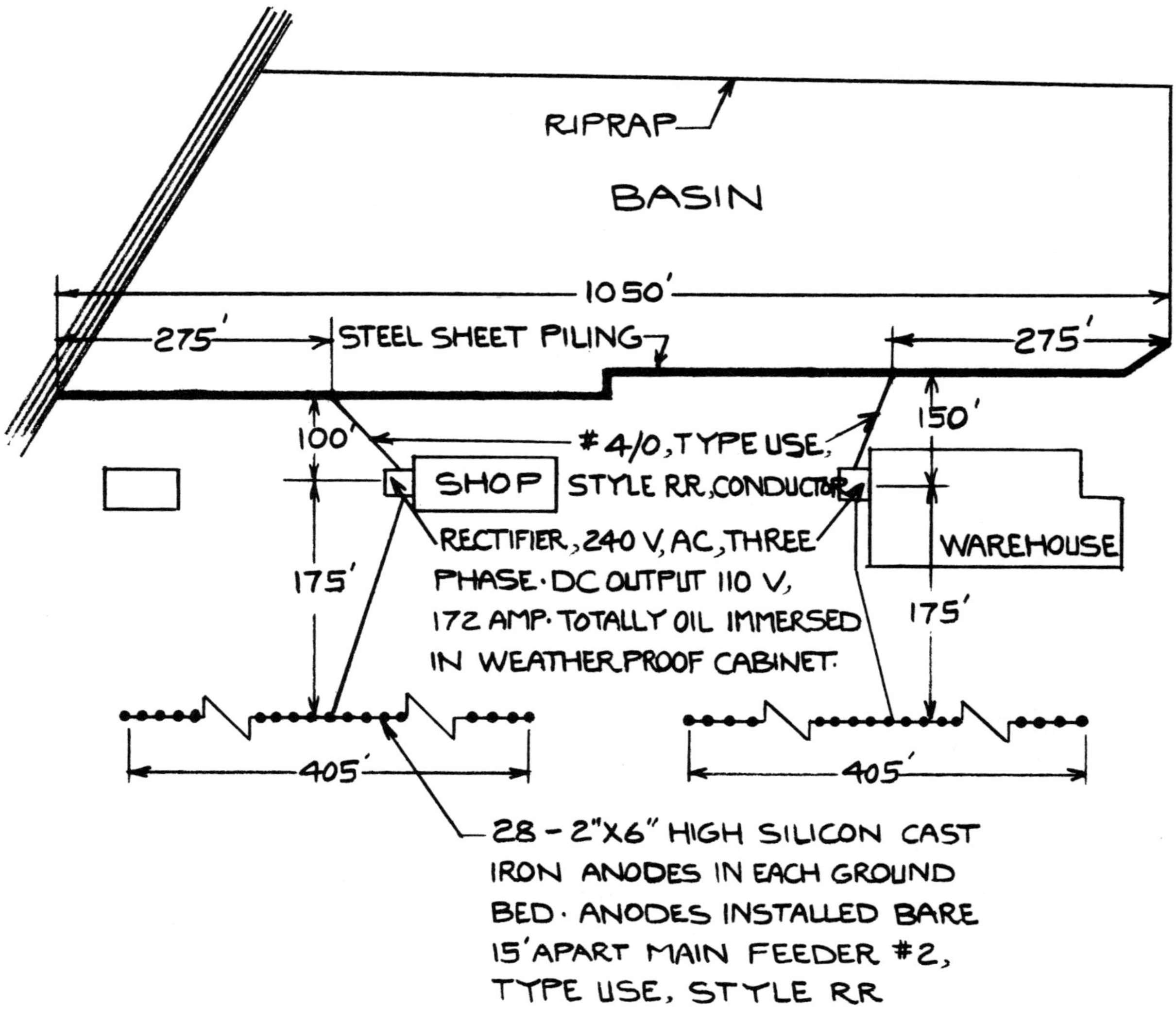

CATHODIC PROTECTION SYSTEM FOR LAND SIDE OF SHEET PILING

NO SCALE

FIGURE 18-10

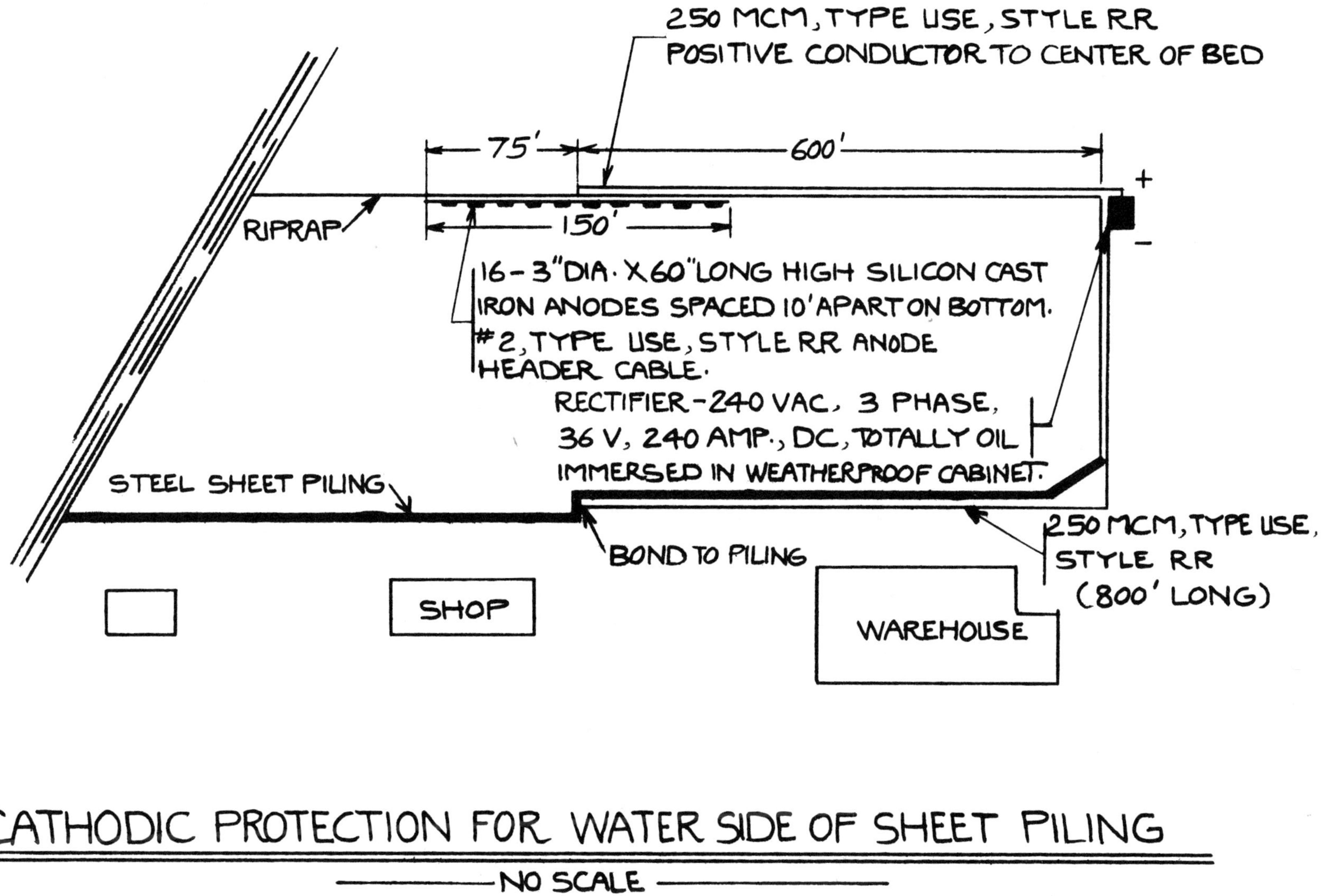
250 MCM, TYPE USE, STYLE RR
POSITIVE CONDUCTOR TO CENTER OF BED
75'
600'
+
-
RIPRAP
150'
16 - 3" DIA. X 60" LONG HIGH SILICON CAST IRON ANODES SPACED 10' APART ON BOTTOM.
#2, TYPE USE, STYLE RR ANODE HEADER CABLE.
RECTIFIER - 240 VAC, 3 PHASE, 36 V, 240 AMP., DC, TOTALLY OIL IMMERSED IN WEATHERPROOF CABINET.
STEEL SHEET PILING
BOND TO PILING
250 MCM, TYPE USE, STYLE RR (800' LONG)
SHOP
WAREHOUSE
CATHODIC PROTECTION FOR WATER SIDE OF SHEET PILING
NO SCALE

FIGURE 18-11

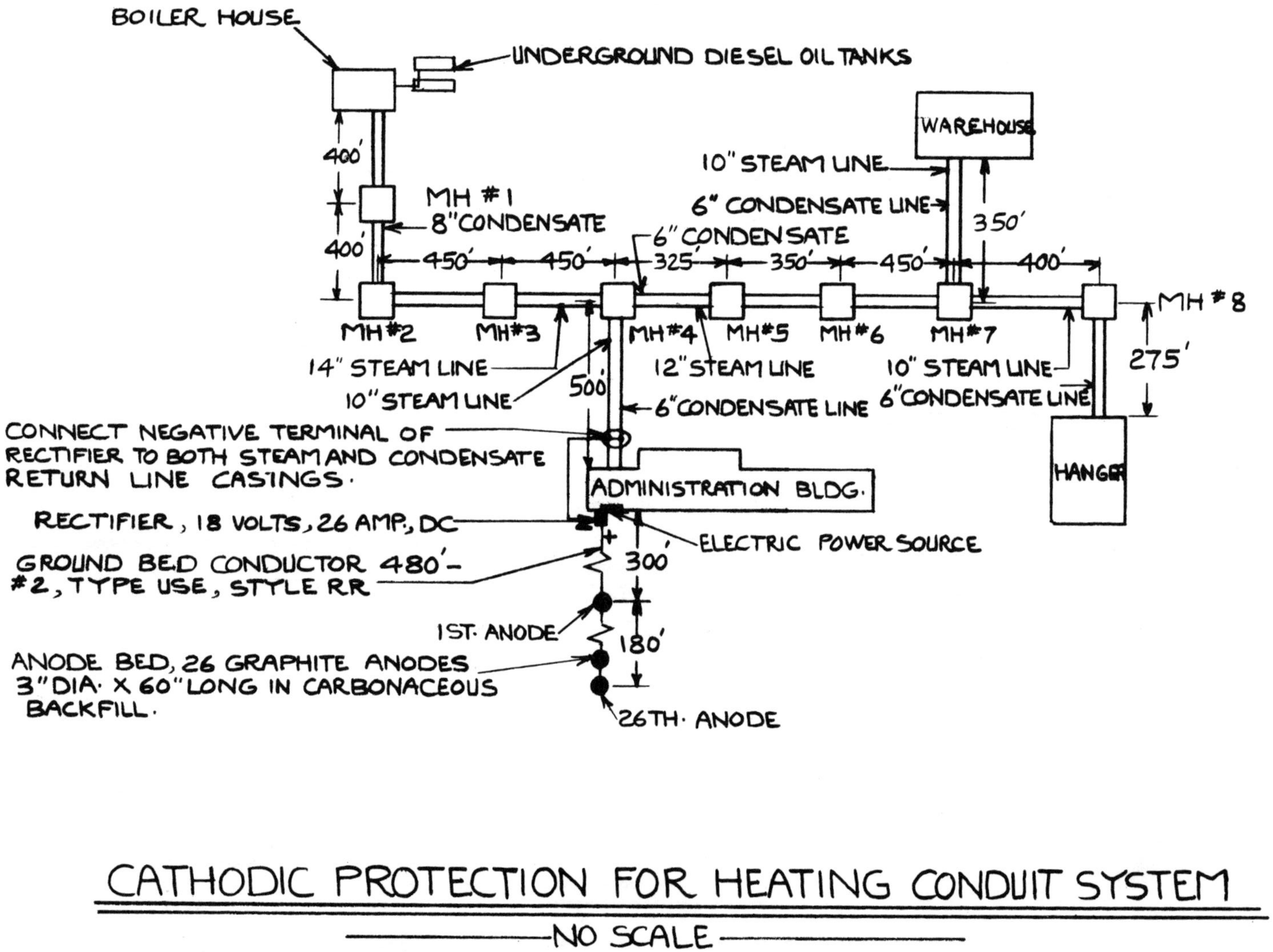
BOILER HOUSE
UNDERGROUND DIESEL OIL TANKS
WAREHOUSE
10" STEAM LINE
6" CONDENSATE LINE
350'
400'
MH #1
8" CONDENSATE
400'
6" CONDENSATE
450'
450'
325'
350'
450'
400'
MH #8
MH #2
MH #3
MH #4
MH #5
MH #6
MH #7
14" STEAM LINE
10" STEAM LINE
500'
12" STEAM LINE
6" CONDENSATE LINE
10" STEAM LINE
6" CONDENSATE LINE
275'
CONNECT NEGATIVE TERMINAL OF RECTIFIER TO BOTH STEAM AND CONDENSATE RETURN LINE CASINGS.
ADMINISTRATION BLDG.
HANGER
RECTIFIER, 18 VOLTS, 26 AMP., DC
ELECTRIC POWER SOURCE
GROUND BED CONDUCTOR 480'-#2, TYPE USE, STYLE RR
300'
1ST. ANODE
180'
ANODE BED, 26 GRAPHITE ANODES 3" DIA. X 60" LONG IN CARBONACEOUS BACKFILL.
26TH. ANODE
CATHODIC PROTECTION FOR HEATING CONDUIT SYSTEM
NO SCALE

FIGURE 18-12

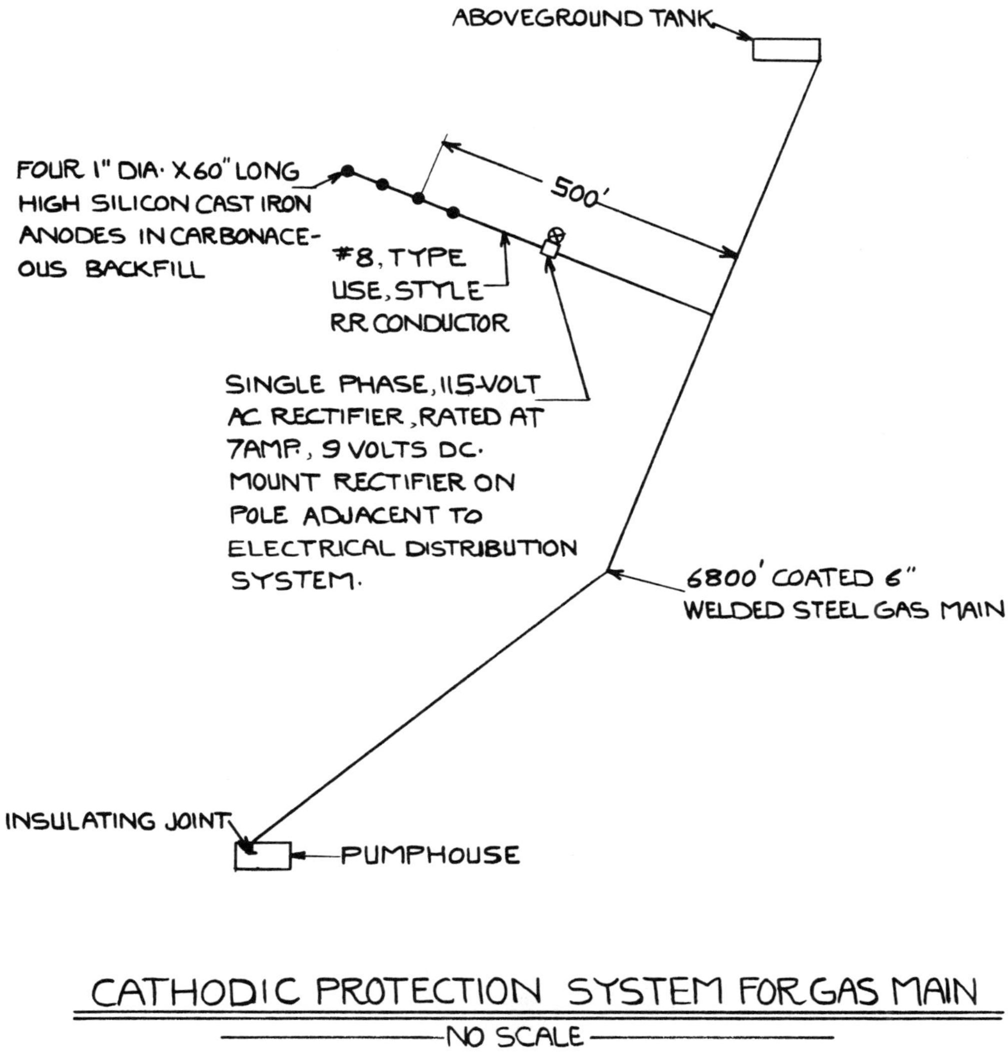
ABOVEGROUND TANK
FOUR 1" DIA. X 60" LONG HIGH SILICON CAST IRON ANODES IN CARBONACEOUS BACKFILL
500'
#8, TYPE USE, STYLE RR CONDUCTOR
SINGLE PHASE, 115-VOLT AC RECTIFIER, RATED AT 7AMP., 9 VOLTS DC. MOUNT RECTIFIER ON POLE ADJACENT TO ELECTRICAL DISTRIBUTION SYSTEM.
6800' COATED 6" WELDED STEEL GAS MAIN
INSULATING JOINT
PUMPHOUSE
CATHODIC PROTECTION SYSTEM FOR GAS MAIN
NO SCALE

FIGURE 18-13

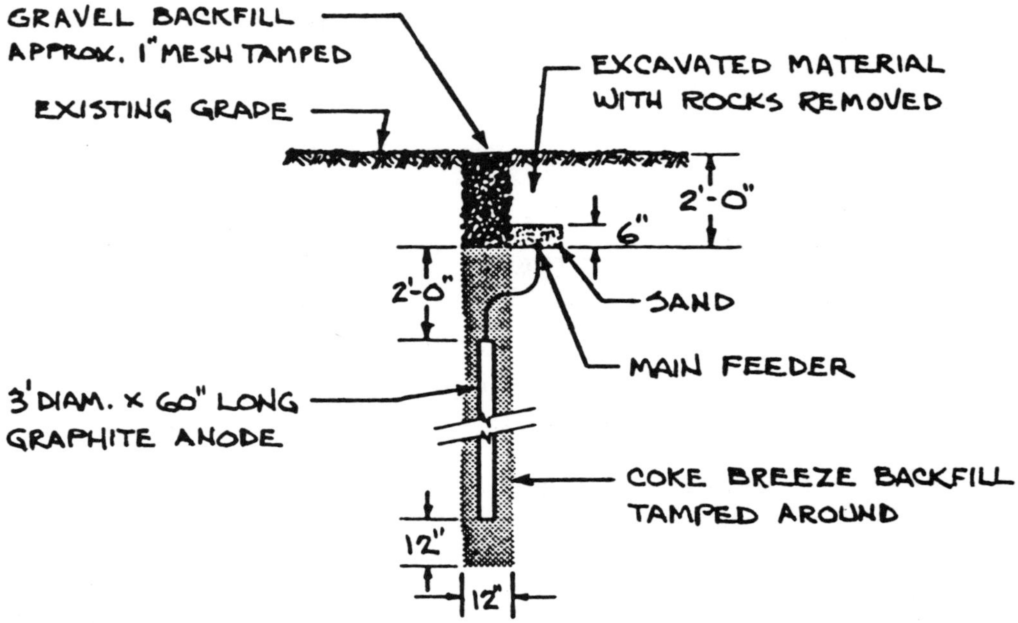

VERTICAL GRAPHITE ANODE

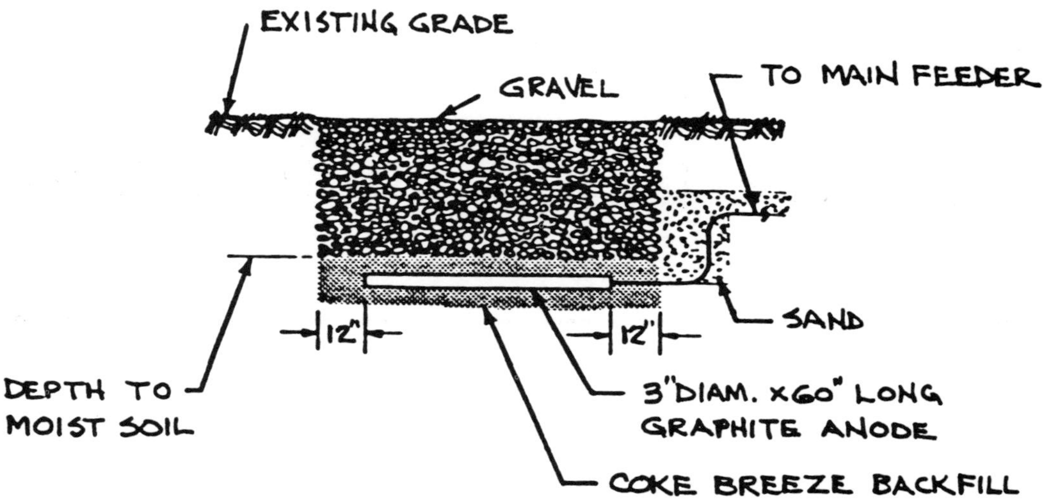

HORIZONTAL GRAPITE ANODE

TYPICAL GRAPHITE ANODE INSTALLATIONS
FOR CATHODIC PROTECTION
— NO SCALE —

FIGURE 18-14

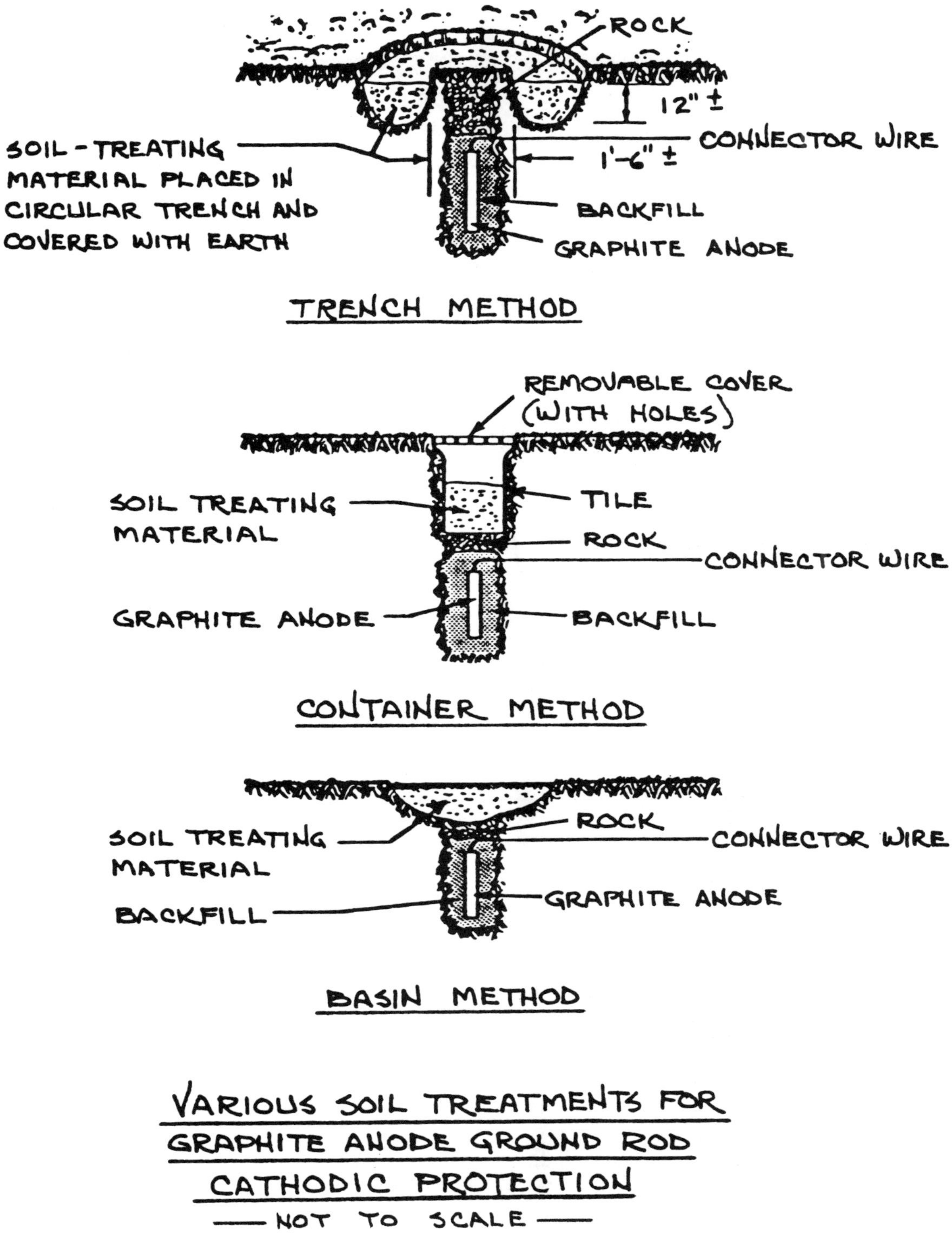

FIGURE 18-15

Anode Installation There are numerous arrangements for anode installation depending on the particular application. The arrangements vary somewhat, but the basic concepts are fairly similar. Figures 18-14 to 18-19 illustrate some of the more common types of installation.

Figures 18-14 to 18-16 delineate carbon, graphite, and magnesium anodes which slowly disintegrate at a rate related to the surrounding environment and the required current density. In general, graphite anodes are economical where long life expectancy is a requirement. Carbon anodes are generally used in *temporary* situations. The preferred installation method is vertical, although properly

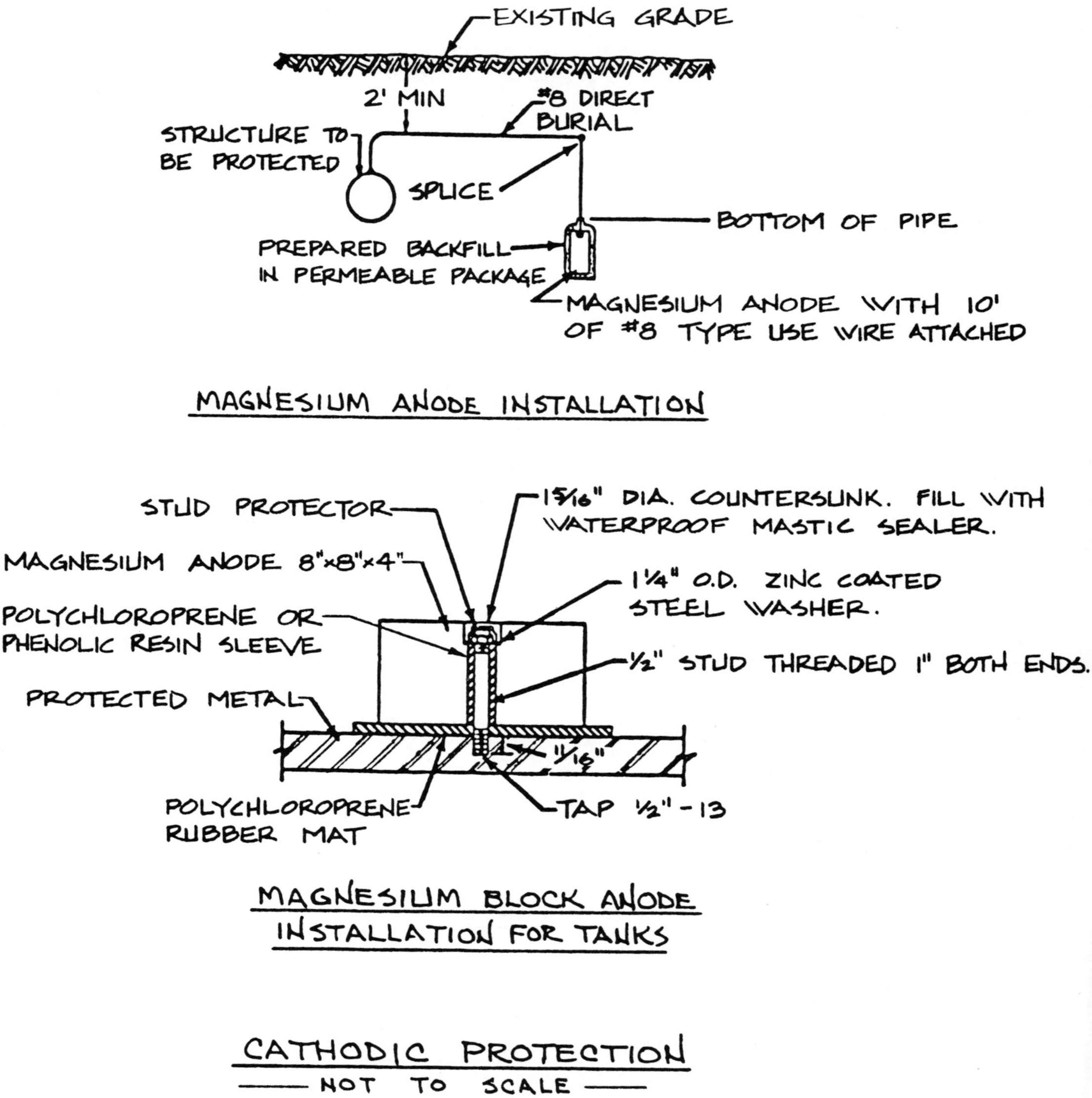

FIGURE 18-16

installed horizontal anodes can be utilized. A moist soil bed is essential. As can be seen, the installation is fairly obvious. The treatment of the soil as shown may be necessary to reduce the ground-bed resistance to undisturbed earth. Treatments may be calcium chloride or magnesium sulfate. The magnesium anode solves a lot of soil problems when furnished complete with its own prepared backfill. In that event your only problem is to be certain of soil conditions. As can be seen, this anode can be used in tanks where rods are not practical. It can be mounted on the surface of the tank separated only by an insulating strip.

Figures 18-17 and 18-18 cover one of the common

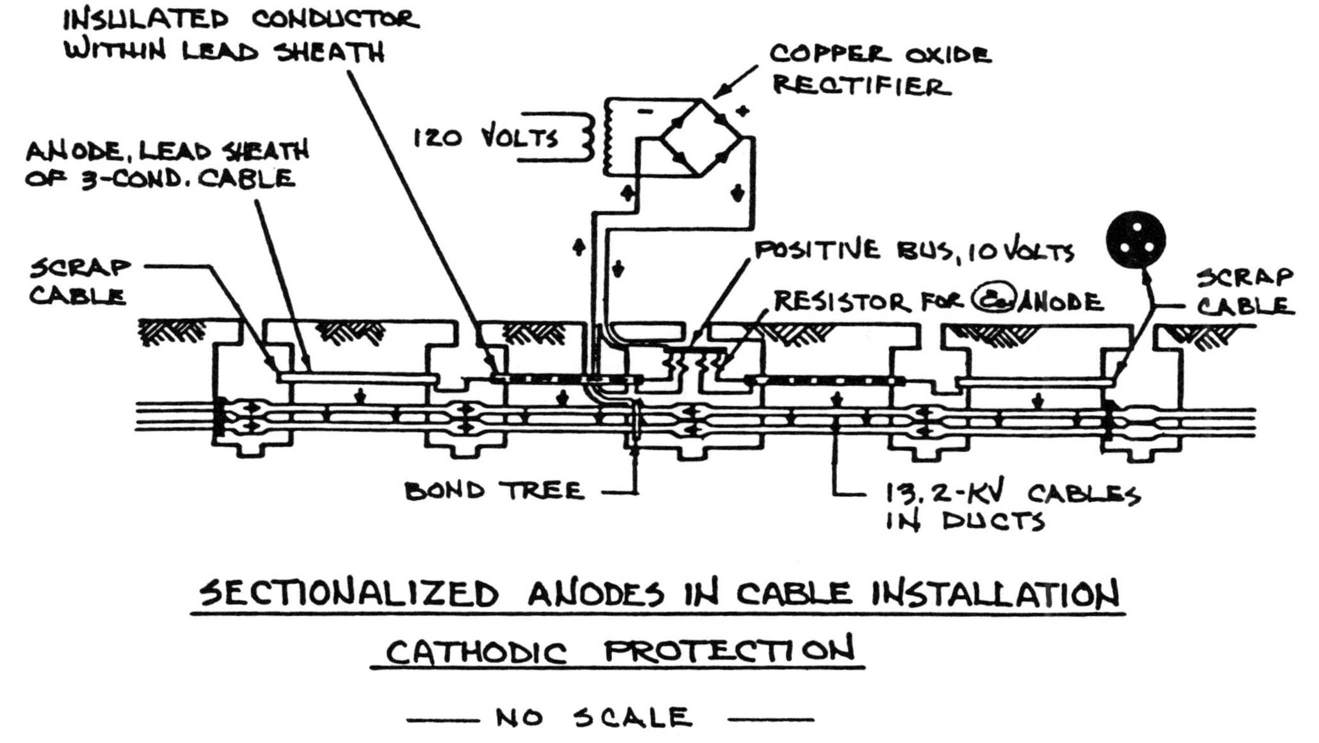

FIGURE 18-17

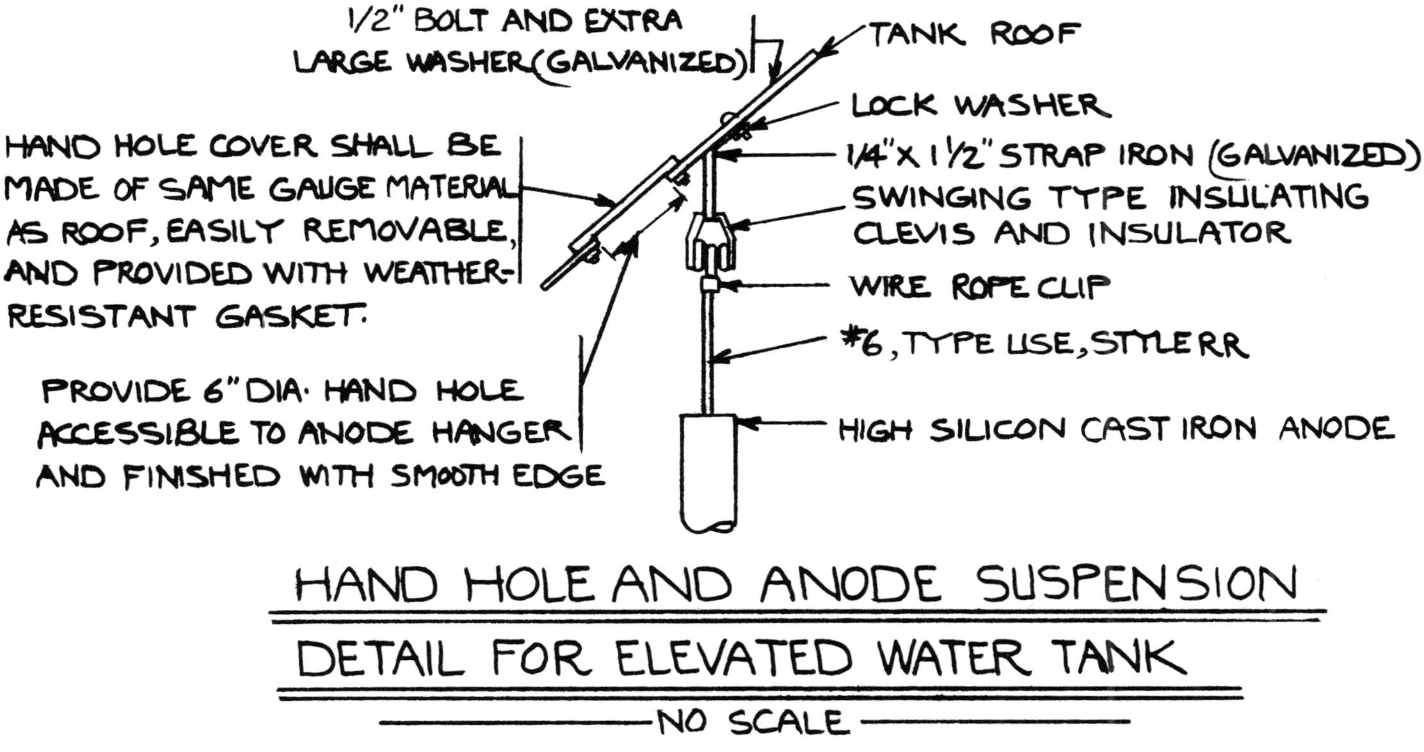

FIGURE 18-18

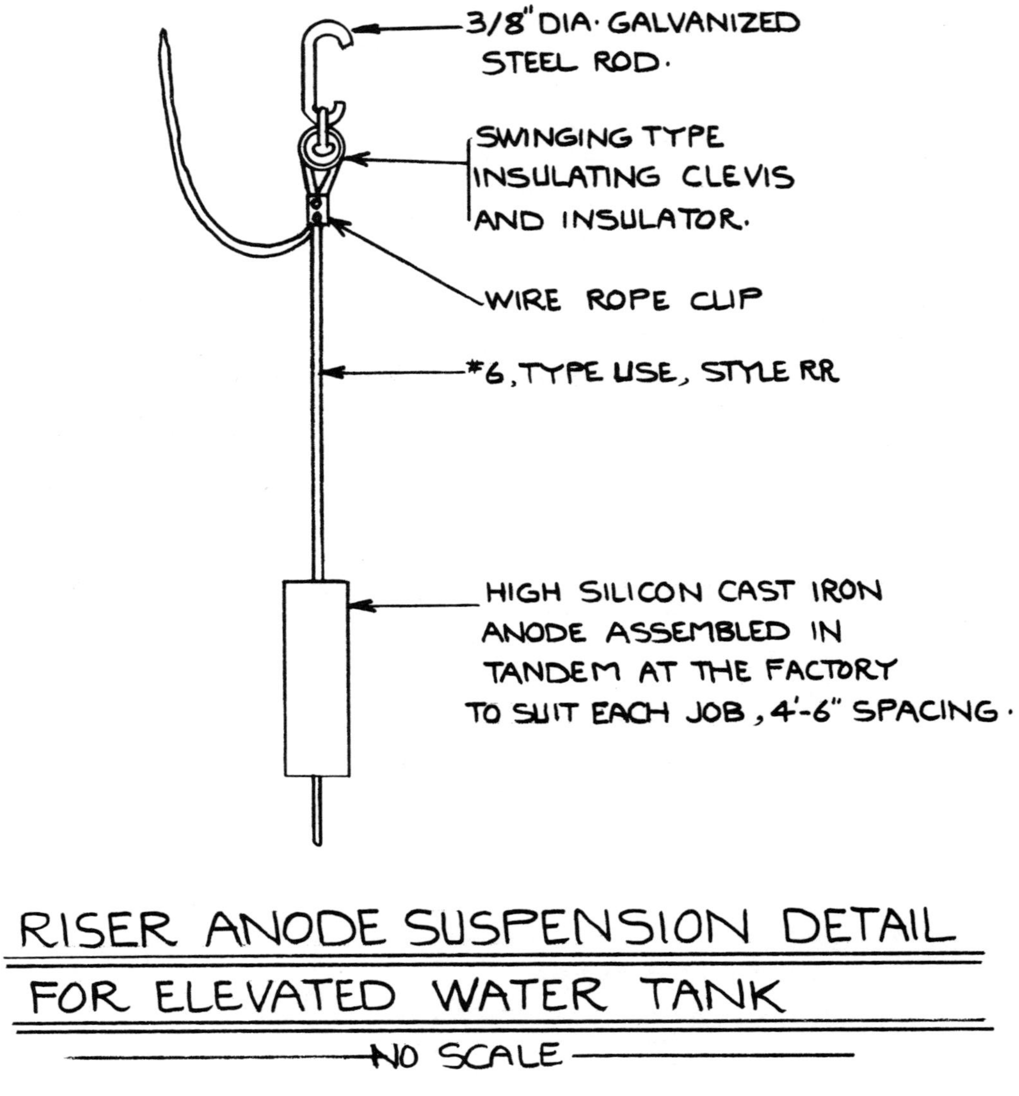

FIGURE 18-19

installations, the cathodic protection of an elevated water tank. The two details are intended, without further explanation, to illustrate clearly how the anodes are suspended in the tank. Note that the anodes are of a different material than all of the others we have shown. They are made of high-silicon cast iron.

Figure 18-19 illustrates cable protection using available materials and a little ingenuity. It shows a portion of the installation of sectionalized anodes with the proper amount of current supplied by control resistors. For the continuous anode section, the lead sheath of an abandoned lead-covered cable in an empty duct not depicted was used. An aluminum cable could have been used. Do not use a bare copper neutral or grounding conductor in a duct bank with lead-covered cables. Instead, use an insulated conductor.

Rectifier Installation A rectifier is an ac-dc converter, and this installation should be detailed similarly to any pad- or pole-mounted transformer. In Figure 18-20 more of the wiring connections are shown for the smaller pole-mounted unit. It should be understood that the wiring connections and associated switches, meters, conduit, and wire apply in all cases.

The smaller pole-mounted rectifier is presumed to be hung on a normal utility-type wooden pole. The unit is contained in a weatherproof housing. This unit is for a small, local cathodic treatment installation.

As the cathodic protection system grows, the rectifier, primarily because of its size rather than weight, becomes an object to be ground-mounted, as shown in the lower part of the detail. It may be installed with or without a fence. Since the ac switch and possible meter have to be

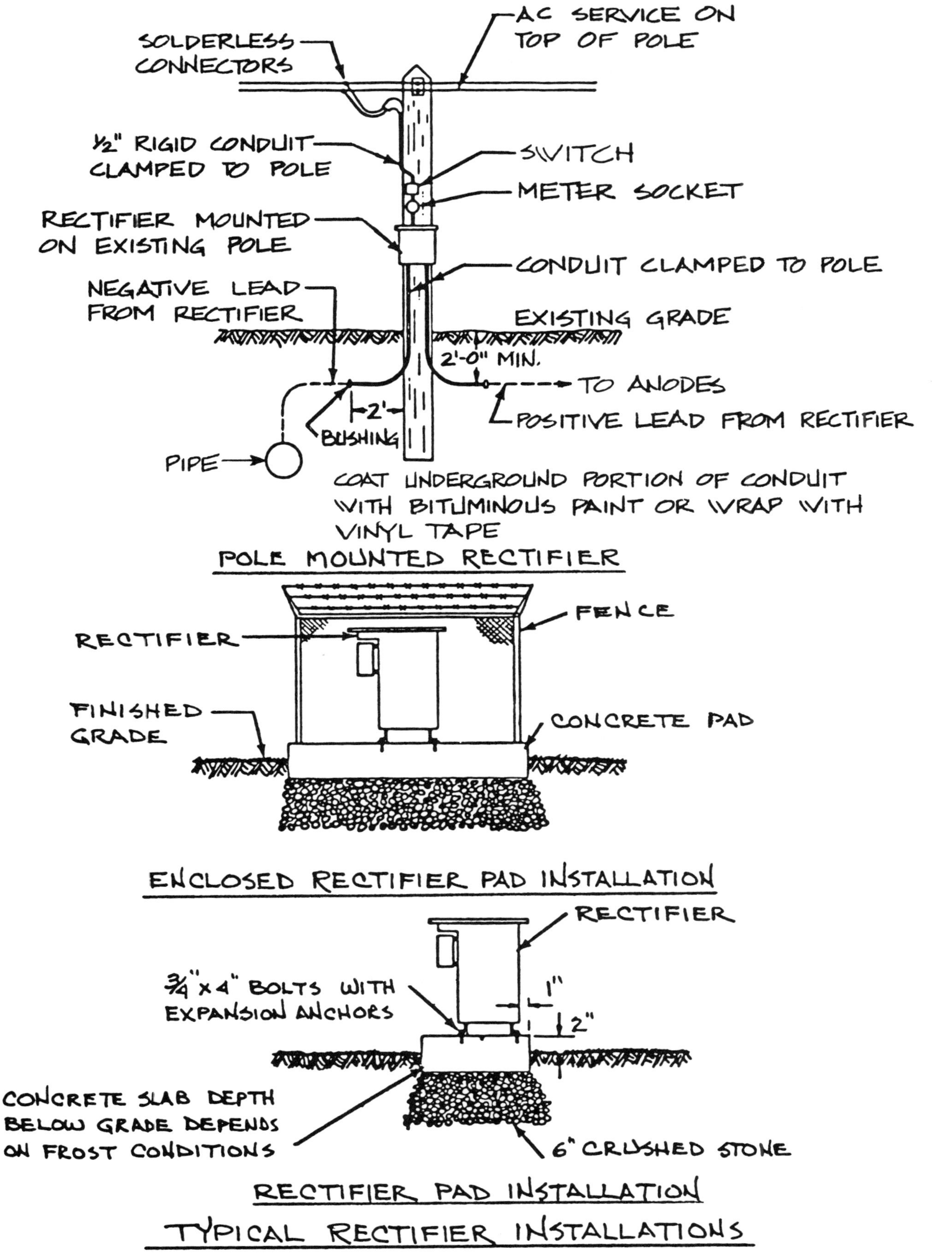

SOLDERLESS CONNECTORS
AC SERVICE ON TOP OF POLE
½" RIGID CONDUIT CLAMPED TO POLE
SWITCH
METER SOCKET
RECTIFIER MOUNTED ON EXISTING POLE
CONDUIT CLAMPED TO POLE
NEGATIVE LEAD FROM RECTIFIER
EXISTING GRADE
2'-0" MIN.
TO ANODES
POSITIVE LEAD FROM RECTIFIER
2'
BUSHING
PIPE
COAT UNDERGROUND PORTION OF CONDUIT WITH BITUMINOUS PAINT OR WRAP WITH VINYL TAPE
POLE MOUNTED RECTIFIER
RECTIFIER
FENCE
FINISHED GRADE
CONCRETE PAD
ENCLOSED RECTIFIER PAD INSTALLATION
RECTIFIER
¾" x 4" BOLTS WITH EXPANSION ANCHORS
1"
2"
CONCRETE SLAB DEPTH BELOW GRADE DEPENDS ON FROST CONDITIONS
6" CRUSHED STONE
RECTIFIER PAD INSTALLATION
TYPICAL RECTIFIER INSTALLATIONS

FIGURE 18-20

installed somewhere, the obvious location for these items is on a fence, a wall, or a building. Presuming the ac supply is underground, a sufficiently large conduit in a concrete base could support the switch. Depending on your application, the rectifier can be located indoors or outdoors. Where hazardous locations are involved, explosion-proof enclosures will be required.

Miter Gate Cathodic protection can be used in a number of atypical situations. To illustrate, Figure 18-21 shows the protection of a hydraulic structure such as a lock or dam gate. Our example illustrates one of the miter gate leaves, which measures some 62 × 33 ft in overall dimensions.

At normal water levels, portions of each gate are above and below the water. In high-water conditions everything may be submerged. The cathodic system controls surface corrosion. Generally, the anodes are silicon cast iron. Although the item is large, it is not, by cathodic protection standards, very unusual. Figure 18-21 is taken from an actual situation but should only be used as a guide. In our calculations, each gate required a standard commercial rectifier rated at 9 V and 27 A. The calculated values were 6.6 V and 18.3 A.

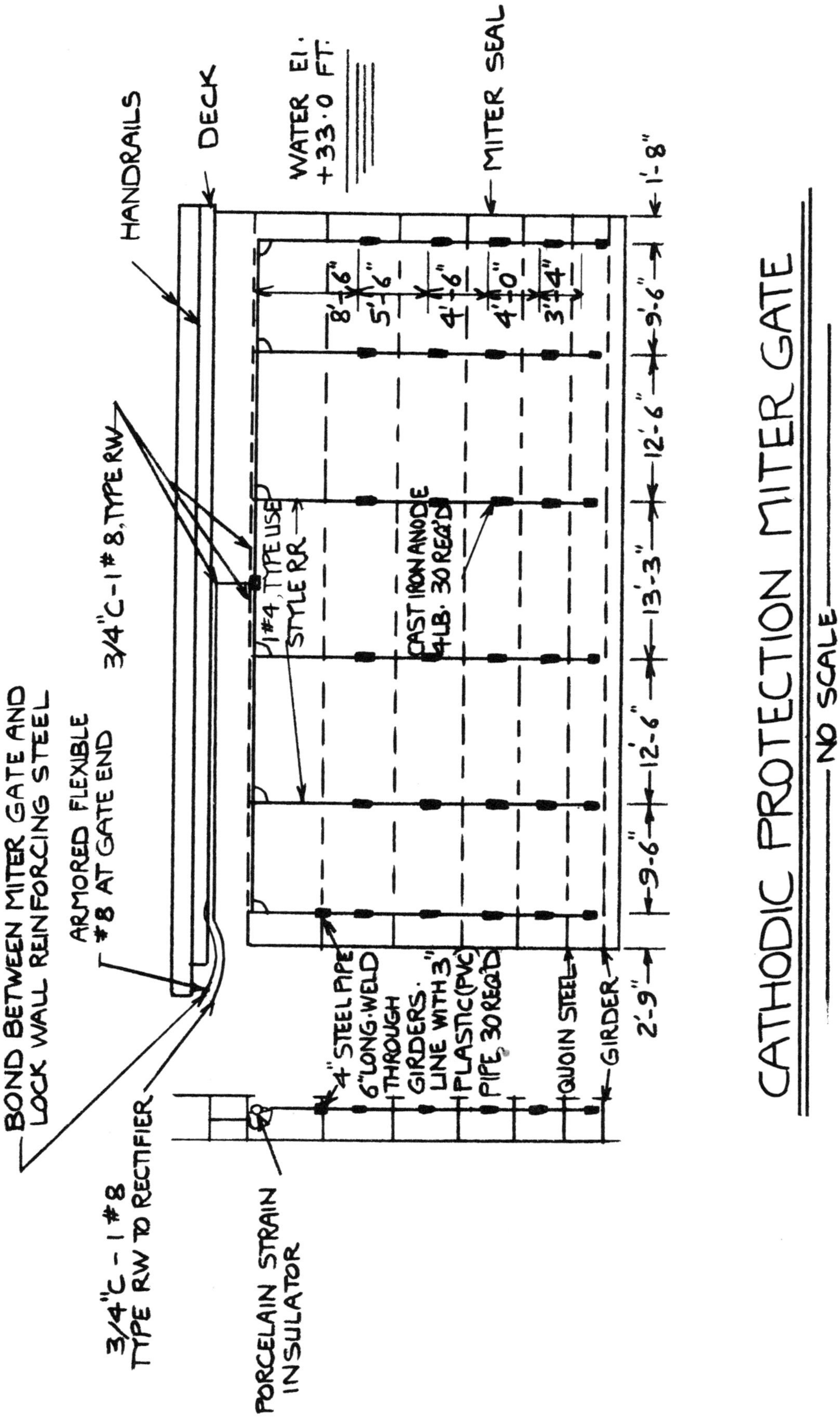

BOND BETWEEN MITER GATE AND LOCK WALL REINFORCING STEEL
ARMORED FLEXIBLE #8 AT GATE END
3/4"C-1 #8, TYPE RW
HANDRAILS
DECK
WATER El. +33.0 FT.
MITER SEAL
1 #4, TYPE USE STYLE RR
CAST IRON ANODE 4 LB. 30 REQ'D
8'-6"
5'-6"
4'-6"
4'-0"
3'-4"
4" STEEL PIPE 6" LONG-WELD THROUGH GIRDERS. LINE WITH 3" PLASTIC (PVC) PIPE, 30 REQ'D
QUOIN STEEL
GIRDER
2'-9"
9'-6"
12'-6"
13'-3"
12'-6"
9'-6"
1'-8"
3/4"C - 1 #8 TYPE RW TO RECTIFIER
PORCELAIN STRAIN INSULATOR
CATHODIC PROTECTION MITER GATE
NO SCALE

FIGURE 18-21

Typical Details Electrical connections are important in the installation of a cathodic protection system. A low-resistance bond between the conductor and the structure being protected is important. Figure 18-22 is a typical example.

The conductor can be either brazed or welded to the structure. Where brazing or welding is considered unsafe, a mechanical connection can be used. It is essential that any bare metal at the bond be well insulated.

In your application any unusual connection should be detailed.

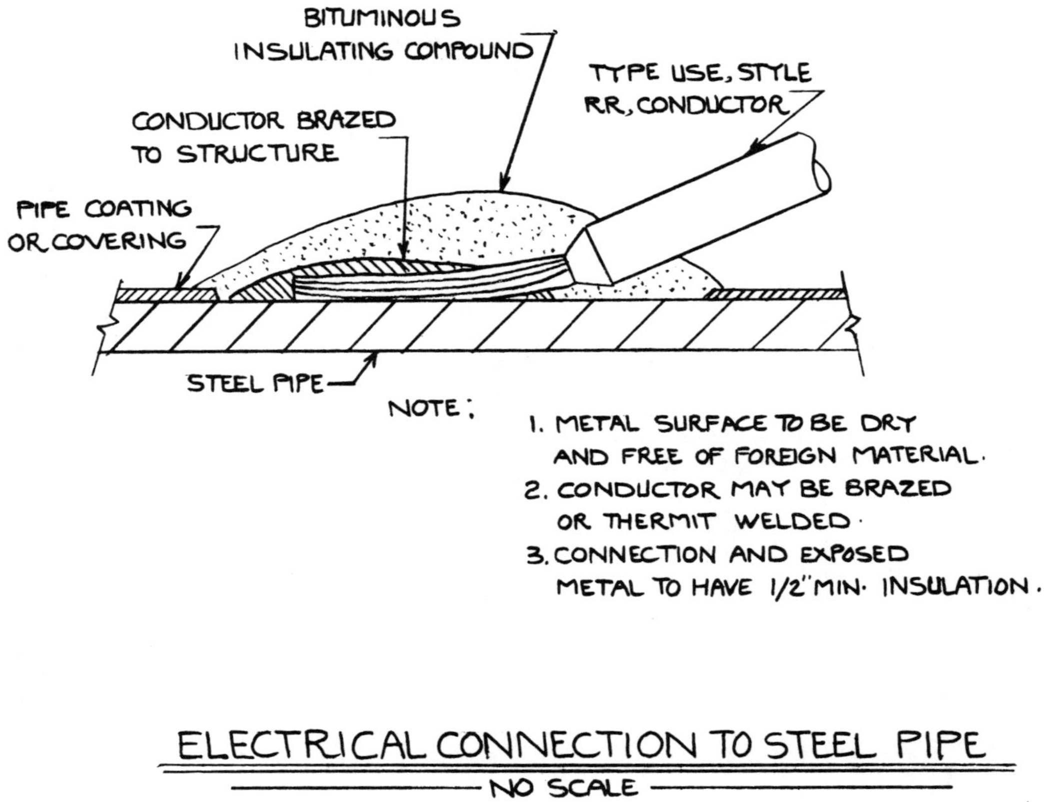

FIGURE 18-22

NINETEEN

Explosionproof Systems

Many engineering firms may practice for years with little or no exposure to explosionproof installation problems. Regardless of the amount of work in this category to which you and your firm are exposed, it is a serious, complex subject. One measure of how complex the subject is can be made by measuring the amount of words and space devoted to explosionproof systems in the NEC and related Code handbooks. About 10 percent of all the material in the Code refers to this subject, beginning with Article 500 and going through Article 517. Considering that the Code starts with Article 90 and goes through Article 820, that may not seem like a large part but reading the Code material quickly changes that opinion. Another simple test is to price an ordinary single-pole switch compared to the same device designed to meet Class I explosionproof conditions. The cost differential is sometimes shocking.

The Code and related handbooks describe all the separate and special provisions covering each explosionproof (or more properly described, hazardous) location better than we can here. But in brief there are three categories with a number of modifications to each.

1. Class I locations, which are those where flammable gases or vapors are or may be present in the air in sufficient quantity to produce an explosive or ignitable mixture.
2. Class II locations, which are considered hazardous because of the presence of combustible duct.
3. Class III locations. which are hazardous because of the presence of easily ignitable fibers or flyings, but where such fibers or flyings are not likely to be in suspension in the air in quantities sufficient to produce ignitable mixtures.

In keeping with the approach used throughout this book, we are avoiding becoming an engineering text. Our objective is to provide helpful information to the people drawing the details.

The above brief excerpted description is from the NEC and is by no means even a part of the Code. Our intention is to alert the detailer who is presumed to be aware of the location of his detail. The above three descriptions form a sort of simplified check procedure for the detailer. If your detail is for a device located in any situation similar to the above description, your standard *nonexplosionproof* details are not applicable or correct. A mistake here can be very expensive for you and your firm. If you have even the slightest doubt, the time to stop and question your superiors is before you start drawing.

An explosionproof device is essentially one that will not trigger an explosion and, if something else does, will survive the resultant explosion. In general the devices (switches, starters, lights, etc.) are basically the same. It is their enclosure which is radically different. Of all the devices frequently omitted or misapplied, the conduit-sealing fitting would be at the top of the list. Our details show seals where they should be installed relative to our details but there are always other locations. Beginning with Code Article 501-5 entitled "Sealing and Drainage," we strongly urge that you read not only the Code but also one of the several explanatory handbooks available from McGraw-Hill or the NFPA.

Certain hazardous situations can be modified or diminished by proper ventilation and drainage. This requires interrelation of mechanical details with electrical details and cross checking of drawings by your project engineer, chief of design, or other designated personnel. Finally, enclosing walls, floors, and ceilings are also frequently a factor to be investigated as they too may modify a particular situation.

In going through the NEC as we have advised, you will

find special situations covered beginning with Article 510 and going through Article 517. Very frequently these are areas where mistakes in detail depiction commonly occur. Because of the many special situations in the typical hospital that require special treatment, we strongly urge that each area of hospital design requiring a detail be specifically discussed with your project engineer or other designated supervisor. No book of details can totally and completely cover all the special situations that may arise. In addition to the NEC, there are local and state fire codes, special hospital construction codes, and other requirements that apply.

It may seem that this portion of our book contains little more than a series of warnings and cautions. To a degree that is correct. Generally your mistake, if any, will be discovered and corrected (sometimes at considerable expense) before it is installed. But where human lives rather than equipment failures are involved, we would prefer to err on the side of safety.

Finally, we have alluded to mistakes being expensive, which may not have impressed you. Following is one example from the author's professional experience. In 1978, when prices were lower, a green ground wire that was required to run from the fifth-floor operating area electrical devices back to a main ground was forgotten. It was discovered at the final punch-list inspection. A mere $11,000 got that one small wire properly installed. The design professional was left with a limited number of choices on how to handle and process this extra.

Power and Lighting Suppose you were asked to put a little schematic diagram on the plans showing a panel with a few lighting circuits and two or three motors and to indicate a typical lighting and motor circuit installation. Showing a rectangle for a panel and three typical lines—one to a motor, starter, and disconnect; one to a typical convenience outlet; and one to a typical lighting fixture—would certainly be a simple task. You probably could sketch the detail as fast as these words could be handwritten.

But, if this detail were to apply to a Class I, Division 1 or Class II, Division 1 explosionproof situation, it could suddenly become a very different task. Figures 19-1 to 19-3 illustrate these two conditions. Admittedly we have gone somewhat overboard in the extent of the detail. A conduit run, for example, could be a single line. A lot of the other devices could have been simplified and indicated by a symbol. Nothing would be wrong with such a presentation as long as you are aware of just how much space these devices require. They are considerably larger than an equal nonexplosionproof device.

To further emphasize, let us take the simple problem of a smoker who drives up to the pump at a neighborhood gasoline station. The attendant filling the tank may or may not tell the driver to put out his or her cigarette. Are you aware that the area within a 20-ft circle of that pump and up to 18 in above ground level is within the Class I Division 2 explosionproof category according to NEC 514-2 parts (b) and (d)? Frequently, to comply with the attendant's request the smoker walks 5 ft from the pump and flips the lighted cigarette on the ground. Either the Code is supersafe or a lot of people are very lucky.

In Figures 19-1 to 19-3 the devices are enclosed in explosionproof housings and the situations seem very similar to each other. Note the locations of all the seals on the various conduit runs. The major difference between Class I, Division 1 and Class I, Division 2 lies in the special requirements for panelboards and junction boxes. In Class II there are many types of cast junction boxes that would be acceptable. In Class II the junction box is a specially constructed explosionproof device.

The Class II panelboard is a fairly normal board enclosed in a bolted explosionproof housing. For Class I every breaker has its own specially designed explosionproof housing. The Class I panelboard resembles a device that might be found on a spaceship to Mars.

We hope we have made several points with these details. Explosionproof devices, except for seals and drains, are not very unusual in detail. The important point is to know where explosionproofing is required. Except in hospitals, other added isolation is generally not required. Since the explosionproof device is special, large, and expensive, so is the contractor's charge when you forget to include explosionproofing when and where it is required.

Hospital Systems The various components of the electrical design of a hospital system have been covered in other sections of this book. The primary problem here is to select properly which of these systems and devices are required to be explosionproof. Since we are not presenting any specific details in this area, we felt it would be helpful to outline some of the major areas that you should consider and further investigate relative to the explosionproof requirements.

As a generalized statement, an explosionproof device is required in a hospital in any location where it is mandated to protect against explosions or fires, electric shock, or mechanical injury from compressed gases or compressed-gas cylinders. This statement is a paraphrase of the NFPA standards 456A to 6-1978. While this may conjure up visions of walls or buildings being destroyed, the actual explosion is frequently not as loud as popping a bubble-gum bubble. The "explosion" occurs in the patient's lungs.

To design properly a hospital electrical system that eliminates hazards, there are two documents that must be referred to and followed precisely. These are the latest copy of NEC Article 517, and Articles 56 A to G of volume 4 of the NFPA's National Fire Codes. Because of the importance of this area of design as well as the possible misinterpretation of the codes, it is mandatory to also have the latest copy of an NEC handbook. There are two excellent versions. One is published by the NFPA and the other by McGraw-Hill.

The following are the four major areas of these books

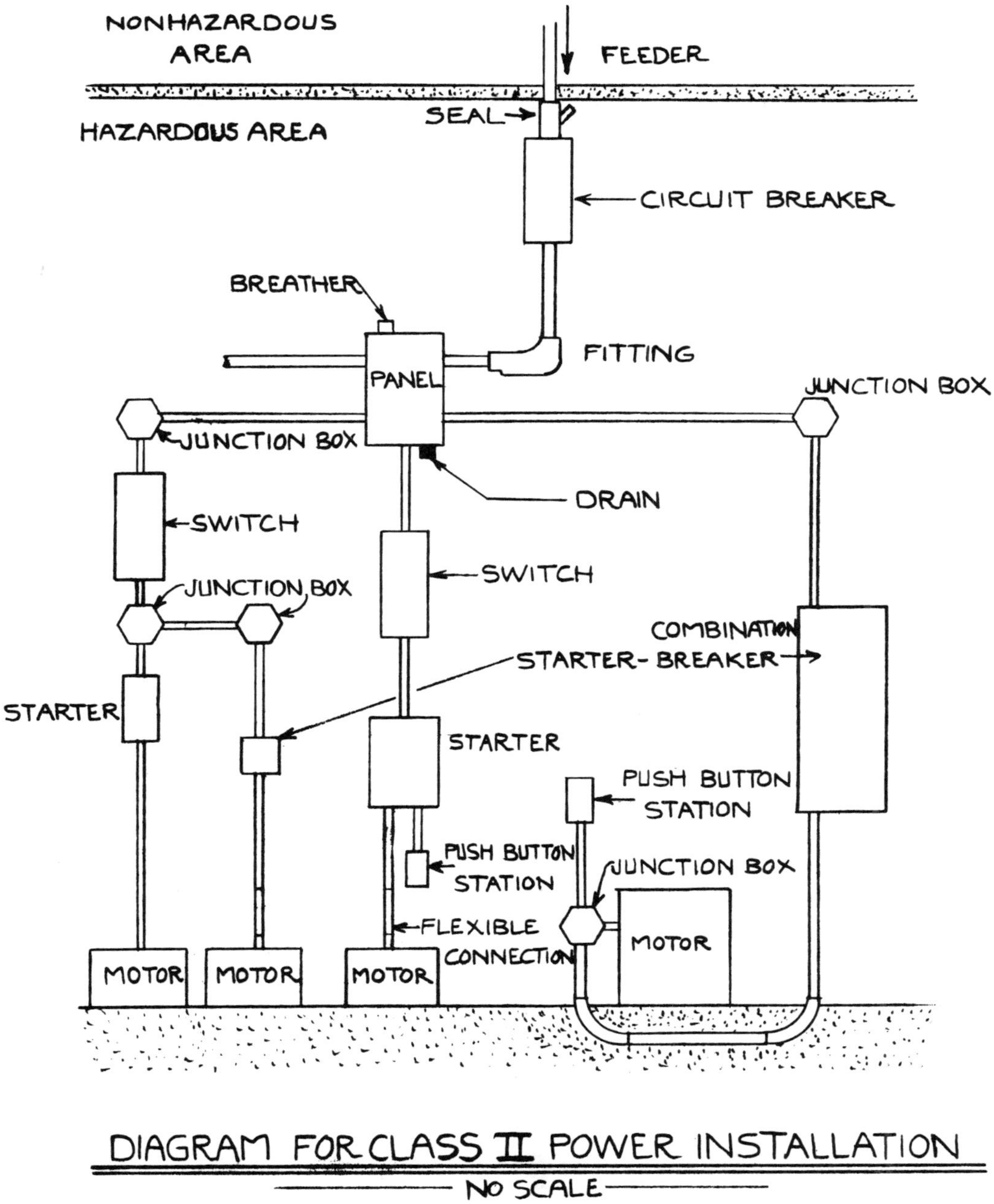
NONHAZARDOUS AREA
FEEDER
SEAL
HAZARDOUS AREA
CIRCUIT BREAKER
BREATHER
FITTING
PANEL
JUNCTION BOX
JUNCTION BOX
DRAIN
SWITCH
SWITCH
JUNCTION BOX
COMBINATION
STARTER- BREAKER
STARTER
STARTER
PUSH BUTTON STATION
PUSH BUTTON STATION
JUNCTION BOX
FLEXIBLE CONNECTION
MOTOR
MOTOR
MOTOR
MOTOR
DIAGRAM FOR CLASS II POWER INSTALLATION
NO SCALE

FIGURE 19-1

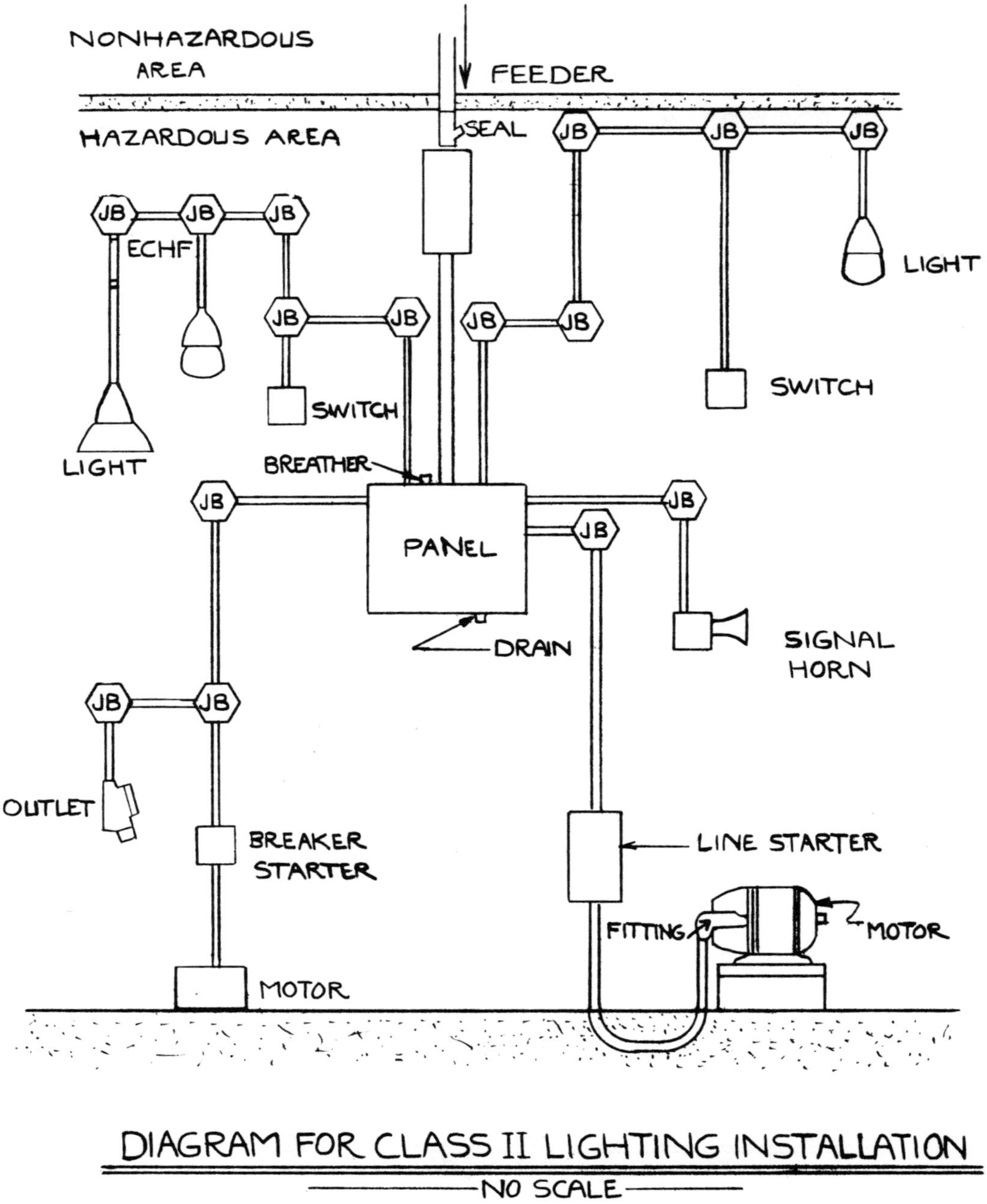

FIGURE 19-2

and codes that directly relate to explosionproof installations. Before our listing, we would like to especially note that explosionproof selection in hospitals as in other areas depends on height above the floor in many cases. Hospital gases are frequently heavier than air and their danger decreases after certain specified heights are achieved.

Ground Fault Protection: In a complete health care facility, NEC Article 517-13 requires that ground fault protection must be applied to every feeder and may be applied further downstream at the option of the health care facility. The definition of what is a health care facility has many ramifications and should be thoroughly checked.

Emergency Systems: These systems in a hospital fall under two definitions called "life safety branch." There are different terms and meanings depending on the type of hospital or health care facility. There are some 40 items in NEC beginning with Article 517-44 that affect decisions in this area.

Wiring in Hazardous Locations: Articles 517-100 through 517-105 of the NEC cover this subject, including

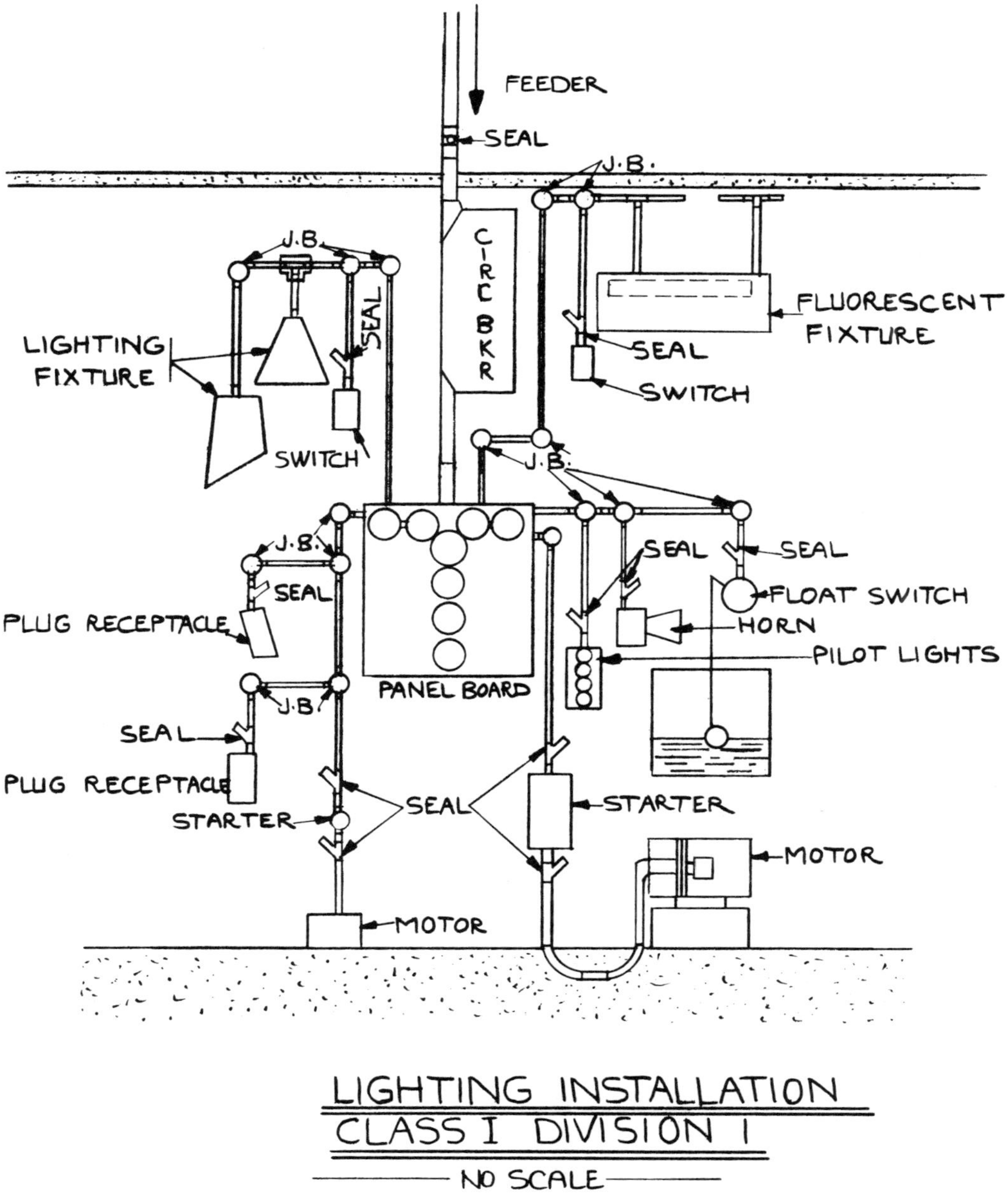

FIGURE 19-3

the frequently misunderstood point of isolated transformers and the requirements of grounded and undergrounded conductors. In general, on a room by room or specific use by use basis, isolation transformers are required to isolate electrical energy from all other areas. In addition there are requirements as explained by NFPA standards 56A to G.

Grounding: This very important item is noted, covered, and referred to throughout NEC Article 517. There are no simple answers in this area. An item-by-item investigation of Code requirements must be made. Improper use of grounding conductors is probably the most common error in hospital electrical system design.

Our discussion may appear as though we are dodging the issue. On the contrary, we are trying to highlight it. Your details are not unusual or complicated; your problem is simply to mark, correctly, which item or items in your detail must be explosionproof.

Seals and Drains To stop the flow of gases, vapors, or flames from one part of the electrical distribution system to another, the conduit runs must be sealed. This precludes

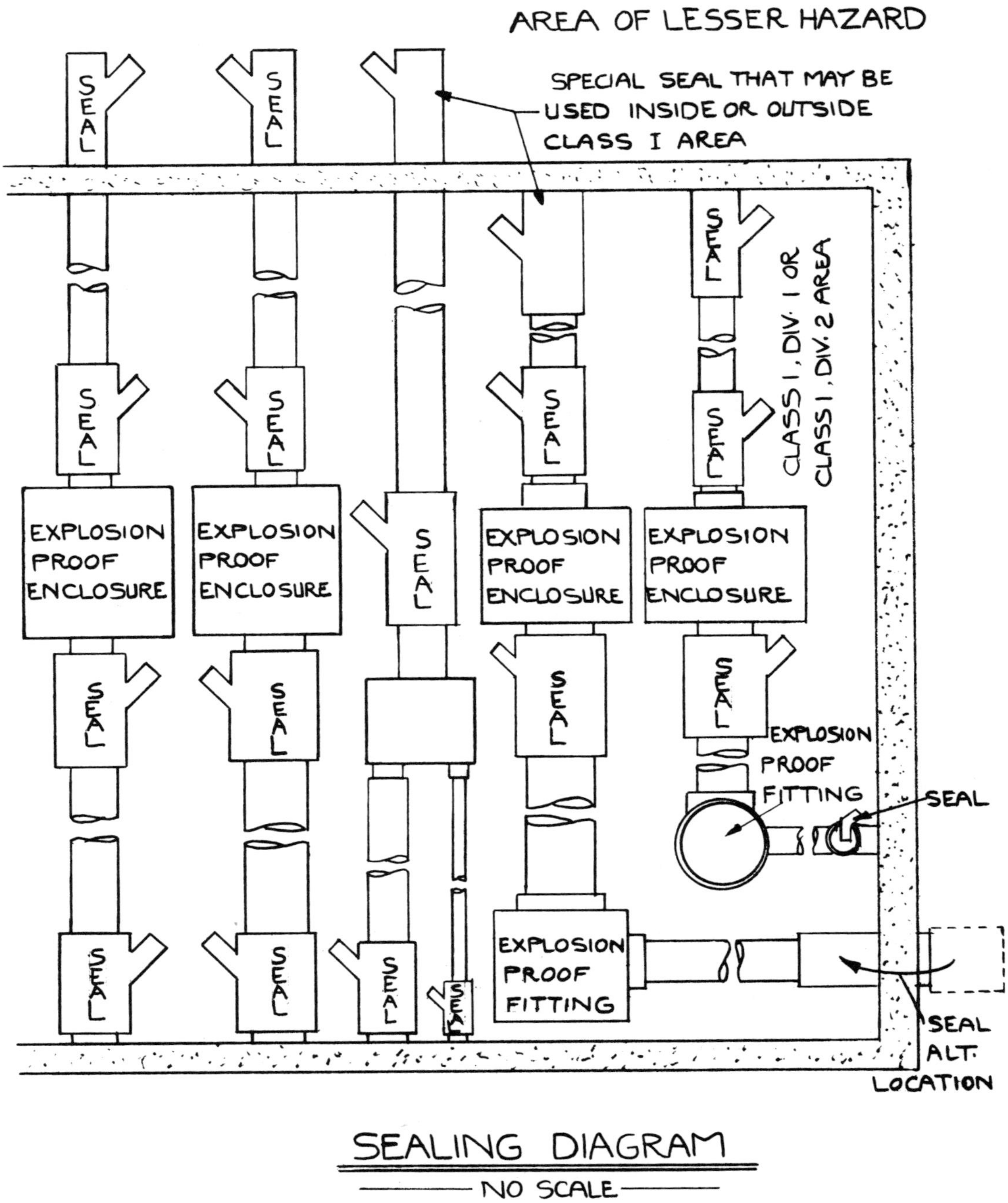

FIGURE 19-4

an occurrence within the conduit system called *pre-compression* or *pressure piling,* which is created when the gases, flames, or vapors pass with terrific speed and power from one location to another. Figures 19-4 to 19-6 show typical installation of seals and drains.

Both Division 1 and Division 2 classifications require seals in each raceway entering enclosures containing arcing, sparking, or high-temperature devices and within 18 in of said enclosure. They are also required in each raceway *leaving* a hazardous area. Division 1 requires seals in conduit runs 2 in and larger that house terminals, splices, or taps, again within 18 in. In Class II Section 502-5 of the Code, certain areas are defined where seals are required.

Where it is probable that water can get into the interiors of enclosures or raceways, the raceways should be pitched

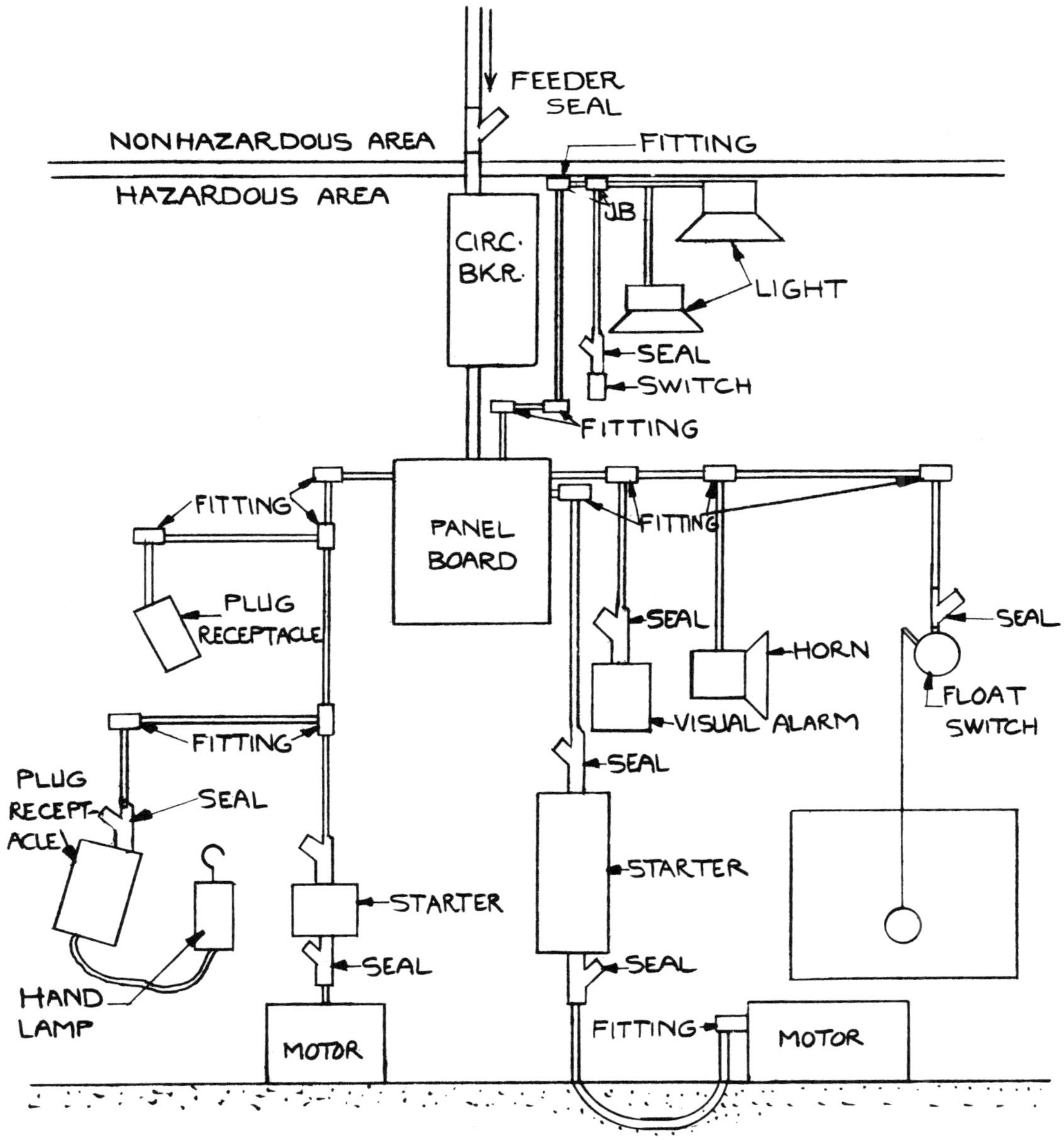

POWER AND LIGHTING DIAGRAM - CLASS I DIVISION 2

NO SCALE

FIGURE 19-5

to low points and special conduit drains should be installed. Frequently, this is far easier said than done. Pitching to a low point may not be possible. In those cases a special drain seal fitting should be installed.

There are very few locations that could be considered as dry. No conduit system is airtight. As the outside temperature and pressure vary over a period of time vapor will collect. Whenever the outside temperature drops below the dew point (saturation temperature) of the vapor, you will have moisture in the raceway. Evaporation of moisture when the exterior temperature rises does not readily occur because of the closed conduit. As a result, moisture in conduits is a far more typical problem then commonly presumed. In view of this physical law, the wisest choice would be to specify drain or inspection seals in all situations, regardless of location.

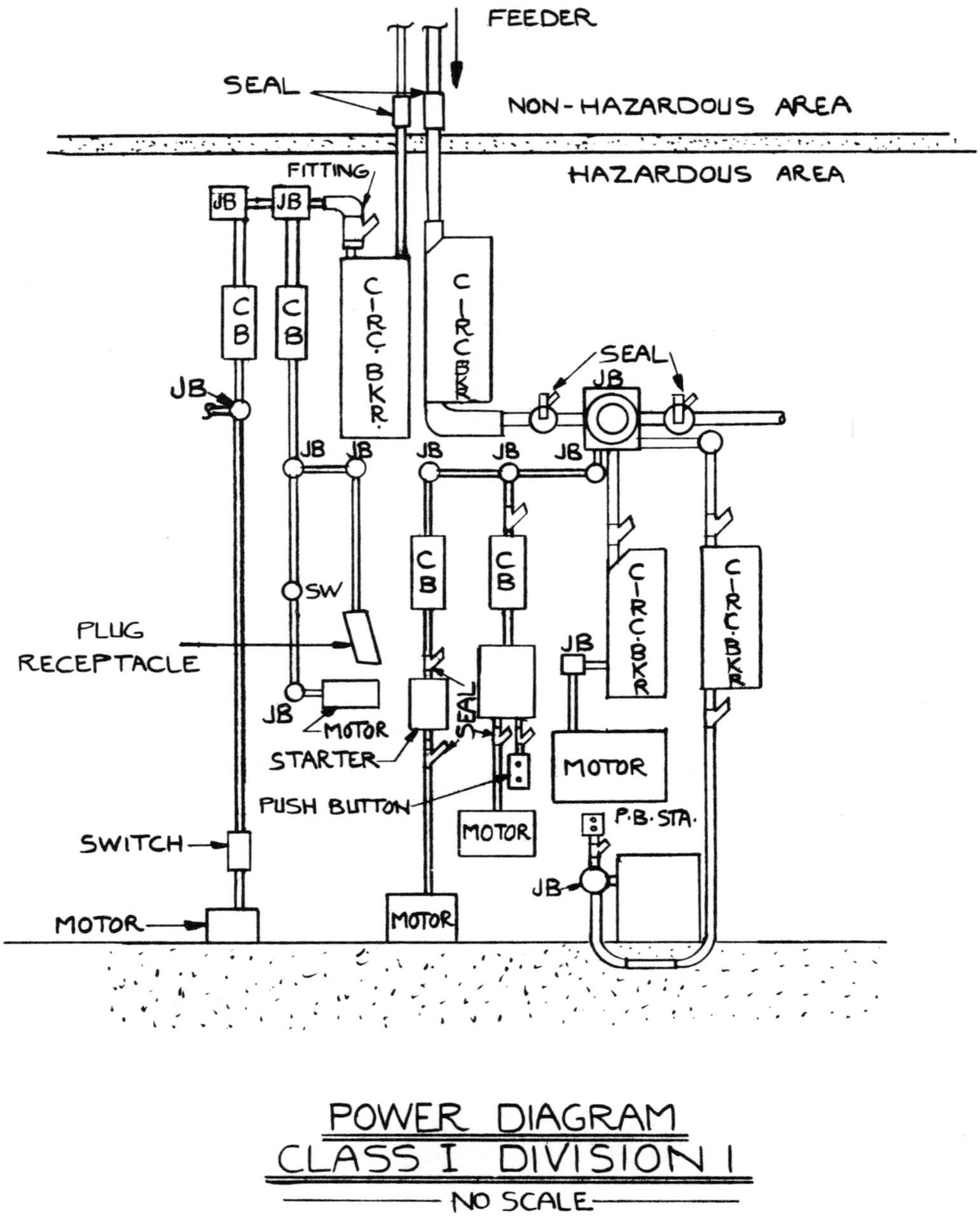
FEEDER
SEAL
NON-HAZARDOUS AREA
HAZARDOUS AREA
FITTING
JB
CB
CIRC. BKR.
SEAL
SW
PLUG
RECEPTACLE
MOTOR
STARTER
PUSH BUTTON
P.B. STA.
SWITCH
MOTOR
POWER DIAGRAM
CLASS I DIVISION I
NO SCALE

FIGURE 19-6

TWENTY

Control Wiring Systems

Control wiring encompasses many varieties and types of systems and applications. As such it does not lend itself very readily to a standard set of details and their applications, which is the basic premise of this book. For the design department of a firm, the design of a control wiring system is seldom a simple repetition of the last project's details. As such, this situation creates problems for the department as well as for the presentation to be made in this book.

The essence of any composite presentation of a control system is the "ladder" diagram. For many, the concept of this diagram is not easily or readily understood. At first glance the ladder diagram seems to indicate that every control is in parallel, and that no one device does anything that relates to or affects another device. Actually the facts are just the opposite. The key to understanding the control diagram is the understanding of the related contactor.

Perhaps another way to bring about an understanding of these diagrams is to know that the actual control device is effectively a switch that closes or opens one or a group of other switches. All controls must have a source of power to operate. What the switches that you push, pull, or turn do is send signals to their related switches called *contactors.*

The contactors usually have two positions, open and closed. These words are used to distinguish between off and on, which normally indicate no flow or flow. A contactor can be set up to operate from either direction.

There is also some confusion about what is meant by closed or open, commonly called "normally closed," which allows current to flow, and "normally open," where current cannot flow, through the contactor.

Another point of confusion concerns the operation of the push button or other device. This is commonly a "momentary" contact. For a very small fraction of time current flows through the push button and opens or closes the related remote contactor. Sometimes there are two buttons—start and stop—or only one button which performs both functions.

How can one button be both start and stop? It is very simple, actually. It can be set up to be sequentially reversing. Press it and current flows; press it again and no current flows, over and over in a continuously repeating cycle.

Generally one button doing two jobs is too confusing in a control situation. So a second device may be incorporated. A pilot light can be added which turns on when current is flowing and turns off when there is no current flow. Even with this feature many users do not find it acceptable. The two-button control with a pilot light on the "on" button is commonly requested.

Finally, an override, or emergency stop device, is frequently demanded or required. Press this button and all power is lost to all controls. The contactors for the emergency stop are at the beginning of the line as well as at the beginning of the ladder diagram. When the emergency stop is activated all current to all control ceases, and nothing operates.

Usually this emergency stop indicates a serious situation. The user (owner) commonly does not want the system restarted unless all problems are resolved. Frequently, the cause is a serious hazard or fire safety problem. To assure this degree of safety, the emergency stop can be reset with a special key controlled by one or a very few individuals.

Not every control situation requires a ladder diagram. Our details will begin with a diagram that is more familiar and then proceed to the seemingly (but not actually) more complex ladder diagrams.

Motor Control Wiring As we have repeatedly stated, this book is not on engineering design but on the application of details. However, to clarify in a limited fashion the whole subject of control wiring, we have decided to show in Figures 20-1 and 20-2 both a control diagram and a ladder diagram.

In each detail we show a more elaborate version of a control diagram similar to a temperature-control diagram. Below this presentation is a ladder diagram of the same thing. When you carefully compare the two modes, we think you will agree that the ladder diagram, once you understand it, is a lot easier to read and to follow.

We have limited our presentation to these few details because control wiring can become a very complex subject where everything about the design becomes unique.

Our two situations are fairly common and frequently occur in normal system design. This list of symbols is by no means complete. While the explanations in the following paragraphs should be sufficient to understand the diagrams, we hope a knowledge of the symbols will make the discussion easier to comprehend.

In the situation shown in Figure 20-1, an overload on any one of the motors will cause all three starters to drop out, that is, to stop the flow of current to the motors. A maintained overload on any of the motors, tripping out the related relays on the affected starter, will drop out all three starters and disconnect all motors from their power source because they are connected and controlled from a single source.

Figure 20-2 illustrates a typical problem where several motors are to be run independently. Some have starters actuated by two-wire pilot devices, and others are started by three-wire devices. A master start and stop push-button station with a control relay is used as an emergency shutdown.

A control relay, noted as CR on the master switch, provides three-wire control to motor M1 which is also controlled by a two-wire device such as a thermostat or pressure switch. Motors M2 and M3 are operated by standard start-stop switches.

Note the auxiliary contacts on M1, M2, and M3 control M4. M4 could be the pump or fan motor. The standard "hold-on" contact on M1 may be used as an auxiliary contact if Y is removed. When a two-wire control device is used, the hold-in contacts are not required. They are designated on the details as A and B.

In Figure 20-2 it is important to remember that the base connections to all starters must be same, that is, L1, L2, and L3 all have the same phased position.

These details should be accompanied by a wiring symbol list, which is usually special and separate from the regular symbol list. Wiring diagram symbols are shown in Figures 20-3 to 20-5.

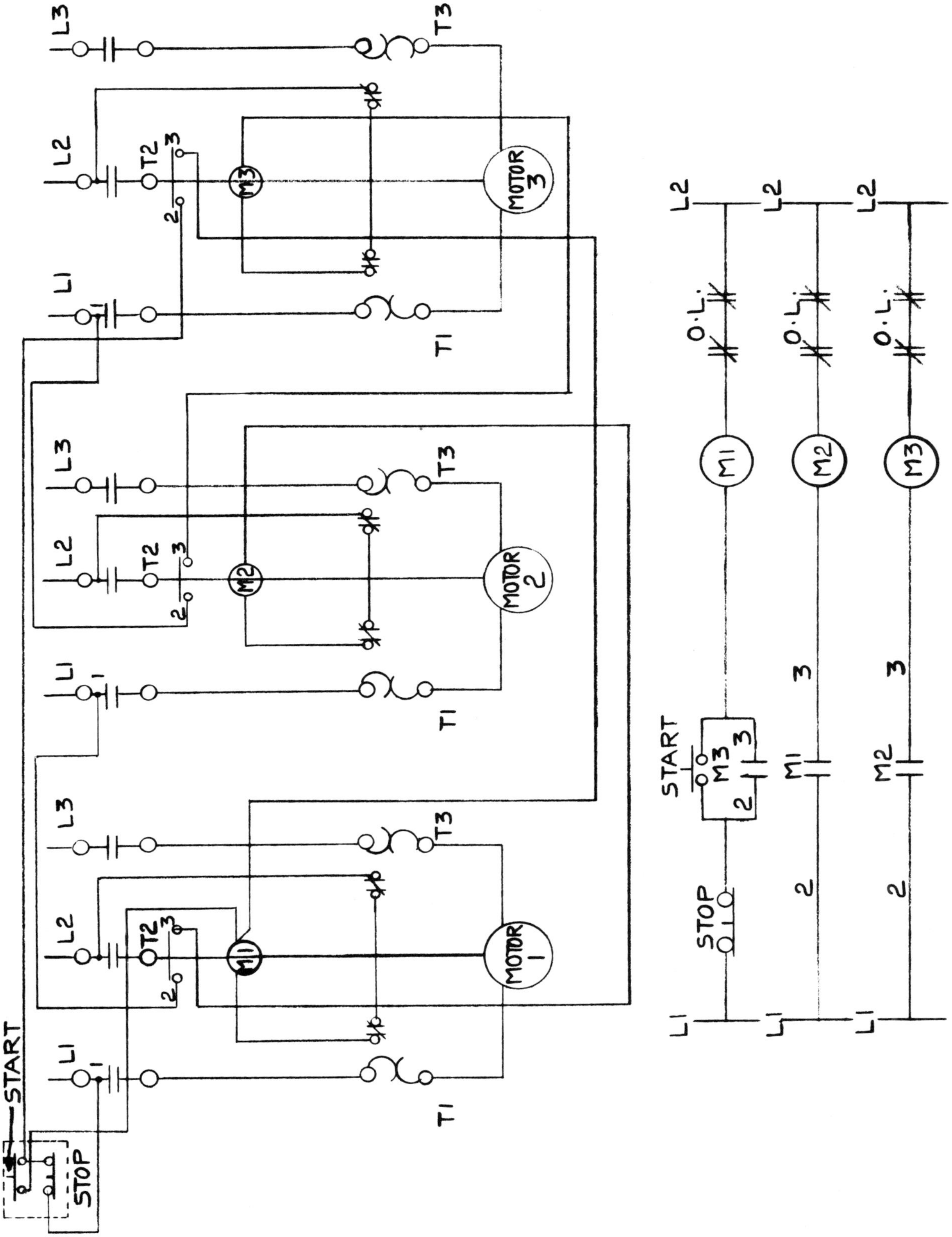

THREE STARTERS ARE OPERATED FROM A SINGLE "START-STOP" STATION

— NOT TO SCALE —

FIGURE 20-1

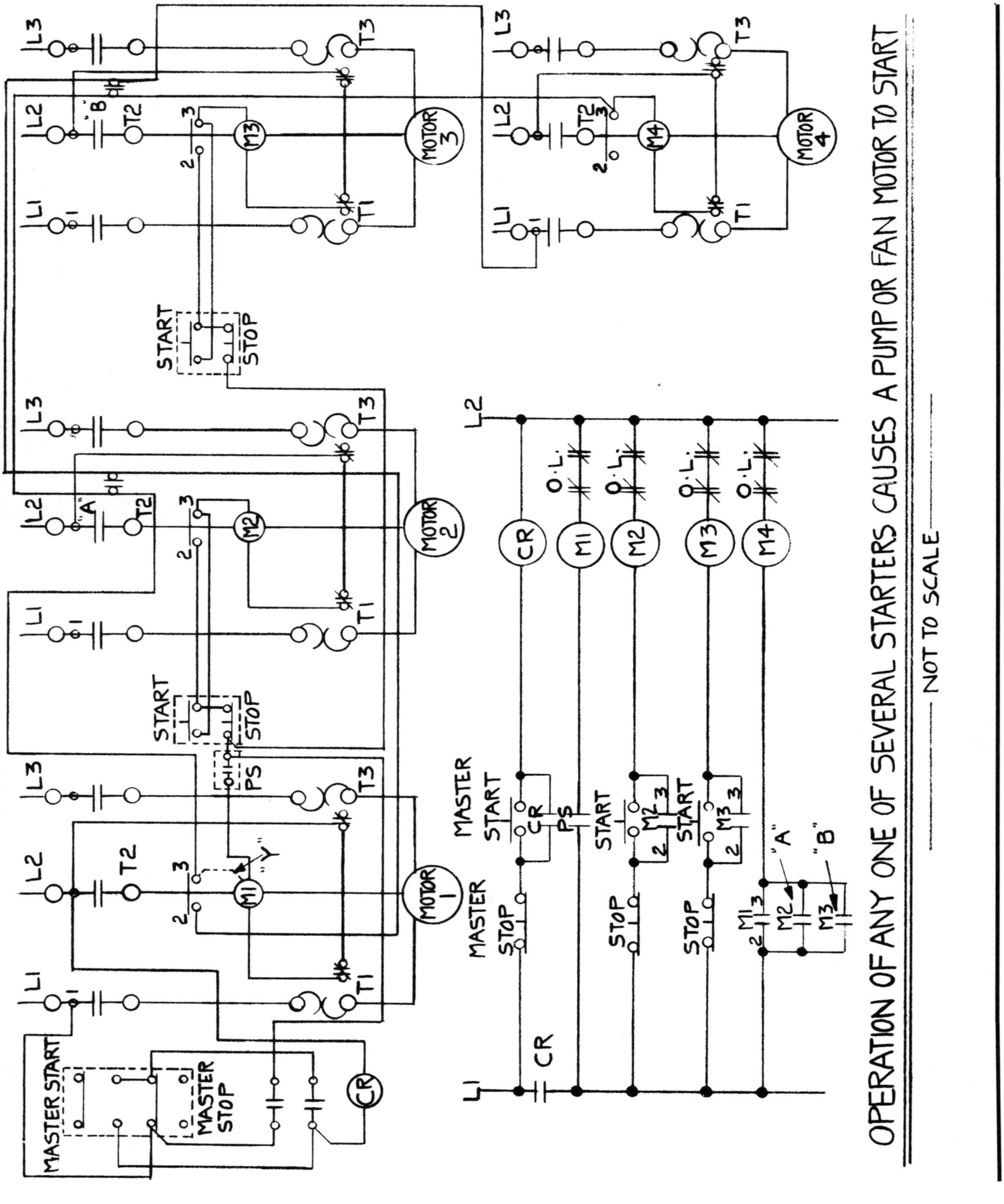

OPERATION OF ANY ONE OF SEVERAL STARTERS CAUSES A PUMP OR FAN MOTOR TO START
NOT TO SCALE
MASTER START
MASTER STOP
CR
L1
L2
L3
T1
T2
T3
MOTOR 1
MOTOR 2
MOTOR 3
MOTOR 4
M1
M2
M3
M4
"A"
"B"
"Y"
PS
START
STOP
MASTER
O.L.

FIGURE 20-2

WIRING DIAGRAM SYMBOLS

DEVICE		SYMBOL
SWITCHES	LIMIT SWITCHES	NORMALLY OPEN; NORMALLY CLOSED; HELD CLOSED; HELD OPEN
	PRESSURE AND TEMPERATURE	WHEN THIS SYMBOL IS USED ON A DRAWING THE SYMBOL SHOULD BE IDENTIFIED AS A "PRESSURE SWITCH" OR A "TEMPERATURE SWITCH".
	PUSH BUTTON STANDARD	NC; NO
	PUSH BUTTON HEAVY DUTY OILTIGHT	MUSHROOM HEAD
	STANDARD DUTY SELECTOR SWITCH	2 POSITION; 3 POSITION
FUSE	GENERAL	

FIGURE 20-3

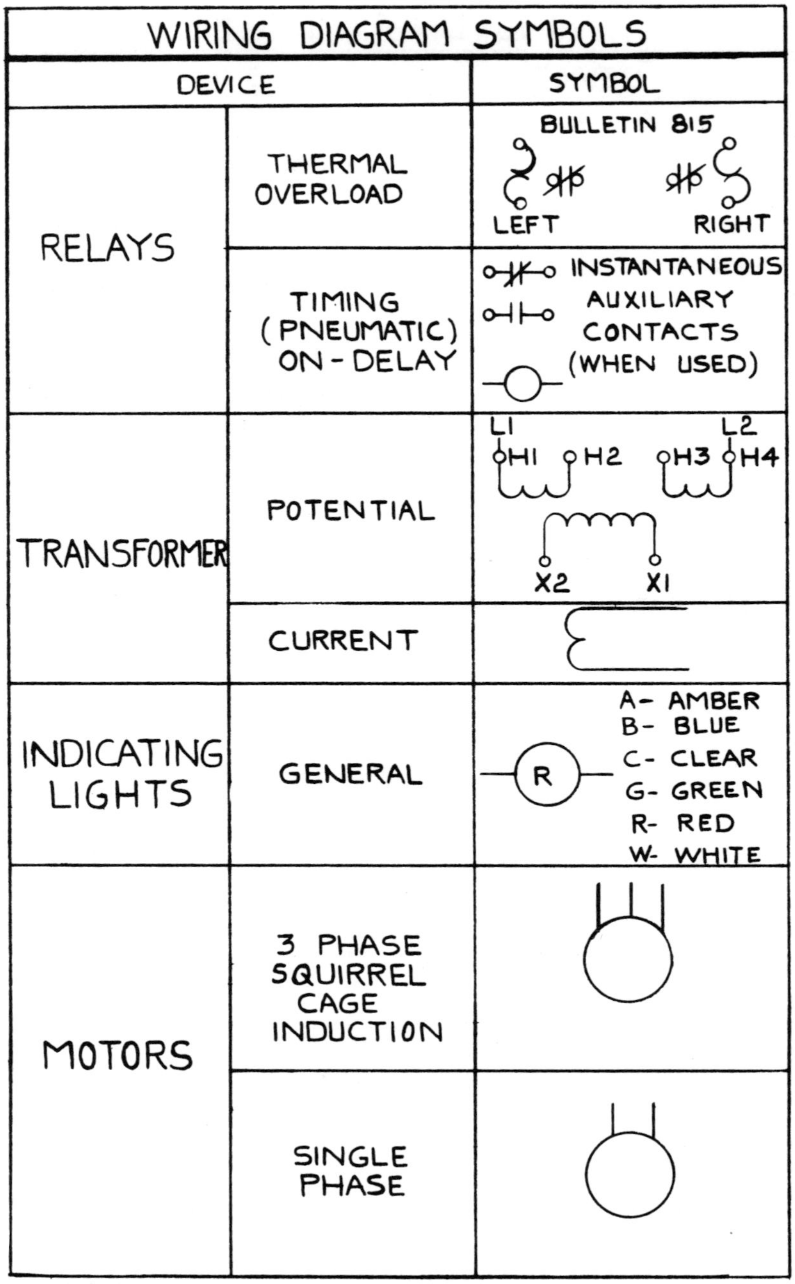
WIRING DIAGRAM SYMBOLS
DEVICE
SYMBOL
RELAYS
THERMAL OVERLOAD
BULLETIN 815
LEFT
RIGHT
TIMING (PNEUMATIC) ON - DELAY
INSTANTANEOUS AUXILIARY CONTACTS (WHEN USED)
TRANSFORMER
POTENTIAL
L1
L2
H1
H2
H3
H4
X2
X1
CURRENT
INDICATING LIGHTS
GENERAL
R
A- AMBER
B- BLUE
C- CLEAR
G- GREEN
R- RED
W- WHITE
MOTORS
3 PHASE SQUIRREL CAGE INDUCTION
SINGLE PHASE

FIGURE 20-4

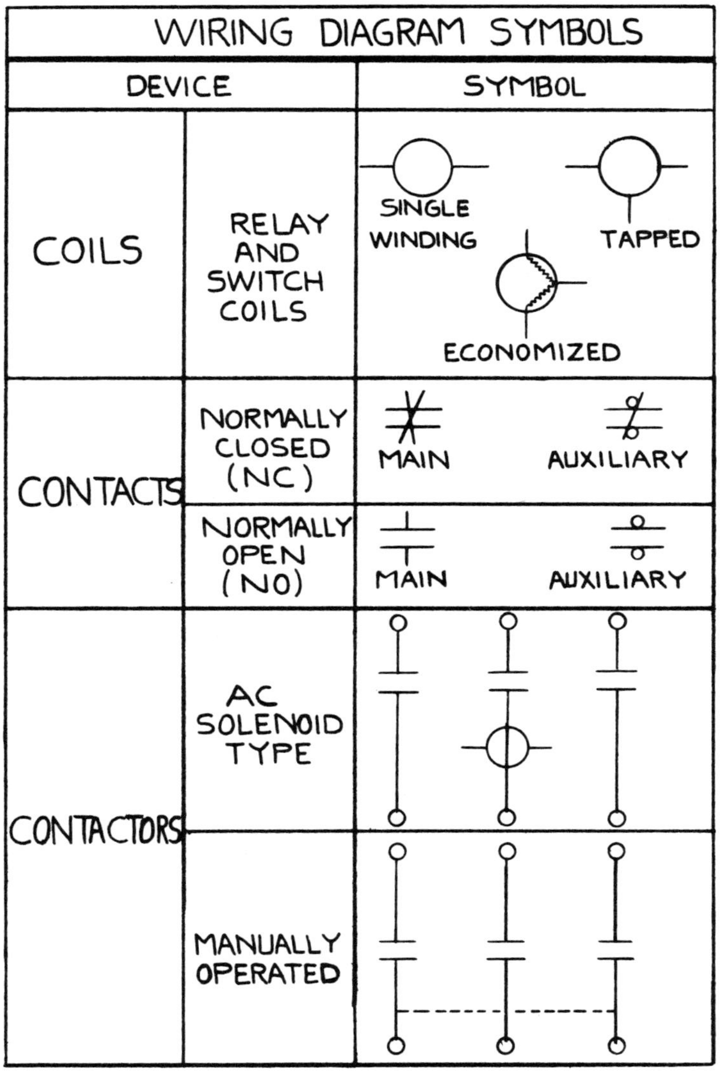
WIRING DIAGRAM SYMBOLS
DEVICE
SYMBOL
COILS
RELAY AND SWITCH COILS
SINGLE WINDING
TAPPED
ECONOMIZED
CONTACTS
NORMALLY CLOSED (NC)
MAIN
AUXILIARY
NORMALLY OPEN (NO)
MAIN
AUXILIARY
CONTACTORS
AC SOLENOID TYPE
MANUALLY OPERATED

FIGURE 20-5

Ladder Diagram Basics Figure 20-6 is only part of a typical ladder diagram. It indicates the beginning of a control system. However, the functions of any control system, even one large enough to cover an entire 28 × 40 inch drawing, would be little more than an elaboration of this simple element of design.

The basic premise is that the process is very important. Allowing it to function is not the casual choice of any employee. The operation of the system and all the controls downstream start in a locked cabinet. Nothing will operate unless the proper system supervisor opens the locked cabinet and energizes the control circuit from the switch contained in the cabinet. But this is not control as such, rather it is control of the control system. Any company wants to preclude operation of its control systems when the plant is not in production.

The real control begins with the emergency stop buttons whose additional contactors are in various parts of the control system. These emergency push-button stops can be, and are, located at various places in the plant. But, regardless of location, their wiring is looped together in series. The ladder diagram, which shows control relationships and not locations, illustrates them as all being together. They are normally closed contacts. Pushing them opens their contacts and breaks the circuit.

Having said that much let us begin at the beginning, keeping clearly in mind that all these push buttons are *momentary* contacts that activate remote contacts. When the switch in the locked cabinet is activated, it closes ES-3 and allows current to flow to all controls. It also opens ES-1, thus shutting off current from the emergency stop light. And, as can be seen, it allows current to flow to a green operating light.

Since none of the stop buttons have been activated, current can also flow through ES-2 and the emergency stop relay. Pushing any one of the emergency stop buttons not only halts the operations interconnected with the button but all others as well. This prevents work from continuing in one location and piling up in another, stopped location. Keep in mind that all three contactors are related to the emergency stop relay.

The emergency stop button causes the closed ES-2 to fall to its normally open position and stop the flow of current to the green operating light. Loss of current to the emergency stop relay is reflected back on ES-3, which falls to its normally open position and stops current to the entire control system. Current is still available at the open ES-1, which is normally closed because its open position depends on both current from the line and from the emergency stop relay which holds it open. Loss of current from the emergency stop relay allows it to fall to its normally closed position and turn on the red emergency light.

Since something serious has happened, restarting is not going to be a casual affair. You need to be a supervisor with a key. Frequently, a security guard may have access to the switch in the control box. The emergency stop can be used as an operating safety control. Regardless of operation of the control, the system cannot be restarted unless the emergency stops are reset at the key-operated reset button. Pressing the button resets the emergency stops and current flows to ES-2 which closes, relights the green light, and causes, through ES (now again seeing current) to open ES-1, turn off the red light, close ES-3, and repower the system.

In the beginning this is a very difficult concept for some to comprehend. The so-called switches, which are usually push buttons, are only switches for an instant of time. In that instant they activate contactors which are similar to a normal switch in that they are on or off. The final part of the concept is to understand that the contacts can be normally on (closed) or normally off (open). Both options have their functions in a control system.

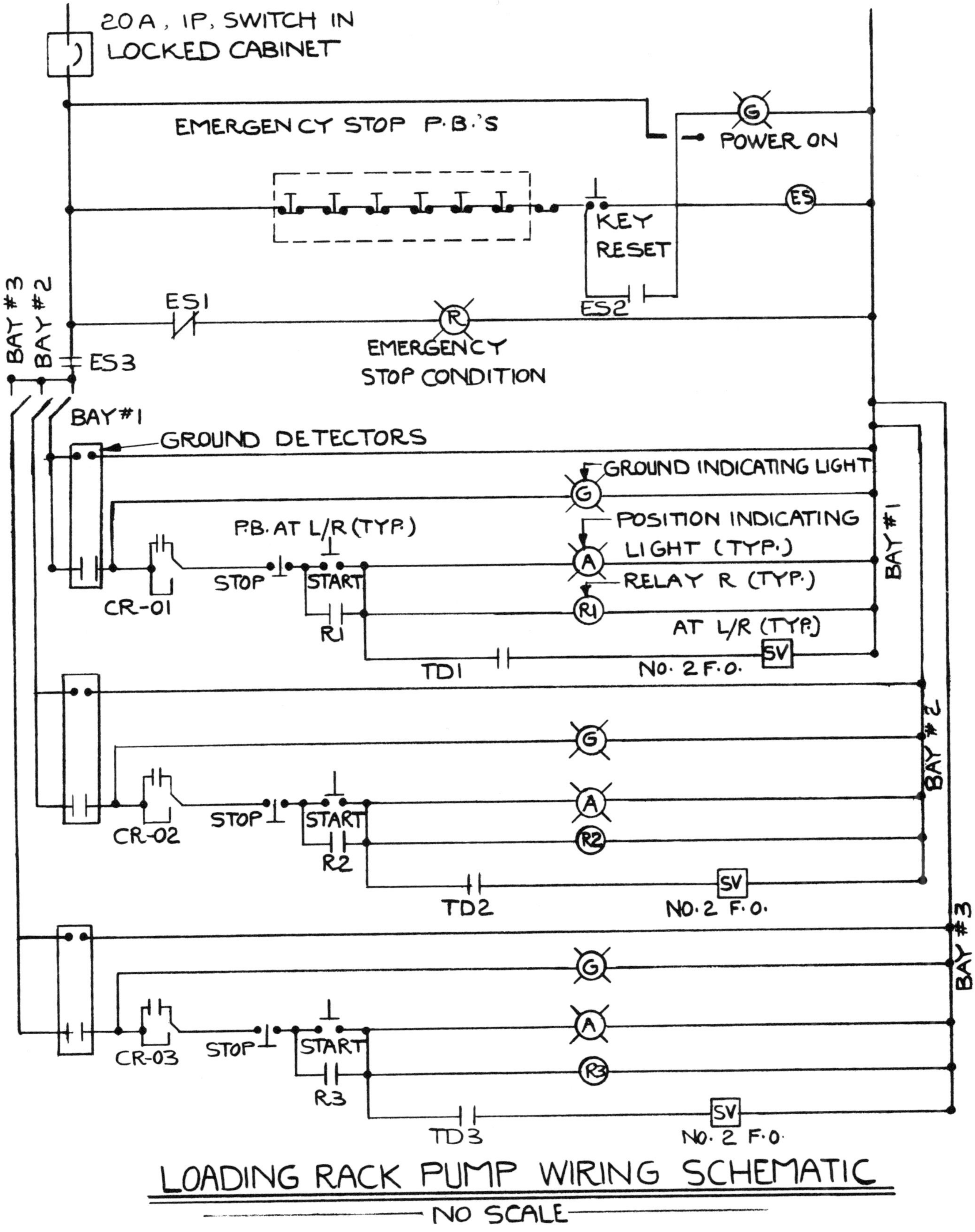
20A, 1P, SWITCH IN LOCKED CABINET
EMERGENCY STOP P.B.'S
POWER ON
G
ES
KEY RESET
ES2
ES1
R
EMERGENCY STOP CONDITION
BAY #3
BAY #2
ES3
BAY #1
GROUND DETECTORS
GROUND INDICATING LIGHT
P.B. AT L/R (TYP.)
POSITION INDICATING LIGHT (TYP.)
RELAY R (TYP.)
AT L/R (TYP.)
STOP
START
CR-01
R1
TD1
NO. 2 F.O.
SV
A
CR-02
R2
TD2
CR-03
R3
TD3
LOADING RACK PUMP WIRING SCHEMATIC
NO SCALE

FIGURE 20-6

CONTROL LEGEND

T1	SPACE MASTER TEMP. SENSOR W/GUARD
T2	MIXED AIR LOW LIMIT CONTROLLER
T3	DISCHARGE AIR SUBMASTER TEMP. SENSOR
T4	OUTDOOR AIR MASTER TEMP. SENSOR & PUMP CONTROLLER
T5	PIPE SUBMASTER TEMP. SENSOR W/WELL
V1	MOD. 3-WAY MIX VALVE ASSEMBLY N.O. TO COIL FLOW
V2	N.C. MOD. STEAM VALVE ASSEMBLIES
M1	MOD. DAMPER MOTOR W/Q605 LINKAGE
HT	DUCT HIGH TEMP. SAFETY CONTROL W/MANUAL RESET
LT	COIL LOW TEMP. SAFETY CONTROL W/MANUAL RESET
ALT	ALTERNATOR

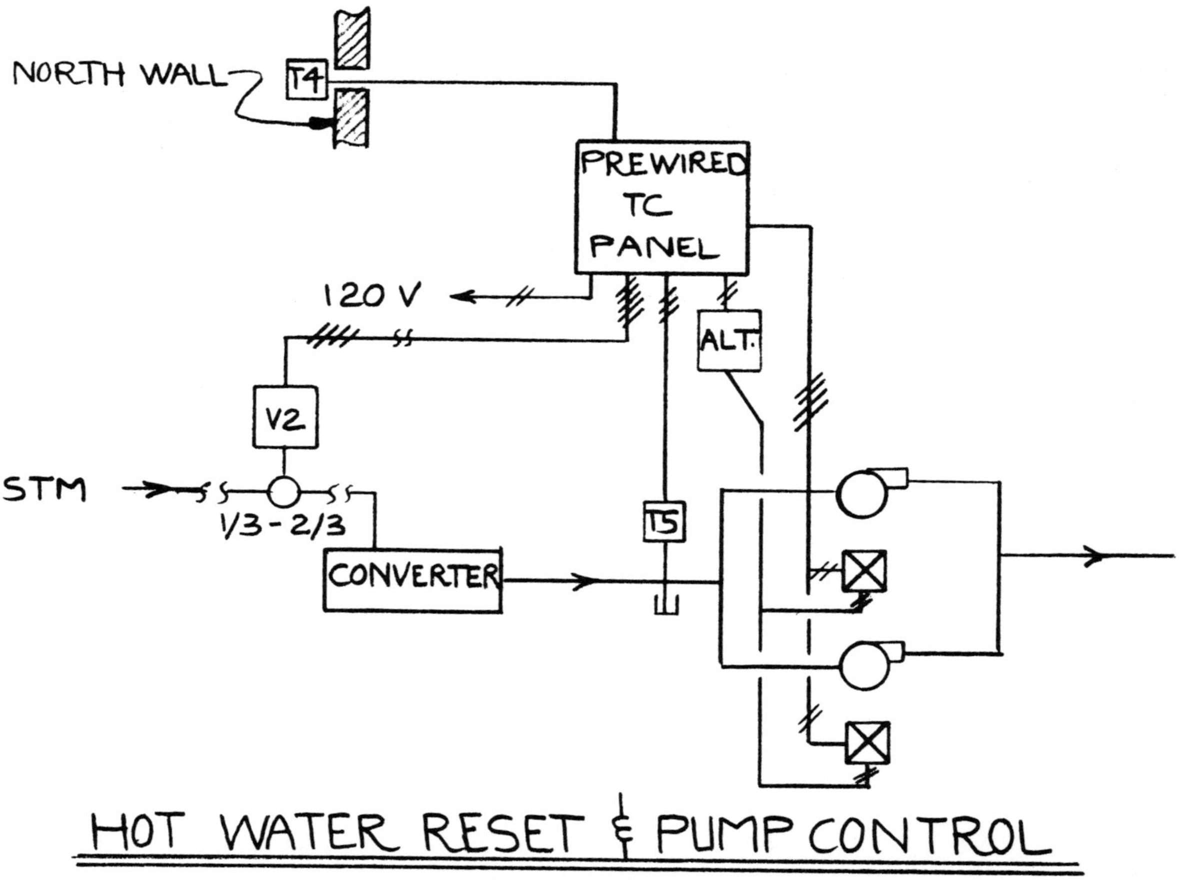

HOT WATER RESET & PUMP CONTROL

NO SCALE

FIGURE 20-7

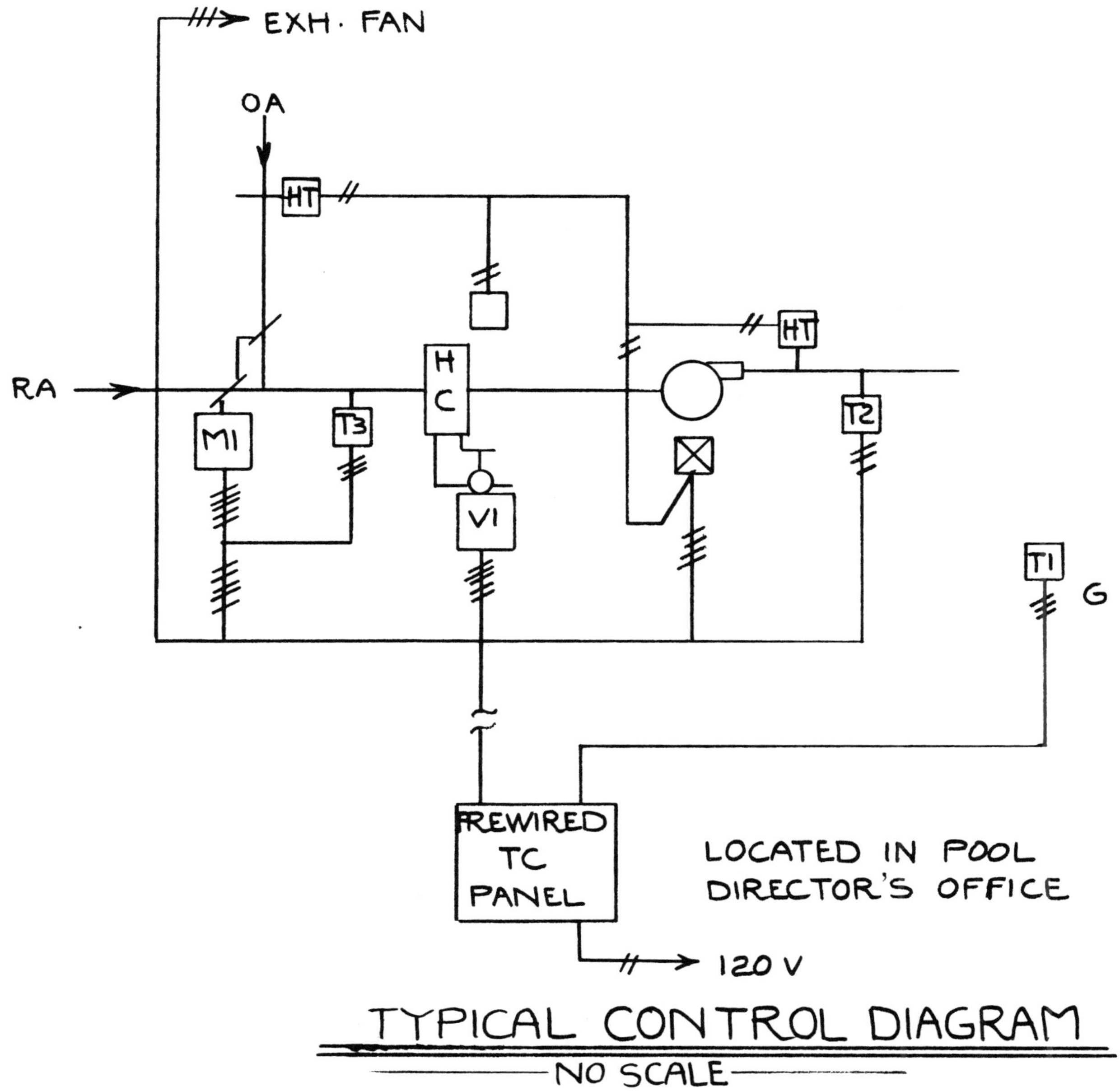

FIGURE 20-8

Temperature-Control Wiring Figures 20-7 to 20-9 cover the installation of a system that controls a series of pool air-handling units plus the controls that relate to the pumps and converter that supply hot water for the coils of the air-handling units. (Note that the control legend for both Figures 20-7 and 20-8 is in Figure 20-7.)

Before describing our details, let us first explore a related control situation that should be fairly familiar—the thermostat in your house. If you have a hot-water heating system, the thermostat controls the circulating pumps. The line voltage to the control devices passes through a control transformer which converts the line voltage to low voltage for your personal safety and then to your thermostat. At a given set point, just like a switch, the bimetallic element in your thermostat bends to make a contact. Current flows and a pump starts. A separate set of controls operates the burner.

In the hot-water system the burner may not start because it operates through another set of controls which govern boiler water temperature. As you know, if you have an air system, the fan that blows the warm air does not come on just because you started the burner (gas or oil). It will only come on when it senses that the burner has heated the air to a degree high enough for comfortable circulation.

Our pool air-handling units have a control system with

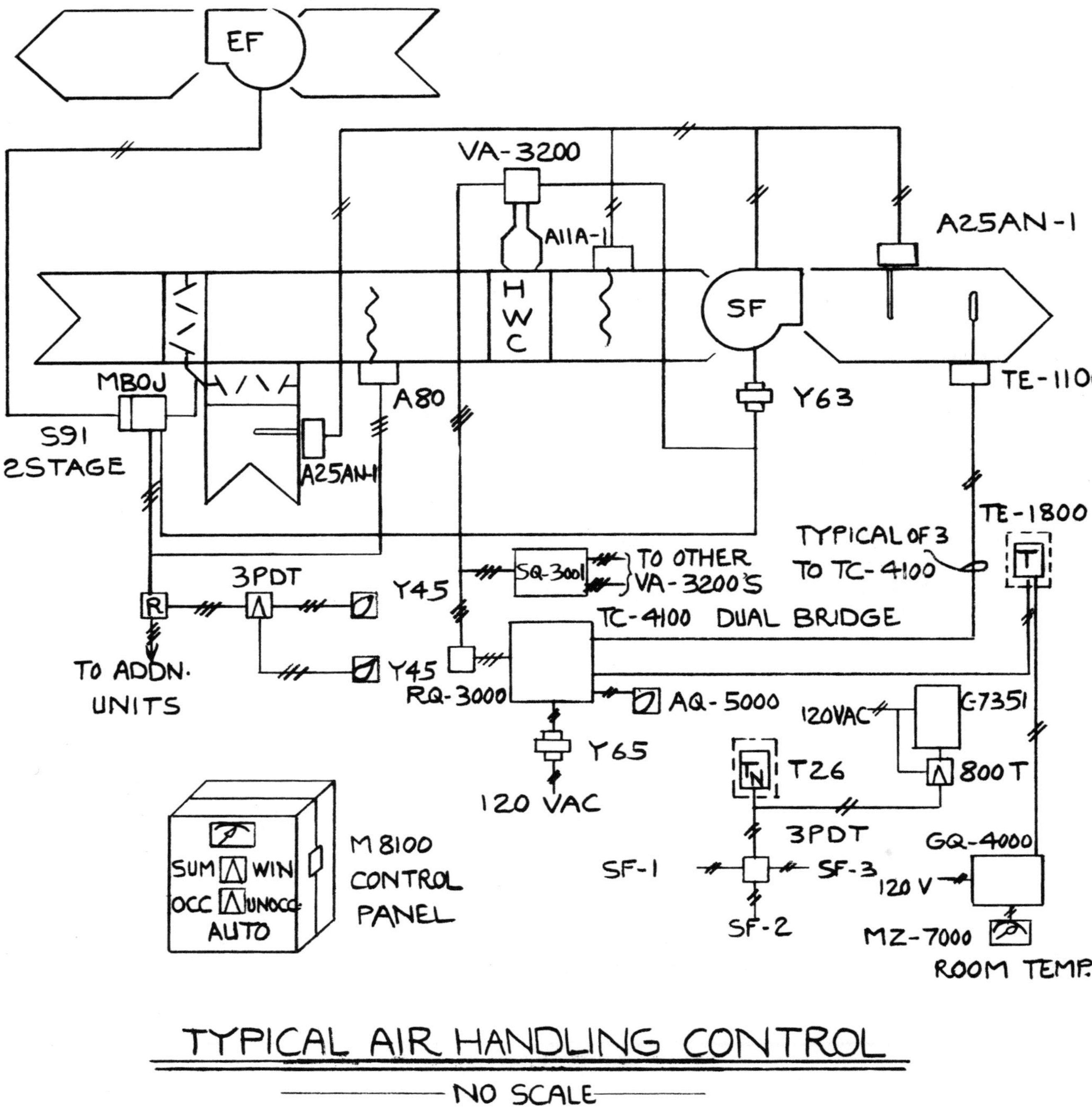

FIGURE 20-9

a number of functions. When the thermostat (T1) calls for heat, the fan starts. The outdoor air damper opens to a preset position modified as required by the mixed air (return and outside air) controller set to maintain the combination of return and outside air above 55° F. This setting value is not shown. It is a part of the specifications.

The system is integrated with a pool exhaust system that operates on low speed in the winter and high speed in the summer through a specified device, not shown, called an *end switch*. When the switch is placed in the summer position, the outdoor damper (OA and MI) opens fully and the return air damper (RA and again MI) closes. The exhaust fan will run on high speed.

T1, the room temperature sensor (thermostat), which is operating through the central control (TC panel) resets the hot-water coil valve (V1) by modulating the degree of opening.

To provide the hot water for the pool fan system from

an existing steam boiler, a device known as a steam-to-water converter was installed in the boiler room. It is also described as a shell-and-tube condenser. Steam is in the shell of the device that heats the water supply contained in tubes in the condenser. The related control diagram is shown on Figure 20-9.

Through a central control (TC) panel, the temperature of the heating-supply water is sensed by a line-immersion thermostat (T5). If the temperature drops, a valve (V2) opens to let more steam in the converter. To preclude unnecessary heating, a special outdoor thermostat (T4) set at 65°F stops all operation when that temperature is reached.

Note that the controls are tied to two pump starters (X) and that a special line called an *alternator* (ALT) is also connected to both starters. This alternator allows only one pump to run, and each time a system signal from the pool TC panel calls for heat, the pump that did not run the last time starts. This is to ensure equal use of both pumps during the heating season.

In lieu of a line diagram the number of wires to operate each device must be known. Once this is known, the devices are simply connected together by lines and the number of wires is indicated by slanted lines in the same manner that other wiring quantities are commonly listed.

We would hasten to caution you that while this is an actual installation, it cannot be simply traced to resolve your problem. Instead it is an illustration of the type of diagram you would *specifically tailor* to solve your problem.

Up to now we have not mentioned Figure 20-9 for a very good reason. There are several schools of thought on how temperature control diagrams should be depicted. In Figure 20-8, which is one version, the specification states that the diagram is typical of three units and that the items are as detailed on the plans. It also describes the control panel and all controls with name, function, and part number. Figure 20-9 takes a second school of thought approach and notes on the drawing the part numbers as can be seen. For specialized switches, it describes the proper function in abbreviated form (3 PDT, which is three-pole, double throw). The specification would be very similar otherwise to that for Figure 20-8. The representation of the fan and duct work is more representative of the actual fan and duct work. The position of dampers, coils, and all temperature sensing devices is claimed to be more understandable. The question really is whether this actually clarifies things for the contractor.

Our position is that we have presented both possibilities for your choice. Since we had two jobs bid almost at the same time where there were two pools and, in each case, three air handling units, we tried Figure 20-8 on one job and Figure 20-9 on the other. In this rather simple test Figure 20-9 got all the compliments. But the systems were both installed without any question, discussion, or debate by the control contractor. Either the contractor knows our firm very well or the effort of Figure 20-9 was not really necessary. We prefer the specification and the depiction of Figure 20-8, which is simpler and clearer and makes it far easier to spot a potential error.

TWENTY-ONE

Energy Conservation

If the total number of words in all magazines, periodicals, and newspapers for the past 5 years were counted, there is little doubt that the subject of energy conservation would be the winner of the largest percentage total. If we also include stories where energy conservation was at least a related subject, the percentage would probably be 35 percent.

Most energy conservation, especially that referring to electrical design, involves slowing or stopping the flow of energy. Since the vast majority of these proposals involve studies, reports, investigations, and flow diagrams or schematic sketches, there is little detailing that will be required or could be presented here.

Ignoring improved operating performance, we can save electricity and conserve energy by turning off its flow. It logically follows that we also save energy by not allowing something to be turned on. Most energy-saving schemes involving mechanical equipment, especially those involving heating, ventilating, air-conditioning, and exhaust equipment, are designed to relate operation to demand. Controlling the on cycle of equipment frequently saves oil and gas as well as electricity.

In the above paragraph we used the phrase "ignoring improved operating performance." For example, except for a hanger detail covered elsewhere in the book, there is no detailing required to illustrate the substitution of high-pressure-sodium luminaires for incandescent luminaires. Yet the savings in this simple fixture substitution can be very large.

At the heart of every controlled energy-saving scheme is the obvious device—the controller. The programmable controller in the form of a 12-hour, 24-hour, or 7-day clock has been around for a long time. And, except for the problem of uncoordinated clocks in innumerable locations, it still is a reliable device. The inference in the above statement should be obvious: you do not need a computer to turn on and off a few lights or motors in a small building.

The applicability of the computer lies in its ability to be programmed to coordinate hundreds of tasks into a total operation. Once the computer is programmed, its beauty is that it does not forget. Through a programmed instruction, which is usually in the form of an "if—then—else" statement, the computer can be set up to react to real-life situations just as a live operator might be expected to do.

A detail related to the situation as outlined above is usually a flow diagram. In the details that follow we will illustrate condensed samples.

Programmable Lighting System Figure 21-1 illustrates the basic elements of a computer-controlled lighting system. The basic elements consist of a computer, transceivers, relays, control wiring, switches, and lighting fixtures. Additive items include a card reader, a printer, a telephone interface (modem), and photorelays.

In its simplest terms a signal from the computer is sent to and processed by the transceiver. From there the signal is sent to relays which energize and deenergize the lighting fixtures. A brief description of each of these elements is contained in the following paragraphs.

The computer is the device that operates the system either by a programmed card reader or by a program written and installed by the user. Generally either system is

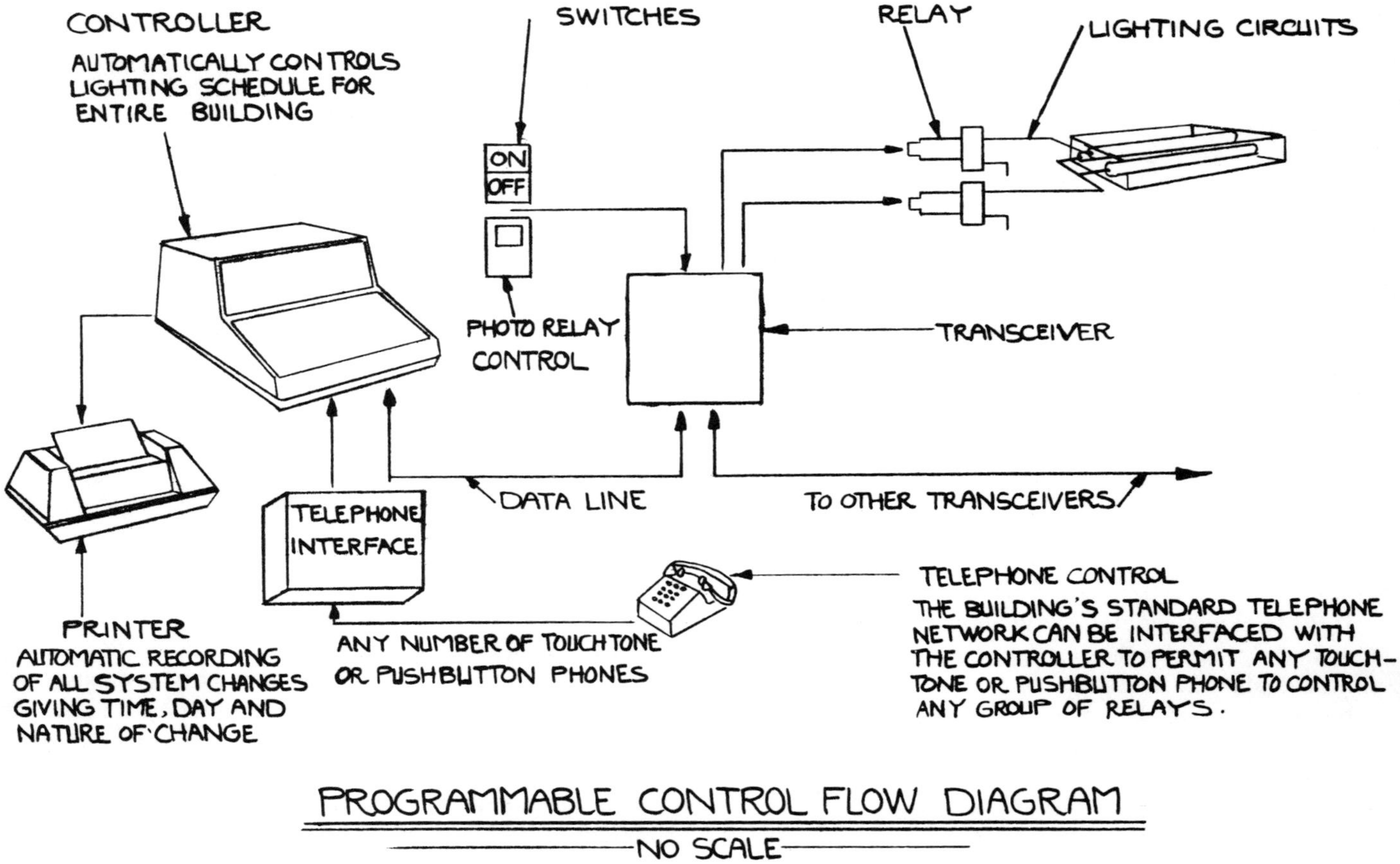
CONTROLLER
AUTOMATICALLY CONTROLS LIGHTING SCHEDULE FOR ENTIRE BUILDING
SWITCHES
ON
OFF
RELAY
LIGHTING CIRCUITS
PHOTO RELAY CONTROL
TRANSCEIVER
TELEPHONE INTERFACE
DATA LINE
TO OTHER TRANSCEIVERS
TELEPHONE CONTROL
THE BUILDING'S STANDARD TELEPHONE NETWORK CAN BE INTERFACED WITH THE CONTROLLER TO PERMIT ANY TOUCH-TONE OR PUSHBUTTON PHONE TO CONTROL ANY GROUP OF RELAYS.
PRINTER
AUTOMATIC RECORDING OF ALL SYSTEM CHANGES GIVING TIME, DAY AND NATURE OF CHANGE
ANY NUMBER OF TOUCHTONE OR PUSHBUTTON PHONES
PROGRAMMABLE CONTROL FLOW DIAGRAM
NO SCALE

FIGURE 21-1

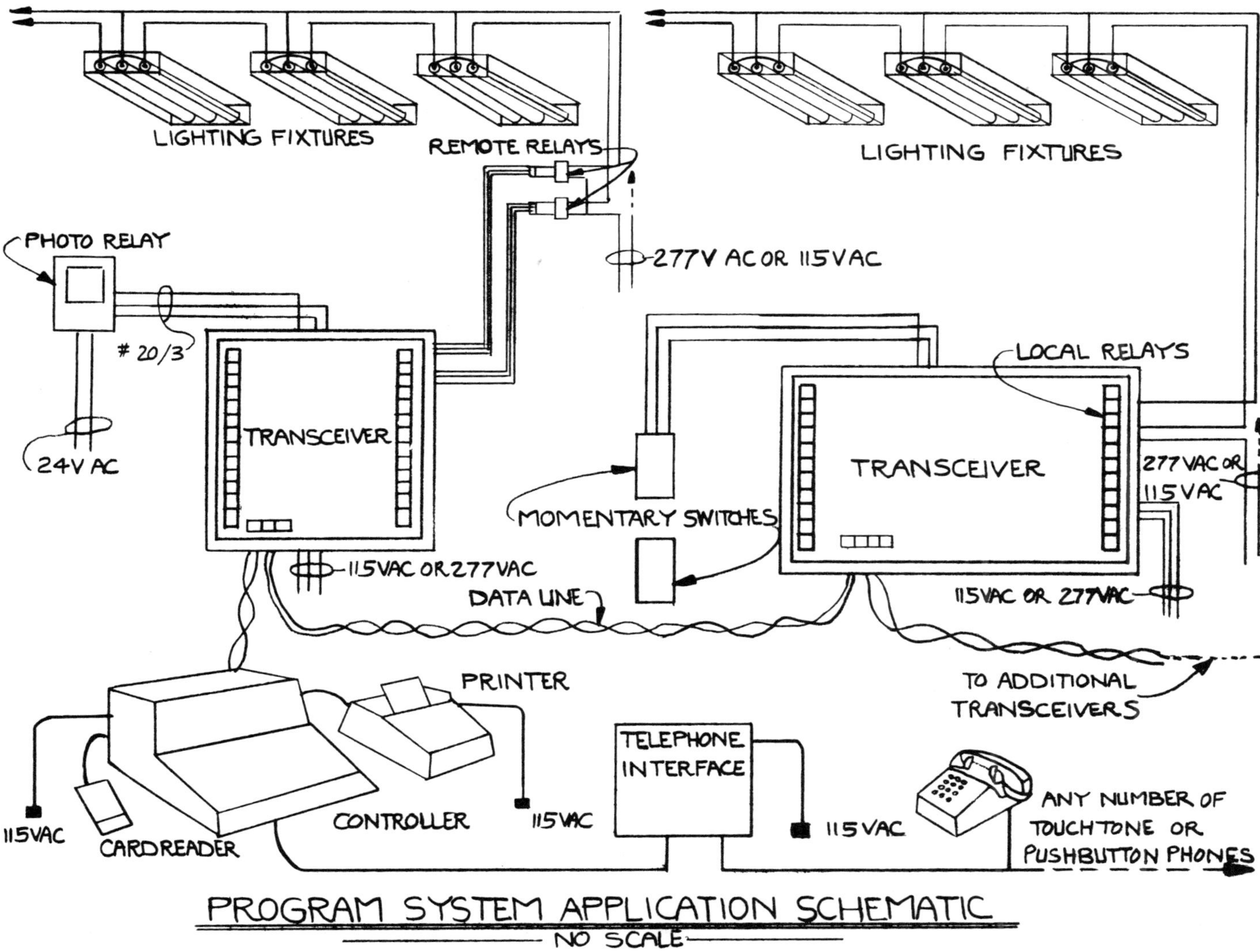

FIGURE 21-2

designed so that no knowledge of programming or programming language is required. There are a large variety of possible outputs available. The most common features of the system are the processor, tape drive, backup battery power, visual and audio display of data, telephone and printer connections, and inputs for optical card readers and data block. Frequently the computer can display and bypass system faults while keeping the rest of the system operative. The transceiver is a microprocessor that also contains the relays shown on the schematic diagram as well as input connections for a limited number of switches. Finally there are the lighting fixtures which are the object of the control system. This is shown as a composite system in Figures 21-2 and 21-3.

The wiring of the devices is also fairly simple. The data line and the connectors between the transceivers and switches are all low-voltage. Wiring from the computer to the transceiver is a pair of #18 wires. Wiring to switches and relays for each device are three #20 wires. The wiring to the opposite side of the relays is 120 or 277 V, depending on fixture voltage. This is also true of the transceiver. The computer and the optional printer and telephone interface (modem) require 120 V each. The card reader merely plugs into its special computer connection.

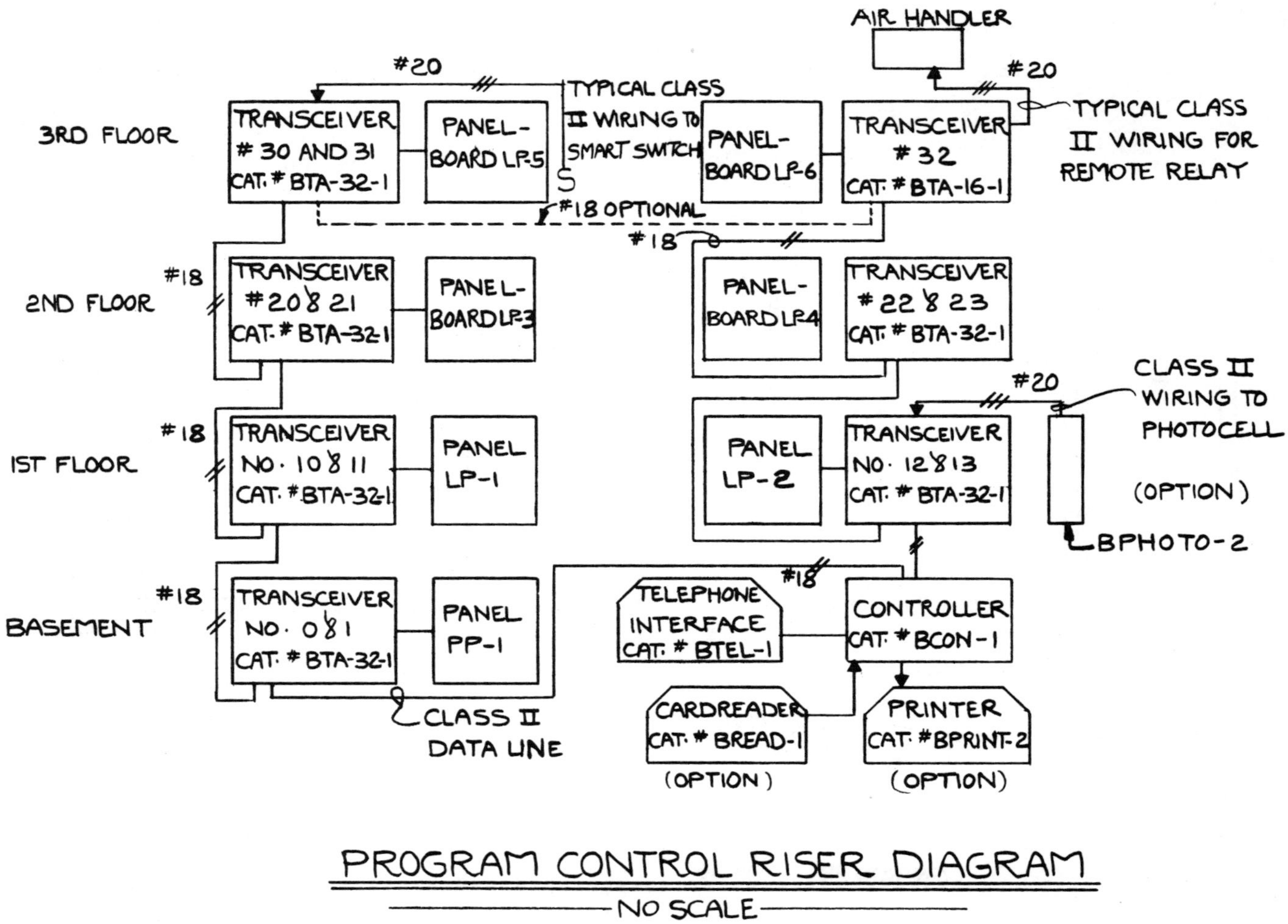

FIGURE 21-3

Switches can be momentary or single-pole double-throw. Where photoelectric relays are used as in Figure 21-4, a time-control device is added so that the photoelectric cell operates in a selected period of time. Many refinements of this system can be added, depending on the requirements of the situation. Our detail merely highlights the essentials. For example, among the additive items shown, the card reader can provide offsite programming and program updating, the printer can provide a record of operations, and the modem can be adjusted from any related, dedicated phone.

Demand-Limiting Devices The demand-limiting device is the second common way to conserve electrical energy. The generalized arrangement is very similar to the situation shown in our previous detail, where the transceiver provided the interface between the computer and the lighting circuits which the computer was programmed to control. In the demand-limiter situation, the demand-limiting device is monitoring (reading) the same demand value that the utility company demand meter is reading.

At this point we depart from details and design and necessarily get into the engineering phase. Since there is no other way to explain this subject properly, we are going to bend our rule and delve, at least superficially, into the basics of this subject.

The demand limiter is a prefabricated black box, so to speak, that contains certain preset conditions. At a certain demand value it simply stops preselected equipment from operating. The key word is "demand," which is a special charge by the utility company created by the power requirements of a large motor or motors. The utility company maintains the capacity to handle any increase in electrical load even though you may, as the user, create this demand on a very infrequent basis. In simple terms, if you want this special service, you are going to have to pay for

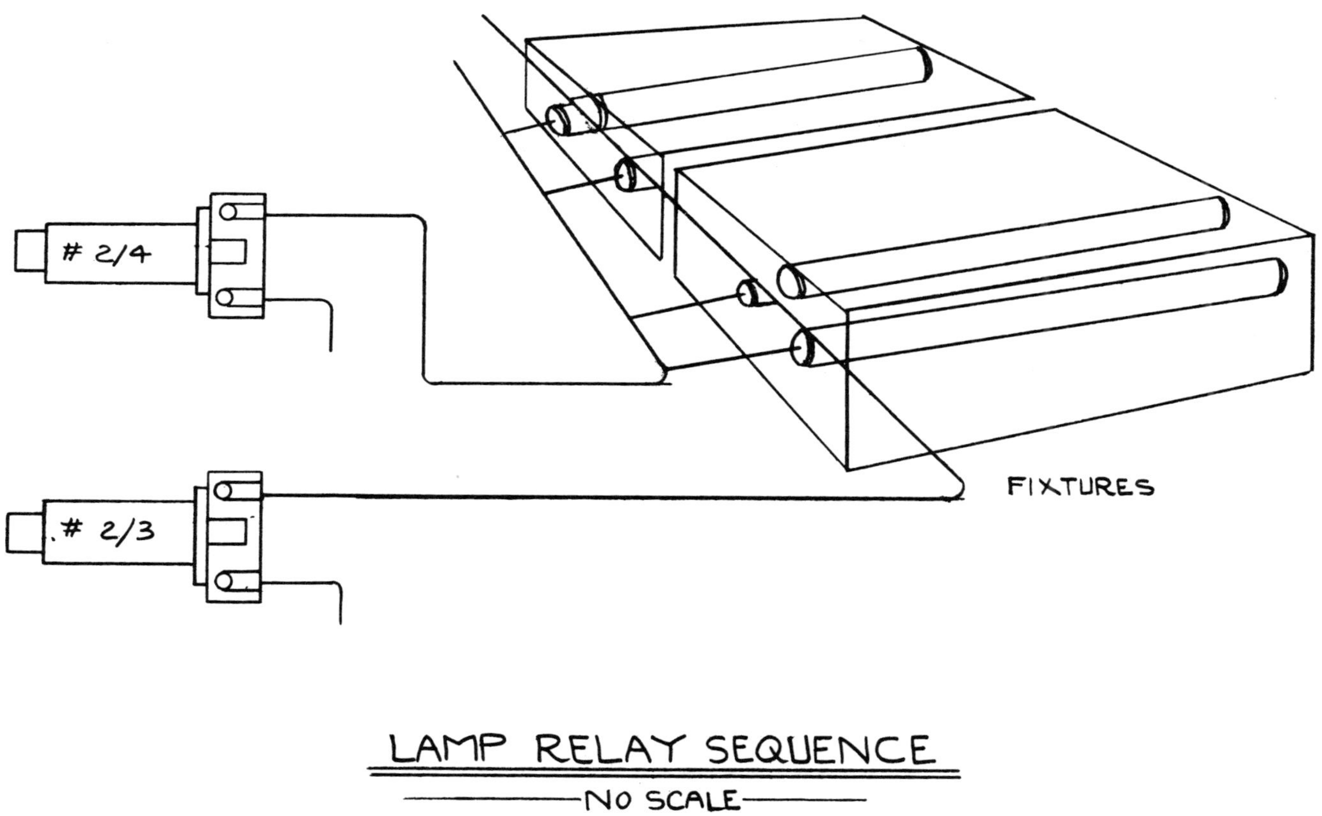

FIGURE 21-4

it. To a degree this situation occurs in a house, usually late in the afternoon, when dinner preparations begin. Since this is fairly predictable, you, as a home owner or renter, pay no demand charge. (Actually you do in the form of an adjusted electric rate.) But, for an office or factory, this demand charge can get to be very large. Demand charges of $6000 to $60,000 over and above basic charges are not uncommon. How can both money and energy be saved? Very simply by controlling demand. Commonly one to five motors or groups of motors create the jump in demand.

The rest of the answer is not so simple. First a survey has to be made to find the big users of electrical energy, then to decide what would really happen if a big user could not start when the operating situation required it and how long this inability could be permitted.

Frequently the air-conditioning machinery is the prime target. As everyone knows air conditioning is often overdone, and a short time of less than perfect conditions can be tolerated. A not uncommon situation would be a 50 percent demand reduction created by merely preventing 30 percent of the major air-conditioning compressors from operating simultaneously.

The demand limiter is wired not only to the demand meter but also to the starting circuits of the sources of electrical energy. When the demand created by other, more vital machinery reaches a preset value, the demand limiter inhibits the controlled motor circuits from operating until system demand drops.

Smaller demand-limiting devices do not necessarily need a computer. Usually they can be internally programmed (like a timer) to control up to 10 load sources. They can be set to stop and start in a user-arranged pattern.

In brief, this is how the demand limiter functions. Your detail is simply a blank square with the title "demand limiter" and wiring going to the demand meter and to selected motor control relays as you did for the programmable lighting controller.

TWENTY-TWO

Legend (Symbol List)

Every set of electrical plans requires a symbol list or legend. There will never be enough room on the plans themselves to provide descriptions beside each electrical symbol, nor is this a reasonable method. The required, and in many cases redundant, explanations would take far too much time.

While the common items such as switches, outlets, and lighting fixtures have fairly standardized symbols, there frequently is some divergence among engineers, designers, and draftspersons over the proper treatment of the more special purpose symbols.

In Figures 22-1 to 22-18 that follow we have used the words "symbol list" and "legend," not to confuse you but rather to make the point that either description is reasonable, logical, and acceptable. We have prefaced the descriptions with the word "electrical." Why use the word "electrical" when the legend or symbol list is usually on a plan labeled "Electrical Details"? Put very simply there is no reason except for filing purposes.

This book presents the legend or symbol list in a manner that conforms to our office draftingroom procedure. Each designer and draftsperson is given, unbound, a photocopied set of symbols on 8½ × 11 inch sheets. He or she can either slide the sheets under the drawing and trace off the appropriate symbols or reduce the detail in size, by hand, to fit in the available space on the drawing. We do not want, or allow, any drawing to be used for bidding and construction purposes that contains symbols not used on the drawing. In our view that is as confusing as specifying products that are not required by the project.

There probably exists somewhere a complete set of symbols that covers every conceivable situation. The list we are presenting is, we freely admit, not perfect. As you peruse our symbols you will inevitably find items not covered that you think should be. For that we make no apology. From time to time we try to add to, alter, or update our list. This is a never-ending task of self-improvement that every firm and individual should take on. We believe the symbols shown make up a pretty good basic collection. We would be the first to agree with you that it could always be improved.

In the area of basic switch and outlet symbols we are sure you can list some added types. Our problem, and yours, is that the list presented already covers some 18 pages. It is getting so long that it almost needs its own table of contents and index. We feel that defeats the purpose of an efficient drafting room tool. If anything, we would like to make it shorter without breaking it up into several separate parts, as we have found it to be far more efficient to keep pieces of information together as complete units.

Finally, you will note that our symbol list does not cover specialties related to bus duct, switchgear, transformers, and high-voltage equipment. In our experience the cost of these specialties is considerable and the possibility of an error can create serious problems. Regardless of the clut-

ELECTRICAL SYMBOL LIST	
SYMBOL	DESCRIPTION
	DISCONNECT SWITCH
⊠	STARTER BY OTHERS REFER TO SECT. 15700 ELECT. SPEC.
	MOTOR
J	JUNCTION BOX
	SURFACE MOUNTED QUADRUPLEX RECEPTACLE
	SURFACE MOUNTED DUPLEX RECEPTACLE
	RECESSED DUPLEX RECEPTACLE
	RECESSED DUPLEX FLOOR RECEPTACLE
S	SMOKE DETECTOR
H	HEAT DETECTOR
$\$_2$	TWO POLE SWITCH RECESSED
$\$_4$	FOUR WAY SWITCH RECESSED
▽	TELEPHONE
	CLOCK
T	THERMOSTAT

FIGURE 22-1

ter, we try to describe every piece with a note on the drawing or detail and, where necessary, provide a brief, custom-made symbol list at or near the detail or drawing. We want every piece of information in these situations to be very closely related and on the same plan or detail sheet so that there is no doubt, debate, or looking through the entire set of plans to find a description.

If our firm specialized only in high-voltage work, or if the project was only concerned with transformers, switchgear, and high voltage, we probably would approach the problem with a specialized symbol list. But nearly all of our work does not fit that category and we are assuming this is also your situation. Thus we feel the symbol list shown is adequate for the vast majority of work which users of this book will perform. That it will not please everyone is a foregone conclusion.

ELECTRICAL SYMBOL LIST

SYMBOL	DESCRIPTION
	EXISTING EXIT SIGN TO BE REMOVED, CUT BACK WIRING & CONDUIT TO ABOVE CEILING & PROVIDE JUNCTION BOX WITH BLANK COVER PLATE.
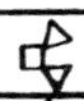	INCANDESCENT LIGHTING FIXTURE
	" " " ON EMERGENCY CIRCUIT
○	" " "
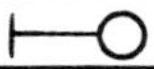	" " "
	" " " DIRECTIONAL AS SHOWN
●	" " " ON EMERGENCY CIRCUIT
○	" " "
	" " " ON EMERGENCY CIRCUIT
■	" " " " " "
□	" " "
	FLUORESCENT LIGHTING FIXTURE
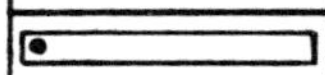	" " "
	" " "
	" " " ON EMERGENCY CIRCUIT
	EXIT SIGN
	" " WALL MOUNTED
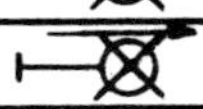	" " " " WITH DIRECTIONAL ARROW
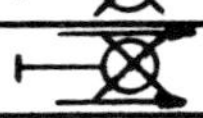	" " " " " " " BOTH SIDES

FIGURE 22-2

ELECTRICAL SYMBOL LIST	
SYMBOL	DESCRIPTION
—//W—	WIREMOLD – NUMBER OF SLASHES DENOTES NUMBER OF WIRES IN RACEWAY.
—//—	BRANCH WIRING IN CONDUIT NUMBER OF SLASHES DENOTES NUMBER OF WIRES IN RACEWAY.
⌒//→	HOME RUN TO PANEL AS NOTED ON PLANS. NUMBER OF SLASHES DENOTES NUMBER OF CIRCUITS TO PANEL. UNLESS OTHERWISE NOTED ALL WIRING SHALL BE #12, THW, COPPER IN 3/4" CONDUIT.
$	SINGLE POLE SWITCH RECESSED.
$ □	" " " SURFACE.
$3	THREE WAY SWITCH RECESSED.
$ □ 3	" " " SURFACE.
$T	THERMAL OVERLOAD SWITCH RECESSED.
$TP	" " " WITH PILOT LIGHT RECESSED.
□┘WP	WATERPROOF DISCONNECT SWITCH.
◪	PANELBOARD
○	LIGHTING OUTLET
Ⓙ	JUNCTION BOX
≡⊖R	RANGE OUTLET
⊙	FLOOR OUTLET

FIGURE 22-3

ELECTRICAL SYMBOL LIST	
SYMBOL	DESCRIPTION
	FIRE ALARM HORN
	FIRE ALARM PULL STATION
	END OF LINE DIODE
GFI	GROUND FAULT INTERRUPTER
C/B	CIRCUIT BREAKER
AFF	ABOVE FINISHED FLOOR
ANN.	ANNUNCIATOR
	SOUND SPEAKER
	ELECTRIC HAND DRYER
	SPECIAL PURPOSE OUTLET
	" " "
$\$_K$	KEY OPERATED SWITCH
$\$_P$	SWITCH AND PILOT LAMP
$\$_{WP}$	WEATHERPROOF SWITCH

FIGURE 22-4

ELECTRICAL LEGEND

CONVENIENCE OUTLETS

SYMBOL	DESCRIPTION
	DUPLEX CONVENIECE OUTLET
1,3	CONVENIENCE OUTLET OTHER THAN DUPLEX 1 = SINGLE, 3 = TRIPLEX, ETC.
WP	WEATHERPROOF CONVENIENCE OUTLET
R	RANGE OUTLET
S	SWITCH AND CONVENIENCE OUTLET
R	RADIO AND CONVENIENCE OUTLET
	SPECIAL PURPOSE OUTLET (DES. IN SPEC.)
	FLOOR OUTLET

SPECIAL OUTLETS

SYMBOL	DESCRIPTION
a,b,c,etc. a,b,c,etc. S a,b,c,etc.	ANY STANDARD SYMBOL AS GIVEN ABOVE WITH THE ADDITION OF A LOWER CASE SUBSCRIPT LETTER MAY BE USED TO DESIGNATE SOME SPECIAL VARIATION OF STANDARD EQUIPMENT OF PARTICULAR INTEREST IN A SPECIFIC SET OF ARCHITECTURAL PLANS. WHEN USED THEY MUST BE LISTED IN THE KEY OF SYMBOLS ON EACH DRAWING AND IF NECESSARY FURTHER DESCRIBED IN THE SPECIFICATIONS.

FIGURE 22-5

ELECTRICAL LEGEND			
SWITCH OUTLETS			
SYMBOL	DESCRIPTION	SYMBOL	DESCRIPTION
S	SINGLE POLE SWITCH	S_{CB}	CIRCUIT BREAKER
S_2	DOUBLE POLE SWITCH	S_{WCB}	WEATHERPROOF CIRCUIT BREAKER
S_3	THREE WAY SWITCH	S_{MC}	MOMENTARY CONTACT SWITCH
S_4	FOUR WAY SWITCH	S_{RC}	REMOTE CONTROL SWITCH
S_D	AUTOMATIC DOOR SWITCH	S_{WP}	WEATHERPROOF SWITCH
S_E	ELECTROLIER SWITCH	S_F	FUSED SWITCH
S_K	KEY OPERATED SWITCH	S_{WF}	WEATHERPROOF FUSED SWITCH
S_P	SWITCH AND PILOT LAMP		

FIGURE 22-6

ELECTRICAL LEGEND		
GENERAL OUTLETS		
CEILING SYMBOL	WALL SYMBOL	DESCRIPTION
○	—○	OUTLET
Ⓑ	—Ⓑ	BLANKED OUTLET
Ⓓ		DROP CORD
Ⓔ	—Ⓔ	ELECTRICAL OUTLET: FOR USE ONLY WHEN CIRCLE USED ALONE MIGHT BE CONFUSED WITH COLUMN, PLUMBING SYM.
Ⓕ	—Ⓕ	FAN OUTLET
Ⓙ	—Ⓙ	JUNCTION BOX
Ⓛ	—Ⓛ	LAMP HOLDER
Ⓛ PS	—Ⓛ PS	LAMP HOLDER WITH PULL SWITCH
Ⓢ	—Ⓢ	PULL SWITCH
Ⓥ	—Ⓥ	OUTLET FOR VAPOR DISCHARGE LAMP
Ⓧ	—Ⓧ	EXIT LIGHT OUTLET
Ⓒ	—Ⓒ	CLOCK OUTLET (SPECIFY VOLTAGE)

FIGURE 22-7

ELECTRICAL LEGEND

PANELS, CIRCUITS, AND MISCELLANEOUS

SYMBOL	DESCRIPTION
	LIGHTING PANEL
	POWER PANEL
	BRANCH CIRCUIT: CONCEALED IN CEILING OR WALL
	BRANCH CIRCUIT: CONCEALED IN FLOOR
	BRANCH CIRCUIT: EXPOSED
	HOME RUN TO PANEL BOARD. INDICATE NUMBER OF CIRCUITS BY NUMBER OF ARROWS. NOTE: ANY CIRCUIT WITHOUT FURTHER DESIGNATION INDICATES A TWO-WIRE CIRCUIT. FOR A GREATER NUMBER OF WIRES INDICATE AS FOLLOWS: (3 WIRES), (4 WIRES)
	FEEDERS. NOTE: USE HEAVY LINES AND DESIGNATE BY NUMBER CORRESPONDING TO LISTING IN FEEDER SCHEDULE.

SYMBOL	DESCRIPTION
	UNDERFLOOR DUCT AND JUNCTION BOX, TRIPLE SYSTEM. NOTE: FOR DOUBLE OR SINGLE SYSTEMS ELIMINATE ONE OR TWO LINES. THIS SYMBOL IS EQUALLY ADAPTABLE TO AUXILIARY SYSTEM LAYOUTS.
G	GENERATOR
M	MOTOR
I	INSTRUMENT
T	POWER TRANSFORMER
	CONTROLLER
	ISOLATING SWITCH

FIGURE 22-8

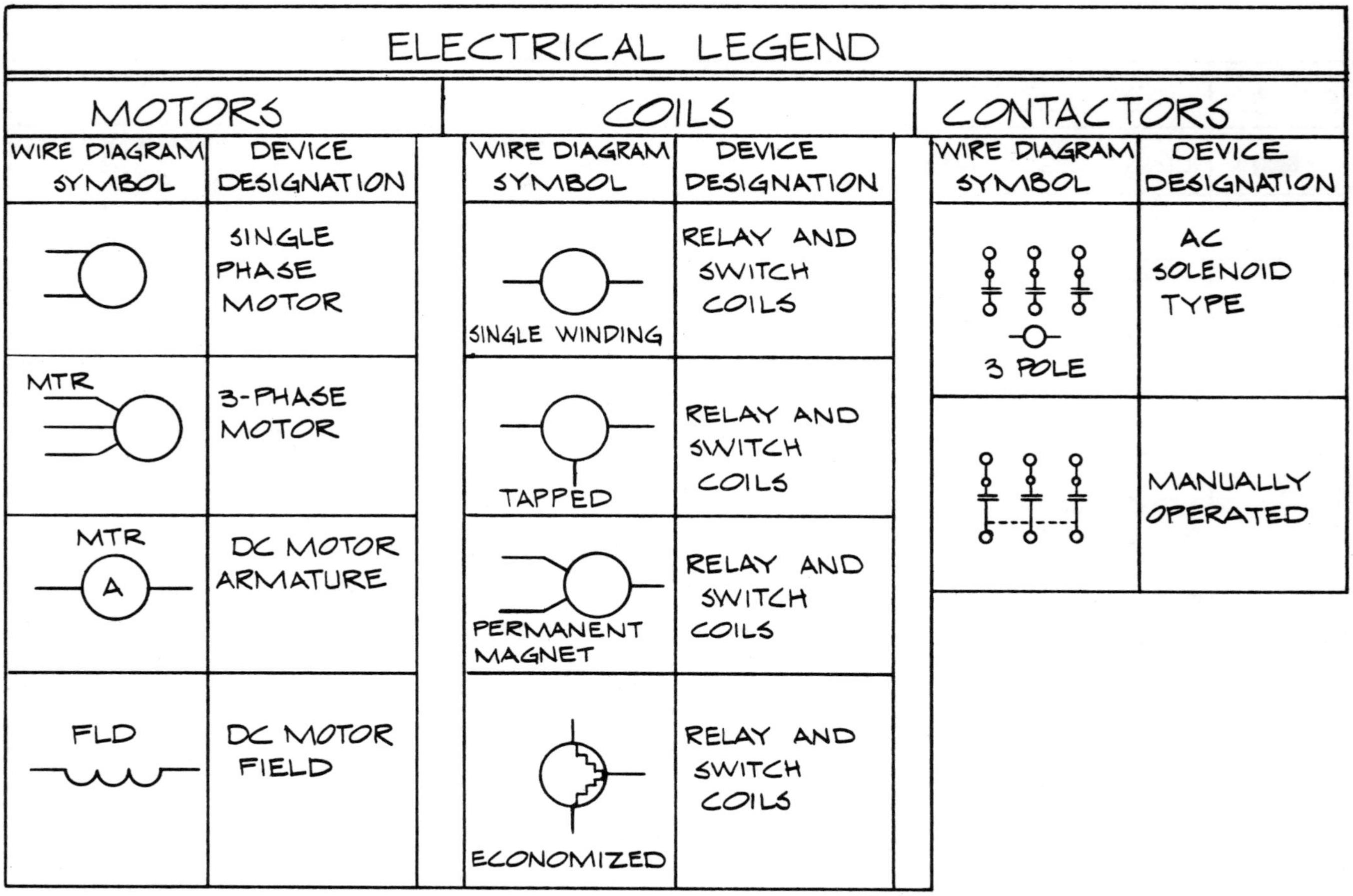
ELECTRICAL LEGEND
MOTORS
WIRE DIAGRAM SYMBOL
DEVICE DESIGNATION
SINGLE PHASE MOTOR
MTR
3-PHASE MOTOR
MTR
A
DC MOTOR ARMATURE
FLD
DC MOTOR FIELD
COILS
WIRE DIAGRAM SYMBOL
DEVICE DESIGNATION
SINGLE WINDING
RELAY AND SWITCH COILS
TAPPED
RELAY AND SWITCH COILS
PERMANENT MAGNET
RELAY AND SWITCH COILS
ECONOMIZED
RELAY AND SWITCH COILS
CONTACTORS
WIRE DIAGRAM SYMBOL
DEVICE DESIGNATION
3 POLE
AC SOLENOID TYPE
MANUALLY OPERATED

FIGURE 22-9

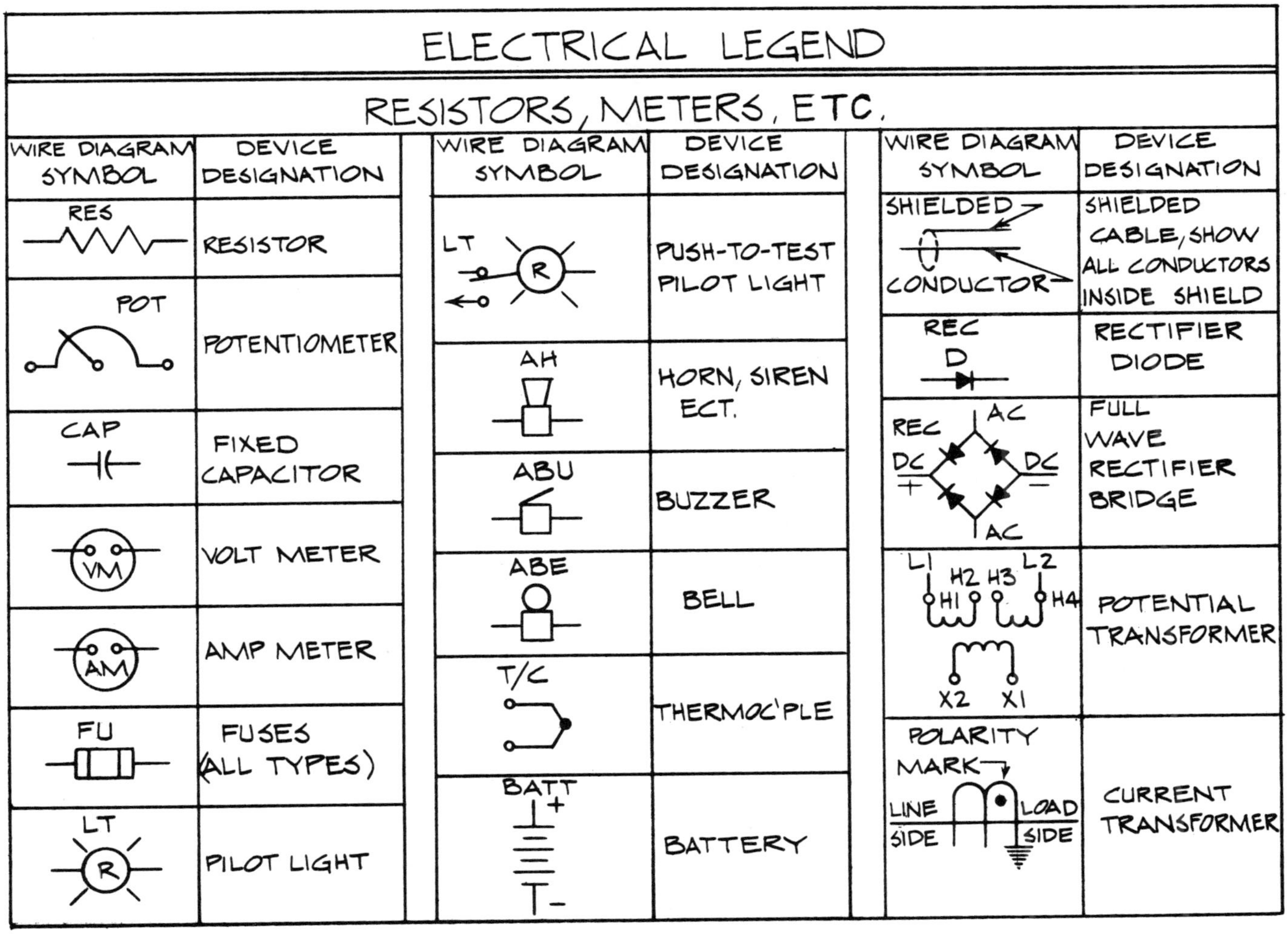
ELECTRICAL LEGEND
RESISTORS, METERS, ETC.
WIRE DIAGRAM SYMBOL
DEVICE DESIGNATION
RES
RESISTOR
POT
POTENTIOMETER
CAP
FIXED CAPACITOR
VM
VOLT METER
AM
AMP METER
FU
FUSES (ALL TYPES)
LT
R
PILOT LIGHT
WIRE DIAGRAM SYMBOL
DEVICE DESIGNATION
LT
R
PUSH-TO-TEST PILOT LIGHT
AH
HORN, SIREN ECT.
ABU
BUZZER
ABE
BELL
T/C
THERMOC'PLE
BATT
+
-
BATTERY
WIRE DIAGRAM SYMBOL
DEVICE DESIGNATION
SHIELDED
CONDUCTOR
SHIELDED CABLE, SHOW ALL CONDUCTORS INSIDE SHIELD
REC
D
RECTIFIER DIODE
REC
AC
DC
+
DC
−
AC
FULL WAVE RECTIFIER BRIDGE
L1
H2 H3
L2
H1
H4
X2 X1
POTENTIAL TRANSFORMER
POLARITY MARK
LINE SIDE
LOAD SIDE
CURRENT TRANSFORMER

FIGURE 22-10

ELECTRICAL LEGEND			
SYMBOLS FOR ONE LINE POWER DIAGRAMS			
SYMBOL	DESCRIPTION	SYMBOL	DESCRIPTION
CT	CURRENT TRANSFORMER		TRANSFORMER
M	CURRENT TRANSFORMER WITH UTILITY METERING		CIRCUIT BREAKER
			DRAWOUT BREAKER
V	VOLT METER		DRAWOUT BREAKER WITH CURRENT LIMITER
VS	VOLT METER SWITCH		FUSE
PT	POTENTIAL TRANSFORMER		SWITCH
			CONTACTOR
A	AMP METER		OVERLOAD RELAY
AS	AMP METER SWITCH		MOTOR
KW	KILOWATT-HOUR METER		
PF	POWER FACTOR METER		
LA	LIGHT ARRESTER		

FIGURE 22-11

ELECTRICAL LEGEND

SWITCHES

WIRE DIAGRAM SYMBOLS	DEVICE DESIGNATION
DISC	DISCONNECT SWITCH
CI	CIRCUIT INTERRUPTER SWITCH
CB	CIRCUIT BREAKER SWITCH
LS	NORMALLY OPEN LIMIT SWITCH
LS	NORMALLY CLOSED LIMIT SWITCH
LS	HELD CLOSED LIMIT SWITCH
LS	HELD OPEN LIMIT SWITCH
LS NP	NEUTRAL POSITION LIMIT SWITCH
LS NP	ACTUATED NEUTRAL POSITION LIMIT SWITCH
LS	MAINTAINED POSITION LIMIT SWITCH
PRS	CLOSED PROXIMITY LIMIT SWITCH
PRS	OPEN PROXIMITY LIMIT SWITCH
FS	NORMALLY OPEN LIQUID LEVEL SWITCH
FS	NORMALLY CLOSED LIQUID LEVEL SWITCH

FIGURE 22-12

ELECTRICAL LEGEND

SWITCHES

WIRE DIAGRAM SYMBOLS	DEVICE DESIGNATION
PS	NORMALLY OPEN VACUUM & PRESSURE SWITCH
PS	NORMALLY CLOSED VACUUM & PRESSURE SWITCH
TAS	NORMALLY OPEN TEMPERATURE SWITCH
TAS	NORMALLY CLOSED TEMPERATURE SWITCH
TGS	TOGGLE SWITCH

WIRE DIAGRAM SYMBOLS	DEVICE DESIGNATION
FLS	NORMALLY OPEN FLOW (AIR, WATER ETC) SWITCH
FLS	NORMALLY CLOSED FLOW (AIR, WATER ETC.) SWITCH
FTS	NORMALLY OPEN FOOT SWITCH
FTS	NORMALLY CLOSED FOOT SWITCH
COS	CABLE OPERATED (EMERG) SWITCH

WIRE DIAGRAM SYMBOL	DEVICE DESIGNATION
PLS F	PLUGGING SWITCH
PLS F R	PLUGGING SWITCH
PLS F R	NON-PLUG SWITCH

FIGURE 22-13

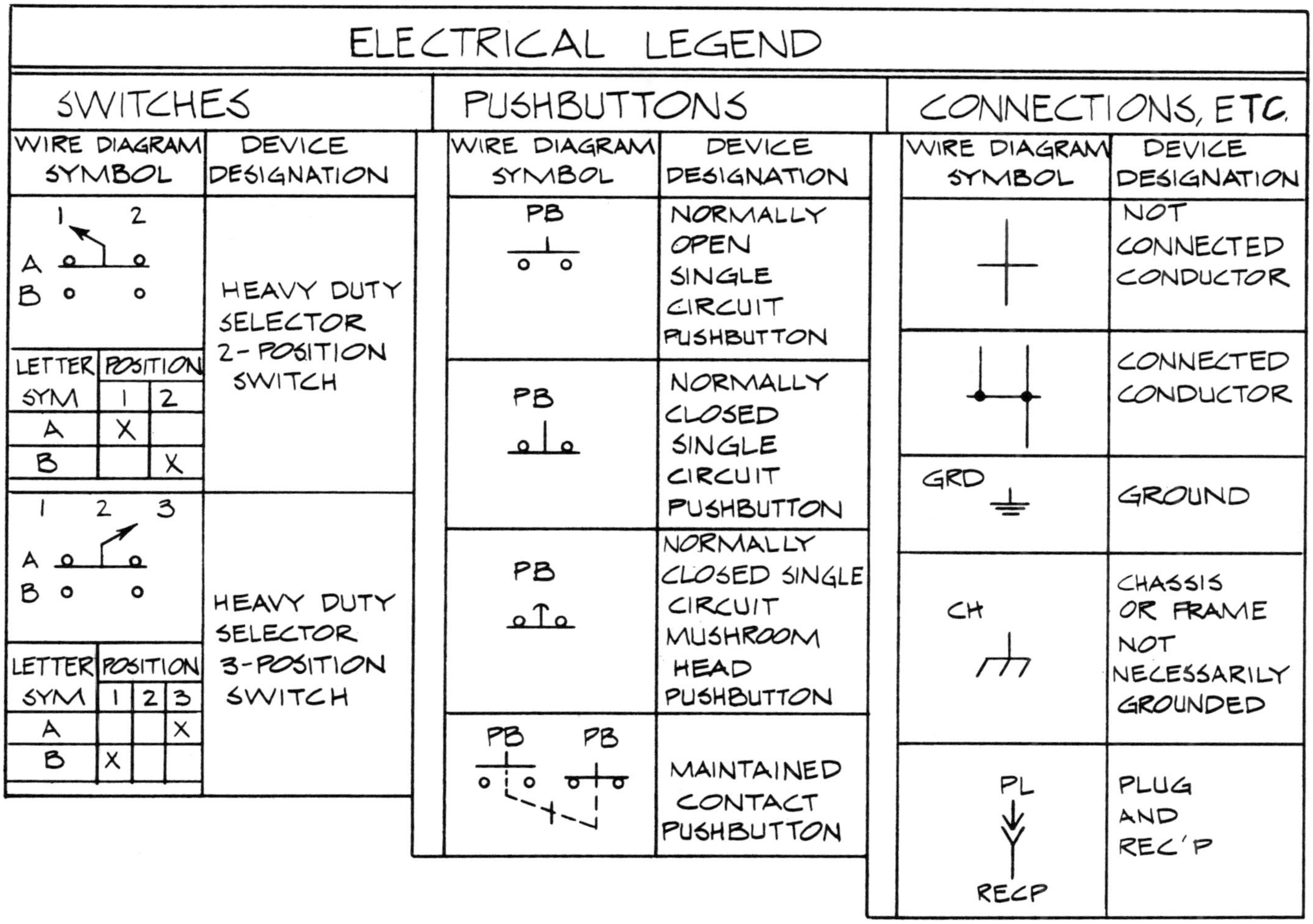
ELECTRICAL LEGEND
SWITCHES
WIRE DIAGRAM SYMBOL
DEVICE DESIGNATION
1 2
A
B
LETTER SYM
POSITION
1 2
A X
B X
HEAVY DUTY SELECTOR 2-POSITION SWITCH
1 2 3
A
B
LETTER SYM
POSITION
1 2 3
A X
B X
HEAVY DUTY SELECTOR 3-POSITION SWITCH
PUSHBUTTONS
WIRE DIAGRAM SYMBOL
DEVICE DESIGNATION
PB
NORMALLY OPEN SINGLE CIRCUIT PUSHBUTTON
PB
NORMALLY CLOSED SINGLE CIRCUIT PUSHBUTTON
PB
NORMALLY CLOSED SINGLE CIRCUIT MUSHROOM HEAD PUSHBUTTON
PB PB
MAINTAINED CONTACT PUSHBUTTON
CONNECTIONS, ETC.
WIRE DIAGRAM SYMBOL
DEVICE DESIGNATION
NOT CONNECTED CONDUCTOR
CONNECTED CONDUCTOR
GRD
GROUND
CH
CHASSIS OR FRAME NOT NECESSARILY GROUNDED
PL
RECP
PLUG AND REC'P

FIGURE 22-14

ELECTRICAL LEGEND			
AUXILIARY SYSTEMS			
SYMBOL	DESCRIPTION	SYMBOL	DESCRIPTION
•	PUSH BUTTON	H	HORN
	BUZZER	N	NURSE'S SIGNAL PLUG
	BELL	M	MAID'S SIGNAL PLUG
	ANNUNCIATOR	R	RADIO OUTLET
	OUTSIDE TELEPHONE	SC	SIGNAL CENTRAL STATION
	INTERCONNECTING TELEPHONE		INTERCONNECTION BOX
	TELEPHONE SWITCHBOARD		BATTERY
T	BELL RINGING TRANSFORMER	————	AUXILIARY SYSTEMS CIRCUITS. NOTE: ANY LINE WITHOUT FURTHER DESIGNATION INDICATES A 2-WIRE SYSTEM FOR A GREATER NUMBER OF WIRES DESIGNATE WITH NUMERALS IN MANNER SIMILAR TO ——— 12-NO. 18 W-3/4" C., OR DESIGNATE BY NUMBER CORRESPONDING TO LISTING IN SCHEDULE.
D	ELECTRIC DOOR OPENER		
F	FIRE ALARM BELL		
F	FIRE ALARM STATION		
	CITY FIRE ALARM STATION		
FA	FIRE ALARM CENTRAL STATION		
FS	AUTOMATIC FIRE ALARM DEVICE		
W	WATCHMAN'S STATION	a,b,c	SPECIAL AUXILIARY OUTLETS SUBSCRIPT LETTERS REFER TO NOTES ON PLANS OR DETAILED DESCRIPTION IN SPECIFICATIONS
W	WATCHMAN'S CENTRAL STATION		

FIGURE 22-15

MATV

ELECTRICAL LEGEND

HEAD-END EQUIPMENT

SYMBOL	DESCRIPTION
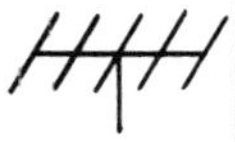	ANTENNA
	BALUN
	ANTENNA JOINER
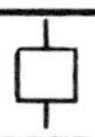	PREAMPLIFIER
	POWER SUPPLY
	HI-LO JOINER
	U/V JOINER
	VARIABLE ATTENUATOR
	FIXED ATTENUATOR
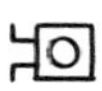	LO BAND SOUND CARRIER REDUCER
	HI BAND SOUND CARRIER REDUCER

SYMBOL	DESCRIPTION
	LO BAND HI "Q" TRAP
	HI BAND HI "Q" TRAP
	LO BAND MIXING UNIT
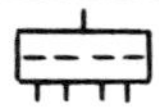	HI BAND MIXING UNIT
	CONVERTER
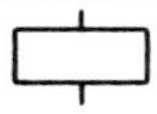	BROADBAND DISTRIBUTION AMPLIFIER
	SINGLE CHANNEL BANDPASS FILTER
	SINGLE CHANNEL AMPLIFIER

FIGURE 22-16

MATV

ELECTRICAL LEGEND

DISTRIBUTION SYSTEM EQUIPMENT

SYMBOL	DESCRIPTION	SYMBOL	DESCRIPTION
	2 WAY SPLITTER		CABLE ADAPTER RG-6
	3 WAY SPLITTER		CABLE ADAPTER RG-11, .412, .500 TO "F"
	4 WAY SPLITTER		CABLE ADAPTER .412, .500 ENTRY MOUNT WITH PIN
	VARIABLE ISOLATION WALL TAP	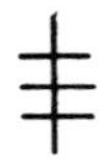	CABLE ADAPTER 5/8" ENTRY TO FEMALE "F"
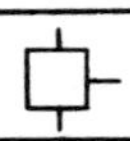	ONE WAY LINE TAP		AUXILIARY POWER SUPPLY
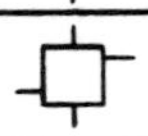	TWO WAY LINE TAP		LINE AMPLIFIER
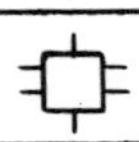	4 WAY LINE TAP		POWER ADDER
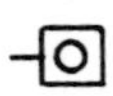	"O" dB WALL OUTLET	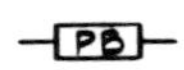	VOLTAGE BLOCK
	TERMINATOR		
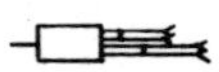	UHF/VHF BAND SEPARATOR		
	MATCHING TRANSFORMER		

FIGURE 22-17

ELECTRICAL LEGEND

CONTACTS

WIRE DIAGRAM SYMBOL	DEVICE DESIGNATION	WIRE DIAGRAM SYMBOL	DEVICE DESIGNATION	WIRE DIAGRAM SYMBOL	DEVICE DESIGNATION
TR 	NORMALLY OPEN ENERGIZED TIME DELAY AFTER COIL CONTACT	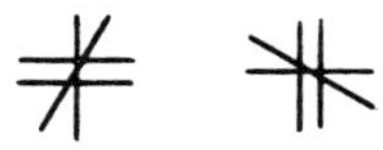 MAIN	NORMALLY CLOSED (N.C.) CONTACT	OL 1OL 	THERMAL OVERLOAD CONTACT
TR 	NORMALLY CLOSED ENERGIZED TIME DELAY AFTER COIL CONTACT	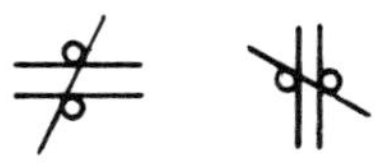 AUXILIARY	NORMALLY CLOSED (N.C.) CONTACT	OL 1OL 	THERMAL OVERLOAD ELEMENT
TR 	NORMALLY OPEN DE-ENERGIZED TIME DELAY AFTER COIL CONTACT	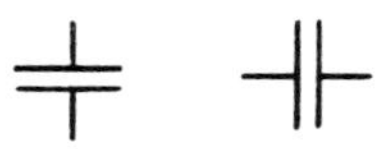 MAIN	NORMALLY OPEN (N.O.) CONTACT	SOL 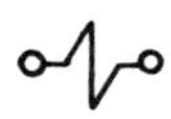	GENERAL
TR 	NORMALLY CLOSED DE-ENERGIZED TIME DELAY AFTER COIL CONTACT	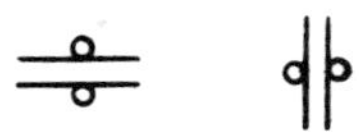 AUXILIARY	NORMALLY OPEN (N.O.) CONTACT		

FIGURE 22-18

Index

ABOUT THE AUTHOR

Jerome F. Mueller has been a practicing consulting engineer in the mechanical and electrical disciplines for over twenty five years. A Professional Engineer in many states, he has designed mechanical and electrical systems for government departments, universities and colleges, cities and towns, the military, architectural firms, and manufacturing companies. Mr. Mueller was an early leader in fluorescent applications lighting and a pioneer in Connecticut's usage of electrical heating systems. He is also the author of the acclaimed text *Standard Mechanical and Electrical Details* (McGraw-Hill).